Cyclic ADP-Ribose and NAADP
Structures, Metabolism and Functions

Cyclic ADP-Ribose and NAADP
Structures, Metabolism and Functions

Editor ***Hon Cheung Lee,*** **Minneapolis, USA**

Cover Designs: Front – the space-filling model on the left is that of cADPR and is based on X-ray crystallography. The molecule on the right is NAADP. The secondary structure of a homo-dimer of the *Aplysia* ADP-ribose cyclase is shown in the middle. The background is derived from a micrograph of cADPR crystals. **Back** – ball-and-stick models of cADPR and NAADP. Graphics designed by HC Lee.

KLUWER ACADEMIC PUBLISHERS
Boston / Dordrecht / London

Distributors for North, Central and South America:
Kluwer Academic Publishers
101 Philip Drive
Assinippi Park
Norwell, Massachusetts 02061 USA
Telephone (781) 871-6600
Fax (781) 681-9045
E-Mail: kluwer@wkap.com

Distributors for all other countries:
Kluwer Academic Publishers Group
Post Office Box 322
3300 AH Dordrecht, THE NETHERLANDS
Telephone 31 786 576 000
Fax 31 786 576 474
E-Mail: services@wkap.nl

Electronic Services <http://www.wkap.nl>

Library of Congress Cataloging-in-Publication Data

Cyclic ADP-Ribose and NAADP: Structures, Metabolism and Functions
Hon Cheung Lee (ed.)
ISBN 1-4020-7281-3

Printed on acid-free paper.

Printed in the United States of America.

Contents

Coordination of Calcium Signaling by cADPR and NAADP in Pancreatic Acinar Cells
Cancela JM, Gif-sur-Yvette, France

Calcium and Calcium-Linked Second Messengers are Main Actors in the Maturation and Fertilization of Starfish Oocytes
Santella L, Nusco GA and Lim D, Naples, Italy

ADP-Ribosylation and CD38 Signaling
Kim UH, Han MK and Yim CY, Chonju, Korea

Preface

In the past decade, we have witnessed the birth and blossoming of a field of research centering on the Ca^{2+} signaling functions of cyclic ADP-ribose (cADPR) and nicotinic acid adenine dinucleotide phoshpate (NAADP). Out of the long shadow cast by the well-celebrated inositol trisphosphate, the first Ca^{2+} messenger identified, emerged a novel cyclic nucleotide. Unlike all other cyclic nucleotides, cADPR is completely circular, formed by head-to-tail linkage of a long linear molecule of NAD. This structure is truly unique among all known compounds. Soon afterward came NAADP, a rather mundanely linear molecule but with a mechanism of action shared by no other Ca^{2+} messengers. More surprising is the fact that these two structurally and functionally different molecules are synthesized by the same members of a new family of multifunctional ADP-ribosyl cyclases. The cover designs of this book are intended both to celebrate this remarkable discovery and to highlight the novel structures of cADPR and NAADP. Much is now known about the catalytic mechanism of these cyclases. Likewise, a decade of progress has brought forth comprehensive understanding of the structures, metabolism and functions of cADPR and NAADP. It is thus timely to assemble these findings in a single volume. The first chapter tells the story behind the emergence of the field. The second chapter summarizes the current understanding of the structure, catalytic mechanism and function of the ADP-ribosyl cyclases. The third chapter is concerned with the regulation of the enzymatic synthesis of cADPR and NAADP. The fourth chapter reviews the gene structure and regulation of human CD38, a mammalian ADP-ribosyl cyclase. The next three chapters describe the methodology developed for and widely used in the field. The rest of the chapters focus on the Ca^{2+} signaling mechanisms and physiological functions of these two novel Ca^{2+} messengers in a wide variety of cellular systems.

All the contributors of this book have played important roles in ushering in and advancing this burgeoning field. This book is dedicated to all of them, to their pioneering spirits, and most of all, to their perseverance against, at times, seemingly overwhelming odds. Looking back, the past fifteen years may seem like a long and winding road, but the journey ahead is now well-mapped by the persistent efforts of all the contributors of this book. Their advances, their accomplishments and their visions are on these pages. There are reasons to believe the future progress of the field will bring about important paradigm shifts that will amend our concept in Ca^{2+} signaling. This book, being the first devoted entirely to the field, represents a cornerstone and is likely to be the first of many to come.

As a personal note, I would like to acknowledge the editorial support of Richard Graeff. I would also like to thank Tim Walseth for the endless hours of stimulating discussion and for all the life-saving help he has provided. I am always grateful to Antonio De Flora for heeding my call from the wilderness and, most of all, for his warm support that has sustained me in this journey.

李漢璋
Hon Cheung Lee
Minneapolis

Cyclic ADP-Ribose and NAADP. Structures, Metabolism and Functions. Lee HC (ed), pp. 1-21, Dordrecht: Kluwer.

Cyclic ADP-Ribose and NAADP

A Story of Two Calcium Messengers

Hon Cheung Lee

Department of Pharmacology, University of Minnesota, Minneapolis, MN 55455, USA.

INTRODUCTION

Cells are surrounded by a plasma membrane, which isolates them from the environment and, in so doing, defines their existence. Internally, cells are likewise compartmentalized by endo-membranes into organelles with diverse and specific functions. Information transfer, or signal transduction, across these membranes is fundamental to coordinating and integrating all cellular activities. Cells have developed various strategies to bypass the isolation imposed by the plasma membrane so that they can detect and adjust to environmental changes. Chemical signals are generally sensed by surface receptors, which upon binding specific ligands, activate intracellular production of second messengers. Cyclic AMP was the first of these second messengers identified by Sutherland and colleagues [1]. In many cases, surface receptor activation can also lead to changes in intracellular Ca^{2+} concentration, which serve as communicating signals to initiate a myriad of further changes, such as enzyme activation, protein phosphorylation and gene expression. The source of Ca^{2+} can be either influx from outside or release from intracellular Ca^{2+} stores. The focus of this book is the latter process, the mobilization of intracellular Ca^{2+} stores.

Ca^{2+} RELEASING EFFECTS OF NAD AND NADP

This story began in 1983. In that year, seminal studies were published by Michael Berridge and colleagues, demonstrating that inositol trisphosphate (IP_3) accumulates rapidly in cells following stimulation by agonists and functions to release Ca^{2+} from intracellular stores [2, 3]. Years of investigation of the metabolism of inositol lipids [4, 5] were brought into focus and established IP_3 as the first Ca^{2+} messenger ever identified.

I had joined the faculty of the University of Minnesota a couple of years earlier and had spent most of my time setting up the laboratory and finishing up a series of studies on sea urchin sperm motility that I started during my postdoctoral study. However, contemplating about the future direction of my budding research career was what occupied most of my thinking. The excitement of the discovery of IP_3 as a Ca^{2+} messenger came at the most opportune time and I was in a particularly receptive frame of mind. My experience with sea urchin eggs during my postdoctoral study at the Hopkins Marine Station, under the mentorship of David Epel, immediately suggested that they may well be a perfect cell system for this kind of investigation.

Sea urchin eggs have long been a model system for investigating the mechanisms of Ca^{2+} signaling. Fertilization initiates a localized elevation of Ca^{2+} at the sperm-egg fusion site. This spark of Ca^{2+} is then amplified and propagates as a Ca^{2+} wave, sweeping across the entire egg and triggering a massive cortical exocytosis in its wake. The consequence is the formation of the fertilization envelope, a protein matrix that serves as a mechanical block to polyspermy. The Ca^{2+} wave also signals the fertilized egg to initiate a multifaceted program that begins with protein and DNA synthesis and culminates in cell division and embryonic development. The critical importance of the Ca^{2+} wave at fertilization in initiating development makes it clear that the eggs must possess robust mechanisms to ensure its successful completion. The choice of sea urchin eggs as a model for investigating the mechanisms of Ca^{2+} mobilization was perhaps the single most important factor that had led us to the discovery of cyclic ADP-ribose (cADPR) and nicotinic acid adenine dinucleotide phosphate (NAADP).

In 1984, David Clapper, a friend and colleague, joined my laboratory and we began in earnest the project of investigating Ca^{2+} mobilization in sea urchin eggs. We first verified that introducing IP_3 into the eggs by electroporation could indeed activate the cortical exocytotic reaction. We then developed an *in vitro* system using egg homogenates to facilitate ready access to the internal Ca^{2+} stores [6]. The eggs possessed abundant endoplasmic reticulum that could be easily isolated by homogenization. Unfractionated homogenates provided a good representation of most, if not all, of the internal Ca^{2+} stores. Incubation with ATP activated the endogenous Ca^{2+} pumps, which sequestered the contaminating Ca^{2+} in the media into the microsomal stores. Subsequent release from the stores could be conveniently followed using fluorescent Ca^{2+} indicators. Liters of egg homogenates could be prepared and stored frozen for years without lost of activity. It was a very simple assay and its stability and robust response made it practically effortless to screen a variety of substances for their ability to release Ca^{2+} from the internal stores of the egg homogenates.

It is, perhaps, logical to think that Ca^{2+} messengers other than IP_3 may well exist, but actually committing to search for a new and unknown second messenger is entirely a different matter. Many would consider it audacious or even foolhardy. If it were not because of the simplicity of the homogenate assay, we too, perhaps, would never have tried either. But try we did, and we found not only one but two substances that were as active in releasing Ca^{2+} from the egg Ca^{2+} stores as IP_3. They were NAD and NADP, well known coenzymes in oxidation-reduction reactions. Neither of them had ever been shown to be participants in Ca^{2+} signaling in cells. As will be described, the substances that were actually active in Ca^{2+} release were not NAD and NADP themselves, but novel metabolites derived from them.

The important roles of NAD in mitochondrial oxidative phosphorylation had been well characterized. It was also known that energized mitochondria use protonmotive force to sequester Ca^{2+}. It was thus possible that exogeneous NAD might interfere with mitochondrial energization in the egg homogenates, resulting in release of Ca^{2+} from the organelles. We quickly ruled this out since the Ca^{2+} release induced by either NAD or NADP was insensitive to mitochondrial inhibitors or uncouplers. Also, fractionation of the egg homogenates showed that the responsive organelles distributed very differently from the mitochondria [7].

We next considered the possibility that NAD and/or NADP might be modulating the IP_3-sensitive release mechanism. Although both NAD and NADP were able to release as much Ca^{2+} from the egg homogenates as IP_3, kinetically there were differences, particularly in the case of NAD. The release induced by IP_3 was immediate, but for NAD, there was an initial delay of a minute or two before the process commenced [7]. We also observed that the egg homogenates showed complete desensitization to repeated additions of IP_3 and would not respond to further challenge with IP_3, and yet NAD was still able to release as much Ca^{2+} from the desensitized homogenates as the naïve one. In fact, the homogenates showed homologous desensitization to repeated additions of all three agonists, IP_3, NAD and NADP, but still respond to the other two heterologous agonists. These results clearly showed that there were three different Ca^{2+} release mechanisms operating in the egg homogenates. A couple of years later, after heparin was shown to be an antagonist of the IP_3-receptor, Peter Dargie, a graduate student, showed that heparin could indeed block Ca^{2+} release induced by IP_3 but not that induced by either NAD or NADP [8]. We were increasingly convinced that what we had stumbled on was something novel and had not been described before.

Our excitement was, however, hampered by the fact that the required concentrations of NAD and NADP were quite high, in the range of tenths of millimolar, which would represent the total concentrations of pyridine nucleotides generally present in cells. Under normal circumstances, it seemed highly unlikely that the cellular levels of NAD or NADP could change sufficiently enough to make what we had observed in the egg homogenates physiologically relevant. Indeed, when Peter Dargie microinjected high concentrations of NAD into live sea urchin eggs, no change in cellular Ca^{2+} concentrations was seen. Certainly, NAD itself could not be a Ca^{2+} messenger. I dreaded the possibility that we were dealing with an artefact of an *in vitro* system. This roller coaster of emotion, from excitement to despair, was typical in the early years.

It finally occurred to us that the initial delay in the Ca^{2+} release induced by NAD could be due to conversion of NAD to an active metabolite and the actual active substance might not be NAD itself. In fact, if enzymatic conversion were required, it would be most appropriate that the substrate NAD concentrations should be in the range of the cytoplasmic concentrations of NAD. A simple experiment to do was, therefore, to incubate NAD with the egg homogenates first to effect the conversion and we predicted that the incubated NAD would then release Ca^{2+} without delay. I was elated that it was indeed the case [7]. We had thus established the existence of a novel enzyme in the egg homogenates that converted NAD to an active metabolite capable of mobilizing intracellular Ca^{2+} stores. This set the stage for the subsequent isolation and identification of the active metabolite as cyclic ADP-ribose.

CYCLIC ADP-RIBOSE (cADPR)

If one had not worked with an unknown substance, one might not truly appreciate how much information we routinely take for granted. Even as simple a task as making up a solution would not be possible if the molecular weight of the substance were not known. Indeed, without knowing the structure of this active metabolite of NAD we could not even refer to it with a specific name and had to resort to calling it E-NAD instead, for enzyme-activated NAD. It was clear to me that if we were to continue with this project, we had to determine the structure of the metabolite. Not trained as a chemist nor having any experience in structure determination, my apprehension was palpable.

With the help of Timothy Walseth, I first set up an HPLC system for purifying the active metabolite. Tim had previously developed an HPLC method for purifying nucleotides that employed trifluoroacetic acid, a volatile acid, for elution [9]. Subsequent concentration of the purified substance and the removal of the solvent could be done easily by simple evaporation. A bioassay employing the egg homogenates was used to monitor the elution of the active metabolite. The simplicity of this bioassay was essential since a large number of fractions needed to be assayed routinely.

Using radioactive NAD of known specific activity and by measuring the radioactivity of the purified metabolite produced after incubation with the egg homogenates, I was able to, first, demonstrate directly that the active metabolite indeed came from NAD and, second, determine accurately the concentration of the active metabolite and thus its extinction coefficient. Finally, I could make routine solutions of known concentrations of the metabolite. Using the same strategy but with the radioactive labels on specific parts of NAD as substrates, I was also able to show that the nicotinamide group of NAD was released during the enzymatic conversion.

To further characterize the structure by biophysical means, much larger quantities of the purified metabolite were needed. With the technical help of Robert Aarhus, we began a laborious process that lasted two years to purify enough of the metabolite for structural determination. Very limited amounts of the active metabolite were produced by the egg homogenates. Even when incubated with millimolar concentrations of NAD, the steady state concentrations of the active metabolite reached were only in the micromolar range, indicating the presence of not only the enzymes for its synthesis but also its degradation. Although it was gratifying to know that a metabolic pathway actively metabolizing the active molecule existed in the egg homogenates, a hallmark of a second messenger, it did pose quite a technical problem for its purification. Liters of egg extracts had to be prepared and scaling up the HPLC procedures to handle large volumes proved to be nontrivial. The use of a volatile acid for elution that Timothy Walseth developed proved to be invaluable in this respect. The highly diluted starting materials also necessitated the use of multiple HPLC columns with diverse chemistries to ensure complete removal of contamination. Many times we thought we had the active metabolite purified, but instead found that it was contaminated by simple substances, such as buffers, obscuring the NMR analyses and requiring yet another column for their removal.

The effort was eventually rewarded. The elation of seeing a clean proton spectrum appearing on the screen of the NMR machine certainly made it worth all the hard work and time invested. The spectrum crystallized the reality of this active metabolite of NAD. It no longer was just a figment of my imagination but something I could see. Gerald Bratt, a spectroscopist collaborating with me on the project, was able to identify and assign the spectral peaks to specific protons of the molecule using a two-dimensional COSY technique. The results showed that the two ribose groups of NAD were unchanged in the metabolite. The NMR results also confirmed that the adenine ring was intact, as shown previously by the radioactive labelling method. It was then clear that the modification must be at the terminal ribose, where the released nicotinamide was originally attached. Satisfying as these results were, they did not provide any clues as to what the actual modification was.

The last piece of the puzzle came from mass spectrometry. After repeated failure to obtain a mass measurement in a local facility, I sent a sample to Roger Hayes at a National Science Foundation facility in Nebraska. The mass of the metabolite was measured to be 540.0526, accurate to better than 2 parts per million. The precision together with what we already knew about the molecule allowed a unique determination of the formula of the molecule as $C_{15}H_{21}N_5O_{13}P_2$. It was close, much closer, but still no structure was revealed.

The actual structure came, most unexpectedly, during a casual conversation with Timothy Walseth. The momentous realization was that the mass of the metabolite was smaller than ADP-ribose, the smallest molecule possible that still retained all parts of NAD except the nicotinamide group. In fact, the mass of the active metabolite was precisely one water molecule less than ADP-ribose. It finally dawned on us that this was exactly analogous to cyclic AMP being one water molecule smaller than AMP. The metabolite had to be ADP-ribose cyclized head-to-tail (Figure 1).

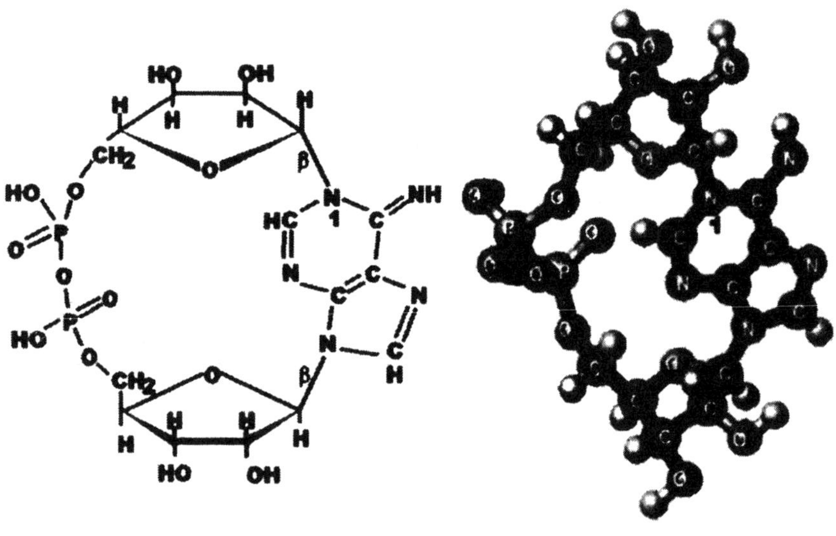

Figure 1. **Structure of cyclic ADP-Ribose.** A color version of the ball-and-stick model of cADPR is shown on the back cover of this book.

A definitive proof of the structure was thus to hydrolyze the metabolite and show that the product was ADP-ribose. Without knowing any specific way to hydrolyze the metabolite, I simply left an acidic solution of the pure metabolite on the bench for two days. Those two days were agonizingly long. Indeed, after two days, about half of the metabolite was hydrolyzed to ADP-ribose.

An ADP-ribose cyclized head-to-tail is such a novel and beautiful structure that I knew at the very moment of realization that it had never been described before. As I was writing this, I relived the euphoria that swept through me more than a decade ago. It was better than a dream came true, for I have never, in my wildest dreams, ever dreamed about determining a structure as unique as this.

I proposed to name it cyclic ADP-ribose (cADPR) [10], which was meant to be descriptive since the exact position where the cyclization occurred was still unknown at the time. The final and definitive resolution of the structure of cADPR was not achieved until five years later. Various approaches, including chemical modifications, isotope labelling and high-resolution NMR techniques, were tried but the results obtained were indirect and in many cases, opened to interpretations. Based on the pH-dependent changes in its UV-spectrum, it was suggested that the linkage site may be at the N1-position of the adenine [11].

Convinced by my colleague, David Levitt, that the only definitive method for solving the structure of a molecule is X-ray crystallography, I finally resolved to crystallize cADPR. Luckily, cADPR crystallized readily by slow evaporation from acidic solutions. The ease of its crystallization most likely related to its compact structure, especially when the charges of the phosphate groups were neutralized under acid conditions. It formed mostly needle shaped crystals when the evaporation was slow, but produced many other amazing shapes when the speed of evaporation was varied. Observing these crystals under the microscope with various lighting conditions produced a dazzling display of birefringence. The background design of the cover of this book was derived from one such micrograph of the cADPR crystals. At last, I was able to see this amazing molecule with my own eyes. It certainly was not a transient artefact with fleeting existence that flew apart under observation. It was real, stable, and crystallizable.

Most importantly, these crystals produced good X-ray diffraction. The final solution of the structure was, for once, straightforward and the cycling sites were clearly shown to be between the nitrogen atom at the 1-position of the adenine and the anomeric carbon of the terminal ribose (Fig. 1). Although anti-climatic as it might seem, the satisfaction of seeing the molecule displayed on the computer screen was truly overwhelming. There it was, with the positions of every atom and every bond definitively determined [12] (cf. right panel of Fig. 1). After seven years of labour, I finally knew every twist and turn of this unique molecule. The crystallographic coordinates of cADPR in Protein Data Base (PDB) format are listed in Table 1.

NICOTINIC ACID ADENINE DINUCLEOTIDE PHOSPHATE (NAADP)

Buoyant by the success of solving the structure of cADPR, I soon began the task of determining the structure of the active metabolite of NADP in January of 1990. Unlike NAD, NADP induced Ca^{2+} release from the egg homogenates without delay, suggesting that it did not require enzymatic

Table 1. Crystallographic coordinates of cADPR in PDB format.

ATOM	1	P1	1	8.087	13.935	7.174
ATOM	2	P2	1	6.089	14.572	5.089
ATOM	3	O1	1	11.300	11.667	6.827
ATOM	4	O2	1	12.813	10.449	3.810
ATOM	5	O3	1	13.830	12.360	5.376
ATOM	6	O4	1	9.631	14.025	6.967
ATOM	7	O5	1	7.578	12.658	7.578
ATOM	8	O6	1	7.811	15.083	8.107
ATOM	9	O7	1	7.531	14.361	5.753
ATOM	10	O8	1	6.382	15.261	3.832
ATOM	11	O9	1	5.198	15.209	6.087
ATOM	12	O10	1	5.580	13.126	4.706
ATOM	13	O11	1	5.092	10.548	3.906
ATOM	14	O12	1	2.432	9.570	4.501
ATOM	15	O13	1	3.949	7.898	5.906
ATOM	16	N1	1	6.828	9.425	5.085
ATOM	17	N3	1	8.776	10.789	5.125
ATOM	18	N6	1	6.720	7.180	5.776
ATOM	19	N7	1	9.712	7.559	6.404
ATOM	20	N9	1	10.666	9.561	6.035
ATOM	21	C2	1	7.519	10.581	4.888
ATOM	22	C4	1	9.376	9.653	5.614
ATOM	23	C5	1	8.778	8.433	5.883
ATOM	24	C6	1	7.389	8.310	5.539
ATOM	25	C8	1	10.794	8.251	6.478
ATOM	26	C9	1	11.671	10.587	5.991
ATOM	27	C10	1	11.928	11.173	4.615
ATOM	28	C11	1	12.461	12.592	4.969
ATOM	29	C12	1	11.641	12.899	6.182
ATOM	30	C13	1	10.343	13.629	5.779
ATOM	31	C14	1	4.731	12.339	5.573
ATOM	32	C15	1	4.080	11.299	4.645
ATOM	33	C16	1	3.293	10.246	5.368
ATOM	34	C17	1	4.404	9.227	5.765
ATOM	35	C18	1	5.378	9.356	4.586
ATOM	36	H1	1	7.027	11.315	4.539
ATOM	37	H2	1	5.841	7.042	5.537
ATOM	38	H3	1	11.611	7.896	6.808
ATOM	39	H4	1	12.485	10.222	6.317
ATOM	40	H5	1	11.099	11.261	4.159
ATOM	41	H6	1	12.437	9.515	3.587
ATOM	42	H7	1	12.367	13.227	4.268
ATOM	43	H8	1	14.221	13.321	5.451
ATOM	44	H9	1	12.143	13.443	6.777
ATOM	45	H10	1	9.799	13.048	5.262
ATOM	46	H11	1	10.558	14.400	5.267
ATOM	47	H12	1	8.234	15.016	9.036
ATOM	48	H13	1	5.465	14.960	7.030
ATOM	49	H14	1	4.069	12.885	5.980
ATOM	50	H15	1	5.246	11.911	6.246
ATOM	51	H16	1	3.511	11.745	4.029
ATOM	52	H17	1	2.838	10.594	6.126
ATOM	53	H18	1	1.601	10.111	4.436
ATOM	54	H19	1	4.829	9.509	6.566
ATOM	55	H20	1	3.279	7.930	6.721
ATOM	56	H21	1	5.274	8.617	3.998
END						

conversion. However, alkaline treatment of NADP greatly increased the activity indicating it was not NADP itself, but a derivative, that was responsible for the Ca^{2+} release activity [7]. This puzzle was solved when I fractionated commercial preparations of NADP by HPLC. It was clear that the Ca^{2+} release activity was due to a small contaminant and not NADP itself. Alkaline treatment of NADP produced more of this substance, indicating that it was derived from NADP. With this crude method of producing the active derivative and with the technical help of Robert Aarhus, we soon isolated enough of the substance for NMR analyses. The result was a complete surprise. The proton NMR spectrum of the active derivative was identical to NADP.

This indicated that all the non-exchangable protons of NADP were conserved in the derivative, including those on the nicotinamide ring. A sample was sent to Roger Hayes in Nebraska for mass spectrometry and the result showed that the derivative and NADP had very similar mass, if not identical. The two molecules must be different in a very subtle way.

Two of the exchangable protons of NADP are in the $-NH_2$ of the amide group in nicotinamide. If they were modified, no changes would be observable in the proton NMR spectrum of the derivative. It soon occurred to me that the simplest modification consistent with the NMR results would be the conversion of the amide group to a carboxyl group. In other words, the $-NH_2$ was converted to $-OH$ and the active derivative was nicotinic acid adenine dinucleotide phosphate (NAADP).

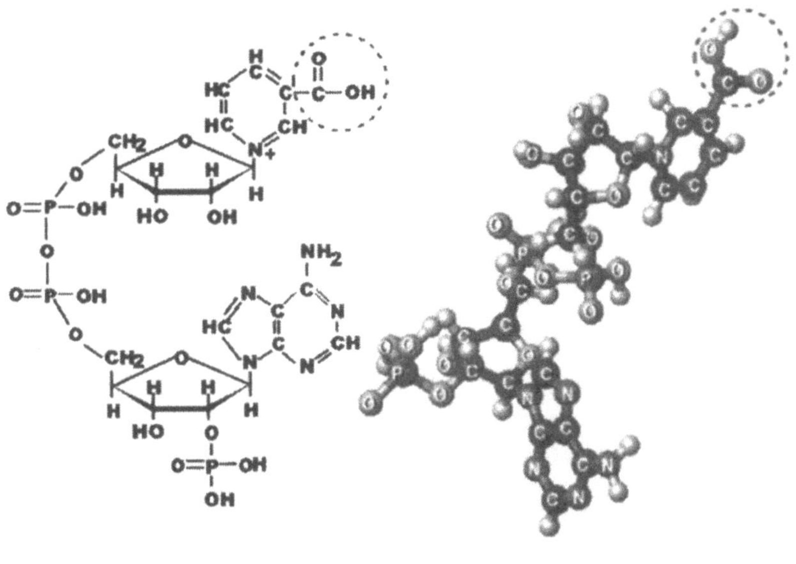

Figure 2. **Structure of nicotinic acid adenine dinucleotide phosphate.** The circles indicate the sole carboxyl group that distinguishes it from NADP. A color version of the ball-and-stick model of cADPR is shown on the back cover of this book.

If this were the case, the active derivative must be exactly one mass unit larger than NADP. A fresh simple was sent to Roger Hayes for high-resolution mass spectrometry. The results returned on March 30 of 1990 definitely confirmed that the active derivative of NADP was NAADP (Fig. 2). In only three months, the structure of the active derivative of NADP was solved, a far cry from the laborious process I spent in solving cADPR.

A manuscript was written up, but I did not submit it for publication until five years later in 1995. That was because of a comment by Nelson Goldberg, a well-respected colleague and a discoverer of cyclic GMP. He thought that the scientific communities might be too skeptical to accept two new Ca^{2+} messengers all at once. After serious consideration of his comment, I decided that it might be strategically more attune to delay the publication of the identification of NAADP until cADPR was better accepted. I did, however, submit all my results and the structure of NAADP to the National Institute of Health in 1991 in a grant proposal, which was received with great enthusiasm and the funding supported many years of my subsequent work on cADPR and NAADP.

In the following years, I focused the effort of my laboratory on investigating the enzymatic synthesis of cADPR and the mechanism of its Ca^{2+} release activity. I was excited to see the gradual popularization and globalization of the interest in cADPR.

I still remember a morning in 1993 when I was surprised by a colleague who congratulated me on the fact that cADPR was found to play a crucial role in mediating insulin secretion. Later, back in my office, I found out that what he was referring to was an article just published in Science by Okamoto's group in Japan. It was most satisfying to know that one's work was not only read, but also taken seriously, by other scientists half way around the world. Indeed, I too was developing international collaborations, most notably, with Antonio De Flora's group in Genova, Italy [13] and Antony Galione in Oxford, England [14]. Locally, I was also collaborating with Thomas Dousa, a nephrologist from the Mayo Clinic in Rochester, Minnesota, who had sent his associates to my laboratory to acquire various techniques used in working with cADPR, including HPLC purification, and preparing and measuring Ca^{2+} release in sea urchin egg homogenates [15].

This growing acceptance of cADPR by scientists with diverse interests was gratifying. It also enticed me to think that the time might be ripe for publishing the structure of NAADP. This was further reinforced by my growing sense of nervousness, since I had not really made any effort to keep the structure of NAADP a secret. Finally, I submitted the manuscript that I had written five years earlier to the Journal of Biological Chemistry, which was published in February of 1995 [16]. Two issues later, in the same journal, Chini *et al.* from Thomas Dousa's laboratory also published a paper identifying NAADP as the active derivative responsible for Ca^{2+} release in sea urchin egg homogenates. Not only the techniques but also the approach and rationale they used in solving the structure were surprisingly similar to mine [17].

Unlike cADPR, NAADP had previously been described. A routine search of the literature using NAADP as a keyword revealed that it was first described as a derivative of NADP in 1980 [18], although its Ca^{2+} releasing function was not known at that time.

ADP-RIBOSYL CYCLASE

How prevalent was the enzyme in the egg homogenates responsible for cyclizing NAD into cADPR? Nancy Rusinko, a graduate student, soon showed that it was widely present in all major mammalian tissues [19]. Not only were the synthesizing enzymes ubiquitous but also the enzymes responsible for hydrolyzing cADPR [20]. We were not merely dealing with peculiar enzymes found only in sea urchin eggs but with a metabolic pathway of general relevance. As described above, cADPR is a very stable molecule. Its spontaneous hydrolysis to ADP-ribose takes days even in highly acidic conditions. If cADPR were to serve as a Ca^{2+} messenger, there must exist enzymes capable of breaking it down and thus terminating its signaling function. We had suspected its existence when we observed that the production of cADPR from NAD in the egg homogenates could only reach a steady state concentration.

It has been generally known for more than half a century that NAD can be hydrolyzed to ADP-ribose by a ubiquitous enzyme, NAD glycohydrolase (NADase). The cellular significance of NADase, however, has never been elucidated. It seems rather wasteful metabolically to hydrolyze a coenzyme so crucial in all cellular oxidoreduction reactions to ADP-ribose, a metabolite that has no specific function.

It should be pointed out that the identification of NADase relied entirely on the accurate identification of its hydrolysis product, ADP-ribose. Although ADP-ribose and cADPR are distinct they also have similar charges and sizes. Many conventional methods cannot, in fact, separate the two. More importantly, cADPR had never been identified before, and there was no reason to even suspect its existence. It was thus entirely possible that, in many cases, enzymes originally believed to be NADases, might in fact be cADPR synthesizing enzymes. The similar widespread occurrence of the cADPR synthesizing enzyme shown by Nancy Rusinko [19] and the previously identified NADase further strengthened such a possibility.

In 1991, I demonstrated directly that a previously purified NADase was, in fact, a cADPR synthesizing enzyme that cyclized NAD to cADPR, producing essentially no ADP-ribose [21]. This enzyme was so efficient in synthesizing cADPR that, in seconds, it produced so much cADPR that it not only totally emptied the cADPR-sensitive Ca^{2+} stores in the egg homogenates but also rendered them completely desensitized. The purified enzyme had a specific activity of about 1.2 mmol cADPR/min/mg [21], corresponding to a turnover rate of about 580/sec, ranking it among the most catalytically active enzymes. It was very clear that this enzyme produced cADPR and not ADP-ribose from NAD, and thus was not a NADase. To avoid further confusion, I proposed to call this enzyme ADP-ribosyl cyclase, a name suggested to me by Nelson Goldberg [21]. The enzyme was originally purified from *Aplysia* ovotestis [22, 23]. It was identified as NADase, because the methods used for assaying the enzymatic product were not capable of separating cADPR from ADP-ribose [22]. Much is now known about this novel enzyme. Its crystal structure has been solved [24], its active site and the catalytic residue have been determined [25], and recombinant enzyme has been produced in large-scale expression in yeast [26]. All these advances and the current understanding of the catalytic mechanism of the cyclase are described in separate chapters in this book.

CD38 - A MAMMALIAN ADP-RIBOSYL CYCLASE

Searching the GenBank database, we identified a mammalian protein, CD38, which shared about 30% sequence identity with the *Aplysia* cyclase, suggesting that it was a mammalian homolog [27]. CD38 was a lymphocyte antigen first defined by monoclonal antibody typing [28]. It was widely used as a marker for lymphocyte differentiation. However, its biological function was unknown, nor was it known that it possessed any enzyme activity.

In 1990, I was contacted by Christopher Grimaldi, an associate of Maureen Howard whose laboratory had succeeded in producing recombinant human CD38. When I tested the recombinant CD38, it was clear that it did have cADPR synthesizing activity, but the activity was transient. Further analyses showed that the recombinant CD38 was capable of not only synthesizing cADPR but also hydrolyzing it to ADP-ribose as well [29]. The cADPR hydrolyzing activity was robust, with a very respectable specific activity of 4-5 μmol/mg/min [30]. To determine if this novel property was peculiar to the recombinant protein, I collaborated with Antonio De Flora in Italy. His group had shown that CD38 was naturally expressed on the surface of human red blood cells [31]. The native CD38 in red blood cells was found to possess the same multifunctionality as the recombinant protein [13]. Kim *et al.* [32] and Takasawa *et al.* [33] independently reported similar results.

The cADPR hydrolyzing activity of CD38 is rather unique and it remains the only enzyme that has so far been reported to be capable of doing so. Many common phosphatases and phosphodiesterases cannot hydrolyze cADPR, a fact consistent with its unique cyclic structure. Much of the cADPR hydrolyzing activity we observed in mammalian tissue extracts [20] may indeed be due to CD38. The concept of a single enzyme using different substrates and producing different products is novel and is at variance with textbook dogma about enzymes in general.

More surprising is that CD38 can, in fact, also use NADP as substrate and catalyze another reaction that exchanges the nicotinamide group of NADP with nicotinic acid. The product is NAADP. The same base-exchange reaction is also catalyzed by the *Aplysia* ADP-ribosyl cyclase to produce NAADP [34]. Much is now known of the catalytic mechanism of CD38 and is the subject of a subsequent chapter. The novelty of these enzymes goes beyond catalyzing multiple reactions. The fact that they can produce two structurally distinct products capable of mobilizing Ca^{2+} stores via independent mechanisms suggests that this class of enzymes plays a crucial role in mediating Ca^{2+} signaling in cells.

CD38, although efficient in hydrolyzing cADPR, does not break down NAADP. Unlike cADPR, the structure of NAADP is similar to many nucleotides and is thus sensitive to various nucleotide hydrolyzing enzymes. For example, alkaline phosphatase, by removing the 2'-phosphate converts NAADP to NAAD, and can thus terminate its Ca^{2+} signaling function [35]. Similarly, nucleotide pyrophosphatase can hydrolyze NAADP into 2'-phospho-AMP and nicotinic acid mononucleotide [36]. The widespread presence of these common hydrolytic enzymes in cells and tissues should ensure effective removal of NAADP once its Ca^{2+} signaling function is completed.

Ca^{2+} MOBILIZATION BY cADPR AND NAADP

The most direct way to show that cADPR could indeed mobilize Ca^{2+} *in vivo* was to directly introduce it into live cells. Sea urchin eggs were particularly suited for such a purpose, as they are large and hardy cells that can tolerate microinjection of various substances. Measuring Ca^{2+} changes in a single egg, however, was not a simple matter at the time. In the mid-nineteen eighties, Ca^{2+} imaging systems were not commercially available. We had to assemble an imaging system from scratch. The prototype used an Apple II+ computer with 48K RAM and a digitizer board designed to take computer portraits. Mark Agre, a graduate student, wrote a program and an assembly routine for the digitization system, which he subsequently "upgraded" to an AT computer and a user friendly program. The system worked so efficiently that we continued to use it for many years, even after commercial systems had become popular.

With this system, Peter Dargie showed that cADPR was indeed active in mobilizing intracellular Ca^{2+} stores in live sea urchin eggs and did so in the nanomolar concentration range [7, 8, 19]. Peter Dargie showed that the injected eggs, in fact, were parthenogenetically activated to produce not only the fertilization envelope but also multiple nuclear cycles and DNA synthesis [8]. Convincing as these results might be, skeptism remained, which we finally laid to rest only years later by synthesizing and using caged cADPR [37]. With it, Ca^{2+} mobilization occurred only after photolysis, eliminating any possibility of the changes being caused by injection artefacts or unknown contaminants in the injection buffer. Using microinjection [16] and a caged analog [35], we showed also that NAADP was equally effective in mobilizing the intracellular Ca^{2+} stores of live eggs.

How did cADPR mobilize the stores? The first clue was provided by Antony Galione in 1991 [14]. Soon after the structure of cADPR was published, I was contacted by Bill Busa, who was interested in a laboratory demonstration of the Ca^{2+} mobilizing effects of cADPR for a course he was teaching at the Marine Biological Laboratory in Woods Hole. The unusual stability of the egg homogenates was ideal for such purpose and I sent him some frozen homogenates we had in storage. Galione, at the time, was doing a postdoctoral study in Bill's laboratory and was also involved in the course. The ease, robust response and reproducibility of the homogenate assay apparently convinced him that cADPR-induced Ca^{2+} mobilization was worth pursuing and he went on to show that cADPR released Ca^{2+} via the Ca^{2+}-induced Ca^{2+} release mechanism [14]. Finally, cADPR, an unknown Ca^{2+} messenger, was related to a well-known mechanism of Ca^{2+} mobilization, the only other known mechanism beside the IP_3-mechansim.

ENDOGENOUS CYCLIC ADP-RIBOSE

We discovered cADPR based on its Ca^{2+} releasing activity *in vitro*. A main concern was whether cADPR was, in fact, endogenously present in cells. In principle, to detect endogenous cADPR, one could label cells with radioactive phosphate or adenine. This kind of technique was popular in the early studies on endogenous IP_3. That cADPR was likely to be only a minute fraction of all the phosphate- or adenine- containing metabolites in cells made it unlikely that the method was applicable in our case. In collaboration with Timothy Walseth, we decided to take advantage of the specificity and reproducibility of the egg homogenates and use it as a bioassay for cADPR. We were gratified to find that cADPR was indeed endogenously present in various mammalian tissues at levels of 1-3 pmol/mg protein, corresponding to cellular concentrations of 0.2-0.6 μM [38].

Assays with higher sensitivity than the bioassay are now available, including a radioimmunoassay [39, 40]. Most recently, Richard Graeff has developed an elegant cycling assay with sensitivity comparable with the radioimmunoassay [41]. The approach Richard uses is to reverse the ADP-ribosyl cyclase reaction, producing NAD from cADPR. The resultant NAD is then measured by a coupled-enzyme cycling assay. The cycling method amplifies NAD a couple of thousands times, enabling ready detection of cADPR in the low nanomolar range. This assay does not require proprietary antibodies and all the reagents are readily available commercially.

The endogenous NAADP has not been measured and no assay for NAADP has been reported. In principle, the egg homogenates can be used as a bioaasay. We have also shown that the egg microsomes bind NAADP with high affinity and specificity and can be used to develop a radio-receptor assay for NAADP [42]. Extending the coupled-enzyme approach, but reversing a different enzyme, nicotinamide-mononucleotide adenylytransferase, Richard Graeff has now developed a cycling assay for NAADP with nanomolar sensitivity [43]. This assay should provide a simple and widely applicable method for measuring endogenous NAADP in cells. The advantages and details of these assays for cADPR and NAADP are described in another chapter in this book.

Elevation of cADPR levels is now shown to accompany cellular activation under various conditions, fulfilling an important and necessary criterion for a second messenger. This is the case during fertilization of sea urchin eggs and the increase in cADPR is found to precede the Ca^{2+} wave, consistent with a causal role for cADPR [44]. Stimulation of a surface receptor, such as the T-cell receptor in lymphocytes, is also accompanied with prolonged elevation of cADPR levels [45]. At variance with the IP_3-system, many of the other agonists that signal through the cADPR-pathway are cell permeant and have their sites of action primarily inside the cell instead of at its surface. These include a gaseous messenger, nitric oxide, metabolic factors such as glucose and vitamin B12, and a transcription factor, retinoic acid (reviewed in [36, 46]). Even cell cycle regulation [47-49], an internally controlled process, as well as physical stimuli such as temperature [50, 51] and drought [52, 53], are all found to involve cADPR. Clearly, the cADPR-pathway is emerging as a highly versatile and universal signaling mechanism.

A DECADE OF EXPLORATION

By the early nineteen nineties, the framework of the cADPR-signaling pathway was sketched. The structure of cADPR had been solved, the class of enzymes responsible for its synthesis and breakdown identified, its effectiveness in live cells had been demonstrated, and its endogenous presence in cells and tissues established. In the following years, we saw major advances in many fronts. A concerted effort by investigators world-wide has gradually but surely transformed a rather esoteric observation, that NAD and NADP can activate Ca^{2+} release in a system unresponsive to IP_3, into a vibrant field of research. Many of these landmark advances are described in the chapters of this book.

We now know that cells responsive to cADPR are not only limited to marine invertebrates such as sea urchin, but span three biological kingdoms:

protist, plant and animal (reviewed in [36, 46]). The cellular functions cADPR is involved in are diverse, from mediating stress response in plants (reviewed in [54]), thermosensing in plants and sponges [50, 51], to insulin secretion [55], neutrophil chemotaxis [56] and clonal expansion of human hemopoietic progenitors [57, 58]. The cADPR-signaling pathway thus has important clinical as well as agricultural relevance.

All the known metabolic enzymes of cADPR and NAADP have now been purified and characterized. Using recombinant techniques, Cyrus Munshi has developed a yeast expression system that can produce them in large quantities [26, 59, 60]. With their ready availability, it is now simply a routine laboratory exercise to synthesize gram quantities of cADPR in days. Compared with the two years of laborious struggle at the beginning just to purify microgram quantities of cADPR for structure determination, we have truly come a long way.

These technical advances allow a large number of specific agonists and antagonists of cADPR and NAADP to be synthesized, providing invaluable tools for pharmacological dissection of the pathways (reviewed in [46, 61]). A comprehensive description of the structures, synthesis and pharmacology of these analogs is provided in a subsequent chapter.

Much is also known about the regulatory mechanisms of the pathway. In many cells, ADP-ribosyl cyclase activity is stimulated by a cGMP-dependent process, most likely by protein phosphorylation [62]. This dependency provides a link between Ca^{2+} mobilization mediated by cADPR and the nitric oxide-signaling pathway. The convergence of these two signaling pathways is found to be important in memory formation in the hippocampus [63] as well as in the activation of defensive genes in plants [64]. It is becoming clear that cADPR-signaling does not operate in isolation but, instead, is interwoven into the tapestry of regulatory pathways of the cell.

Accumulating evidence indicates the target of cADPR is the ryanodine receptor [65-67], an intracellular Ca^{2+} channel responsible for Ca^{2+}-induced Ca^{2+} release. The action of cADPR is to increase the sensitivity of the release mechanism to Ca^{2+} [14, 68, 69]. In this mode of action, cADPR serves as a modulator, amplifying Ca^{2+} signals initiated by other means, such as influx or mobilization of other Ca^{2+} stores. Large increases in cellular cADPR induced by stimuli can sensitize the release mechanism so much that even basal Ca^{2+} levels in cells are sufficient to activate Ca^{2+}-induced Ca^{2+} release. In this alternate mode, cADPR is functioning operationally as a Ca^{2+} messenger [70, 71]. This dual mode of action allows cADPR to signal via various forms of Ca^{2+} changes, such as sparks, waves and oscillations [72-74], making it a highly versatile Ca^{2+} signaling mechanism.

The finding that NAADP activates a release mechanism totally independent of cADPR and IP_3 has opened a new page in Ca^{2+} signaling. The general relevance of the NAADP-mechanism has been established (reviewed in [46, 75]) and a wide variety of cells, including invertebrate [16, 76], mammalian [77] and plant cells [78], are known to be responsive to NAADP. Its unique nature is evidenced by the fact that not only the release mechanism is independent from all that are known but also the Ca^{2+} stores sensitive to NAADP are physically separable from the endoplasmic reticulum containing the cADPR- and the IP_3-sensitive stores [16, 73]. It is becoming clear that the existence of multiple Ca^{2+} stores sensitive to distinct Ca^{2+} messengers may well be a necessary prerequisite for cells to differentiate and respond to diverse stimuli that signal through Ca^{2+} mobilization.

This story of cADPR and NAADP has now spanned more than a decade. It is an exciting story that has brought forth many unexpected findings, amending our views and understanding of Ca^{2+} signaling in cells. The ending of this story is not in sight. It has become an evolving saga propelled forward by self-sustaining discoveries that raise more questions than they have answered. It has acquired a force of its own. Like all good stories, one just cannot wait to see what the next chapter will bring.

REFERENCES

1. Rall TW and Sutherland EW. 1958. Formation of a cyclic adenine ribonucleotide by tissue particles. *J. Biol. Chem.* 232: 1065-1076.
2. Berridge MJ. 1983. Rapid accumulation of inositol trisphosphate reveals that agonists hydrolyse polyphosphoinositides instead of phosphatidylinositol. *Biochem. J.* 212: 849-858.
3. Streb H, Irvine RF, Berridge MJ and Schulz I. 1983. Release of Ca^{2+} from a nonmitochondrial intracellular store in pancreatic acinar cells by inositol-1,4,5-trisphosphate. *Nature* 306: 67-69.
4. Hokin MR and Hokin LE. 1953. Enzyme secretion and the incorporation of P^{32} into phospholipides of pancreas slices. *J. Biol. Chem.* 203: 967-977.
5. Michell RH. 1975. Inositol phospholipids and cell surface receptor function. *Biochim. Biophys. Acta* 415: 81-147.
6. Clapper DL and Lee HC. 1985. Inositol trisphosphate induces Ca^{2+}- release from non-mitochondrial stores in sea urchin egg homogenates. *J. Biol. Chem.* 260: 13947-13954.
7. Clapper DL, Walseth TF, Dargie PJ and Lee HC. 1987. Pyridine nucleotide metabolites stimulate calcium release from sea urchin egg microsomes desensitized to inositol trisphosphate. *J. Biol. Chem.* 262: 9561-9568.
8. Dargie PJ, Agre MC and Lee HC. 1990. Comparison of Ca^{2+} mobilizing activities of cyclic ADP-ribose and inositol trisphosphate. *Cell Regul.* 1: 279-290.
9. Axelson JT, Bodley JW and Walseth TF. 1981. A volatile liquid chromatography system for nucleotides. *Anal. Biochem.* 116: 357-360.
10. Lee HC, Walseth TF, Bratt GT, Hayes RN and Clapper DL. 1989. Structural determination of a cyclic metabolite of NAD^+ with intracellular Ca^{2+}-mobilizing activity. *J. Biol. Chem.* 264: 1608-1615.
11. Kim H, Jacobson EL and Jacobson MK. 1993. Position of cyclization in cyclic ADP-ribose. *Biochem. Biophys. Res. Commun.* 194: 1143-1147.
12. Lee HC, Aarhus R and Levitt D. 1994. The crystal structure of cyclic ADP-ribose. *Nature Struct. Biol.* 1: 143-144.

13. Lee HC, Zocchi E, Guida L, Franco L. Benatti U and De Flora A. 1993. Production and hydrolysis of cyclic ADP-ribose at the outer surface of human erythrocytes. *Biochem. Biophys. Res. Commun.* 191: 639-645.
14. Galione A, Lee HC and Busa WB. 1991. Ca^{2+}-induced Ca^{2+} release in sea urchin egg homogenates: modulation by cyclic ADP-ribose. *Science* 253: 1143-1146.
15. Beers KW, Chini EN, Lee HC and Dousa TP. 1995. Metabolism of cyclic ADP-ribose in opossum kidney renal epithelial cells. *Am. J. Physiol.* 268: C741-C746.
16. Lee HC and Aarhus R. 1995. A derivative of NADP mobilizes calcium stores insensitive to inositol trisphosphate and cyclic ADP-ribose. *J. Biol. Chem.* 270: 2152-2157.
17. Chini EN, Beers KW and Dousa TP. 1995. Nicotinate adenine dinucleotide phosphate (NAADP) triggers a specific calcium release system in sea urchin eggs. *J. Biol. Chem.* 270: 3216-3223.
18. Bernofsky C. 1980. Nicotinic acid adenine dinucleotide phosphate (NAADP). *Meth. Enzymol.* 66: 105-112.
19. Rusinko N and Lee HC. 1989. Widespread occurrence in animal tissues of an enzyme catalyzing the conversion of NAD^+ into a cyclic metabolite with intracellular Ca^{2+}-mobilizing activity. *J. Biol. Chem.* 264: 11725-11731.
20. Lee HC and Aarhus R. 1993. Wide distribution of an enzyme that catalyzes the hydrolysis of cyclic ADP-ribose. *Biochim. Biophys. Acta* 1164: 68-74.
21. Lee HC and Aarhus R. 1991. ADP-ribosyl cyclase: an enzyme that cyclizes NAD^+ into a calcium-mobilizing metabolite. *Cell Regul.* 2: 203-209.
22. Hellmich MR and Strumwasser F. 1991. Purification and characterization of a molluscan egg-specific NADase, a second-messenger enzyme. *Cell Regul.* 2: 193-202.
23. Glick DL, Hellmich MR, Beushausen S, Tempst P, Bayley H and Strumwasser F. 1991. Primary structure of a molluscan egg-specific NADase, a second-messenger enzyme. *Cell Regul.* 2: 211-218.
24. Prasad GS, McRee DE, Stura EA, Levitt DG, Lee HC and Stout CD. 1996. Crystal structure of Aplysia ADP ribosyl cyclase, a homologue of the bifunctional ectozyme CD38. *Nature Struct. Biol.* 3: 957-964.
25. Munshi C, Thiel DJ, Mathews II, Aarhus R, Walseth TF and Lee HC. 1999. Characterization of the active site of ADP-ribosyl cyclase. *J. Biol. Chem.* 274: 30770-30777.
26. Munshi C and Lee HC. 1997. High-level expression of recombinant Aplysia ADP-ribosyl cyclase in Pichia Pastoris by fermentation. *Prot. Express. Purif.* 11: 104-110.
27. States DJ, Walseth TF and Lee HC. 1992. Similarities in amino acid sequences of *Aplysia* ADP-ribosyl cyclase and human lymphocyte antigen CD38. *Trends Biochem. Sci.* 17: 495.
28. Reinherz EL, Kung PC, Goldstein G, Levey RH and Schlossman SF. 1980. Discrete stages of human intrathymic differentiation: analysis of normal thymocytes and leukemic lymphoblasts of T-cell lineage. *Proc. Natl. Acad. Sci. USA* 77: 1588-1592.
29. Howard M, Grimaldi JC, Bazan JF, Lund FE, Santos-Argumedo L, Parkhouse RM, Walseth TF and Lee HC. 1993. Formation and hydrolysis of cyclic ADP-ribose catalyzed by lymphocyte antigen CD38. *Science* 262: 1056-1059.
30. Munshi C, Aarhus R, Graeff R, Walseth TF, Levitt D and Lee HC. 2000. Identification of the enzymatic active site of CD38 by site-directed mutagenesis. *J. Biol. Chem.* 275: 21566-21571.
31. Zocchi E, Franco L, Guida L, Benatti U, Bargellesi A, Malavasi F, Lee HC and De Flora A. 1993. A single protein immunologically identified as CD38 displays NAD^+ glycohydrolase, ADP-ribosyl cyclase and cyclic ADP-ribose hydrolase activities at the outer surface of human erythrocytes. *Biochem. Biophys. Res. Commun.* 196: 1459-1465.
32. Kim H, Jacobson EL and Jacobson MK. 1993. Synthesis and degradation of cyclic ADP-ribose by NAD glycohydrolases. *Science* 261: 1330-1333.

33. Takasawa S, Tohgo A, Noguchi N, Koguma T, Nata K, Sugimoto T, Yonekura H and Okamoto H. 1993. Synthesis and hydrolysis of cyclic ADP-ribose by human leukocyte antigen CD38 and inhibition of the hydrolysis by ATP. *J. Biol. Chem.* 268: 26052-26054.
34. Aarhus R, Graeff RM, Dickey DM, Walseth TF and Lee HC. 1995. ADP-ribosyl cyclase and CD38 catalyze the synthesis of a calcium-mobilizing metabolite from NADP. *J. Biol. Chem.* 270: 30327-30333.
35. Lee HC, Aarhus R, Gee KR and Kestner T. 1997. Caged nicotinic acid adenine dinucleotide phosphate - Synthesis and use. *J. Biol. Chem.* 272: 4172-4178.
36. Lee HC. 1997. Mechanisms of calcium signaling by cyclic ADP-ribose and NAADP. *Physiol. Rev.* 77: 1133-1164.
37. Aarhus R, Gee K and Lee HC. 1995. Caged cyclic ADP-ribose - Synthesis and use. *J. Biol. Chem.* 270: 7745-7749.
38. Walseth TF, Aarhus R, Zeleznikar RJ, Jr. and Lee HC. 1991. Determination of endogenous levels of cyclic ADP-ribose in rat tissues. *Biochim. Biophys. Acta* 1094: 113-120.
39. Takahashi K, Kukimoto I, Tokita K, Inageda K, Inoue S, Kontani K, Hoshino S, Nishina H, Kanaho Y and Katada T. 1995. Accumulation of cyclic ADP-ribose measured by a specific radioimmunoassay in differentiated human leukemic HL-60 cells with all-trans-retinoic acid. *FEBS Lett.* 371: 204-208.
40. Graeff RM, Walseth TF and Lee HC. 1997. A radio-immunoassay for measuring endogenous levels of cyclic ADP-ribose in tissues. *Meth. Enzymol.* 280: 230-241.
41. Graeff R and Lee HC. 2002. A novel cycling assay for cellular cyclic ADP-ribose with nanomolar sensitivity. *Biochem. J.* 361: 379-384.
42. Aarhus R, Dickey DM, Graeff RM, Gee KR, Walseth TF and Lee HC. 1996. Activation and inactivation of Ca^{2+} release by $NAADP^{+}$. *J. Biol. Chem.* 271: 8513-8516.
43. Graeff R and Lee HC. 2002. A novel cycling assay for NAADP with nanomolar sensitivity. *Biochem. J.* (in press).
44. Kuroda R, Kontani K, Kanda Y, Katada T, Nakano T, Satoh Y-i, Suzuki N and Kuroda H. 2001. Increase of cGMP, cADP-ribose and inositol 1,4,5-trisphosphate preceding Ca^{2+} transients in fertilization of sea urchin eggs. *Development* 128: 4405-4414.
45. Guse AH, da Silva CP, Berg I, Skapenko AL, Weber K, Heyer P, Hohenegger M, Ashamu GA, Schulze-Koops H, Potter BVL and Mayr GW. 1999. Regulation of calcium signalling in T lymphocytes by the second messenger cyclic ADP-ribose. *Nature* 398: 70-73.
46. Lee HC. 2001. Physiological functions of cyclic ADP-ribose and NAADP as calcium messengers. *Ann. Rev. Pharmacol. Toxicol.* 41: 317-345.
47. Masuda W, Takenaka S, Inageda K, Nishina H, Takahashi K, Katada T, Tsuyama S, Inui H, Miyatake K and Nakano Y. 1997. Oscillation of ADP-ribosyl cyclase activity during the cell cycle and function of cyclic ADP-ribose in a unicellular organism, Euglena Gracilis. *FEBS Lett.* 405: 104-106.
48. Zocchi E, Daga A, Usai C, Franco L, Guida L, Bruzzone S, Costa A, Marchetti C and Deflora A. 1998. Expression of CD38 increases intracellular calcium concentration and reduces doubling time in HeLa and 3T3 cells. *J. Biol. Chem.* 273: 8017-8024.
49. Franco L, Zocchi E, Usai C, Guida L, Bruzzone S, Costa A and De Flora A. 2001. Paracrine roles of NAD^{+} and cyclic ADP-ribose in increasing intracellular calcium and enhancing cell proliferation of 3T3 fibroblasts. *J. Biol. Chem.* 276: 21642-21648.
50. Zocchi E, Carpaneto A, Cerrano C, Bavestrello G, Giovine M, Brozzone S, Guida L, Franco L and Usai C. 2001. The temperature-signaling cascade in sponges involves a heat-gated cation channel, abscisic acid and cyclic ADP-ribose. *Proc. Natl. Acad. Sci. USA* 98: 14859-14864.
51. Sangwan V, Foulds I, Singh J and Dhindsa RS. 2001. Cold-activation of Brassica napus BN115 promoter is mediated by structural changes in membranes and cytoskeleton, and requires Ca^{2+} influx. *Plant J.* 27: 1-12.

52. Wu Y, Kuzma J, Marechal E, Graeff R, Lee HC, Foster R and Chua NH. 1997. Abscisic acid signaling through cyclic ADP-ribose in plants. *Science* 278: 2126-2130.
53. Leckie CP, McAinsh MR, Allen GJ, Sanders D and Hetherington AM. 1998. Abscisic acid-induced stomatal closure mediated by cyclic ADP-ribose. *Proc. Natl. Acad. Sci. USA* 95: 15837-15842.
54. Lee HC. 2002. Cyclic ADP-ribose and NAADP. Roles in cellular response to stress. In *Sensing, Signalling and Cell Adaptation*, Storey KB and Storey JM (eds), vol. 3, pp. 91-104, Elsevier.
55. Okamoto H. 1999. The CD38-cyclic ADP-ribose signaling system in insulin secretion. *Mol. Cell. Biochem.* 193: 115-118.
56. Partida-Sanchez S, Cockayne D, Monard S, Jacobson EL, Oppenheimer N, Garvy B, Kusser K, Goodricj S, Howard M, Harmsen A, Randall T and Lund FE. 2001. Cyclic ADP-ribose production by CD38 regulates intracellular calcium release, extracellular calcium influx and chemotaxis in neutrophils and is required for bacterial clearance in vivo. *Nature Med.* 7: 1209-1216.
57. Podesta M, Zocchi E, Pitto A, Usai C, Franco L, Bruzzone S, Guida L, Bacigalupo A, Scadden DT, Walseth TF, De Flora A and Daga A. 2000. Extracellular cyclic ADP-ribose increases intracellular free calcium concentration and stimulates proliferation of human hemopoietic progenitors. *FASEB J.* 14: 680-690.
58. Zocchi E, Podesta M, Pitto A, Usai C, Bruzzone S, Franco L, Guida L, Bacigalupo A and De Flora A. 2001. Paracrinally stimulated expansion of early human hemopoietic progenitors by stroma-generated cyclic ADP-ribose. *FASEB J.* 15: 1610-1612.
59. Munshi CB, Fryxell KB, Lee HC and Branton WD. 1997. Large scale production of human CD38 in yeast by fermentation. *Meth. Enzymol.* 280: 318-330.
60. Hussain AMM, Lee HC and Chang CF. 1998. Functional expression of secreted mouse BST-1 in yeast. *Prot. Express. Purif.* 12: 133-137.
61. Lee HC, Munshi C and Graeff R. 1999. Structures and activities of cyclic ADP-ribose, NAADP and their metabolic enzymes. *Mol. Cell. Biochem.* 193: 89-98.
62. Galione A, White A, Willmott N, Turner M, Potter BV and Watson SP. 1993. cGMP mobilizes intracellular Ca^{2+} in sea urchin eggs by stimulating cyclic ADP-ribose synthesis. *Nature* 365: 456-459.
63. Reyes-Harde M, Potter BVL, Galione A and Stanton PK. 1999. Induction of hippocampal LTD requires nitric-oxide-stimulated PKG activity and Ca^{2+} release from cyclic ADP-ribose-sensitive stores. *J. Neurophysiol.* 82: 1569-1576.
64. Durner J, Wendehenne D and Klessig DF. 1998. Defense gene induction in tobacco by nitric oxide, cyclic GMP, and cyclic ADP-ribose. *Proc. Natl. Acad. Sci. USA* 95: 10328-10333.
65. Perez CF, Marengo JJ, Bull R and Hidalgo C. 1998. Cyclic ADP-ribose activates caffeine-sensitive calcium channels from sea urchin egg microsomes. *Am. J. Physiol.* 274: C430-C439.
66. Lokuta AJ, Darszon A, Beltran C and Valdivia HH. 1998. Detection and functional characterization of ryanodine receptors from sea urchin eggs. *J. Physiol.* 510.1: 155-164.
67. Li P-L, Tang W-X, Valdivia HH, Zou A-P and Campbell WB. 2001. cADP-ribose activates reconstituted ryanodine receptors from coronary arterial smooth muscle. *Am. J. Physiol.* 280: H208-H215.
68. Lee HC. 1993. Potentiation of calcium- and caffeine-induced calcium release by cyclic ADP-ribose. *J. Biol. Chem.* 268: 293-299.
69. Lee HC, Aarhus R, Graeff R, Gurnack ME and Walseth TF. 1994. Cyclic ADP ribose activation of the ryanodine receptor is mediated by calmodulin. *Nature* 370: 307-309.
70. Lee HC, Aarhus R and Graeff RM. 1995. Sensitization of calcium-induced calcium release by cyclic ADP-ribose and calmodulin. *J. Biol. Chem.* 270: 9060-9066.
71. Lee HC. 1996. Modulator and messenger functions of cyclic ADP-ribose in calcium signaling. *Re. Prog. Horm. Res.* 51: 355-388.

72. Cui Y, Galione A and Terrar DA. 1999. Effects of photoreleased cADP-ribose on calcium transients and calcium sparks in myocytes isolated from guinea-pig and rat ventricle. *Biochem. J.* 342: 269-273.
73. Lee HC and Aarhus R. 2000. Functional visualization of the separate but interacting calcium stores sensitive to NAADP and cyclic ADP-ribose. *J. Cell Sci.* 113: 4413-4420.
74. Churchill GC and Galione A. 2001. NAADP induces Ca^{2+} oscillations via a two-pool mechanism by priming IP_3- and cADPR-sensitive Ca^{2+} stores. *EMBO J.* 20: 2666-2671.
75. Lee HC. 2000. NAADP: An emerging calcium signaling molecule. *J. Memb. Biol.* 173: 1-8.
76. Albrieux M, Lee HC and Villaz M. 1998. Calcium signaling by cyclic ADP-ribose, NAADP, and inositol trisphosphate are involved in distinct functions in Ascidian oocytes. *J. Biol. Chem.* 273: 14566-14574.
77. Cancela JM, Churchill GC and Galione A. 1999. Coordination of agonist-induced Ca^{2+}-signalling patterns by NAADP in pancreatic acinar cells. *Nature* 398: 74-76.
78. Navazio L, Bewell MA, Siddiqua A, Dickinson GD, Galione A and Sanders D. 2000. Calcium release from the endoplasmic reticulum of higher plants elicited by the NADP metabolite nicotinic acid adenine dinucleotide phosphate. *Proc Natl Acad Sci USA* 97: 8693-8698.

Hon Cheung Lee, Department of Pharmacology, University of Minnesota,
321 Church St., SE, 4-145 Jackson Hall, Minneapolis, MN 55455, USA.
Tel: 612-625-7120 (office), 612-625-4641 (lab)
Fax: 612-625-0991
e-mail: leehc@tc.umn.edu
Web site: http://www.tc.umn.edu/~leehc

Cyclic ADP-Ribose and NAADP. Structures, Metabolism and Functions. Lee HC (ed), pp. 23-43, Dordrecht: Kluwer.

ADP-Ribosyl Cyclases - A Family of cADPR and NAADP Metabolizing Enzymes

Hon Cheung Lee, Cyrus B. Munshi and Richard Graeff
Department of Pharmacology, University of Minnesota, Minneapolis, MN 55455, USA.

INTRODUCTION

ADP-ribosyl cyclase cyclizes NAD to produce cyclic ADP-ribose (cADPR) and releases nicotinamide in the process [1]. It is a ubiquitous enzyme widely present in animal cells [2, 3] and has even been identified in *Euglena*, a protozoan [4]. The first cyclase purified was from *Aplysia* ovotestis [1, 5]. It is a soluble protein of approximately 30 kDa. Sequence comparison reveals that CD38, a lymphocyte antigen, is a mammalian homolog [6]. CD38 is a transmembrane glycoprotein of approximately 45 kDa [7]. It catalyzes not only the synthesis of cADPR but also its hydrolysis to ADP-ribose [8-11]. Additionally, both CD38 and the *Aplysia* cyclase use NADP as a substrate and catalyze the exchange of the nicotinamide group with nicotinic acid, producing nicotinic acid adenine dinucleotide phosphate (NAADP) [12]. The *Aplysia* cyclase has been crystallized and its structure solved [13, 14]. Current knowledge of the structures of these enzymes and the novel multiplicity of their catalytic activities are described in this chapter. Since both of their products are Ca^{2+} messengers, this class of enzymes is crucial in mediating Ca^{2+} signaling functions in cells. Some of their physiological functions have now been revealed by gene ablation and are summarised here.

A NEW FAMILY OF ENZYMES

In addition to the *Aplysia* cyclase and CD38, CD157 (BST-1), another antigen, is also a homolog [15, 16]. The three proteins share 25-30% sequence identity (see below). More importantly, they share the same function, cyclizing NAD to produce cADPR. Another member of this growing family is an ADP-ribosyl cyclase purified from *Euglena*, a membrane–bound enzyme of 40 kDa [4].

Enzymes are generally classified by the specific products they produce. Since cADPR is a newly identified metabolite, it is appropriate to group all the cADPR synthesizing enzymes together in a new family of ADP-ribosyl cyclases. Two of the members of this family, CD38 and CD157 have very similar catalytic properties, possessing both cADPR synthesizing and hydrolyzing activities [9, 15], while the other two members, the *Aplysia* and *Euglena* cyclases, lack the hydrolase activity [1, 4]. The dominant enzymatic product of CD38 and CD157 from NAD is thus ADP-ribose, with cADPR being only a minor component. The overall reaction is similar to classical NADases. In contrast, the *Aplysia* and *Euglena* cyclases, are clearly different from NADase as they produce cADPR instead of ADP-ribose.

A simple method has been devised to operationally distinguish ADP-ribosyl cyclase from NADase by using nicotinamide guanine dinucleotide (NGD) as a substrate instead of NAD. All members of the cyclase family cyclize NGD to cyclic GDP-ribose (cGDPR), a fluorescent product [17, 18]. The reason that cGDPR is fluorescent while cADPR is not is because the site of cyclization is at the 7-nitrogen of the purine ring instead of the 1-nitrogen as shown in Figure 1 [19]. The cyclization product of nicotinamide hypoxanthine dinucleotide (NHD), cIDPR, is likewise fluorescent because the site of cyclization is also at the 7-nitrogen [19].

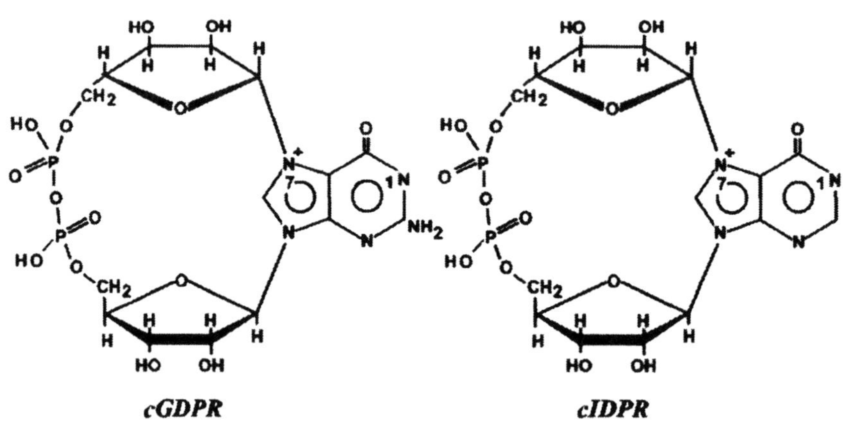

Figure 1. **Structures of cyclic GDP-ribose and cyclic IDP-ribose.**

NADase, such as that in *Neurospora*, cannot cyclize NGD but instead hydrolyzes it to GDP-ribose, a non-fluorescent product [19-21]. ADP-ribosyl transferases, which also possess intrinsic NADase activity [22], likewise cannot cyclize NGD or NHD to produce the fluorescent products. Using this simple fluorimetric assay, ADP-ribosyl cyclase can be easily distinguished from other NADases and the cyclases can be operationally classified.

THE MULTI-FUNCTIONALITY OF THE ADP-RIBOSYL CYCLASES

Figure 2 lists three classes of reactions, cyclization, hydrolysis and base-exchange, that are catalyzed by many of the enzymes in the cyclase family. All members of the family catalyze the cyclization reaction. The hydrolysis reaction is catalyzed by CD38 and CD157, but not by the *Aplysia* nor the *Euglena* cyclases. Both CD38 and the *Aplysia* cyclase catalyze the base-exchange reaction. Of all the products listed in Fig. 2, only cADPR and NAADP have Ca^{2+} signaling functions. Two of the products of the cyclization reaction, cGDPR and cIDPR, are fluorescent.

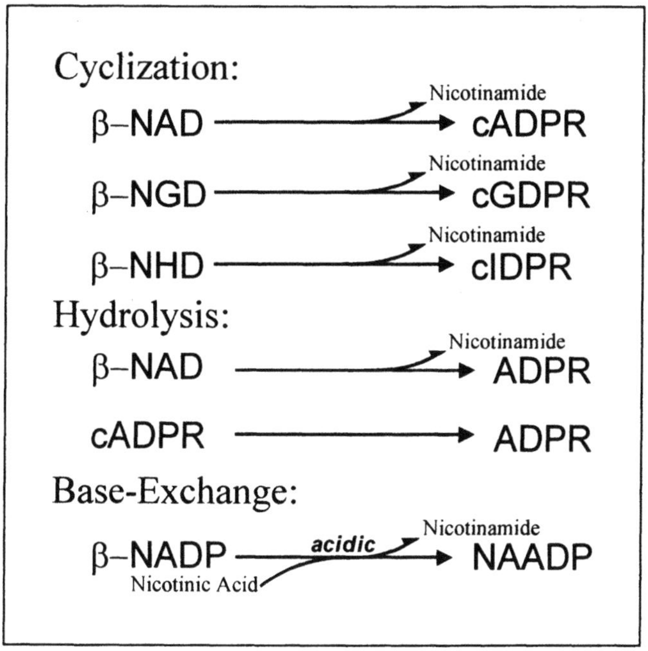

Figure 2. **Reactions catalyzed by the cyclase family of enzymes.**

Many other closely related structural analogs of NAD, such as those with various substitutions at the 8-position of the adenine ring [23, 24], are also cyclized, providing a versatile method for preparing a variety of cADPR analogs, including specific antagonists and agonists [23-27]. The reaction is, however, stereo specific and α-NAD is not a substrate [1].

Concerning the hydrolysis reaction, CD38 is the only enzyme reported so far that can efficiently hydrolyze cADPR. Many of the common nucleotidases, phosphatases or phosphodiesterases are not capable of doing so. This unique stability of cADPR toward common hydrolytic enzymes has been exploited in developing the cycling assay for endogenous cADPR and is described in a later chapter. In contrast to the cyclization reaction, the hydrolysis reaction shows a higher degree of substrate specificity. Many of the close structural analogs of cADPR, such as cGDPR [17, 19, 20], 7-deaza-8-Br-cADPR [28] or 3-deaza-cADPR [27], are resistant to hydrolysis by CD38 and are useful pharmacological tools for investigating cell systems that are metabolically active.

The base-exchange reaction also shows a certain degree of structural specificity with respect to the substrate, nicotinic acid. The 2-position of the pyridine ring is most crucial. Substitutions at that position render the substrate incapable of exchanging with the nicotinamide group of NADP [29]. On the other hand, neither the negative charge nor the position of the 3-carboxyl group of nicotinic acid is required for the exchange reaction.

An important factor governing the catalysis of the cyclases is pH. At neutral or alkaline pH, the cyclization and the hydrolysis reaction are dominant, while the base-exchange reaction is preferred at acidic pH [12]. This unusual acidic requirement has led to the proposal that the product of the exchange reaction, NAADP, may be important in mediating Ca^{2+} signaling in the endocytic or lyosomal pathways [30, 31], since the organelles involved are acidic.

STRUCTURAL HOMOLOGY OF THE ADP-RIBOSYL CYCLASE FAMILY

Despite sequence identity of 25-30%, the members of the cyclase family are a diverse group of proteins. The *Aplysia* cyclase is a soluble protein, while the rest are membrane-bound. As illustrated in Figure 3, CD38 is a type II glycoprotein with a single stretch of transmembrane sequence (M, Fig. 3) near the amino-terminus [7]. It has four potential glycosylation sites (diamonds, Fig. 3). CD157, is also a glycoprotein, but it is an glycosyl-phosphatidylinositol-anchored protein with the anchoring site (GPI, Fig. 3) at the carboxyl-terminus [16].

Two main similarities shared by all three proteins are the presence of a highly conserved sequence (C, Fig. 3) of 18 residues in the middle of the molecules and the positions of the cysteines (black lines, Fig. 3). Structural evidence indicates the conserved sequence is part of the active site of the cyclase and CD38 [32, 33]. The similarity of the conserved sequence is higher between the *Aplysia* cyclase and CD38, but is less so with CD157. Indeed, compared to the cyclase and CD38, the enzyme activities of CD157 are, in fact, also very much lower [15, 34]. Site directed mutagensis analyses identify three critical residues, Trp77, Trp140 and Glu179, in the *Aplysia* cyclase [32]. In CD38, the corresponding residues, Trp125, Trp189 and

Glu226, as well as Glu146, are likewise important for its enzymatic activities [33].

Another notable similarity in the three proteins is the location of the cysteines. *Aplysia* cyclase has ten cysteines and all can be aligned perfectly with those of the CD38 and CD157. The crystal structure of the cyclase shows that all cysteines are paired by disulfide linkages (s-s, Fig. 3) [14]. Homology modelling of CD38 indicates that the ten corresponding cysteines as well as the extra pair, Cys119 and Cys201 (dashed line labelled s-s, Fig. 3), are all close enough to form disulfide bonds also [14, 33].

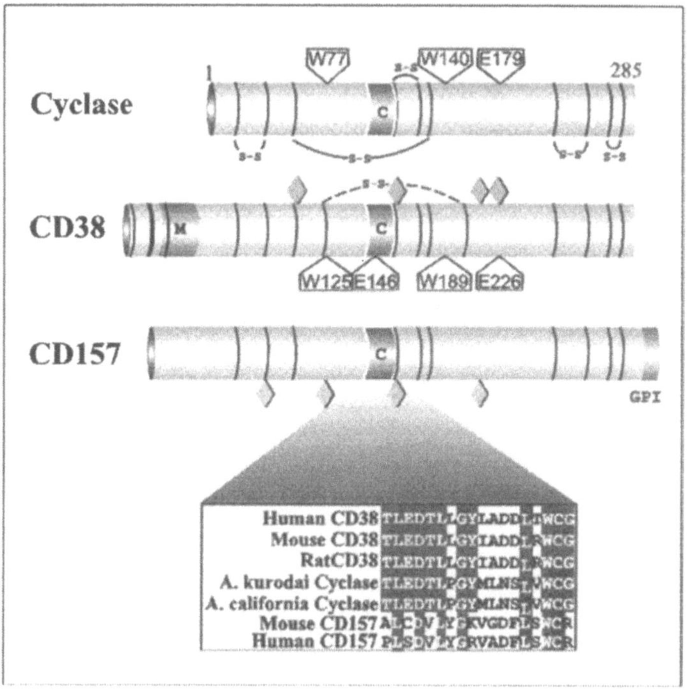

Figure 3. **Structural homology of the ADP-ribosyl cyclase family.**

DISTRIBUTION OF ADP-RIBOSYL CYCLASES IN CELLS AND TISSUES

When expressed on the cell surface, both CD38 and CD157 are ecto-enzymes with their catalytic domains located outside the cell [8, 15, 35]. How these ecto-enzymes can produce intracellular signaling is a topological paradox that has recently been elucidated [36]. It is found that CD38 can function as a transporter during catalysis, transporting the enzymatic product, cADPR, across the plasma membrane and into cells, where it functions as a Ca^{2+} messenger to mobilize the intracellular Ca^{2+} stores [37].

This novel function of CD38 accounts for the observations that NAD added externally can effect internal Ca^{2+} mobilization in several cell types, including fibroblasts [38], astrocytes [39], osteoclasts [40] and cerebellar granule cells [41]. Additionally, CD38 can be induced to internalize by its substrate, NAD, and exert its signaling function internally [38, 42, 43]. The topology of CD38 signaling is described in more detail in the chapter by De Flora *et al.* in this volume.

In many cells, CD38 is normally an intracellular enzyme expressed on various organelles, as revealed by immuno-electronmicrospy [44]. Of particular interest is its presence in the inner membrane of the nucleus [45-47]. It has been shown that the nuclear envelope itself is a functioning Ca^{2+} store mobilizable by cADPR [45, 46, 48]. The presence of both the metabolic enzyme of cADPR and the Ca^{2+} release mechanism sensitive to the messenger in the nucleus suggests that cADPR may play a direct role in mediating nuclear activation and gene expression.

Although CD38 was first identified as a lymphocyte antigen by antibody typing [49], it should not be regarded as lymphocyte specific. It is now known to be widely expressed not only in a variety of other blood cells, including even red blood cells [8, 50], but is also ubiquitous in other tissues, including brain [44], liver [46, 51], retina [52] and pancreas [53, 54]. Indeed, its presence in pancreatic β-cells is found to play a significant role in mediating insulin secretion [53, 54].

CD157 was similarly identified first as an antigen on bone marrow stromal cells but later shown to be present in various blood cells and other human tissues, such as small intestine and liver (reviewed in [55]). Transgenic mice deficient in CD157 show impaired local mucosal immune response to oral antigenic challenge [55, 56]. It is clear that both CD38 and CD157 have a wide range of distribution as well as functions, and their ecto-expression on the cell surface is just one of them.

The *Aplysia* cyclase is an intracellular enzyme first purified from ovotestis, in which large quantities are present [1, 5]. Immuno-staining localizes the cyclase inside granules of the oocytes [5]. It is also present in the presynatic neurons in the buccal ganglion of the animal and is believed to be involved in regulating neurotransmitter release at the cholinergic synapse [57]. In fact, neuronal acetylcholine release is found to be modulated by cADPR and NAADP; both are products of the cyclase [58].

The *Euglena* cyclase is also an intracellular enzyme [4]. The amino acid sequence of this membrane-bound enzyme has not been determined. It is not known if it has higher homology with the soluble *Aplysia* cyclase or with the membrane-bound CD38. The activity of the *Euglena* cyclase and the cellular levels of its product, cADPR, both exhibit periodic increases preceding and correlating with the cell cycle [59]. Since the *Euglena* Ca^{2+} stores are also sensitive to cADPR, its periodic increases can function as causal signals for and/or time keeping mechanism of the cell cycle [60].

STRUCTURE OF THE *APLYSIA* ADP-RIBOSYL CYCLASE

The structures of cADPR, NAADP and their metabolic enzymes, the *Apylsia* cyclase and CD38, are shown in Figure 4. The structures are rendered using the program MolMol [61]. Cyclic ADP-ribose is a cyclic molecule and is very compact. The space-filling model shown in Fig. 4A is determined by X-ray crystallography [62]. In contrast, NAADP is a long linear molecule. The crystal structure of NAADP has not been determined. The structure shown in Fig. 4B is modeled using the crystal coordinates of NADP bound to isocitrate dehydrogenase [63], which is then changed to NAADP and its energy minimized using the program ChemBuilder3D. It is remarkable that despite their clearly disparate structures, both cADPR and NAADP are synthesized by the same enzyme [12].

The *Aplysia* ADP-ribosyl cyclase is a homo-dimer as revealed by X-ray crystallography [14]. It is also a dimer in solution as shown by chemical cross-linking and dynamic laser light scattering studies [64]. Fig. 4C depicts the dimer with one monomer rendered with the van der Waals contact surfaces, while the other shows the secondary structures. The two halves of each monomer are distinct, with four large β-sheets (green) in the carboxyl-domain while the amino-domain is composed mainly of α-helixes (red). The formation of the dimer is very stable and involves three of the ten α-helixes of the molecule. The contacting amino acids include acidic, basic as well as hydrophobic residues [31], indicating the dimer is stablized by both ionic and hydrophobic interactions. The five pairs of disulfide linkages of the molecule (cf. Fig. 3) are shown in cyan, two in the carboxyl-domain and the rest in the amino-domain.

The conserved sequence (TLEDTL, cf. Fig. 3) is shown in yellow in Fig. 4C and is located near the central cleft separating the carboxyl- and amino-domains of the molecule. It, in fact, forms part of the base (yellow) of the active site pocket, which is shown in close-up view in Fig. 4D. Site-directed mutagenesis studies identify three critical residues at the active site [32]. Glu179 is most likely the catalytic residue since its modification renders the cyclase inactive [32]. The residue is located deep inside the active site pocket (Fig. 4D).

The two tryptophans, Trp77 and Trp140, located at the rim of the pocket, are also important. Mutagenizing them results in loss of both the ADP-ribosyl cyclase and the base-exchange activities, but the mutants retain some GDP-ribosyl cyclase activity. These two residues are likely to function in coordinating the substrate NAD and molding it into a folded conformation, so that the two ends of the molecule can be cyclized. The retention of GDP-ribosyl cyclase activity of the mutants is attributable to the alternate cyclization site in cGDPR. As described above, the cyclization of NGD to produce cGDPR occurs at N-7 instead of N-1 in cADPR, requiring rotation

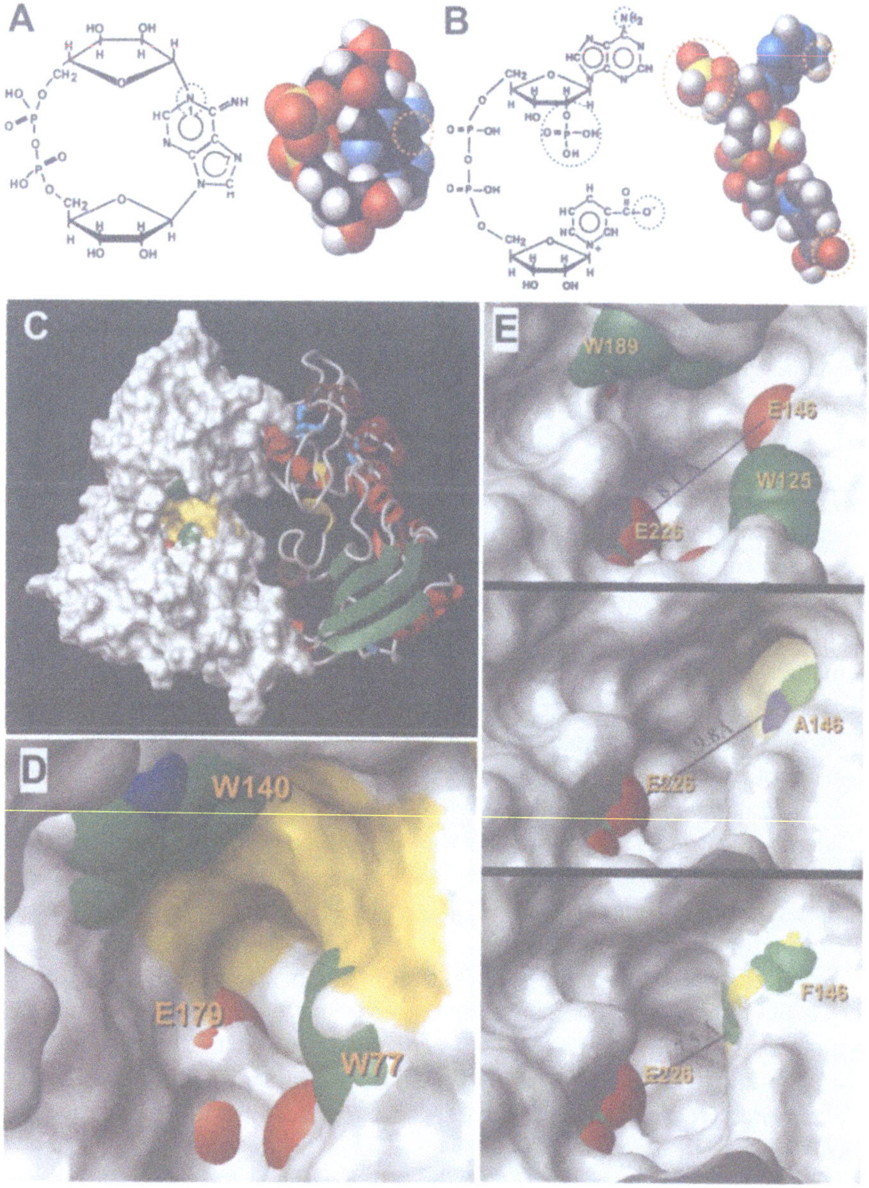

Figure 4. **Structures of cADPR (A), NAADP (B), ADP-ribosyl cyclase (C), and the active sites of both the cyclase (D) and CD38 (E).** The circle in (A) indicates the N1-cyclization site. Circles in (B) indicate determinants important for Ca^{2+} releasing activity of NAADP. Models of the active sites of two of the CD38 mutants, A146 and F146 are also shown in E. The color code for atoms is: oxygen, red; nitrogen, blue; phosphorous, yellow; carbon, black (A, B) or green (C, D, E) and hydrogen, white (A, B) or yellow (E).

of the guanine ring around the glycosidic bond between N-9 and the ribose. The alternate orientation of the base could be less affected by changes in the coordinating tryptophans, allowing the retention of some cyclizing activity when NGD is used as substrate.

STRUCTURE OF CD38

The crystal structure of CD38 has not been determined. However, it is possible to model the structure of its extra-membrane domain using the crystal coordinates of the *Aplysia* cyclase because of the sequence homology between the two proteins. The resulting energy-minimized structure reveals that the two extra pairs of cysteines, Cys119 and Cys201 (cf. Fig. 3) are close enough to form a disulfide bond [14, 33].

More importantly, it also reveals that the residues in CD38, Glu226, Trp125 and Trp189, corresponding to the three critical amino acids of the cyclase are all clustered in a pocket near the middle of the molecule, very similar to the active site pocket of the *Aplysia* cyclase. This is remarkable considering that these residues spread over a third of the sequence of the molecule (cf. Fig. 3). Mutagenizing Glu226 to even highly conserved residues, such as aspartate, results in complete loss of all enzymatic activities [33]. This is not due to non-specific structural change induced by the mutation since the mutated proteins retain the ability to bind NAD and show essentially no change in secondary structure as determined by circular dichroism. That Glu226 is indeed the catalytic amino acid is further supported by the finding that the same residue is covalently ribosylated during catalysis when an alternative substrate analog is used [65].

Similar to the *Aplysia* cyclase, the two tryptophans appear to also serve a substrate coodinating function, since their mutagenesis results in large decrease in all enzyme activities except the GDP-ribosyl cyclase activity [33]. The top panel of Fig. 4E depicts the active site pocket of CD38 with the critical residues identified.

Another critical residue shown in Fig. 4E is Glu146. It is part of the conserved sequence (cf. Fig. 3). As described above, CD38 catalyzes multiple reactions and when incubated with NAD mainly hydrolyzes it to ADP-ribose. The amount of cADPR produced is minute. Mutagenizing Glu146 to alanine greatly increases the cyclase activity and depresses its NAD hydrolyzing activity to the extent that the former is 3-4 fold higher than the latter [66]. The GDP-ribosyl cyclase activity of the mutant protein is likewise increased, while the cADPR hydrolase activity is decreased. A series of mutants of Glu146 has been produced. When the enzymatic activities of all of them are considered, the two cyclase activities and the two hydrolase activities are linearly correlated, indicating that a single residue is responsible for controlling both the cyclizing and hydrolyzing activities of CD38 [66]. The Glu146 replacements, however, only minimally affect the base-exchange activity that is responsible for synthesizing NAADP.

As shown in the top panel of Fig. 4E, Glu146 is located directly across from the catalytic residue, Glu226. It is not situated as deep inside the active site pocket as Glu226 and is closer to the position of the coordinating tryptophan, Trp189, which is at the rim of the pocket. The distance between Glu226 and Glu146 is about 8Å, which is about the width of a molecule of cADPR.

Homology modelling is used to assess the structural changes at the active site caused by the mutations [66]. A model of the active site of the alanine mutant (E146A) is shown in the middle panel of Fig. 4E. Replacement of Glu146 with alanine greatly increases the cyclase activity and depresses the hydrolase activity. The side chain of alanine is smaller than glutamate. The distance between Ala146 and Glu226 is thus larger, at 9.8 Å, making the opening of the active site pocket slightly wider. In addition, the region around the alanine (tinted yellow) appears less restricted as compared to the wildtype. The net effect of these changes likely allows the substrate to bind deeper into the pocket.

Another type of substitution that results in enhanced cyclase activity is increasing hydrophobicity. For example, substituting Glu146 of CD38 with phenylalanine (E146F) enhances the cyclase activity as much as the E146A mutant. The main difference between the two mutants is that E146F has higher hydrolase activity than E146A. A model depicting the structural change at the active site of the E146F mutant is shown in the lower panel of Fig. 4E. The size of the side chain of the phenylalanine is larger than glutamine and the distance between Phe146 and Glu226 is now closer at 7.5 Å. The substitution is likely to reduce access to the catalytic residue. A model accounting for many of the structural and catalytical properties of the ADP-ribosyl cyclase is described below.

In addition to the residues in the active site pocket, three other have also been shown to be important for catalytic activities. Lysine 129 is identified by chemical labelling and site-directed mutagenesis as part of the binding site for ATP, which at high concentration (5-10 mM), can inhibit cADPR hydrolysis activity of CD38 [67]. Lys129 is quite close to active site pocket and changing it to either alanine or arginine eliminates the cADPR hydrolysis activity of CD38, although the enzyme remains active using NAD as substrate [67]. In addition, two cysteine residues, Cys119 and Cys201, are also important for the cADPR hydrolysis activity [68]. These are the two extra cysteine residues that are not conserved in the *Aplysia* cyclase, but are likely to be close enough to form a disulfide linkage as indicated by homology modeling (cf. Fig. 3). Mutagenizing Cys 119 to lysine and/or Cys201 to glutamate preferentially inhibits the cADPR hydrolyzing activity. Changing them to other residues, such as alanine, totally eliminates all activities [68]. If these two cysteine residues are indeed linked by a disulfide, changing them would break the linkage and could induce structural changes in the whole molecule.

STRUCTURE OF CD157/BST-1

The extracellular domain of CD157 has been produced as soluble proteins in different expression systems and is shown to possess ADP-ribosyl cyclase activity [34, 69]. The crystal structure of the extracellular domain of CD157 has recently been solved [70]. Similar to the *Aplysia* cyclase, CD157 also crystallizes as a homo-dimer. In fact, the secondary structure elements of the two proteins are superimposable, with a root-mean-square deviation of only 3.8 Å for the main polypeptide chains [70]. All ten of the conserved cysteines are paired as disulfide linkages, as in the *Aplysia* cyclase. The remarkable similarity extends to the active site, which, in the case of CD157, is identified by co-crystallization with various substrates. Not only are the locations of the active site pockets of the two proteins the same, but also the critical residues, the catalytic glutamate and the two substrate coordinating tryptophans. The three dimensional positions of the three critical residues are, in fact, superimposable between the two proteins, with a root-mean-square deviation of only 0.2 Å. Considering the primary sequences of the two proteins share only about 27% identity, the similarity of the three-dimensional structures is truly remarkable.

A UNIFIED MODEL OF CATALYSIS OF ADP-RIBOSYL CYCLASE

The multi-functionality of the enzymes in the cyclase family is not only catalytically novel, it is also biologically important since two of the products, cADPR and NAADP, are functioning as Ca^{2+} messengers in a wide variety of cells. A unified model of catalysis for these enzymes has been proposed [31, 66, 71]. Accordingly, the first step in the process is the binding of the substrate to the active site pocket, whose structure is such that the bound substrate is molded into a folded conformation so that the two ends can be linked (Fig. 5). The two critical tryptophans, Trp77 and Trp140 in the *Aplysia* cyclase or the Trp125 and Trp189 in CD38, can serve such a function. Co-crystallization of CD157 with various substrates provides direct evidence that substrate recognition is indeed predominantly through van der Waals interactions with the two tryptophans [70]. The active site does not recognize the 2'-position of the ribose and can use either NAD (R is OH, Fig. 5) or NADP (R is P_i, Fig. 5) as substrate.

The next step is the release of the nicotinamide group (hexagon labelled n, Fig. 5) from the substrate and the rest of the molecule then forms an activated intermediate with the enzyme. The nature of this intermediate has not been elucidated, but recent evidence suggests the intermediate is likely to be covalent ribosylation of the catalytic residue, Glu179 in the *Aplysia* cyclase or Glu226 in CD38 [65]. Alternatively, an oxocarbonium ion has also been suggested in analogy with other NAD utilizing enzymes [72]. Structural evidence obtained in CD157 indicates the N-glycosidic bond of the substrate, NAD, is exposed near the catalytic glutamate residue and its carboxyl side-chain stabilizes the catalytic intermediate of the S_N-1 type, more consistent with a non-covalent intermediate [70]. It should be noted that CD157 is catalytically much less active than either the *Aplysia* cyclase or CD38, which may well be related to the apparent differences in the nature of the enzyme intermediate. In any case, kinetic analyses provide support for a single intermediate being responsible for all of the reaction chemistries of CD38 [73].

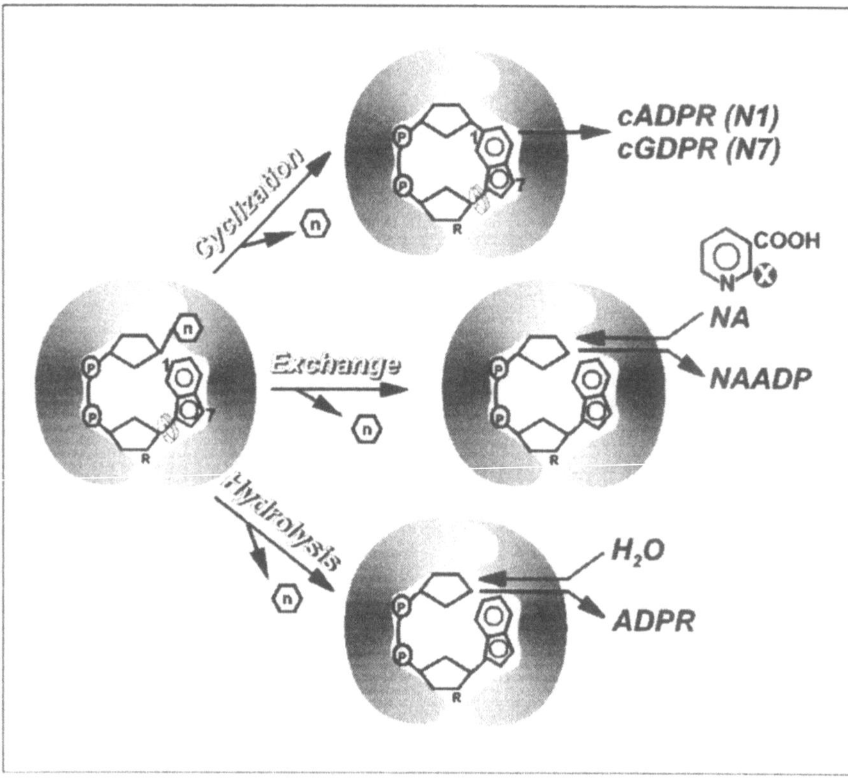

Figure 5. **A unified catalytic model for the family of ADP-ribosyl cyclases.**

Which course the subsequent reaction will take is dependent on the reaction conditions. In the absence of other nucleophilic substrates and at neutral or alkaline conditions, intra-molecular attack on the activated anomeric carbon of the ribose by the 1-nitrogen of the adenine ring would result in cyclization, producing cADPR. All members of the cyclase family also catalyze the cyclization of NGD to cGDPR, which occurs at the 7-nitrogen of the guanine ring instead of the 1-position. This requires a rotation of the base around the glycosidic bond linking to the ribose (cf. Fig. 5). The substrate coordination of the active site must thus support both orientations of the base. Such flexibility could be provided by the presence of two critical tryptophans lining the active site pocket; each could function to coordinate the substrate via hydrophobic interaction with the purine ring of the substrate.

In the presence of a nucleophile such as nicotinic acid (NA, structure in Fig. 5) and under acidic conditions, reaction of the activated intermediate with the nucleophilic substrate dominates over the intra-molecular linkage with the 1-nitrogen. The result is the exchange reaction, producing NAADP. The 2-position of nicotinic acid is crucial (marked by an X on its structure in Fig. 5) and a substitution at that position prevents access to and/or reaction with the activated enzyme intermediate [29].

The third class of reaction, hydrolysis, occurs when water gains access to the active site and reacts with the activated intermediate, producing ADP-ribose. Which one of the three classes of reaction dominates would thus depend on the competition between intra-molecular attack and exogenous nucleophiles, such as nicotinic acid and water, for the activated intermediate. Accordingly, the main difference between the *Aplysia* cyclase and CD38 is that the activated intermediate in the former case is inaccessible to water, making cyclization overwhelmingly predominate.

Accessibility of water can also account for the conversion of CD38 to a cyclase-like enzyme by mutating Glu146. Substitution with alanine reduces the bulkiness of the sidechain, allowing the substrate to bind deeper into the active site pocket (middle panel in Fig. 5E). The net effect would be to occlude the opening and thereby restrict access of water into the active site. Intra-molecular cyclization is thus enhanced with concomitant decrease in hydrolysis, as is observed [66]. In the case of the E146F mutation, the hydrophobic nature of the phenyl group tends to reduce water access to the active site and thus promote cyclization. The enhancing effect is, however, counter-balanced to a certain extent by the bulkiness of the side chain, which reduces the size of the mouth of the active pocket. The distance between Glu226 and Phe146 is now only 7.5 Å (lower panel of Fig. 5), smaller than the width of a molecule of cADPR. This would require the substrate to tilt and insert sideways into the active side instead. The less than perfect fit could result in leakage of water into the site, accounting for the higher hydrolase activity of E146F than E146A [66].

The affinity of the active site for cADPR is also a determining factor for the hydrolysis activity. In the case of the *Aplysia* cyclase, the affinity is likely to be low and once cADPR is formed from NAD it leaves the active site immediately. This would greatly reduce the chance of the activated intermediate reacting with water. The result is high cyclase activity and low cADPR hydrolysis. The low affinity also accounts for the lack of cADPR hydrolyzing activity of the *Aplysia* cyclase, since cADPR would not be able to bind to the active site and thus could not serve as a substrate. The opposite is true for CD38. With high affinity, cADPR binds avidly to the active site and is converted to the activated intermediate. Subsequent reaction with water then results in hydrolysis of cADPR to ADP-ribose. The enzyme thus catalyzes hydrolysis much better than cyclization. This catalytic model provides a simple but rational scheme accounting for many of the functional as well as structural features of this novel family of enzymes.

PHYSIOLOGICAL FUNCTIONS OF CD38

CD38 was first thought to be a lymphocyte antigen and had widely been used as a marker for lymphocyte differentiation [49]. The initial investigations thus focused on its functioning as an antigen. Consistent with this idea is that ligation of surface CD38 by specific antibodies against CD38 can induce a variety of responses, from proliferation to apoptosis, in lymphocytes as well as in hematopoietic cells (reviewed in [74, 75]). However, not all antibodies against CD38 are active and only certain specific antibodies, such as IB4, can produce these effects [76]. Since ligation by antibodies is clearly an artificial intervention, searches have been conducted for possible natural ligands of CD38. CD31 has now been identified as such a ligand, albeit with rather low affinity [77]. Although binding of CD31 to CD38 can elicit Ca^{2+} changes in leukocytes, so far evidence is lacking that it can indeed produce all the effects observed with antibody ligation [78].

It should be noted that CD38 expressed on the cell surface is known to associate with other receptors, such as CD3/TCR in T-cells or BCR in B-cells ([79], reviewed in [78]). Ligation of CD38 by antibodies can potentially activate other cell surface receptors through indirect interactions. It is not entirely clear that this kind of indirect or, perhaps, non-specific activation, should be considered as true signaling through CD38 as an antigen. Indeed, the effects of antibody ligation of CD38 are lost in cells with defective CD3/TCR [76]. Similarly, in natural killer cells that are defective in CD16, their response to ligation by anti-CD38 is also lost [80]. It thus appears that CD38 does not have an antigenic function of its own, but relies on association with other antigens for signaling.

Furthermore, antibody ligation of CD38 so far has not been reported to affect its enzymatic activities. Conversely, cells transfected with CD38 defective in enzymatic activities remain responsive to ligation [81]. It is thus fair to conclude that if antibody ligation can be considered as a true signaling function of CD38, then this function is unlikely to be related to cADPR-signaling.

On the other hand, recent results provide strong and direct evidence that enzymatic production of cADPR by CD38 indeed has important physiological functions in cells and in animals. In 3T3 and HeLa cells, expression of CD38 is accompanied with consumption of cellular NAD, production of cADPR, depletion of intracellular Ca^{2+} stores and elevation of cytosolic Ca^{2+} concentration [82]. The biological consequence is reduction in both the duration of the S phase and the cell doubling time.

In osteoclasts, cells responsible for bone resorption, CD38 is present at the plasma membrane [40]. Incubation with NAD elicits intracellular Ca^{2+} elevation through mobilization of Ca^{2+} stores. Pre-treatment with antagonistic antibodies against CD38, which blocks its ADP-ribosyl cyclase activity, inhibits the NAD-effect [40]. Conversely, agonistic antibodies enhance the NAD-induced Ca^{2+} changes, which in turn inhibit bone resorption.

A more direct way to block the activity of CD38 and assess the biological consequences is to produce knockout mice, which has now been accomplished by two independent groups [54, 83]. The NADase activity in all tissues examined in these mice show dramatic decrease, indicating most, if not all, of mammalian NADase is attributable to CD38, which, as described above, is more appropriately classified as a member of the cyclase family. The cADPR levels in many tissues of the knockout mice are also depressed [54, 83]. Notable exceptions are in the heart and brain, where cADPR levels are similar to wildtype mice [84], indicating the presence of another, as yet unidentified, ADP-ribosyl cyclase in these tissues.

The CD38 knockout mice exhibit several phenotypic defects. In pancreatic acinar cells of the knockout mice, muscarinic Ca^{2+} signaling is found to be aberrant [85]. Wildtype acinar cells are responsive to acetylcholine and show concentration-dependent elevation of endogenous cADPR levels. This response is totally abolished in the knockout mice. The CD38-knockout acinar cells also fail to produce the typical Ca^{2+} spiking in response to acetylcholine as observed in the wildtype cells [85]. Both wildtype and knockout cells respond similarly when cADPR is directly infused into the cells. Therefore, the defect in the knockout cells is not in the process of cADPR-induced Ca^{2+} mobilization but, instead, is because of the failure of acetylcholine to activate cADPR production, due to the lack of CD38.

Not only are the acinar cells defective, the β-cells in the pancreas of the knockout mice also show impairments [54]. It has previously been shown that cADPR plays an important role in mediating glucose-induced insulin secretion in the islets [86, 87]. This topic will be described in more detail in the chapter by Takasawa and Okamoto in this volume. Normally, treatment of the wildtype pancreatic islets with glucose activates production of cADPR and elevation of intracellular Ca^{2+} concentration, resulting in increased insulin secretion [54]. In the CD38-knockout islets, cADPR production fails to increase following the glucose treatment. As a consequence, both the Ca^{2+} changes and the insulin secretion are depressed [54]. Correlating with these cellular defects, the serum levels of insulin following a glucose challenge are depressed as compared with the wildtype mice. The blood glucose of the knockout mice also shows prolonged elevation after the initial glucose challenge, indicating impairments in glucose tolerance [54].

Another notable phenotype in the CD38 knockout mice is the susceptibility to bacterial infection, which is traced to defects in neutrophil chemotaxis [84]. In wildtype mice, bacterial infection induced migration of neutrophils to the infection site, where they confine and combat the infecting bacteria. In so doing, the neutrophils function to prevent the spread of the bacteria to the rest of the body. Neutrophil migration is believed to be chemotactically directed by bacterial products, mainly N-formylpeptides, which, in wildtype neutrophils, induce mobilization of intracellular Ca^{2+} stores. The depletion of the stores then activates capacitative Ca^{2+} influx [84]. The whole process is mediated by cADPR since blocking the mobilization of Ca^{2+} stores with a cADPR antagonist, 8-Br-cADPR, effectively inhibits the store-depletion stimulated Ca^{2+} influx and blocks the peptide-induced neutrophil chemotaxis [84].

Neutrophils isolated from the knockout mice lack CD38 and cannot produce cADPR. They exhibit a much depressed chemotactic response to N-formylpeptides. The peptides also fail to elicit the Ca^{2+}-influx associated with chemotaxis in the mutant neutrophils. The biological consequence is that CD38-knockout neutrophils fail to accumulate at the site of bacterial infection, allowing the infecting bacteria to dissimulate rapidly into the blood of the mutant mice. The bacterial dose producing 50% fatality in the knockout mice is thus ten times lower than in the wildtype mice [84].

CONCLUDING REMARKS

The *in vivo* studies described above establish a clear and causal relationship between CD38 and cADPR-signaling. This is a major advance, which like all important advances, raises further unanswered questions. Principal among them are the exact mechanisms by which the enzymatic activities of CD38 are regulated by external ligands, such as N-formylpeptides or acetylcholine, in cells. More intriguing is the possibility that another, perhaps, redundant ADP-ribosyl cyclase may also exist, as is suggested by the high endogenous cADPR levels measured in some tissues of the CD38 knockout mice [84]. In any case, the studies described demonstrate unequivocally the importance of the cADPR-pathway in the normal functioning of cells and tissues and, in the case of bacterial infection, it is, in fact, crucial to the survival of the animals.

REFERENCES

1. Lee HC and Aarhus R. 1991. ADP-ribosyl cyclase: an enzyme that cyclizes NAD^+ into a calcium-mobilizing metabolite. *Cell Regul.* 2: 203-209.

2. Rusinko N and Lee HC. 1989. Widespread occurrence in animal tissues of an enzyme catalyzing the conversion of NAD^+ into a cyclic metabolite with intracellular Ca^{2+}-mobilizing activity. *J. Biol. Chem.* 264: 11725-11731.
3. Lee HC and Aarhus R. 1993. Wide distribution of an enzyme that catalyzes the hydrolysis of cyclic ADP-ribose. *Biochim. Biophys. Acta* 1164: 68-74.
4. Masuda W, Takenaka S, Tsuyama S, Inui H, Miyatake K and Nakano Y. 1999. Purification and characterization of ADP-ribosyl cyclase from Euglena gracilis. *J. Biochem.* 125: 449-453.
5. Hellmich MR and Strumwasser F. 1991. Purification and characterization of a molluscan egg-specific NADase, a second-messenger enzyme. *Cell Regul.* 2: 193-202.
6. States DJ, Walseth TF and Lee HC. 1992. Similarities in amino acid sequences of *Aplysia* ADP-ribosyl cyclase and human lymphocyte antigen CD38. *Trends Biochem. Sci.* 17: 495.
7. Jackson DG and Bell JI. 1990. Isolation of a cDNA encoding the human CD38 (T10) molecule, a cell surface glycoprotein with an unusual discontinuous patern of expression during lymphocyte differentiation. *J. Immunol.* 144: 2811-2815.
8. Lee HC, Zocchi E, Guida L, Franco L, Benatti U and De Flora A. 1993. Production and hydrolysis of cyclic ADP-ribose at the outer surface of human erythrocytes. *Biochem. Biophys. Res. Commun.* 191: 639-645.
9. Howard M, Grimaldi JC, Bazan JF, Lund FE, Santos-Argumedo L, Parkhouse RM, Walseth TF and Lee HC. 1993. Formation and hydrolysis of cyclic ADP-ribose catalyzed by lymphocyte antigen CD38. *Science* 262: 1056-1059.
10. Kim H, Jacobson EL and Jacobson MK. 1993. Synthesis and degradation of cyclic ADP-ribose by NAD glycohydrolases. *Science* 261: 1330-1333.
11. Takasawa S, Tohgo A, Noguchi N, Koguma T, Nata K, Sugimoto T, Yonekura H and Okamoto H. 1993. Synthesis and hydrolysis of cyclic ADP-ribose by human leukocyte antigen CD38 and inhibition of the hydrolysis by ATP. *J. Biol. Chem.* 268: 26052-26054.
12. Aarhus R, Graeff RM, Dickey DM, Walseth TF and Lee HC. 1995. ADP-ribosyl cyclase and CD38 catalyze the synthesis of a calcium-mobilizing metabolite from NADP. *J. Biol. Chem.* 270: 30327-30333.
13. Prasad GS, Levitt DG, Lee HC and Stout CD. 1996. Crystallization of ADP-ribosyl cyclase from Aplysia californica. *Proteins* 24: 138-140.
14. Prasad GS, McRee DE, Stura EA, Levitt DG, Lee HC and Stout CD. 1996. Crystal structure of Aplysia ADP ribosyl cyclase, a homologue of the bifunctional ectozyme CD38. *Nature Struct. Biol.* 3: 957-964.
15. Hirata Y, Kimura N, Sato K, Ohsugi Y, Takasawa S, Okamoto H, Ishikawa J, Kaisho T, Ishihara K and Hirano T. 1994. ADP ribosyl cyclase activity of a novel bone marrow stromal cell surface molecule, BST-1. *FEBS Lett.* 356: 244-248.
16. Itoh M, Ishihara K, Tomizawa H, Tanaka H, Kobune Y, Ishikawa J, Kaisho T and Hirano T. 1994. Molecular cloning of murine BST-1 having homology with CD38 and *Aplysia* ADP-ribosyl cyclase. *Biochem. Biophys. Res. Commun.* 203: 1309-1317.
17. Graeff RM, Walseth TF, Fryxell K, Branton WD and Lee HC. 1994. Enzymatic synthesis and characterizations of cyclic GDP-ribose. A procedure for distinguishing enzymes with ADP-ribosyl cyclase activity. *J. Biol. Chem.* 269: 30260-30267.
18. Graeff RM, Mehta K and Lee HC. 1994. GDP-ribosyl cyclase activity as a measure of CD38 induction by retinoic acid in HL-60 cells. *Biochem. Biophys. Res. Commun.* 205: 722-727.
19. Graeff RM, Walseth TF, Hill HK and Lee HC. 1996. Fluorescent analogs of cyclic ADP-ribose: synthesis, spectral characterization, and use. *Biochemistry* 35: 379-386.
20. Lee HC, Graeff R and Walseth TF. 1995. Cyclic ADP-ribose and its metabolic enzymes. *Biochimie* 77: 345-355.

21. Graeff RM, Franco L, Deflora A and Lee HC. 1998. Cyclic GMP-dependent and -independent effects on the synthesis of the calcium messengers cyclic ADP-ribose and nicotinic acid adenine dinucleotide phosphate. *J. Biol. Chem.* 273: 118-125.
22. Takada T, Iida K and Moss J. 1994. Expression of NAD glycohydrolase activity by rat mamary adenocarcinoma cells transformed with rat T cell alloantigen RT6.2. *J. Biol. Chem.* 269: 9420-9423.
23. Walseth TF and Lee HC. 1993. Synthesis and characterization of antagonists of cyclic-ADP-ribose-induced Ca^{2+} release. *Biochim. Biophys. Acta* 1178: 235-242.
24. Walseth TF. Aarhus R, Gurnack ME, Wong L, Breitinger H-GA, Gee KR and Lee HC. 1997. Preparation of cyclic ADP-ribose antagonists and caged cyclic ADP-ribose. *Meth. Enzymol.* 280: 294-305.
25. Ashamu GA, Sethi JK, Galione A and Potter BVL. 1997. Roles for adenosine ribose hydroxyl groups in cyclic adenosine 5'-diphosphate ribose-mediated Ca^{2+} release. *Biochemistry* 36: 9509-9517.
26. Bailey VC. Sethi JK, Galione A and Potter BVL. 1997. Synthesis of 7-deaza-8-bromo cyclic adenosine 5'-diphosphate ribose - The first hydrolysis resistant antagonist at the cADPR receptor. *Chem. Commun.* : 695-696.
27. Wong L, Aarhus R, Lee HC and Walseth TF. 1999. Cyclic 3-deaza-adenosine diphosphoribose: a potent and stable analog of cyclic ADP-ribose. *Biochim. Biophys. Acta* 1472: 555-564.
28. Sethi JK. Empson RM, Bailey VC, Potter BVL and Galione A. 1997. 7-Deaza-8-bromo-cyclic ADP-ribose, the first membrane-permeant, hydrolysis-resistant cyclic ADP-ribose antagonist. *J. Biol. Chem.* 272: 16358-16363.
29. Lee HC and Aarhus R. 1997. Structural determinants of nicotinic acid adenine dinucleotide phosphate important for its calcium-mobilizing activity. *J. Biol. Chem.* 272: 20378-20383.
30. Lee HC. 1997. Mechanisms of calcium signaling by cyclic ADP-ribose and NAADP. *Physiol. Rev.* 77: 1133-1164.
31. Lee HC. 2000. Enzymatic functions and structures of CD38 and homologs. *Chem. Immunol.* 75: 39–59.
32. Munshi C, Thiel DJ, Mathews II, Aarhus R, Walseth TF and Lee HC. 1999. Characterization of the active site of ADP-ribosyl cyclase. *J. Biol. Chem.* 274: 30770-30777.
33. Munshi C, Aarhus R, Graeff R, Walseth TF, Levitt D and Lee HC. 2000. Identification of the enzymatic active site of CD38 by site-directed mutagenesis. *J. Biol. Chem.* 275: 21566-21571.
34. Hussain AMM, Lee HC and Chang CF. 1998. Functional expression of secreted mouse BST-1 in yeast. *Prot. Express. Purif.* 12: 133-137.
35. Kontani K, Nishina H, Ohoka Y, Takahashi K and Katada T. 1993. NAD glycohydrolase specifically induced by retinoic acid in human leukemic HL-60 cells. *J. Biol. Chem* 268: 16895-16898.
36. De Flora A, Franco L, Guida L, Bruzzone S, Usai C and Zocchi E. 2000. Topology of CD38. *Chem. Immunol.* 75: 79-98.
37. Franco L, Guida L, Bruzzone S. Zocchi E, Usai C and Flora AD. 1998. The transmembrane glycoprotein CD38 is a catalytically active transporter responsible for generation and influx of the second messenger cyclic ADP-ribose across membranes. *FASEB J.* 12: 1507-1520.
38. Franco L, Zocchi E, Usai C, Guida L, Bruzzone S, Costa A and De Flora A. 2001. Paracrine roles of NAD^+ and cyclic ADP-ribose in increasing intracellular calcium and enhancing cell proliferation of 3T3 fibroblasts. *J. Biol. Chem.* 276: 21642-21648.
39. Verderio C, Bruzzone S, Zocchi E, Fedele E, Schenk U, De Flora A and M. M. 2001. Evidence of a role for cyclic ADP-ribose in calcium signalling and neurotransmitter release in cultured astrocytes. *J. Neurochem.* 78: 646-657.

40. Sun L, Adebanjo OA, Moonga BS, Corisdeo S, Anandatheerthavarada HK, Biswas G, Arakawa T, Hakeda Y, Koval A, Sodam B, Bevis PJR, Moser AJ, Lai FA, Epstein S, Troen BR, Kumegawa M and Zaidi M. 1999. CD38/ADP-ribosyl cyclase: A new role in the regulation of osteoclastic bone resorption. *J. Cell Biol.* 146: 1161-1171.
41. Deflora A, Guida L, Franco L, Zocchi E, Pestarino M, Usai C, Marchetti C, Fedele E, Fontana G and Raiteri M. 1996. Ectocellular in vitro and in vivo metabolism of cADP-ribose in cerebellum. *Biochem. J.* 320: 665-671.
42. Zocchi E, Franco L, Guida L, Piccini D, Tacchetti C and Deflora A. 1996. NAD^+-dependent internalization of the transmembrane glycoprotein CD38 in human Namalwa β cells. *FEBS Lett.* 396: 327-332.
43. Zocchi E, Usai C, Guida L, Franco L, Bruzzone S, Passalacqua M and De Flora A. 1999. Ligand-induced internalization of CD38 results in intracellular Ca^{2+} mobilization: role of NAD^+ transport across cell membranes. *FASEB J.* 13: 273-283.
44. Yamada M, Mizuguchi M, Otsuka N, Ikeda K and Takahashi H. 1997. Ultrastructural localization of CD38 immunoreactivity in rat brain. *Brain Res.* 756: 52-60.
45. Adebanjo OA, Anandatheerthavarada HK, Koval AP, Moonga BS, Biswas G, Sun L, Sodam BR, Bevis PJR, Huang CLH, Epstein S, Lai FA, Avadhani NG and Zaidi M. 1999. A new function for CD38/ADP-ribosyl cyclase in nuclear Ca^{2+} homeostasis. *Nature Cell Biol.* 1: 409-414.
46. Khoo KM, Han M-K, Park JB, Chae SW, Kim U-H, Lee HC, Bay BH and Chang CF. 2000. Localization of the cyclic ADP-ribose-dependent calcium signaling pathway in hepatocyte nucleus. *J. Biol. Chem.* 275: 24807-24817.
47. Khoo KM and Chang CF. 2001. Identification and characterization of nuclear CD38 in the rat spleen. *Int. J. Biochem. Cell Biol.* 34: 43-54.
48. Santella L, Deriso L, Gragnaniello G and Kyozuka E. 1998. Separate activation of the cytoplasmic and nuclear calcium pools in maturing starfish oocytes. *Biochem. Biophys. Res. Commun.* 252: 1-4.
49. Reinherz EL, Kung PC, Goldstein G, Levey RH and Schlossman SF. 1980. Discrete stages of human intrathymic differentiation: analysis of normal thymocytes and leukemic lymphoblasts of T-cell lineage. *Proc. Natl. Acad. Sci. USA* 77: 1588-1592.
50. Zocchi E, Franco L, Guida L, Benatti U, Bargellesi A, Malavasi F, Lee HC and De Flora A. 1993. A single protein immunologically identified as CD38 displays NAD^+ glycohydrolase, ADP-ribosyl cyclase and cyclic ADP-ribose hydrolase activities at the outer surface of human erythrocytes. *Biochem. Biophys. Res. Commun.* 196: 1459-1465.
51. Fernandez JE, Deaglio S, Donati D, Svoboda-Beusan I, Corno F, Aranega A, Forni M, Falini B and Malavasi F. 1998. Analysis of the distribution of human CD38 and of its ligand CD31 in normal tissues. *J. Biol. Regul. Homeostatic Agents* 12: 81-91.
52. Khoo KM and Chang CF. 1999. Characterization and localization of CD38 in the vertebrate eye. *Brain Res.* 821: 17-25.
53. Kato I, Takasawa S, Akabane A, Tanaka O, Abe H, Takamura T, Suzuki Y, Nata K, Yonekura H, Yoshimoto T and et al. 1995. Regulatory role of CD38 (ADP-ribosyl cyclase/cyclic ADP-ribose hydrolase) in insulin secretion by glucose in pancreatic beta cells. Enhanced insulin secretion in CD38-expressing transgenic mice. *J. Biol. Chem.* 270: 30045-30050.
54. Kato I, Yamamoto Y, Fujimura M, Noguchi N, Takasawa S and Okamoto H. 1998. CD38 disruption impairs glucose-induced increases in cyclic ADP-ribose, $[Ca^{2+}]_i$ and insulin secretion. *J. Biol. Chem.* 274: 1869-1872.
55. Ishihara K and Hirano T. 2000. BST-1/CD157 regulates the humoral immune responses in vivo. *Chem. Immunol.* 75: 235-255.
56. Itoh M, Ishihara K, Hiroi T, Ok Lee B, Maeda H, Iijima H, Yanagita M, Kiyono H and Hirano T. 1998. Deletion of bone marrow stromal cell antigen-1 (CD157) gene Impaired systemic thymus independent-2 antigen-induced IgG3 and mucosal TD antigen-elicited IgA responses. *J Immunol* 161: 3974-3983.

57. Mothet JP, Fossier P, Meunier FM, Stinnakre J, Tauc L and Baux G. 1998. Cyclic ADP-ribose and calcium-induced calcium release regulate neurotransmitter release at a cholinergic synapse of Aplysia. *J. Physiol.* 507.2: 405-414.
58. Chameau P, Van De Vrede Y, Fossier P and Baux G. 2001. Ryanodine-, IP_3- and NAADP-dependent calcium stores control acetylcholine release. *Pflugers Arch.* 443: 289-296.
59. Masuda W, Takenaka S, Inageda K, Nishina H, Takahashi K, Katada T, Tsuyama S, Inui H, Miyatake K and Nakano Y. 1997. Oscillation of ADP-ribosyl cyclase activity during the cell cycle and function of cyclic ADP-ribose in a unicellular organism, Euglena Gracilis. *FEBS Lett.* 405: 104-106.
60. Masuda W, Takenaka S, Tsuyama S, Tokunaga M, Yamaji R, Inui H, Miyatake K and Nakano Y. 1997. Inositol 1,4,5-trisphosphate and cyclic ADP-ribose mobilize Ca^{2+} in a protist, *Euglena Gracilis*. *Comp. Biochem. Physiol.* 118: 279-283.
61. Koradi R, Billeter M and Wuthrich K. 1996. MOLMOL: a program for display and analysis of macromolecular structures. *J. Mol. Graphics* 14: 51-55.
62. Lee HC, Aarhus R and Levitt D. 1994. The crystal structure of cyclic ADP-ribose. *Nature Struct. Biol.* 1: 143-144.
63. Bolduc JM, Dyer DH, Scott WG, Singer P, Sweet RM, Koshland DE and Stoddard BL. 1995. Mutagenesis and Laue structures of enzyme intermediates-Isocitrate dehydrogenase. *Science* 268: 1312-1318.
64. Munshi C, Baumann C, Levitt D, Bloomfield VA and Lee HC. 1998. The homo-dimeric form of ADP-ribosyl cyclase in solution. *Biochim. Biophys. Acta* 1388: 428-436.
65. Sauve AA, Deng HT, Angeletti RH and Schramm VL. 2000. A covalent intermediate in CD38 is responsible for ADP-ribosylation and cyclization reactions. *J. Am. Chem. Soc.* 122: 7855-7859.
66. Graeff R, Munshi C, Aarhus R, Johns M and Lee HC. 2001. A single residue at the active site of CD38 determines its NAD cyclizing and hydrolyzing activities. *J. Biol. Chem.* 276: 12169-12173.
67. Tohgo A, Munakata H, Takasawa S, Nata K, Akiyama T, Hayashi N and Okamoto H. 1997. Lysine 129 of CD38 (ADP-ribosyl cyclase/cyclic ADP-ribose hydrolase) participates in the binding of ATP to inhibit the cyclic ADP-ribose hydrolase. *J. Biol. Chem.* 272: 3879-3882.
68. Tohgo A, Takasawa S, Noguchi N, Koguma T, Nata K, Sugimoto T, Furuya Y, Yonekura H and Okamoto H. 1994. Essential cysteine residues for cyclic ADP-ribose synthesis and hydrolysis by CD38. *J. Biol. Chem.* 269: 28555-28557.
69. Yamamoto-Katayama S, Sato A, Ariyoshi M, Suyama M, Ishihara K, Hirano T, Nakamura H, Morikawa K and Jingami H. 2001. Site-directed removal of N-glycosylation sites in BST-1/CD157: effects on molecular and functional heterogeneity. *Biochem. J.* 357: 385-392.
70. Yamamoto-Katayama S, Ariyoshi M, Ishihara K, Hirano T, Jingami H and Morikawa K. 2002. Crystallographic studies on human BST-1/CD157 with ADP-ribosyl cyclase and NAD glycohydrolase activities. *J. Mol. Biol.* 316: 711-723.
71. Lee HC. 1999. A unified mechanism for enzymatic synthesis of two calcium messengers, cyclic ADP-ribose and NAADP. *Biol. Chem.* 380: 785-793.
72. Muller-Steffner HM, Augustin A and Schuber F. 1996. Mechanism of cyclization of pyridine nucleotides by bovine spleen NAD^+ glycohydrolase. *J. Biol. Chem.* 271: 23967-23972.
73. Sauve AA, Munshi C, Lee HC and Schramm VL. 1998. The reaction mechanism for CD38. A single intermediate is responsible for cyclization, hydrolysis, and base-exchange chemistries. *Biochemistry* 37: 13239-13249.
74. Deterre P, Berthelier V, Bauvois B, Dalloul A, Schuber F and Lund F. 2000. CD38 in T- and B-cell functions. *Chem. Immunol.* 75: 146-168.

75. Campana D, Suzuki T, Todisco E and Kitanaka A. 2000. CD38 in hematopoiesis. *Chem. Immunol.* 75: 169-188.
76. Morra M, Zubiaur M, Terhorst C, Sancho J and Malavasi F. 1998. CD38 is functionally dependent on the TCR/CD3 complex in human T cells. *FASEB J.* 12: 581-592.
77. Deaglio S, Morra M, Mallone R, Ausiello CM, Prager E, Garbarino G, Dianzani U, Stockinger H and Malavasi F. 1998. Human CD38 (ADP-ribosyl cyclase) is a counter-receptor of CD31, an Ig superfamily member. *J. Immunol.* 160: 395-402.
78. Deaglio S, Mallone R, Baj G, Arnulfo A, Surico N, Dianzani U, Mehta K and Malavasi F. 2000. CD38/CD31, a receptor/ligand system ruling adhesion and signaling in human leukocytes. *Chem. Immunol.* 75: 99-120.
79. Funaro A, De Monte LB, Dianzani U, Forni M and Malavasi F. 1993. Human CD38 is associated to distinct molecules which mediate transmembrane signaling in different lineages. *Eur. J. Immunol.* 23: 2407-2411.
80. Deaglio S, Zubiaur M, Gregorini A, Bottarel F, Ausiello CM, Dianzani U, Sancho J and Malavasi F. 2002. Human CD38 and CD16 are functionally dependent and physically associated in natural killer cells. *Blood* 99: 2490-2498.
81. Lund FE, Muller-Steffner HM, Yu NX, Stout CD, Schuber F and Howard MC. 1999. CD38 signaling in B lymphocytes is controlled by its ectodomain but occurs independently of enzymatically generated ADP-ribose or cyclic ADP-ribose. *J. Immunol.* 162: 2693-2702.
82. Zocchi E, Daga A, Usai C, Franco L, Guida L, Bruzzone S, Costa A, Marchetti C and Deflora A. 1998. Expression of CD38 increases intracellular calcium concentration and reduces doubling time in HeLa and 3T3 cells. *J. Biol. Chem.* 273: 8017-8024.
83. Cockayne DA, Muchamuel T, Grimaldi JC, Mullersteffner H, Randall TD, Lund FE, Murray R, Schuber F and Howard MC. 1998. Mice deficient for the ecto-nicotinamide adenine dinucleotide glycohydrolase CD38 exhibit altered humoral immune responses. *Blood* 92: 1324-1333.
84. Partida-Sanchez S, Cockayne D, Monard S, Jacobson EL, Oppenheimer N, Garvy B, Kusser K, Goodricj S, Howard M, Harmsen A, Randall T and Lund FE. 2001. Cyclic ADP-ribose production by CD38 regulates intracellular calcium release, extracellular calcium influx and chemotaxis in neutrophils and is required for bacterial clearance in vivo. *Nature Med.* 7: 1209-1216.
85. Fukushi Y, Kato I, Takasawa S, Sasaki T, Ong BH, Sato M, Ohsaga A, Sato K, Shirato K, Okamoto H and Maruyama Y. 2001. Identification of cyclic ADP-ribose-dependent mechanisms in pancreatic muscarinic Ca^{2+} signaling using CD38 knockout mice. *J. Biol. Chem.* 276: 649-655.
86. Takasawa S, Nata K, Yonekura H and Okamoto H. 1993. Cyclic ADP-ribose in β cells. *Science* 262: 585.
87. Okamoto H. 1999. The CD38-cyclic ADP-ribose signaling system in insulin secretion. *Mol. Cell. Biochem.* 193: 115-118.

Hon Cheung Lee, Department of Pharmacology,
University of Minnesota, Minneapolis, MN 55455, USA.
e-mail: leehc@tc.umn.edu

Cyclic ADP-Ribose and NAADP. Structures, Metabolism and Functions. Lee HC (ed), pp. 45-64, Dordrecht: Kluwer.

Regulation of Synthesis of cADPR and NAADP

Antony Galione
Department of Pharmacology, Oxford University, Mansfield Road, Oxford OX1 3QT, UK.

INTRODUCTION

The discovery of the Ca^{2+} mobilizing properties of cyclic adenosine diphosphate-ribose (cADPR), a cyclic metabolite of NAD [1], and product of enzyme activities termed ADP-ribosyl cyclases [2], has engendered a number of studies aimed at investigating the regulation of synthesis of this molecule. This is important since for cADPR to satisfy the criteria of a second messenger as originally laid down by Sutherland [3], its intracellular levels must be regulated and, in particular, by signals acting at cell surface receptors [4]. Our understanding of the regulation of cADPR synthesis, although still incomplete, has come from two types of approach. The first is the biochemical studies of regulation of ADP-ribosyl cyclase activities, and the second from the development of techniques to quantify changes in cell and tissue levels of cADPR in intact cells and tissues [4]. Endogenous levels in cells and tissues are approximately in the tens of pmol/mg protein, which coincide broadly with the concentration-response relationship range for cADPR-induced Ca^{2+} release. The first evidence for clues as to possible regulatory mechanisms of cADPR synthesis came from studies investigating the pathway for cGMP-induced Ca^{2+} mobilization in sea urchin eggs [5]. In this chapter we will review the role of the nitric oxide-cGMP signaling pathway in regulating cADPR-mediated Ca^{2+} signaling, and also the interactions with the other major cyclic nucleotide second messenger, cAMP. The interactions between these different intracellular messenger pathways at present have given the greatest insight into the regulation of cADPR synthesis [6, 7], and indeed the only clue as to possible modes of modulating synthesis of the other Ca^{2+} mobilizing pyridine nuclotide metabolite, nicotinic acid adenine dinucleotide phosphate (NAADP) [8].

ENZYMOLOGY OF cADPR PRODUCTION

The discovery of cADPR as a Ca^{2+} mobilizing agent came from studies looking at the effects of pyridine nucleotides on Ca^{2+} release in sea urchin egg homogenates [9]. NAD was found to release Ca^{2+} in this cell-free system dependent on the presence of ADP-ribosyl cyclase activities which cyclize NAD^+ to cADPR. ADP-ribosyl cyclases have been demonstrated in a wide range of cellular systems [2]. The first to be characterized at the molecular level was a soluble form found in the granules of *Aplysia* ovotestes [10, 11]. This enzyme has now been crystallized and its structure solved [12]. An important finding based on sequence homologies [13] was that the type II membrane glycoprotein CD38 expresses ADP-ribosyl cyclase activity [14, 15]. This molecule is a multifunctional protein, expressing both ADP-ribosyl cyclase activity and activities leading to the hydrolysis of cADPR to ADP-ribose, termed cADPR hydrolases [16]. CD38 has now been reported to be present in many tissues along with related molecules with similar enzyme activities such as BST-1. They are largely present as ectoenzymes with catalytic domains facing away from the cytoplasm which poses a problem for a potential intracellular messenger generating enzyme. However, evidence has been presented that extracellular NAD or cADPR may regulate intracellular Ca^{2+} signals involving transport across the plasma membrane [17]. In contrast, extracellular production of cADPR was found to be unrelated to cADPR-induced Ca^{2+} release in T cells [18]. Studies from CD38 knockout mice, however, have suggested that CD38 and intracellular cADPR may play an important role in regulating insulin secretion [19], muscarinic receptor-mediated Ca^{2+} signaling in pancreatic acinar cells [20], and in chemotaxis in response to f-met-leu-phe in neutrophils [21]. In addition, evidence is emerging for intracellular locations of CD38, for example, in the vertebrate eye [22] and liver nuclei [23] and other endomembranes [24]. As yet uncharacterized ADP-ribosyl cyclase activities have been demonstrated in cytosols of sea urchin eggs [8], brain [25], and T-cells [26] as well as associated with mitochondria [27, 28], and the sarcoplasmic reticulum [29]. The first clue as to how cADPR synthesis might be controlled came from studies exploring messengers at fertilization in the sea urchin egg.

REGULATION OF cADPR SYNTHESIS BY cGMP

Following the discovery that cADPR evokes Ca^{2+} release from intracellular stores through activation of ryanodine-sensitive Ca^{2+} release channels (RYRs) [30, 31], intense activity has indicated that cADPR is a widespread regulator of RYR-based Ca^{2+} signaling [2, 4]. It has been hypothesized that by analogy to the role of inositol trisphosphate (IP_3) in Ca^{2+} signaling, cADPR levels are regulated by a variety of cellular stimuli [32], and elevations of cADPR lead to the triggering of Ca^{2+}-induced Ca^{2+} release by sensitizing RYRs to activation by Ca^{2+} ions (see the Chapter by Morgan and Galione in this book). The first proposal suggesting a mechanism for regulating cADPR synthesis came from the study of how cGMP mobilizes Ca^{2+} in sea urchin eggs.

The Ca^{2+} signal at fertilization in sea urchin eggs has several distinct features [33]. At sperm egg fusion there is a brief but global Ca^{2+} influx into the egg, the "cortical flash", followed by a latent period of a number of seconds before a larger but more localized Ca^{2+} rise is detected, after which it sweeps across the cell as a regenerative wave. In addition, the Ca^{2+} signal is prolonged with duration of typically several minutes. A number of messengers have been proposed to mediate this event, whose intracellular application should faithfully mimic the fertilization Ca^{2+} signals in terms of its features just described. The first candidate, IP_3, produces a rapid but short lived transient, but cGMP induces a prolonged Ca^{2+} signal and importantly with a long latency, following microinjection into eggs [34]. Importantly, as for fertilization [33], block of IP_3-sensitive Ca^{2+} release channels (IP_3Rs) with heparin, did not abolish the cGMP-evoked Ca^{2+} signal. Further investigation of the mechanism of cGMP-induced Ca^{2+} release was carried out in sea urchin homogenate systems [5]. Using dilute egg homogenates, cGMP was unable to trigger Ca^{2+} release unless homogenates were supplemented with NAD. In the presence of micromolar NAD levels, cGMP mobilized Ca^{2+} from microsomal stores, with a pronounced latency that was inversely related to the cGMP concentration. cGMP-evoked Ca^{2+} release was blocked by cADPR antagonists such as 8-amino-cADPR but not by heparin, and in intact eggs by RYR inhibitors. Furthermore using a thin layer chromatography analysis, cGMP was shown to stimulate NAD metabolism. These data were taken to indicate that cGMP mobilizes Ca^{2+} in sea urchin eggs by initiating a signaling pathway promoting cADPR synthesis and consequent activation of RYRs [5, 7, 32]. Furthermore, inhibitors of ADP-ribosyl cyclase such as nicotinamide, and inhibitors of cGMP-dependent protein kinases inhibited cGMP-induced Ca^{2+} release [35]. These studies were extended to show that nitric oxide, by virtue of its activation of soluble guanylyl cyclase and hence cGMP production, could also evoke Ca^{2+} signals in sea urchin eggs via the cADPR signaling pathway [36]. (see Fig. 1). The proposed signalling pathway is illustrated in Fig. 2.

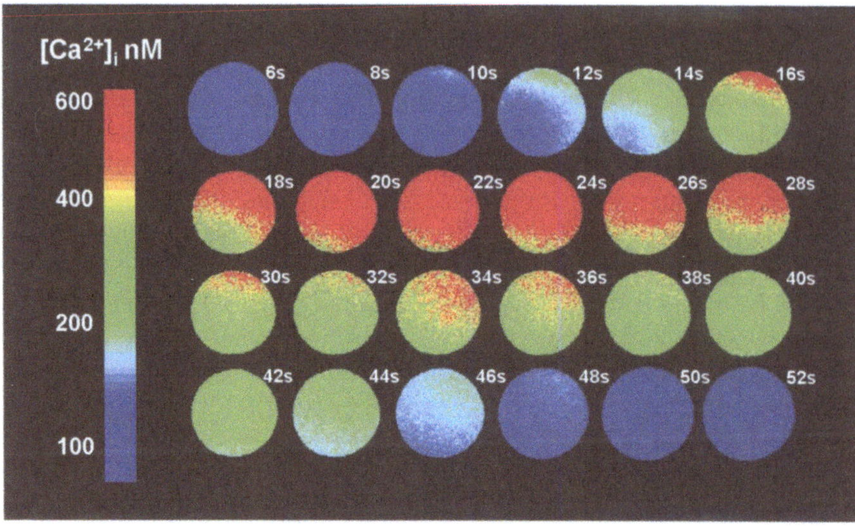

Figure 1. **Nitric oxide induces a Ca^{2+} wave in the sea urchin egg.** Digital ratio images measured with fura-2 are displayed at different times after exposure to sea water containing NO. The images shown were measured 3 s apart. There was a latency of at least 14 s before a Ca^{2+} rise was detected, and the Ca^{2+} often increased in one region of the egg before spreading across the cell at a velocity of around 15 μm / s. This response to nitric oxide persisted in Ca^{2+}-free media, but was abolished by prior microinjection of the cADPR antagonist, 8-NH_2-cADPR into the egg [36].

Further analysis in sea urchin egg homogenates (*Strongylocentrotus purpuratus*) distinguished between two types of ADP-ribosyl cyclase. The first, a soluble form, is regulated in a cGMP-dependent manner, probably via a sea urchin specific cGMP-dependent protein kinase [37], while a second membrane-associated form is cGMP-independent. In a different study using homogenates of *Lytechinus pictus* eggs, the predominant cADPR synthesizing activity was found to be in the soluble fraction, but the NAADP-sythesising activity was mainly membrane-associated. Interestingly, NAADP synthesis was enhanced by cAMP but not by cGMP suggesting a differential regulation of the two synthetic activities by the different cyclic nucleotides [8].

A Role for cGMP at Fertilization

Despite over a century of investigation, the mechanisms of egg activation by sperm are still poorly understood. Three major theories for sperm-induced Ca^{2+} signals have been proposed. The first, the contact hypothesis proposes that sperm interacts with a cell surface receptor linked to signal transduction

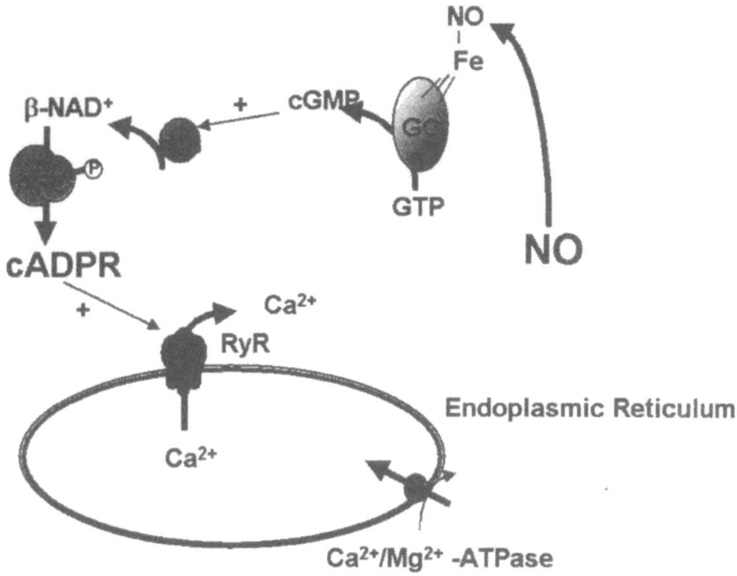

Figure 2. **Pathway for interaction of nitric oxide/cGMP and cADPR signaling pathways.** Nitric oxide activates soluble guanylyl cyclase resulting in an increase in cGMP. cGMP acts via a G-kinase to stimulate cADPR synthesis from NAD by activating ADP-ribosyl cyclase (ARC) activities. The site of regulation is unclear with either phosphorylation of ARC or a regulatory protein that interacts with this enzyme proposed. cADPR then mobilizes Ca^{2+} from endoplasmic reticulum stores by activation of RYRs.

pathways [38]. The second is the conduit hypothesis that upon sperm-egg fusion, Ca^{2+} flows into the egg through channels in the sperm [39] to overload Ca^{2+} stores resulting in Ca^{2+} release. The third is the content hypothesis in which sperm-egg fusion allows the diffusion of a cytosolic sperm factor into the egg which then initiates events leading to the mobilization of Ca^{2+} [40]. Most of the studies have focused on the role of IP_3 in mediating the fertilization Ca^{2+} transient. For example, it seems likely that a major sperm factor is a sperm-specific phospholipase C, which after introduction into the egg, hydrolyses phosphatidylinositol 4,5 bisphosphate to produce IP_3 and hence leads to Ca^{2+} mobilization. However, in a number of eggs there is a clear role for cADPR in shaping the fertilization Ca^{2+} transient [41, 42] and mediating other fertilization-associated events [43-45]. How sperm-egg interactions lead to an increase in egg cADPR levels is at present unknown, but most studies have begun to focus on a possible role for cGMP and nitric oxide in mediating this event in view of the work described in the section above. At fertilization of sea urchin eggs, early increases in cGMP have been reported [46, 47]. In a recent study, a primary messenger role for cGMP at sea urchin egg fertilization was assumed since increases in its levels preceded that of cADPR or IP_3 [47]. Nitric oxide (NO) may also play an important role: not only might it initiate signals, but since NO synthesis is Ca^{2+}-dependent, it may help elicit regenerative Ca^{2+} signals. Direct interactions of nitric oxide with RYRs are also reported which are discussed in the Chapter by Morgan and Galione in this book. Using a fluorescent NO indicator, Epel and his colleagues found that fertilization of sea urchin eggs was accompanied by a small early increase in nitric oxide production, followed by a later and larger production [48]. In addition, egg activation during insemination was inhibited by prior microinjection of haemoglobin, a nitric oxide scavenger. However, a role for cGMP at fertilization has not been fully accepted. Inhibitors of the cGMP/ADP-ribosyl cyclase pathway were found not to block the cADPR-component of the sea urchin fertilization-induced Ca^{2+} signal [49], and a role for NO in chordate fertilization was apparently discounted [50]. Work from our laboratory suggests that the NO/cGMP/cADPR pathway may play a role in later events by prolonging the fertilization Ca^{2+} transient (Leckie, M, Empson, R, Churchill, GC, Whitaker, M, and Galione, A unpublished observations). It is clear that more work is needed to ascertain the mechanisms of cADPR synthesis at fertilization.

A link between NO, cGMP, cADPR and RYR-mediated Ca^{2+} release has also been investigated in a number of mammalian cells and tissues. In the neurosecretory cell line PC12, NO mobilizes Ca^{2+} if the type 2 RYR is present in cells [51]. NO donors caused a two-fold increase in cADPR levels, an effect that was reduced by cGMP-dependent protein kinase inhibitors. A role for the cGMP/cADPR pathway has also been proposed in the generation of α_1-adrenergic receptor-mediated Ca^{2+} mobilization in rat lacrimal gland cells [52], and in parotid glands, NO may regulate secretion through a cGMP/cADPR signaling pathway [53]. In mammalian hippocampal slices, the phenomenon of long-term depression (LTD) is thought to involve NO as a retrograde messenger. LTD was abolished by cGMP-dependent protein kinase inhibitors [54], and by a membrane permeant antagonist of cADPR [55], 7-deaza-8-bromo-cADPR [56]. Furthermore, induction of LTD was associated by an increase in cADPR levels in the brain slices [55]. A role for the NO/cGMP/cADPR pathway has been proposed in the induction of synaptic plasticity.

In plant systems, the NO/cGMP/cADPR pathway may be involved in plant defense responses. Resistant plants responded to infection by tobacco mosaic virus and NO donors by the induction of two genes, those encoding phenylalanine ammonia lyase and pathogenesis related protein [57]. These genes were also induced by cADPR and cGMP, but induction was reduced by cGMP-dependent protein kinase inhibitors in the case of cGMP, and the RYR blocker, ruthenium red for both cGMP and cADPR. Thus a cascade involving NO, cGMP, and cADPR-mediated Ca^{2+} release via RYRs was proposed in plant defense signaling [58].

Finally, decreases in cGMP levels may be associated with decreases in cADPR synthesis. In melanocytes, melatonin negatively regulates cADPR signaling. Melatonin has been shown to cause a decrease in both cGMP and cADPR levels in melanocytes, which results in decreases in protein secretion [59, 60].

cAMP INTERACTIONS WITH cADPR SIGNALING PATHWAYS

In the above sections we have focused on the regulation of cADPR synthesis by cGMP. There is also evidence that the cAMP signaling pathway may also regulate cADPR synthesis in other cells. As described above, in the sea urchin system, NAADP synthesis through catalysis of base exchange of nicotinic acid for the nicotinamide moiety of NADP, may be regulated by cAMP [8]. cAMP appears to lower the K_m for the NADP substrate, an effect that is probably mediated via cAMP-dependent protein kinases. In other systems, there is evidence that cAMP may also regulate cADPR synthesis. In bovine chromaffin cells, acetylcholine was found to stimulate ADP-ribosyl cyclase, as did raised extracellular potassium [61]. Ca^{2+} influx was required indicating a possible role for Ca^{2+} in regulating cADPR synthesis, as has also been proposed in gastrointestinal longitudinal smooth muscle [62]. In addition, a role for cAMP-dependent protein kinase in regulating cADPR synthesis was also proposed, in chromaffin cells, since activators of this kinase such as 8-bromo-cAMP, or adenylyl cyclase activators such as forskolin, increased the initial velocity of the ADP-ribosyl cyclase activity [61]. These effects which were blocked by H-89 and Rp-cAMPS, inhibitors of cAMP-dependent protein kinases. As with the effects of cGMP-dependent protein kinase, it is still not clear whether direct phosphorylation of the cyclase enzyme occurs or whether these effects are mediated by possible regulatory proteins. Recent work in our laboratory has shown that the effects of the β-adrenoceptor agonist, isoprenaline, in increasing cADPR levels in guinea pig heart (Fig. 3), are partially mediated by a cAMP-signaling cascade involving cAMP-dependent protein kinase (Thomas and Galione, unpublished observations; Fig. 3). These findings complement the data from Higashida's laboratory that this drug enhances ADP-ribosyl cyclase activities in cardiac tissue [63]. Likewise relaxation of pulmonary smooth muscle in response to isoprenaline involves cADPR-evoked Ca^{2+} release via RYRs that hyperpolarizes the sarcolemma through activation of BK_{Ca} channels, and this effect is dependent on cAMP-dependent protein kinases (Biottin, Galione and Evans, unpublished observations).

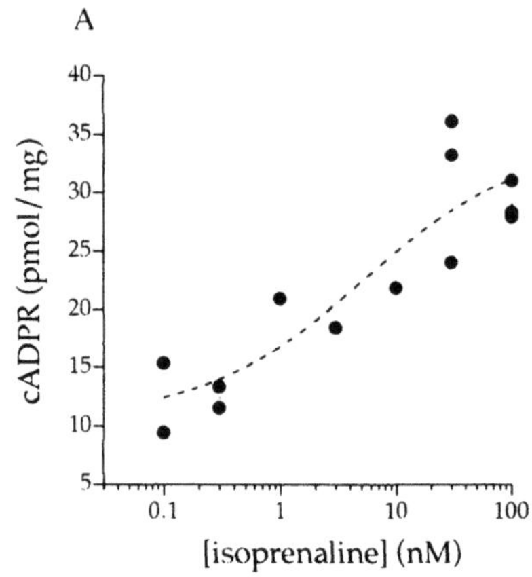

Figure 3. **Concentration-dependent effect of isoprenaline on cADPR levels in guinea pig ventricular tissue.** Following perfusion for 3 minutes with varying concentrations of isoprenaline, acutely isolated hearts were snap frozen in liquid nitrogen. Following, acid extraction, absolute cADPR levels were assessed using the sea urchin based radioreceptor assay. Each point represents the mean (*n*=3-5 determinations) cADPR value (pmol/mg protein) from 1 heart.

The effects of cAMP in regulating cADPR synthesis are not confined to mammalian systems. A recent report examining heat-stress responses in sponges has shown that heat stress causes the opening of a cation channel and ion fluxes lead to the production of abscisic acid which activates cADPR synthesis through a mechanism dependent on the activation of cAMP-dependent protein kinases [64].

OTHER WAYS OF REGULATING cADPR SYNTHESIS

In this chapter we have emphasized the role of cyclic nucleotide-regulation of cADPR synthesis in cells and tissues. However, other modes of regulation of ADP-ribosyl cyclase activities have been proposed, some of them perhaps more direct. Some examples of these are depicted in Fig. 4.

Tyrosine Kinase-Linked Receptors

In T-cell lines, cADPR has been demonstrated as having a second messenger role linking T-cell receptor (TCR/CD3) activation to Ca^{2+} mobilization via the type 3 RYR [26]. The finding that the tyrosine phosphatase selective inhibitor dephostin enhances cADPR synthesis when incubated with cytosolic fractions, implicates a cytosolic non-receptor tyrosine kinase in the regulation of T-cell ADP-ribosyl cyclase. In addition, phosphorylation of RYRs by fyn kinases enhanced cADPR-induced Ca^{2+} release in these cells [65].

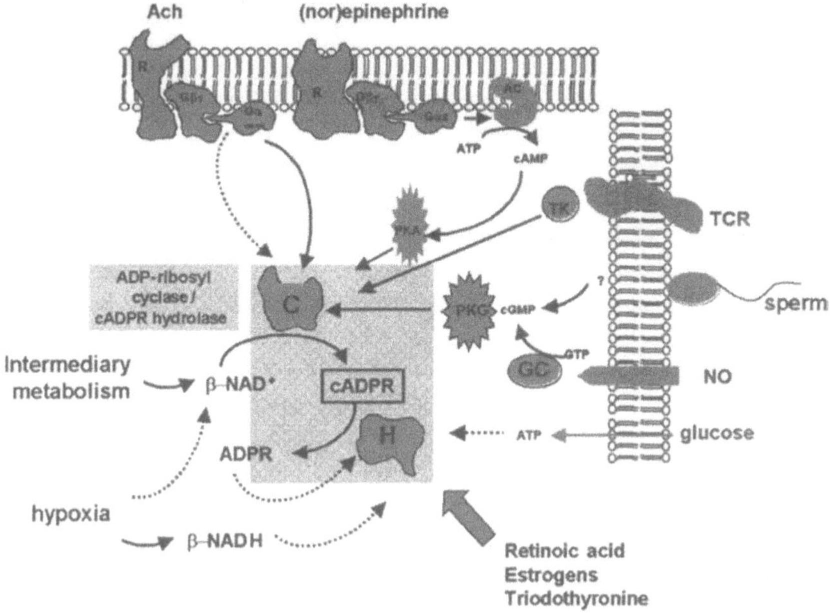

Figure 4. **Summary of some of the mechanisms proposed for regulation of cADPR levels in cells by a variety of different stimuli.** Examples of stimuli are given for each of the mechanisms depicted. The gray central box represents the enzyme activities of a CD38-like enzyme with both ADP-ribosyl cyclase and cADPR hydrolase activities that are expressed on the same protein. In some cells these may be separate and either soluble or membrane-bound for cyclases. Solid arrows represent stimulation, while dashed arrows signify inhibition. Chronic regulation of levels may be regulated by hormones modulating enzyme levels (solid gray arrow). See text for details.

cADPR Synthesis Linked to Cellular Metabolic Changes

Since the substrate for cADPR synthesis NAD plays a central role in metabolic pathways it is not surprising that changes in cADPR synthesis have been linked to various metabolic states. In addition, cADPR synthesis has been shown to be regulated by changes in other pyridine nucleotide metabolites such as NADH and ADP-ribose. The latter may have a signaling role in its own right by modulating cation influx channels at the plasma membrane [66, 67], and some of these channels may also catalyse the metabolism of ADP-ribose as well as being gated by it [68]. These metabolites and substrates as well as a central player in cellular metabolism, ATP, may directly regulate enzyme activities of multifunctional ADP-ribosyl cyclase/cADPR hydrolase enzymes.

Regulation by ATP–A role for cADPR has been proposed during stimulus secretion-coupling in beta pancreatic cells [69]. New evidence is now emerging to show directly that cADPR mobilizes Ca^{2+} from intracellular stores in beta cell lines [70, 71] after some years of much debate [72]. Evidence from Okamoto's laboratory has shown that under certain conditions, glucose elevates cADPR levels in endocrine pancreatic cells and tissues [73]. A role for CD38 as the primary ADP-ribosyl cyclase activity responsible has come from studies in beta cells derived from CD38 knockout mice [19] or transgenic animals overexpressing CD38 [74]. It has been proposed that the intracellular rise in ATP as a consequence of glucose stimulation, inhibits the cADPR hydrolase activity of CD38 thereby enhancing cellular increases in cADPR [75, 76], leading to Ca^{2+} mobilization and insulin secretion.

Regulation by NADH–Pulmonary smooth muscle is unusual in that it constricts in response to hypoxia, and is thought to play a physiological role in matching perfusion to ventilation in the lung. This hypoxic pulmonary vasoconstriction (HPV) is thought to occur in part by Ca^{2+} release from the sarcoplasmic reticulum via RYRs [77-79]. An analysis of ADP-ribosyl cyclase activities in different vascular smooth muscle homogenates indicated a substantially higher enzyme activity in pulmonary as opposed to systemic smooth muscle tissue [80]. Hypoxia induced a large increase in cADPR synthesis in intact muscle, an effect that was mimicked in muscle extracts by increasing the NADH/NAD ratio, as occurs during hypoxia. It was found that raised NADH levels inhibited cADPR hydrolase activity providing one explanation for hypoxia-induced increases in cADPR levels [80].

Regulation by NAD Levels–A recent study has linked changes in NAD levels to modulation of cADPR levels [24]. In osteoblasts CD38 was found to be colocalized with RYRs to the endoplasmic reticulum. Extracellular NAD was found to mobilize Ca^{2+} in CD38 expressing NIH3T3 cells but after a delay of several seconds. This delay was proposed to be due to transport of NAD across both the plasma and endoplasmic reticulum membranes, where it is converted to cADPR luminally. cADPR is then translocated across the endoplasmic reticulum membrane where it activates RYRs and induces Ca^{2+} release. The authors of this study propose that cADPR-based Ca^{2+} signaling in this way is linked to intermediary metabolism via CD38 which acts as an NAD sensor. However, it should be noted that many of the transport steps, particularly for membrane transport of NAD and cADPR have yet to be elucidated. The recent findings that cADPR synthesis may occur in the mitochondria may suggest an additional role for cADPR signaling in coupling the metabolic state of cell to the generation of Ca^{2+} signals [27, 28].

Regulation by ADP-ribose–ADP-ribose (ADPR), the molecule formed by the hydrolysis of cADPR by cADPR hydrolases, has been found to have two interesting properties with regard to Ca^{2+} signaling. The first is that it may, by end product inhibition. inhibit cADPR hydrolase and thus enhance levels of cADPR [81]. The second is that it may act as a messenger in its own right to regulate cation (including Ca^{2+}) influx across the plasma membrane. This phenomenon was initially observed in ascidian eggs were nano-molar concentrations of ADP-ribose, but not cADPR, gated a cation current associated with egg activation at fertilization [66]. A recent report has shown that a cation channel of the TRP family, TRPC7, is not only gated by ADP-ribose, but is also itself a ADPR-pyrophosphatase. This represents a new family of proteins which appear to express both channel and enzyme activities [82].

Changes in Enzyme Levels

Chronic changes in cADPR levels may be associated with changes in levels of CD38 and other ADP-ribosyl cyclase activities. Indeed in pancreatic beta cells, overexpression of CD38 is associated with elevated basal cADPR levels as well as larger increases upon stimulation with glucose [74]. A number of hormones have been shown to enhance the expression of CD38-like molecules. In HL60 cells, retinoic acid leads to an increase in ADP-ribosyl cyclase activity and resting levels of cADPR [83], while in the rat uterus oestrogen enhances ADP-ribosyl cyclase activities [84, 85]. In vascular smooth muscle, retinoic acid and tri-iodothyronine increase cyclase activities presumably by increasing enzyme levels [86].

G Protein-Linked Receptors and cADPR Signaling

G-protein coupled receptors represent one of the largest families of cellular proteins. It is not surprising therefore that the largest number of receptors that have been proposed to signal through the cADPR pathway are of this family. Most studies implicating G protein coupled receptors in the regulation of cADPR signaling pathways have come from experiments using inhibitors of cADPR-induced Ca^{2+} release [4]. However, there are reports of such receptors modulating ADP-ribosyl cyclase activities, although detailed descriptions in molecular terms have not been reported. The best studied systems in this respect are the neuroblastoma cell line, NG-108, and cardiac myocytes. Using a TLC analysis, activation of m1/m3 muscarinic receptors by acetylcholine in NG108 cell membranes enhanced the metabolism of exogenously added NAD concomitant with cADPR production [87]. In contrast, activation of m2/m4 mucarinic receptors inhibited ADP-ribosyl cyclase activities and reduced cADPR levels [87]. Both these effects were mimicked and potentiated by GTP, and stimulation was blocked by cholera toxin, while inhibition was blocked by pertussis toxin. Interestingly, these membrane effects were not affected by cGMP, although this does not exclude a cytoplasmic based modulation by this nucleotide on soluble enzymes as described above. In cardiac tissue both activation of β_1-adrenergic receptors [63] and angiotensin II receptors [88] in cardiac cell membranes are associated with enhanced ADP-ribosyl cyclase activities. As mentioned above and as depicted in Fig. 4, the β-adrenergic agonist, isoprenaline, increases cADPR levels in isolated guinea pig hearts in a concentration-dependent manner, although an indirect regulation of synthesis via cAMP-dependent protein kinase is proposed. In pancreatic acinar cells, muscarinic receptor activation was found to induce large increases in cADPR levels in a concentration-dependent manner [20], and effect that was not apparent in acini from CD38 knockout mice. Finally, in testicular myoid cells, endothelin B receptors were found to couple to activation of ADP-ribosyl cyclase activities [89].

NAADP SYNTHESIS

Although the sibling messenger of cADPR, NAADP has been proposed as an intracellular messenger in a number of cells (see [90] for review), there are no data at present on physiological changes in the intracellular levels of this molecule in response to cellular stimuli. The only pathway described for NAADP synthesis involves a base-exchange reaction catalysed by ADP-ribosyl cyclases, including CD38 [91]. As mentioned above, in the sea urchin egg homogenate, NAADP synthesis was only detected in the presence of NADP and nicotinic acid as substrates implicating a base-exchange reaction. Such in vitro studies have required high concentrations of nicotinic acid for NAADP production, and it is possible that other synthetic routes including NAAD kinases or NADP deaminases may be important, although such enzyme activities have not been reported. Indeed, a recent report studying NAADP synthesis in tissues from CD38 knockout mice [92], suggests that CD38 is the major NAADP synthesizing enzyme in mammalian tissues. However, non-base exchange mechanisms were not investigated. To date the only insight into the regulation of NAADP synthesis has come from the findings that NAADP synthesis by base change mechanisms is favoured by acidic pH [91], and may in the sea urchin egg, be regulated by cAMP [8].

CONCLUSIONS

The development of assays for measuring ADP-ribosyl cyclase activities and for measuring tissue levels of cADPR (see [93] for review) has begun to allow us to study the mechanisms regulating cADPR synthesis. Although much work in this area remains to be done, it is likely that a simpler high through put assay for cADPR level determination in tissues will facilitate these studies [94].

The diversity of stimuli that mobilize Ca^{2+} via the cADPR signaling pathway means that there may be a variety of coupling mechanisms which modulate ADP-ribosyl cyclase/cADPR hydrolase activities (Table 1). These may involve more direct modulation by G protein subunits, tyrosine phosphorylation or indirect modulation through other messengers controlling kinases such as cGMP or cAMP.

These studies have been hampered by the current lack of identity of the enzymes involved in cADPR metabolism in many cells, so precise molecular interactions remain unclear. In the case of CD38, which has been well characterized at the molecular level, only inhibition of the cADPR hydrolase activity of this protein by ATP is understood to any extent at the molecular level [75].

Table 1. **Stimuli that mobilize Ca^{2+} utilising cADPR as an intracellular messenger.**

Stimulus	Receptor	Coupling	Cell	cADPR levels	Method	Response	Ref.
1-Me-adenine	1-Me-adenine	G-protein	Starfish oocyte	ND	Fluorimetry	maturation	[95]
1S, 3R-ACPD	mGluR	G-protein	DRG	ND	Electro-physiology	Ca-gcation /Cl^-	[96]
ABA	ABA	?	Tomato	+	bioassay	Gene expression	[97]
ACh	Muscar-inic	G-protein	NG-108	+/-	TLC	Membrane currents	[87]
ACh	Nicotinic	Ca^{2+} influx	Chroma-ffin	ND	NGD	Catechol-amine secretion	[61]
ACh	Muscar-inic	G-protein	Tracheal smooth muscle	ND	Fluorimetry	contraction	[98]
ACh	Muscar-inic	G-protein	Pancreas: acinar	20x	RIA	Ca release	[20]
Adrenal-ine	α_1-adrenergic	G-protein	Lacrimal gland	ND	Fluorimetry	Ca release	[99]
Anti-CD3 mAb	TCR/CD3	Tyrosine phosphory-lation	Jurkat	4x +	HPLC	T cell activation	[26]
ATP	Purinergic		Macro-phage	ND	Electro-physiology	Ca-g K^+	[100]
CCK	CCK	G-protein	Longi-tudinal smooth muscle	+	Bioassay	Contractility	[62]

Table 1- continued

Stimulus	Receptor	Coupling	Cell	cADPR levels	Method	Response	Ref
CCK	CCK_A	G-protein	Pancreas: acinar cell	ND	Electrophysiology	Secretion	[101]
cGMP	G-kinase	G-kinase	Sea urchin egg	ND	TLC	Ca^{2+} release	[5]
Estrogen	E2 receptor		Myometrium	ND	NGD	Contractility	[85]
Estrogen	Steroid	Protein synthesis	Rat uterus	ND	Bioassay		[84]
Glucose	Glucokinase	ATP	Pancreas: β cell	4x +	RIA	Insulin secretion	[102]
Glucose	Glucokinase	ATP	Pancreas: β cell	14x	RIA	Secretion	[103]
Heat	ABA	A-kinase	Sponge	ND	NGD		[64]
Hypoxia	ADP-ribosyl cyclase	NADH	Lung smooth muscle	10x	Radioreceptor	Contraction	[80]
IL-8	Cytokine	?	NK	ND	Ca release		[104]
Isoprenaline	β_1-adrenergic	G-protein	Cardiac myocyte	+	TLC	Contractility	[63]
Isoprenaline	β_1-adrenergic	A-kinase	Cardiac myocyte	ND	Fluorimetry 8-NH_2-cADPR	Ca^{2+} waves	[105]
Laminin	α6β1-integrin	Tyrosine kinase	Jurkat T cell	ND	7-deaza-8-Br-cADPR	Ca^{2+} release	[106]
Melatonin	Melatonin receptor	cGMP	Melanocyte	-		Reduced protein secretion	[59]
NO	Guanylyl cyclase	G-kinase	Sea urchin egg	+	Bioassay	Ca^{2+} release	[36]
NO	Guanylyl cyclase	Phosphorylation	PC12	ND	Fluorimetry	Ca^{2+} release	[51]
NO	Guanylyl cyclase	Phosphorylation	Brain slice	1.5x +	Radioreceptor	LTD	[55]
Retinoic acid	RA receptor	Protein synthesis	HL60	20-30x +	RIA		[83]
Sperm	Fusion	?	Sea urchin egg	ND	Fluorimetry	Exocytosis	[14]
Sperm	Fusion	?	Sea bream egg		HPLC	Fertilization	[45]

Abbreviations used: Ca-gcation/Cl^-, Ca^{2+}-activated nonselective cation/Cl^- channels; Ca-gK^+, Ca^{2+} activated K^+ channels; ACh, acetylcholine; CCK, cholecystokinin; ABA, abscisic acid; LTD, long-term depression; TLC, thin layer chromatography; RIA, radioimmunoassay; A-kinase, cAMP-activated protein kinase; G-kinase, cGMP-activated protein kinase; NO, nitric oxide.

REFERENCES

1. Lee HC, Walseth TF, Bratt GT, Hayes RN and Clapper DL. 1989. Structural determination of a cyclic metabolite of NAD with intracellular Ca^{2+}-mobilizing activity. *J. Biol. Chem.* 264: 1608-1615.
2. Lee HC. 2001. Physiological functions of cyclic ADP-ribose and NAADP as Ca^{2+} messengers. *Annu. Rev. Pharmacol. Toxicol.* 41: 317-345
3. Robison GA, Butcher R and Sutherland EW. 1971. *Cyclic AMP*. pp 531. New York and London: Academic Press.
4. Galione A and Churchill GC. 2000. Cyclic ADP-ribose as a Ca^{2+} mobilizing messenger. *Science STKE* www.stke.org/cgi/content/full/OC_sigtrans;2000/41/pe1: 1-6
5. Galione A, White A, Willmott N, Turner M, Potter BVL and Watson SP. 1993. cGMP mobilizes intracellular Ca^{2+} in sea urchin eggs by stimulating cyclic ADP-ribose synthesis. *Nature* 365: 456-459.
6. Galione A. 1994. Cyclic ADP-ribose, the ADP-ribosyl cyclase pathway and Ca^{2+} signalling. *Mol. Cell Endocrinol.* 98: 125-131.
7. Lee HC. 1994. A signaling pathway involving cyclic ADP-ribose, cGMP, and nitric oxide. *News Physiol. Sci.* 9: 134-137.
8. Wilson HL and Galione A. 1998. Differential regulation of nicotinic acid-adenine dinucleotide phosphate and cADP-ribose production by cAMP and cGMP. *Biochem. J.* 331: 837-843.
9. Clapper DL, Walseth TF, Dargie PJ and Lee HC. 1987. Pyridine nucleotide metabolites stimulate Ca^{2+} release from sea urchin egg microsomes desensitized to inositol trisphosphate. *J. Biol. Chem.* 262: 9561-9568.
10. Glick DL, Hellmich MR, Beushausen S, Tempst P, Bayley H and Strumwasser F. 1991. Primary structure of a molluscan egg-specific NADase, a second-messenger enzyme. *Cell Regul.* 2: 211-218.
11. Hellmich MR and Strumwasser F. 1991. Purification and characterization of a molluscan egg-specific NADase, a second-messenger enzyme. *Cell Regul.* 2: 193-202.
12. Prasad GS, Levitt DG, Lee HC and Stout CD. 1996. Crystallization of ADP-ribosyl cyclase from *Aplysia californica. Proteins Structure Function and Genetics* 24: 138-140.
13. States DJ, Walseth TF and Lee HC. 1992. Similarities in amino acid sequences of *Aplysia* ADP-ribosyl cyclase and human lymphocyte antigen CD38. *Trends Biochem. Sci.* 17: 495.
14. Howard M, Grimaldi JC, Bazan JF and Lund FE, Santosargumedo L, *et al.* 1993. Formation and hydrolysis of cyclic ADP-ribose catalyzed by lymphocyte antigen-CD38. *Science* 262: 1056-1059.
15. Summerhill RJ, Jackson DG and Galione A. 1993. Human lymphocyte antigen CD38 catalyzes the production of cyclic ADP- ribose. *FEBS Lett.* 335: 231-233.
16. Lee HC and Aarhus R. 1993. Wide distribution of an enzyme that catalyzes the hydrolysis of cyclic ADP-ribose. *Biochim. Biophys. Acta* 1164: 68-74.
17. De Flora A, Guida L, Franco L and Zocchi E. 1997. The CD38/cyclic ADP-ribose system: A topological paradox. *Int. J. Biochem. Cell Biol.* 29: 1149-1166.
18. da Silva CP, Schweitzer K, Heyer P, Malavasi F, Mayr GW and Guse AH. 1998. Ectocellular CD38-catalyzed synthesis and intracellular Ca^{2+}-signalling activity of cyclic ADP-ribose in T-lymphocytes are not functionally related. *FEBS Lett.* 439: 291-296.
19. Kato I, Yamamoto Y, Fujimura M, Noguchi N, Takasawa S and Okamoto H. 1999. CD38 disruption impairs glucose-induced increases in cyclic ADP-ribose, $[Ca^{2+}]_i$, and insulin secretion. *J. Biol. Chem.* 274: 1869-1872.
20. Fukushi Y, Kato I, Takasawa S, Sasaki T, Ong BH, *et al.* 2001. Identification of cyclic ADP-ribose-dependent mechanisms in pancreatic muscarinic Ca^{2+}-signaling using CD38 knockout mice. *J. Biol. Chem.* 276: 649-655.

21. Partida-Sanchez S, Cockayne DA, Monard S, Jacobson EL, Oppenheimer N, *et al.* 2001. Cyclic ADP-ribose production by CD38 regulates intracellular Ca^{2+} release, extracellular Ca^{2+} influx and chemotaxis in neutrophils and is required for bacterial clearance in vivo. *Nature Med.* 7: 1209-1216.
22. Khoo KM and Chang CF. 1999. Characterization and localization of CD38 in the vertebrate eye. *Brain Res.* 821: 17-25.
23. Adebanjo OA, Anandatheerthavarada HK, Koval AP, Moonga BS, Biswas G, *et al.* 1999. A new function for CD38/ADP-ribosyl cyclase in nuclear Ca^{2+} homeostasis. *Nature Cell Biol* 1: 409-414.
24. Sun L, Adebanjo OA, Koval A, Anandatheerthavarada HK, Iqbal J, *et al.* 2002. A novel mechanism for coupling cellular intermediary metabolism to cytosolic Ca^{2+} signaling via CD38/ADP-ribosyl cyclase, a putative intracellular NAD^{+} sensor. *FASEB J.* 16: 302-314.
25. Matsumura N and Tanuma S. 1998. Involvement of cytosolic NAD^{+} glycohydrolase in cyclic ADP-ribose metabolism. *Biochem. Biophys. Res. Commun.* 253: 246-252.
26. Guse AH, da Silva CP, Berg I, Skapenko AL, Weber K, *et al.* 1999. Regulation of Ca^{2+} signalling in T lymphocytes by the second messenger cyclic ADP-ribose. *Nature* 398: 70-73.
27. Liang M, Chini EN, Cheng J and Dousa TP. 1999. Synthesis of NAADP and cADPR in mitochondria. *Arch. Biochem. Biophys.* 371: 317-325.
28. Ziegler M, Jorcke D and Schweiger M. 1997. Identification of bovine liver mitochondrial NAD^{+} glycohydrolase as ADP- ribosyl cyclase. *Biochem. J.* 326: 401-405.
29. Meszaros LG, Wrenn RW and Varadi G. 1997. Sarcoplasmic reticulum-associated and protein kinase C-regulated ADP- ribosyl cyclase in cardiac muscle. *Biochem. Biophys. Res. Commun.* 234: 252-256.
30. Galione A, Lee HC and Busa WB. 1991. Ca^{2+}-induced Ca^{2+} release in sea urchin egg homogenates: modulation by cyclic ADP-ribose. *Science* 253: 1143-1146.
31. Galione A. 1992. Ca^{2+}-induced Ca^{2+} release and its modulation by cyclic ADP-ribose. *Trends Pharmacol. Sci.* 13: 304-306.
32. Berridge MJ. 1993. Cell signalling. A tale of two messengers. *Nature* 365: 388-389.
33. Whitaker MJ and Swann K. 1993. Lighting the fuse at fertilization. *Development* 117: 1-12.
34. Whalley T, McDougall A, Crossley I, Swann K and Whitaker M. 1992. Internal Ca^{2+} release and activation of sea urchin eggs by cGMP are independent of the phosphoinositide signaling pathway. *Mol. Biol. Cell* 3: 373-383.
35. Sethi JK, Empson RM and Galione A. 1996. Nicotinamide inhibits cyclic ADP-ribose-mediated Ca^{2+} signalling in sea urchin eggs. *Biochem. J.* 319: 613-617.
36. Willmott N, Sethi J, Walseth TF, Lee HC, White AM and Galione A. 1996. Nitric oxide induced mobilization of intracellular Ca^{2+} via the cyclic ADP-ribose signalling pathway. *J. Biol. Chem.* 271: 3699-3705.
37. Graeff RM, Franco L, De Flora A and Lee HC. 1998. Cyclic GMP-dependent and -independent effects on the synthesis of the Ca^{2+} messengers cyclic ADP-ribose and nicotinic acid adenine dinucleotide phosphate. *J. Biol. Chem.* 273: 118-125.
38. Kline D, Simoncini L, Mandel G, Maue RA, Kado RT and Jaffe LA. 1988. Fertilization events induced by neurotransmitters after injection of mRNA in Xenopus eggs. *Science* 241: 464-467.
39. Creton R and Jaffe LF. 1995. Role of Ca^{2+} influx during the latent period in sea urchin eggs. *Develop. Growth Differ.* 37: 703-709
40. Swann K. 1990. A cytosolic sperm factor stimulates repetitive Ca^{2+} increases and mimics fertilization in hamster eggs. *Development* 110: 1295-1302.
41. Galione A, McDougall A, Busa WB, Willmott N, Gillot I and Whitaker M. 1993. Redundant mechanisms of Ca^{2+}-induced Ca^{2+}-release underlying Ca^{2+} waves during fertilization of sea-urchin eggs. *Science* 261: 348-352.

42. Lee HC, Aarhus R and Walseth TF. 1993. Ca^{2+} mobilization by dual receptors during fertilization of sea urchin eggs. *Science* 261: 352-355.
43. Ayabe T, Kopf GS and Schultz RM. 1995. Regulation of mouse egg activation: presence of ryanodine receptors and effects of microinjected ryanodine and cyclic ADP ribose on uninseminated and inseminated eggs. *Development* 121: 2233-2244.
44. Albrieux M, Lee HC and Villaz M. 1998. Ca^{2+} signaling by cyclic ADP-ribose, NAADP, and inositol trisphosphate are involved in distinct functions in ascidian oocytes. *J. Biol. Chem.* 273: 14566-14574.
45. Polzonetti V, Cardinali M, Mosconi G, Natalini P, Meiri I and Carnevali O. 2002. Cyclic ADPR and Ca^{2+} signaling in sea bream (Sparus aurata) egg fertilization. *Mol. Reprod. Dev.* 61: 213-217.
46. Ciapa B and Epel D. 1996. An early increase in cGMP follows fertilization of sea urchin eggs. *Biochem. Biophys. Res. Commun.* 223: 633-636.
47. Kuroda R, Kontani K, Kanda Y, Katada T, Nakano T, *et al.* 2001. Increase of cGMP, cADP-ribose and inositol 1,4,5-trisphosphate preceding Ca^{2+}- transients in fertilization of sea urchin eggs. *Development* 128: 4405-4414.
48. Kuo RC, Baxter GT, Thompson SH, Stricker SA, Patton C, *et al.* 2000. NO is necessary and sufficient for egg activation at fertilization. *Nature* 406: 633-636.
49. Lee SJ, Christenson L, Martin T and Shen SS. 1996. The cyclic GMP-mediated Ca^{2+} release pathway in sea urchin eggs is not required for the rise in Ca^{2+} during fertilization. *Dev. Biol.* 180: 324-335.
50. Hyslop LA, Carroll M, Nixon VL, McDougall A and Jones KT. 2001. Simultaneous measurement of intracellular nitric oxide and free Ca^{2+} levels in chordate eggs demonstrates that nitric oxide has no role at fertilization. *Dev. Biol.* 234: 216-230.
51. Clementi E, Riccio M, Sciorati C, Nistico G and Meldolesi J. 1996. The type 2 ryanodine receptor of neurosecretory PC12 cells is activated by cyclic ADP-ribose. Role of the nitric oxide/cGMP pathway. *J. Biol. Chem.* 271: 17739-17745.
52. Jorgensen TD, Dissing S and Gromada J. 1996. Cyclic GMP potentiates phenylephrine but not cyclic ADP-ribose-evoked Ca^{2+} release from rat lacrimal acinar cells. *FEBS Lett.* 391: 117-120.
53. Looms D, Tritsaris K, Nauntofte B and Dissing S. 2001. Nitric oxide and cGMP activate Ca^{2+}-release processes in rat parotid acinar cells. *Biochem. J.* 355: 87-95.
54. Reyes-Harde M, Potter BV, Galione A and Stanton PK. 1999. Induction of hippocampal LTD requires nitric-oxide-stimulated PKG activity and Ca^{2+} release from cyclic ADP-ribose-sensitive stores. *J. Neurophysiol.* 82: 1569-1576.
55. Reyes-Harde M, Empson R, Potter BV, Galione A and Stanton PK. 1999. Evidence of a role for cyclic ADP-ribose in long-term synaptic depression in hippocampus. *Proc. Natl. Acad. Sci. USA* 96: 4061-4066.
56. Sethi JK, Empson RM, Bailey VC, Potter BVL and Galione A. 1997. 7-Deaza-8-bromo-cyclic ADP-ribose, the first membrane-permeant, hydrolysis-resistant cyclic ADP-ribose antagonist. *J. Biol. Chem.* 272: 16358-16363.
57. Durner J, Wendehenne D and Klessig DF. 1998. Defense gene induction in tobacco by nitric oxide, cyclic GMP, and cyclic ADP-ribose. *Proc. Natl. Acad. Sci. USA* 95: 10328-10333.
58. Klessig DF, Durner J, Noad R, Navarre DA, Wendehenne D, *et al.* 2000. Nitric oxide and salicylic acid signaling in plant defense. *Proc. Natl. Acad. Sci. USA* 97: 8849-8855.
59. Bubis M and Zisapel N. 1998. A role for NAD^+ and cADP-ribose in melatonin signal transduction. *Mol. Cell Endocrinol.* 137: 59-67.
60. Bubis M and Zisapel N. 1999. Involvement of cGMP in cellular melatonin responses. *Biol. Cell* 91: 45-49.
61. Morita K, Kitayama S and Dohi T. 1997. Stimulation of cyclic ADP-ribose synthesis by acetylcholine and its role in catecholamine release in bovine adrenal chromaffin cells. *J. Biol. Chem.* 272: 21002-21009.

62. Kuemmerle JF and Makhlouf GM. 1995. Agonist-stimulated cyclic ADP ribose. Endogenous modulator of Ca^{2+}- induced Ca^{2+}- release in intestinal longitudinal muscle. *J. Biol. Chem.* 270: 25488-25494.
63. Higashida H, Egorova A, Higashida C, Zhong ZG, Yokoyama S, *et al.* 1999. Sympathetic potentiation of cyclic ADP-ribose formation in rat cardiac myocytes. *J. Biol. Chem.* 274: 33348-33354.
64. Zocchi E, Carpaneto A, Cerrano C, Bavestrello G, Giovine M, *et al.* 2001. The temperature-signaling cascade in sponges involves a heat-gated cation channel, abscisic acid, and cyclic ADP-ribose. *Proc. Natl. Acad. Sci. USA.* 98: 14859-14864.
65. Guse AH, Tsygankov AY, Weber K and Mayr GW. 2001. Transient tyrosine phosphorylation of human ryanodine receptor upon T cell stimulation. *J. Biol. Chem.* 276: 34722-34727.
66. Wilding M, Russo GL, Galione A, Marino M and Dale B. 1998. ADP-ribose gates the fertilization channel in ascidian oocytes. *Am. J. Physiol.* 275: C1277-C1283.
67. Li PL, Zhang DX, Ge ZD and Campbell WB. 2002. Role of ADP-ribose in 11,12-EET-induced activation of K(Ca) channels in coronary arterial smooth muscle cells. *Am. J. Physiol.* 282: H1229-H1236.
68. Perraud AL, Fleig A, Dunn CA, Bagley LA, Launay P, *et al.* 2001. ADP-ribose gating of the Ca^{2+}-permeable LTRPC2 channel revealed by Nudix motif homology. *Nature* 411: 595-599.
69. Takasawa S, Nata K, Yonekura H and Okamoto H. 1993. Cyclic ADP-ribose in insulin secretion from pancreatic beta cells. *Science* 259: 370-3
70. Mitchell KJ, Pinton P, Varadi A, Tacchetti C, Ainscow EK, *et al.* 2001. Dense core secretory vesicles revealed as a dynamic Ca^{2+} store in neuroendocrine cells with a vesicle-associated membrane protein aequorin chimaera. *J. Cell Biol.* 155: 41-51.
71. Varadi A and Rutter GA. 2002. Dynamic imaging of endoplasmic reticulum Ca^{2+} concentration in insulin-secreting MIN6 cells using recombinant targeted cameleons: roles of sarco(endo)plasmic reticulum Ca^{2+}-ATPase (SERCA)-2 and ryanodine receptors. *Diabetes* 51 Suppl 1: S190-S201.
72. Islam MS and Berggren PO. 1997. Cyclic ADP-ribose and the pancreatic beta cell: where do we stand? *Diabetologia* 40: 1480-1484.
73. Okamoto H, Takasawa S and Nata K. 1997. The CD38-cyclic ADP-ribose signalling system in insulin secretion: molecular basis and clinical implications. *Diabetologia* 40: 1485-1491.
74. Kato I, Takasawa S, Akabane A, Tanaka O, Abe H, *et al.* 1995. Regulatory role of CD38 (ADP-ribosyl cyclase/cyclic ADP-ribose hydrolase) in insulin secretion by glucose in pancreatic beta cells. Enhanced insulin secretion in CD38-expressing transgenic mice. *J. Biol. Chem.* 270: 30045-30050.
75. Takasawa S, Tohgo A, Noguchi N, Koguma T, Nata K, *et al.* 1993. Synthesis and hydrolysis of cyclic ADP-ribose by human leukocyte antigen CD38 and inhibition of the hydrolysis by ATP. *J. Biol. Chem.* 268: 26052-26054.
76. Tohgo A, Takasawa S, Noguchi N, Koguma T, Nata K, *et al.* 1994. Essential cysteine residues for cyclic ADP-ribose synthesis and hydrolysis by CD38. *J. Biol. Chem.* 269: 28555-28557.
77. Jabr RI, Toland H, Gelband CH, Wang XX and Hume JR. 1997. Prominent role of intracellular Ca^{2+} release in hypoxic vasoconstriction of canine pulmonary artery. *Br. J. Pharmacol.* 122: 21-30.
78. Dipp M and Evans AM. 2001. Cyclic ADP-ribose is the primary trigger for hypoxic pulmonary vasoconstriction in the rat lung in situ. *Circ. Res.* 89: 77-83.
79. Morio Y and McMurtry IF. 2002. Ca^{2+} release from ryanodine-sensitive store contributes to mechanism of hypoxic vasoconstriction in rat lungs. *J. Appl. Physiol.* 92: 527-534.

80. Wilson HL, Dipp M, Thomas JM, Lad C, Galione A and Evans AM. 2001. ADP-ribosyl cyclase and cyclic ADP-ribose hydrolase act as a redox sensor: a primary role for cADPR in hypoxic pulmonary vasoconstriction. *J. Biol. Chem.* 276: 11180-11188.
81. Genazzani AA, Bak J and Galione A. 1996. Inhibition of cADPR-Hydrolase by ADP-ribose potentiates cADPR synthesis from beta-NAD$^+$. *Biochem. Biophys. Res. Commun.* 223: 502-507.
82. Cahalan MD. 2001. Cell biology: Channels as enzymes. *Nature* 411: 542-543.
83. Takahashi K, Kukimoto I, Tokita K, Inageda K, Inoue S, *et al.* 1995. Accumulation of cyclic ADP-ribose measured by a specific radioimmunoassay in differentiated human leukemic HL-60 cells with all- trans-retinoic acid. *FEBS Lett.* 371: 204-208.
84. Chini EN, de Toledo FG, Thompson MA and Dousa TP. 1997. Effect of estrogen upon cyclic ADP ribose metabolism: beta-estradiol stimulates ADP ribosyl cyclase in rat uterus. *Proc. Natl. Acad. Sci. USA.* 94: 5872-5876.
85. Dogan S, White TA, Deshpande DA, Murtaugh MP, Walseth TF and Kannan MS. 2002. Estrogen Increases CD38 gene expression and leads to differential regulation of adenosine diphosphate (ADP)-ribosyl cyclase and cyclic ADP-ribose hydrolase activities in rat myometrium. *Biol. Reprod.* 66: 596-602
86. de Toledo FG, Cheng J and Dousa TP. 1997. Retinoic acid and triiodothyronine stimulate ADP-ribosyl cyclase activity in rat vascular smooth muscle cells. *Biochem. Biophys. Res. Commun.* 238: 847-850.
87. Higashida H, Yokoyama S, Hashii M, Taketo M, Higashida M, *et al.* 1997. Muscarinic receptor-mediated dual regulation of ADP-ribosyl cyclase in NG108-15 neuronal cell membranes. *J. Biol. Chem.* 272: 31272-31277.
88. Higashida H, Zhang JS, Hashii M, Shintaku M, Higashida C and Takeda Y. 2000. Angiotensin II stimulates cyclic ADP-ribose formation in neonatal rat cardiac myocytes. *Biochem. J.* 352: 197-202.
89. Barone, F, Genazzani, AA, Conti, A., Churchill, GC, Palombi, F, Ziparo, E, Sorrentino, V, Galione A and Filippini, A 2002. A pivotal role for cADPR-mediated Ca^{2+} signaling: regulation of endothelin-induced contraction in peritubular smooth muscle cells. *FASEB J.* 16: 697-705.
90. Patel S, Churchill GC and Galione A. 2001. Coordination of Ca^{2+} signalling by NAADP. *Trends Biochem. Sci.* 26: 482-489.
91. Aarhus R, Graeff RM, Dickey DM, Walseth TF and Lee HC. 1995. ADP-ribosyl cyclase and CD38 catalyze the synthesis of a Ca^{2+}- mobilizing metabolite from NADP. *J. Biol. Chem.* 270: 30327-30333.
92. Chini EN, Chini CC, Kato I, Takasawa S and Okamoto H. 2002. CD38 is the major enzyme responsible for synthesis of nicotinic acid- adenine dinucleotide phosphate in mammalian tissues. *Biochem. J.* 362: 125-130.
93. Galione A, Cancela J-M, Churchill G, Genazzani A, Lad C, *et al.* 2000. Methods in cADPR and NAADP research. In *Methods in Ca^{2} Signaling*, JW Putney, Jr (ed) pp. 249-296. Boca Raton: CRC Press.
94. Graeff R and Lee HC. 2002. A novel cycling assay for cellular cADP-ribose with nanomolar sensitivity. *Biochem. J.* 361: 379-384.
95. Santella L and Kyozuka K. 1997. Effects of 1-methyladenine on nuclear Ca^{2+} transients and meiosis resumption in starfish oöcytes are mimicked by the nuclear injection of inositol 1,4,5-trisphosphate and cADP-ribose. *Cell Calcium* 22: 11-20.
96. Pollock J, Crawford JH, Wootton JF, Seabrook GR and Scott RH. 1999. Metabotropic glutamate receptor activation and intracellular cyclic ADP- ribose release Ca^{2+} from the same store in cultured DRG neurones. *Cell Calcium* 26: 139-148.
97. Wu Y, Kuzma J, Marechal E, Graeff R, Lee HC, *et al.* 1997. Abscisic acid signaling through cyclic ADP-ribose in plants. *Science* 278: 2126-2130.

98. Prakash YS, Kannan MS, Walseth TF and Sieck GC. 1998. Role of cyclic ADP-ribose in the regulation of $[Ca^{2+}]_i$ in porcine tracheal smooth muscle. *Am. J. Physiol.* 274: C1653-C1660.
99. Gromada J, Jorgensen TD and Dissing S. 1995. The release of intracellular Ca^{2+} in lacrimal acinar cells by alpha-, beta-adrenergic and muscarinic cholinergic stimulation: the roles of inositol triphosphate and cyclic ADP-ribose. *Pflügers Arch.* 429: 751-761.
100. Ebihara S, Sasaki T, Hida W, Kikuchi Y, Oshiro T, *et al.* 1997. Role of cyclic ADP-ribose in ATP-activated potassium currents in alveolar macrophages. *J. Biol. Chem.* 272: 16023-16029.
101. Cancela JM and Petersen OH. 1998. The cyclic ADP ribose antagonist 8-NH_2-cADP-ribose blocks cholecystokinin-evoked cytosolic Ca^{2+} spiking in pancreatic acinar cells. *Pflügers Arch.* 435: 746-748.
102. Takasawa S, Akiyama T, Nata K, Kuroki M, Tohgo A, *et al.* 1998. Cyclic ADP-ribose and inositol 1,4,5-trisphosphate as alternate second messengers for intracellular Ca^{2+} mobilization in normal and diabetic beta-cells. *J. Biol. Chem.* 273: 2497-2500.
103. An NH, Han MK, Um C, Park BH, Park BJ, *et al.* 2001. Significance of ecto-cyclase activity of CD38 in insulin secretion of mouse pancreatic islet cells. *Biochem. Biophys. Res. Commun.* 282: 781-786.
104. Inngjerdingen M, Al-Aoukaty A, Damaj B and Maghazachi AA. 1999. Differential utilization of cyclic ADP-ribose pathway by chemokines to induce the mobilization of intracellular Ca^{2+} in NK cells. *Biochem. Biophys. Res. Commun.* 262: 467-472.
105. Rakovic S, Cui Y, Iino S, Galione A, Ashamu GA, *et al.* 1999. An antagonist of cADP-ribose inhibits arrhythmogenic oscillations of intracellular Ca^{2+} in heart cells. *J. Biol. Chem.* 274: 17820-17827.
106. Schottelndreier H, Potter BVL, Mayr GW and Guse AH. 2001. Mechanisms involved in alpha6beta1-integrin mediated Ca^{2+}signalling. *Cell Signal.* 13: 895-899.

Antony Galione, Department of Pharmacology,
Oxford University, Mansfield Road,
Oxford OX1 3QT, UK.
e-mail: antony.galione@pharmacology.oxford.ac.uk

Cyclic ADP-Ribose and NAADP. Structures, Metabolism and Functions. Lee HC (ed), pp. 65-79, Dordrecht: Kluwer.

A Natural History of the Human CD38 Gene

Enza Ferrero and Fabio Malavasi
Laboratory of Immunogenetics; Department of Genetics, Biology and Biochemistry; Faculty of Medicine, University of Turin and Experimental Medicine Research Center; S. Giovanni Battista Hospital, Turin, Italy.

INTRODUCTION

In the beginning, CD38 (gene symbol *CD38*) was but one of a handful of thymic lymphocyte surface markers [1]. It was subsequently found to be expressed also by B cells, monocytes and natural killer cells, as detailed in [2]. In immunocytes, it emerged that engagement of CD38 by specific monoclonal antibodies could command a variety of responses such as activation and cytokine release in T cells [3, 4], modulation of B cell antigen receptor responses [4], B and myeloid cell apoptosis [6, 7], modulation of the superantigen response in monocytes [7] and induction of NK cell activation and cytotoxicity [9,10]. In addition, CD38 can influence immunocyte adhesion to endothelium [10] through a counterreceptor identified as CD31 [11], leaving no doubts about the receptorial nature of CD38. The intracellular pathways activated via CD38 are being characterized and numerous cytoplasmic substrates have been identified that undergo phosphorylation following CD38 signaling [12-16].

However, the most surprising turn of events was sparked by the 1992 finding of amino acid sequence similarity between CD38 and ADP ribosyl cyclase, a soluble enzyme found in the mollusk *Aplysia* [17]. CD38 was found to have a complex enzymatic activity that can synthesize cyclic ADP ribose (cADPR) and nicotinic acid adenine dinucleotide phosphate (NAADP), two calcium-releasing messengers, or adenonsine 5' diphospho-ribose (ADPR) [18], which plays crucial roles in ion channel gating [19] and post-translational protein modification [20]. Novel biological activities mediated by CD38 enzymatic activity emerged such as the regulation of insulin secretion [21] and neutrophil clearance of bacteria in sepsis [22].

As experimental evidence independently supports the receptor-mediated mechanism of action and the enzymatic activity, we can add CD38 to the growing list of two-in-one proteins that, aside from many leukocyte receptor-ectoenzymes, includes the TRP (transient receptor potential) cation channel LTRPC2 [19], which is both ion channel and protein kinase, and migration inhibitory factor (MIF), a secreted macrophage protein that is a cytokine with tautomerase enzymatic activity [23].

Molecular phylogenetics should give us some clues toward understanding the biological role of CD38 by suggesting *how* and *why* the receptorial and enzymatic activities co-evolved, and will be dealt with in the final part of this chapter. However, the beginning is dedicated to the structure and regulation of the human CD38 gene in view of its increasing repercussions in human physiology and biochemistry in health and disease.

THE GENESCAPE

With the availability of the first draft of the human genome the CD38 neighborhood, located at the 4p15 chromosomal region [24], is now reasonably well delineated. The organization of the *CD38* locus and its surroundings, presented in Fig. 1, is based on the nucleotide sequence of a 576,121 base pair (bp) genomic segment from the GenBank database (accession number NT_027004). Proceeding from telomere to centromere, upstream of *CD38* lies an important neighbor because *CD157* (BST-1, BP3] is a paralog of *CD38*. These two genes have exquisitely similar intron-exon organization as a consequence of their origin by gene duplication [25-27]. *CD38* lies appoximately 50 kb downstream of *CD157*, with both genes having the same 5'→3' orientation. However, *CD157* and *CD38* may not be contiguous since, according to *in silico* analysis, another gene lies between them, identified only as LOC132699. The hypothetical product of this gene is a protein (accession number MGC5629) which is very similar to high-mobility group protein 17 (HMG-17), a non-histone chromosomal protein that binds to nucleosomes and enhances transcription [26].

About 86 kb downstream of *CD38* are another couple of paralogous genes: *HBP17* and *Ksp37*. HBP17 (heparin-binding growth factor binding protein), is a 17 kDa protein first identified in conditioned medium of human epidermoid carcinoma cells [27] encoded by a small, two-exon gene that is followed ~21 kb downstream by the similarly organized gene encoding *Ksp37*. This is a recently-described secreted protein found in human serum whose expression is limited to cytotoxic lymphocytes, be they NK or T cells [28].

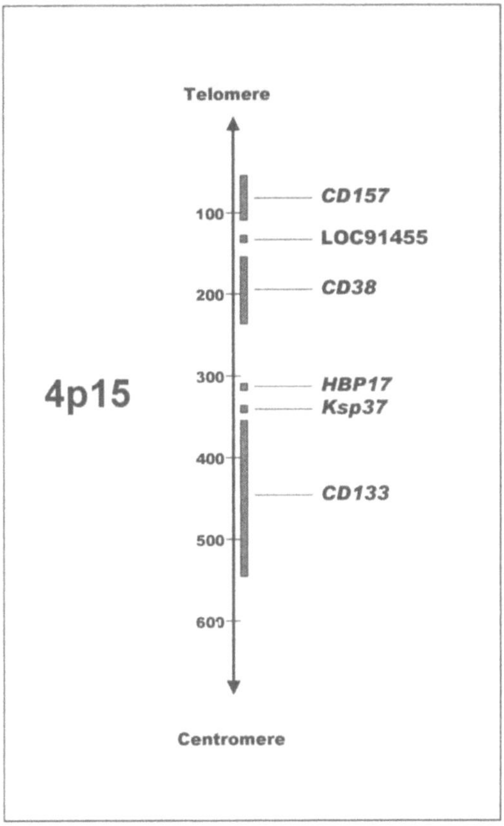

Figure 1. **Chromosomal neighbourhood of CD38.** Schematic diagram of a 600,000 basepair segment of a subregion of the the human 4p15 chromosomal band. The five established genes (*CD157, CD38, HBP17, Ksp37* and *CD133*) are indicated by their gene symbol. LOC91455 is a hypothetical gene identified *in silico*. The approximate size and location of each gene is indicated by a vertical gray bar. The map is based on the genomic sequence with accession number NT_027004.

CD133 (AC133, Prominin-like 1, PROML1 and hematopoietic stem cell antigen) is a large, 23-exon gene distributed over 50 kb that lies just 5 kb downstream of *Ksp37*. CD133 is a 120 kDa five-transmembrane domain glycoprotein with a highly restricted pattern of expression in hematopoietic stem cells and progenitors in bone marrow which overlaps with that of the older, well-known stem cell marker CD34 [29].

Two observations are striking from this glance at the immediate chromosomal vicinity of *CD38*. Firstly, we note the presence of two sets of duplicated genes. Based on the fact that the two paralogous sets of genes form adjacent couples organized as (A1-A2) (B1-B2) and not (A1-B1) (A2-B2) suggests that two independent duplications events occurred in this region to give rise to the *CD157/CD38* and *HBP17/Ksp37* pairs of genes. This is not unusual and the human genome is littered with clusters of related genes. The destiny of a duplicate varies from redundancy and inactivation to divergence and acquisition of new functions, and evidently the latter has prevailed in the case of *CD157* and *CD38*.

The second observation is that CD157, CD38, Ksp37 and CD133 are all expressed in highly-specific stages of development in blood cells, forming a sort of mini-cluster of hematopoietic antigens. For example, witness the myeloid-restricted expression of CD157 [30], the discontinuous expression of CD38 during lymphoid ontogenesis, the killer-cell-specific expression of Ksp37 and the limitation of CD133 expression to hematopoietic stem cells and progenitors. CD38 expression in bone marrow precursors is also exemplary: as CD34 (or CD133^{+}) progenitor cells differentiate they acquire CD38 expression [31].

HUMAN CD38: GENE STRUCTURE AND EXPRESSION

In 1990, Jackson and Bell cloned the cDNA for human CD38 [32], predicting that the product of the human CD38 gene was a polypeptide of 300 amino acids with one membrane-spanning portion close to the N-terminus, a topology characteristic of type II membrane proteins. CD38 was biochemically characterized by Alessio *et al.* as a glycoprotein with a Mw of ~45 kDa in its monomeric form [33].

Cloning of the CD38 gene was reported by two laboratories in 1997 [25, 34]. As shown in Fig. 2, it consists of 8 relatively small exons that vary in size from 86 to 446 bp. However, the exons are dotted over approximately 70 kb of genomic DNA so that the seven intervening sequences constitute 98.2% of the gene. The intron lengths vary from 334 bp (intron 6) to around 40 kb for the surprisingly long first intron.

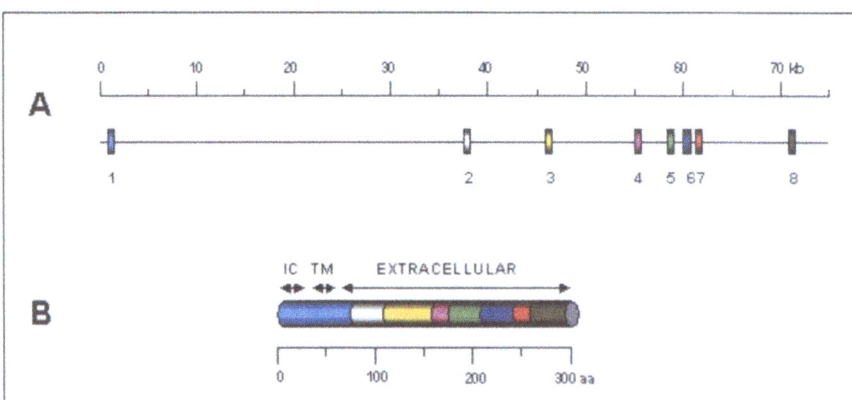

Figure 2. **Organization of human CD38 gene and protein. A.** Schematic diagram of the anatomy of the human CD38 gene, highlighting the position of the eight exons over a continuous 70 kb span of genomic DNA. **B.** The 300 amino acid CD38 protein is illustrated, with relative composition in eight exon-derived blocks. Arrows above the cylindrical representation of the protein indicate the intracellular (IC), transmembrane (TM) and extracellular portions.

The ATG initiation codon is located in exon 1. As the exon with the most coding capacity, exon 1 encodes the 21 amino acid intracytoplasmic portion, the 23 amino acid transmembrane portion and the first 34 juxtamembrane amino acids of the extracellular domain. The remainder of the extracellular domain is encoded by exons 2 to 8.

This straightforward description of the CD38 molecule as a type II membrane protein with an extracellular portion is incomplete without the description of a soluble form of the molecule, present in human serum, derived from the membrane form by proteolytic cleavage [35]. To complicate matters, however, CD38 antigen and/or CD38-type enyzmatic activities are associated with the cytosol [36], mitochondria [37] and nucleus [38, 39] in diverse cell types in non-human animal models.

Table 1. **Expression of CD38 analyzed *in silico* or by immunohistochemistry in human tissues**

Tissue	EST	Antigen	IHC detail
Brain	+	+	Purkinje cells, neurofibrillary tangles
Cardiac muscle	NT	+	Sarcolemma
Digestive tract	NT	+	*Lamina propria* lymphocytes
Germ cells	+	NT	
Germ cell tumor	+	NT	
Germinal center B cells	+	+	Lymph node, spleen
Head-neck	+	NT	
Hepatocellular carcinoma	+	NT	
Hypothalamus	+	NT	
Fetal liver-spleen	+	NT	
Kidney	NT	+	Proximal convoluted tubules
Lymph node	NT	+	Germinal center B cells
Pancreas	+	+	Islets of Langerhans
Pediatric Pre-B ALL.	+	NT	
Prostate	+	+	Epithelium
Skeletal muscle	+	+	Sarcolemma
Skin	NT	-	
Spleen	+	+	Germinal centers, red pulp blood cells
Testis	+	NT	
THP-1 myeloid + DMSO	+	NT	
Thymus	NT	+	Paracortical and medullary areas
Tonsillar B cells	+	+	
Umbilical cord	NT	-	

Abbreviations: IHC, immunohistochemistry; NT, not tested; ALL, acute lymphocytic leukemia. Details of all human CD38 ESTs are available at: http://www.ncbi.nlm.nih.gov/UniGene.

Proceeding beyond the cell to expression of CD38 in the human organism *in toto*, the 1980's gave us details of the distribution of CD38 in cells from blood and bone marrow, obtained mostly by immunofluorescence. In 1998, Fernandez *et al.* used immunohistochemistry (IHC) to survey the distribution of the CD38 antigen in various human organs and tissues [39]. Not surprisingly, 2000 gave us a computational tool for mRNA expression analysis in the form of the expressed sequence tag (EST) database. This aims to collect all the mRNAs expressed in an organism. Each mRNA is represented by a unique short sequenced fragment or tag. A list of the human RNA sources that harbour CD38 ESTs is shown in Table I where, for comparison, the data is flanked by the CD38 antigen distribution obtained by IHC. The picture that emerges is one of quite diffuse expression. Note that in the mouse, EST analysis is far more informative as it currently offers a more complete representation of the murine body, and murine CD38 ESTs are present in almost every adult tissue or organ examined.

HUMAN CD38: GENE REGULATION AND TRANSCRIPTS

How then does one reconcile the quite widespread expression of CD38 in adult tissues with the intricate on/off expression in blood and bone marrow? As with any gene, CD38 expression is probably regulated through a combination of chromatin structure, methylation, and control of transcription at the promoter, elongation or termination stages.

Promoter Control–The sites where RNA polymerase II (pol II) starts transcribing the CD38 gene have been analysed in different laboratories using different templates - Jurkat (T cell-acute lymphocytic leukemia cell line), HL-60 (acute promyelocytic leukemia cell line), pancreas (endocrine) and thymus [25, 34, 40]. The results concord on the existence of multiple transcription start sites (TSSs) mostly situated from 75 to 102 bp upstream of the first ATG (Fig. 3). No canonical TATA box is found in the vicinity of the transcription start sites. However, TATA-less promoters may have alternative start sites such as: (i) an initiator (Inr) sequence, loosely defined as PyPyA_{+1}NT/APyPy or (ii) binding sites for the PU.1 transcription factor may function as the core promoter element, especially in myeloid and B cells [41]. In *CD38*, a potential Inr sequence (5'- TTAAACT-3') is located 205 bp upstream of the first ATG and a pair of potential binding sites for PU.1 are present in the promoter. On the other hand, the combination of (i) multiple TSSs; (ii) a high G/C promoter content; (iii) a CpG island associated with the exon 1/intron 1; and (iv) promoter and intronic potential binding sites for the ubiquitous transcription factor Sp1 give the *CD38* promoter a "housekeeping" (meaning constitutively active) look about it [41].

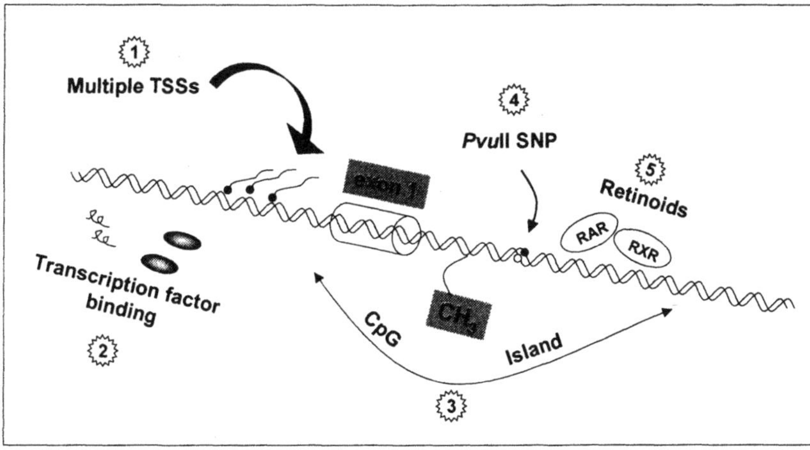

Figure 3. **Regulation and polymorphism of human CD38. 1.** Multiple transcription start sites are indicated as pol II (•) and nascent transcripts (black ribbons). **2.** Hypothetical transcription factors that bind the promoter. Potential transcription binding sites are listed in ref. 34. **3.** A 900 bp CpG associated with exon 1 and the 5' end of intron 1 is shown. **4.** Location of the *Pvu*II polymorphic site at the 5' end of intron 1, with black and white dots representing the two alleles. **5.** Intronic retinoic acid responsive element that binds RAR-RXR complexes.

According to the literature, the 1.5 kb CD38 upstream sequence would seem to be transcriptionally active in uninduced HL-60 cells [40]. Therefore, we decided to make a reporter construct consisting of this upstream sequence and the coding region of the enhanced green fluorescence protein. As expected, fluorescence was detected in CD38-surface expressing cells but also in five human cell lines that do not have detectable CD38 either on the cell surface or by RT-PCR (E. Ferrero, manuscript in preparation].

The importance of cis-acting transcription factors (TFs) remains to be experimentally verified. Potential binding factors are listed in [34] and includes numerous immunologically-relevant TFs and an estrogen responsive element which warrants further investigation in view of the recent reports of estrogen-mediated induction of CD38 expression or simil-CD38 enzymatic activity [44, 45]. Details of other CD38 inducers such as retinoids, vitamin D_3 and interferon-γ are reviewed in [43].

Elongation and Termination–The hypothesis that control of transcriptional elongation may play a major role in the regulation of CD38 derives from the in silico and subsequently physical cloning of a novel CD38 transcript which is truncated in intron 1 (E. Ferrero, manuscript in preparation). Now detected in several tissues, this transcript is consistently truncated at the same spot in intron 1, suggesting the presence of a STOP or GO decision point for continuing transcription. This adds to our previous interest in intron 1, raised by the description there of a functional retinoic acid responsive element by Kishimoto *et al.* [40].

Transcripts–It is intuitive from the dimensions of the gene that the production and processing of CD38 transcripts must require considerable time and effort. If the average elongation rate of RNA polymerase II (pol II) is 2.4 kb min^{-1} [44] then almost 30 minutes are required to complete transcription of the gene. It is not known whether the primary transcript is spliced during or after synthesis is complete, *i.e.*, co- or post-transcriptionally. This is not an arcane point because analysis of *CD38* transcripts by Northern blot reveals numerous hybridizing bands (of 1.4, 1.7, 2.4, 3.7, 4.5, 5.2, 6.3, 7.0 and 8.5 kb). Their contents are mostly unknown and some may represent accumulated splice products. Two *CD38* transcripts have been reported so far: the full-length and *CD38 alt* which is the only known alternative splice product [25]. In *CD38 alt*, exon 2 is spliced to exon 4, excluding exon 3, leading to a reading frame shift which translates into a premature stop codon in exon 4. As mentioned above, a third transcript, *tCD38* (truncated CD38), is currently being analysed in our laboratory.

HUMAN CD38 : POLYMORPHISM

A further reason for interest in *CD38* intron 1 is the presence of a single nucleotide polymorphism (SNP), a C→G nucleotide change at bp 184 of intron 1, which creates a *Pvu*II polymorphism [45]. The *CD38*B* allele (C at +184, presence of the *Pvu*II site) is the most common, with a frequency of 0.86 in a cohort of 60 healthy Italian adults, while *CD38*A* (G at +184, loss of the PvuII site) has an allelic frequency of 0.14. The frequency of heterozygotes (0.25) is close to that predicted by the Hardy-Weinberg law (0.24). *CD38*B* homozygotes have the most common genotype while only 1% of the population is expected to be homozygous for *CD38*A*.

It is not yet clear if the polymorphism has biological significance. Analysis of 37 individuals with type 1 diabetes and 115 type 2 diabetics failed to reveal differences in allelic frequencies with respect to controls [46]. We are currently investigating the significance of RT-PCR findings that suggest higher transcription of the G allele (E. Ferrero, unpublished results) and the possibility that E2A, a basic helix-loop-helix DNA-binding protein required for B and T lymphocyte differentiation, influences CD38 expression as the PvuII recognition sequence is part of a potential binding site for E2A (5' RCAGNTG 3').

Interestingly, the intronic *Pvu*II site is present in the Old World monkey, *Macaca fasicularis* (F. Saccucci, S. Crovella, unpublished result). However, analysis of genomic DNA from 30 macaques showed that the *Pvu*II site was present in all 60 copies of the gene analysed. Since it is estimated that macaque and Homo sapiens last shared a common ancestor 25 million years ago [47], this suggests that the SNP evolved quite "recently".

CD38 IN PHYLOGENESIS

CD38 is part of a 'superfamily' of ADP-ribosyl cyclases along with the CD157 family and *ADP-ribosyl cyclase* (from the mollusk *Aplysia*). The constituent genes of this superfamily show such high conservation over most of their gene structure to leave little doubt about their common evolutionary origin. Exon-by-exon comparison of human *CD38* (8 exons) [25, 34], human and murine *CD157* (9 exons) [51, 52] and *ADP-ribosyl cyclase* (9 exons) [49] reveals a quite spectacular conservation of exon size and intron positioning from exons 2 to 8. One noteworthy difference is that exons 3 and 4 in *CD38* and *CD157* form a single exon in the *cyclase*, thus the intron insertion (or intron excision in the *cyclase* gene?) could be a useful molecular marker.

Although the genes are highly homologous and their deduced proteins share 25-30% amino acid sequence, each family member forms a distinct protein entity: the cyclase is a soluble molecule [50], CD38 is membrane-bound with a membrane-spanning portion that probably evolved through modification of a signal peptide, and CD157 is membrane bound by a GPI anchor. Note that in consequence of their different mechanisms of membrane attachment, the two proteins have opposite membrane topology: CD38 is C-terminus out, CD157 is N-terminus out.

Among the conserved amino acids, the most striking are represented by a 5-disulfide forming, 10-cysteine (Cys) array, the true protein signature of this family. In addition, the cyclase and CD38 share another pair of features that distinguish them from the CD157 family: a additional disulfide-forming pair of Cys residues and a six amino acid stretch Thr144- Leu-Glu-Asp-Thr-Leu149 (TLEDTL) that contains one of the critical glutamates for enzymatic activity [51]. The other, Glu226 (human CD38 numbering] [52] is found in all members of the superfamily. The complete conservation of the 10-Cys array throughout 750 million years of evolution (~when mollusks and human last shared an ancestor) suggests that disulfide bonds play an essential role in the structure-function of these molecules.

Armed with these molecular landmarks, we attempted to trace the origins of CD38 in phylogenesis by molecular sequence analysis of the genome drafts available to date (Fig. 4). Plenty of mammalian orthologs were found as CD38 cDNAs have been cloned in numerous species such as man, rat, mouse, dog, pig, rabbit and cow, and we have recently cloned full-length cDNAs from macaque (E. Ferrero, manuscript in preparation) and chicken (E. Ferrero, unpublished results). ESTs from likely CD38 orthologs are

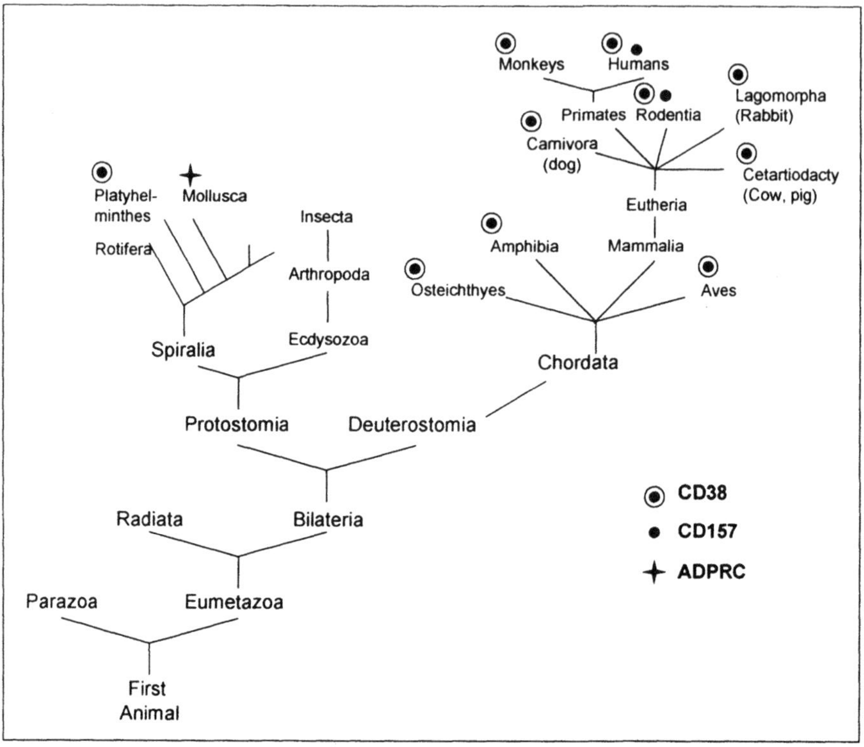

Figure 4. **Charting the animal evolutionary history of the CD38/CD157/ADPRC gene family.** Complete CD38 (bull's-eye) orthologs are found in the primates *Homo sapiens* and *Macaca fascicularis*. the rodents mouse and rat, rabbit (order *Lagomorpha*), dog (order *Carnivora*), cow and pig (infraorder *Cetartiodacty*), and first among non-mammalian vertebrates, in the chicken (class *Aves*). Suspected CD38 orthologs are found in the liver fluke *Schistosoma mansoni* (Phylum *Platyhelminthes*), in the frog *Xenopus* laevis (Class *Amphibia*), and in the puffer fish *Teraodon nigroviridis* (Class *Osteichthyes*). CD157 (black dot) has been cloned in man, rat and mouse. ADP ribosyl cyclase (ADPRC, four-pointed star) is so far unique to the Phylum *Mollusca*, genus *Aplysia*.

present in the puffer fish (*Tetraodon nigroviridis*), the frog (*Xenopus laevis*) and in the flatworm (*Schistosoma mansoni*), which is quite baffling. The schistosome candidate is only a partial clone and the deduced protein sequence available has most similarity to canine and other CD38s: the order of similarity is CD38>ADPRC>CD157. As the protein sequence corresponds to exons 3-5, it is impossible to know whether it is membrane-bound or soluble. The TLEDTL signature of mammalian CD38 is SLETTL in the *schistosome* protein, with conservation of the glutamate active site. We hope that future genome sequence information will become available to confirm whether or not this is the most ancient CD38 ortholog known so far. Surprisingly, no CD38 homologs were found by computational searches in

Drosophila, *C. elegans*, *S. cerevisiae* or *S. pombe*, at least at the time of writing.

In conclusion, this tale of how a routine marker blossomed into a beautiful multifuctional molecule is a scientific fairy story come true that will not end until the biological role of CD38 is completely unfurled, which seems far away at the moment. We have always favoured the hypothesis that CD38 is an enzyme that gained membrane attachment and was subsequently recruited for signaling duties [53], apparently deriving a role in both the innate and adaptive immune sytems where it seems to have overcome the shortness of its cytoplasmic tail by 'parasitizing' professional signal transducing molecules on immunocytes, such as the T cell receptor [54], the B cell receptor [4] and CD16 [55]. But perhaps the true role of CD38 is exerted elsewhere, seeing as it is expressed in other subcellular locations. New roles for CD38 also emerge as analysis of the CD38 knockout mice proceeds [22, 56, 57], while in humans, CD38 is gaining in practical value in human diseases, most notably in B cell chronic lymphocytic leukemia [58], while its roles in acute promyelocytic leukemia [59] and other forms of cancer [60], AIDS [61], autoimmunity [46, 62-64] and diabetes [21] are currently being verified experimentally. The story continues.....

ACKNOWLEDGEMENTS

We are grateful to Dr. Alfredo Brusco (University of Turin] for carefully reading the manuscript. This work was supported by grants from AIRC, Special Project AIDS (Istituto Superiore di Sanità), Biotechnology (CNR/MURST), and Cofinanziamento (MURST) to FM. EF is supported by the Fondazione per le Biotecnologie. Funds were also provided by the International Foundation for Research in Experimental Medicine (FIRMS) in Torino, Italy.

REFERENCES

1. Reinherz EL, Kung PC, Goldstein G, Levey RH and Schlossman SF. 1980. Discrete stages of human intrathymic differentiation: analysis of normal thymocytes and leukemic lymphoblasts of T-cell lineage. *Proc. Natl. Acad. Sci. USA* 77: 1588-1592.
2. Malavasi F, Funaro A, Alessio M, DeMonte LB, Ausiello CM, Dianzani U, Lanza F, Magrini E, Momo M and Roggero S. 1992. CD38: a multi-lineage cell activation molecule with a split personality. *Int. J. Clin. Lab. Res.* 22: 73-80.
3. Funaro A, Spagnoli GC, Ausiello CM, Alessio M, Roggero S, Delia D, Zaccolo M and Malavasi F. 1990. Involvement of the multilineage CD38 molecule in a unique pathway of cell activation and proliferation. *J. Immunol.* 145: 2390-2396.
4. Lund FE, Yu N, Kim KM, Reth M and Howard MC. 1996. Signaling through CD38 augments B cell antigen receptor (BCR) responses and is dependent on BCR expression. *J. Immunol.* 157: 1455-1467.
5. Zupo S, Rugari E, Dono M, Taborelli G, Malavasi F and Ferrarini M. 1994. CD38 signaling by agonistic monoclonal antibody prevents apoptosis of human germinal center B cells. *Eur. J. Immunol.* 24: 1218-1222.

6. Kumagai M, Coustan-Smith E, Murray DJ, Silvennoinen O, Murti KG, Evans WE, Malavasi F and Campana D. 1995. Ligation of CD38 suppresses human B lymphopoiesis. *J. Exp. Med.* 181: 1101-1110.
7. Zilber MT, Gregory S, Mallone R, Deaglio S, Malavasi F, Charron D and Gelin C. 2000. CD38 expressed on human monocytes: a coaccessory molecule in the superantigen-induced proliferation. *Proc. Natl. Acad. Sci. USA* 97: 2840-2845.
8. Mallone R, Funaro A, Zubiaur M, Baj G, Ausiello CM, Tacchetti C, Sancho J, Grossi C and Malavasi F. 2001. Signaling through CD38 induces NK cell activation. *Int. Immunol.* 13: 397-409.
9. Sconocchia G, Titus JA, Mazzoni A, Visintin A, Pericle F, Hicks SW, Malavasi F and Segal DM. 1999. CD38 triggers cytotoxic responses in activated human natural killer cells. *Blood* 94: 3864-3871.
10. Dianzani U, Funaro A, DiFranco D, Garbarino G, Bragardo M, Redoglia V, Buonfiglio D and De Monte LB, Pileri A and Malavasi F. 1994. Interaction between endothelium and $CD4^{+}CD45RA^{+}$ lymphocytes. Role of the human CD38 molecule. *J. Immunol.* 153: 952-959.
11. Deaglio S, Morra M, Mallone R, Ausiello CM, Prager E, Garbarino G, Dianzani U, Stockinger H and Malavasi F. 1998. Human CD38 (ADP-ribosyl cyclase) is a counter-receptor of CD31, an Ig superfamily member. *J. Immunol.* 160: 395-402.
12. Kikuchi Y, Yasue T, Miyake K, Kimoto M and Takatsu K. 1995. CD38 ligation induces tyrosine phosphorylation of Bruton tyrosine kinase and enhanced expression of interleukin 5-receptor alpha chain: synergistic effects with interleukin 5. *Proc. Natl. Acad. Sci. USA* 92: 11814-11818.
13. Silvennoinen O, Nishigaki H, Kitanaka A, Kumagai M, Ito C, Malavasi F, Lin Q, Conley ME and Campana D. 1996. CD38 signal transduction in human B cell precursors. Rapid induction of tyrosine phosphorylation, activation of syk tyrosine kinase, and phosphorylation of phospholipase C-gamma and phosphatidylinositol 3- kinase. *J. Immunol.* 156: 100-107.
14. Zubiaur M, Izquierdo M, Terhorst C, Malavasi F and Sancho J. 1997. CD38 ligation results in activation of the Raf-1/mitogen-activated protein kinase and the CD3-zeta/zeta-associated protein-70 signaling pathways in Jurkat T lymphocytes. *J. Immunol.* 159: 193-205.
15. Inoue S, Kontani K, Tsujimoto N, Kanda Y, Hosoda N, Hoshino S, Hazeki O and Katada T. 1997. Protein-tyrosine phosphorylation by IgG1 subclass CD38 monoclonal antibodies is mediated through stimulation of the Fc-gamma-II receptors in human myeloid cell lines. *J. Immunol.* 159: 5226-5232.
16. Yasue T, Nishizumi H, Aizawa S, Yamamoto T, Miyake K, Mizoguchi C, Uehara S, Kikuchi Y and Takatsu K. 1997. A critical role of Lyn and Fyn for B cell responses to CD38 ligation and interleukin 5. *Proc. Natl. Acad. Sci. USA* 94: 10307-10312.
17. States DJ, Walseth TF and Lee HC. 1992. Similarities in amino acid sequences of Aplysia ADP-ribosyl cyclase and human lymphocyte antigen CD38. *Trends Biochem. Sci.* 17: 495.
18. Lee HC. 2001. Physiological functions of cyclic ADP-ribose and NAADP as calcium messengers. *Annu. Rev. Pharmacol. Toxicol.* 41: 317-345.
19. Perraud AL, Fleig A, Dunn CA, Bagley LA, Launay P, Schmitz C, Stokes AJ, Zhu Q, Bessman MJ, Penner R, Kinet JP and Scharenberg AM. 2001. ADP-ribose gating of the calcium-permeable LTRPC2 channel revealed by Nudix motif homology. *Nature* 411: 595-599.
20. Koch-Nolte F, Reche P, Haag F and Bazan F. 2001. ADP-ribosyltransferases: plastic tools for inactivating protein and small molecular weight targets. *J. Biotechnol.* 92: 81-87.
21. Okamoto H, Takasawa S, Nata K, Kato I, Tohgo A and Noguchi N. 2000. Physiological and pathological significance of the CD38/cyclic ADP-ribose signaling system. *Chem. Immunol.* 75: 121-145.

22. Partida-Sanchez S, Cockayne DA, Monard S, Jacobson EL, Oppenheimer N, Garvy B, Kusser K, Goodrich S, Howard M, Harmsen A, Randall TD and Lund FE. 2001. Cyclic ADP-ribose production by CD38 regulates intracellular calcium release, extracellular calcium influx and chemotaxis in neutrophils and is required for bacterial clearance in vivo. *Nature Med.* 7: 1209-1216.
23. Rosengren E, Aman P, Thelin S, Hansson C, Ahlfors S, Bjork P, Jacobsson L and Rorsman H. 1997. The macrophage migration inhibitory factor MIF is a phenylpyruvate tautomerase. *FEBS Lett.* 417: 85-88.
24. Nakagawara K, Mori M, Takasawa S, Nata K, Takamura T, Berlova A, Tohgo A, Karasawa T, Yonekura H, Takeuchi T, *et al.* 1995. Assignment of CD38, the gene encoding human leukocyte antigen CD38 (ADP-ribosyl cyclase/cyclic ADP-ribose hydrolase), to chromosome 4p15. *Cytogenet. Cell Genet.* 69: 38-39.
25. Nata K, Takamura T, Karasawa T, Kumagai T, Hashioka W, Tohgo A, Yonekura H, Takasawa S, Nakamura S and Okamoto H. 1997. Human gene encoding CD38 (ADP-ribosyl cyclase/cyclic ADP-ribose hydrolase): organization, nucleotide sequence and alternative splicing. *Gene* 186: 285-292.
26. Landsman D, Soares N, Gonzalez FJ and Bustin M. 1986. Chromosomal protein HMG-17. Complete human cDNA sequence and evidence for a multigene family. *J. Biol. Chem.* 261: 7479-7484.
27. Okamoto T, Tanaka Y, Kan M, Sakamoto A, Takada K and Sato JD. 1996. Expression of fibroblast growth factor binding protein HBp17 in normal and tumor cells. *In Vitro Cell Dev. Biol. Anim.* 32: 69-71.
28. Ogawa K, Tanaka K, Ishii A, Nakamura Y, Kondo S, Sugamura K, Takano S, Nakamura M and Nagata K. 2001. A novel serum protein that is selectively produced by cytotoxic lymphocytes. *J. Immunol.* 166: 6404-6412.
29. Kratz-Albers K, Zuhlsdorp M, Leo R, Berdel WL, Buchner T and Serve H. 1998. Expression of a AC133, a novel stem cell marker, on human leukemic blasts lacking CD34-antigen and on a human $CD34^+$ leukemic line: MUTZ-2. *Blood* 92: 4485-4487.
30. Ortolan E, Capobianco A, Armando E, Crivellin F, Horenstein A and Malavasi F. 2002. CD157, the Janus of CD38 but with a unique personailty. *Cell Biochem. Funct.* 20: 1-14.
31. Terstappen LW, Huang S, Safford M, Lansdorp PM and Loken MR. 1991. Sequential generations of hematopoietic colonies derived from single nonlineage-committed $CD34^+CD38^-$ progenitor cells. *Blood* 77: 1218-1227.
32. Jackson DG and Bell JI. 1990. Isolation of a cDNA encoding the human CD38 (T10) molecule, a cell surface glycoprotein with an unusual discontinuous pattern of expression during lymphocyte differentiation. *J. Immunol.* 144: 2811-2815.
33. Alessio M, Roggero S, Funaro A, De Monte LB, Peruzzi L, Geuna M and Malavasi F. 1990. CD38 molecule: structural and biochemical analysis on human T lymphocytes, thymocytes, and plasma cells. *J. Immunol.* 145: 878-884.
34. Ferrero E and Malavasi F. 1997. Human CD38, a leukocyte receptor and ectoenzyme, is a member of a novel eukaryotic gene family of nicotinamide adenine dinucleotide-converting enzymes: extensive structural homology with the genes for murine bone marrow stromal cell antigen 1 and aplysian ADP-ribosyl cyclase. *J. Immunol.* 159: 3858-3865.
35. Funaro A, Horenstein AL, Calosso L, Morra M, Tarocco RP, Franco L, De Flora A and Malavasi F. 1996. Identification and characterization of an active soluble form of human CD38 in normal and pathological fluids. *Int. Immunol.* 8: 1643-1650.
36. Yamauchi J and Tanuma S. 1994. Occurrence of an NAD^+ glycohydrolase in bovine brain cytosol. *Arch. Biochem. Biophys.* 308: 327-329.
37. Moser B, Winterhalter KH and Richter C. 1983. Purification and properties of a mitochondrial NAD^+ glycohydrolase. *Arch. Biochem. Biophys.* 224: 358-364.
38. Khoo KM and Chang CF. 2002. Identification and characterization of nuclear CD38 in the rat spleen. *Int. J. Biochem. Cell Biol.* 34: 43-54.

39. Fernandez JE, Deaglio S, Donati D, Beusan IS, Corno F, Aranega A, Forni M, Falini B and Malavasi F. 1998. Analysis of the distribution of human CD38 and of its ligand CD31 in normal tissues. *J. Biol. Regul. Homeost. Agents* 12: 81-91.
40. Kishimoto H, Hoshino S, Ohori M, Kontani K, Nishina H, Suzawa M, Kato S and Katada T. 1998. Molecular mechanism of human CD38 gene expression by retinoic acid. Identification of retinoic acid response element in the first intron. *J. Biol. Chem.* 273: 15429-15434.
41. Ernst P and Smale ST. 1995. Combinatorial regulation of transcription. I: General aspects of transcriptional control. *Immunity* 2: 311-9.
42. Chini EN, de Toledo FG, Thompson MA and Dousa TP. 1997. Effect of estrogen upon cyclic ADP ribose metabolism: beta-estradiol stimulates ADP ribosyl cyclase in rat uterus. *Proc. Natl. Acad. Sci. USA* 94: 5872-5876.
43. Funaro A, Ferrero E, Mehta K and Malavasi F. 2000. Schematic portrait of human CD38 and related molecules. *Chem. Immunol.* 75: 256-273.
44. Tennyson CN, Klamut HJ and Worton RG. 1995. The human dystrophin gene requires 16 hours to be transcribed and is cotranscriptionally spliced. *Nature Genet.* 9: 184-190.
45. Ferrero E, Saccucci F and Malavasi F. 1999. The human CD38 gene: polymorphism, CpG island, and linkage to the CD157 (BST-1) gene. *Immunogenet.* 49: 597-604.
46. Mallone R, Ortolan E, Baj G, Funaro A, Giunti S, Lillaz E, Saccucci F, Cassader M, Cavallo-Perin P and Malavasi F. 2001. Autoantibody response to CD38 in Caucasian patients with type 1 and type 2 diabetes: immunological and genetic characterization. *Diabetes* 50: 752-762.
47. Goodman M, Porter CA, Czelusniak J, Page SL, Schneider H, Shoshani J, Gunnell G and Groves CP. 1998. Toward a phylogenetic classification of Primates based on DNA evidence complemented by fossil evidence. *Mol. Phylogenet. Evol.* 9: 585-598.
48. Muraoka O, Tanaka H, Itoh M, Ishihara K and Hirano T. 1996. Genomic structure of human BST-1. *Immunol. Lett.* 54: 1-4.
49. Nata K, Sugimoto T, Tohgo A, Takamura T, Noguchi N, Matsuoka A, Numakunai T, Shikama K, Yonekura H, Takasawa S, *et al.* 1995. The structure of the Aplysia kurodai gene encoding ADP-ribosyl cyclase, a second-messenger enzyme. *Gene* 158: 213-218.
50. Glick DL, Hellmich MR, Beushausen S, Tempst P, Bayley H and Strumwasser F. 1991. Primary structure of a molluscan egg-specific NADase, a second- messenger enzyme. *Cell Regul.* 2: 211-218.
51. Graeff R, Munshi C, Aarhus R, Johns M and Lee HC. 2001. A single residue at the active site of CD38 determines its NAD cyclizing and hydrolyzing activities. *J. Biol. Chem.* 276: 12169-12173.
52. Munshi C, Aarhus R, Graeff R, Walseth TF, Levitt D and Lee HC. 2000. Identification of the enzymatic active site of CD38 by site-directed mutagenesis. *J. Biol. Chem.* 275: 21566-21571.
53. Ferrero E and Malavasi F. 1999. The metamorphosis of a molecule: from soluble enzyme to the leukocyte receptor CD38. *J. Leukoc. Biol.* 65: 151-161.
54. Morra M, Zubiaur M, Terhorst C, Sancho J and Malavasi F. 1998. CD38 is functionally dependent on the TCR/CD3 complex in human T cells. FASEB J. 12: 581-592.
55. Deaglio S ZM, Gregorini A, Bottarel, Ausiello C, Dianzani U, Sancho J and Malavasi F. 2002. Human CD38 and CD16 are functionally dependent and physicaly associated in natural killer cells. *Blood* 90: (in press).
56. Cockayne DA, Muchamuel T, Grimaldi JC, Muller-Steffner H, Randall TD, Lund FE, Murray R, Schuber F and Howard MC. 1998. Mice deficient for the ecto-nicotinamide adenine dinucleotide glycohydrolase CD38 exhibit altered humoral immune responses. *Blood* 92: 1324-1333.
57. Kato I, Yamamoto Y, Fujimura M, Noguchi N, Takasawa S, Okamoto H. 1999. CD38 disruption impairs glucose-induced increases in cyclic ADP-ribose, $[Ca^{2+}]_i$, and insulin secretion. *J. Biol. Chem.* 274: 1869-1872.

58. Hamblin TJ, Orchard JA, Ibbotson RE, Davis Z, Thomas PW, Stevenson FK and Oscier DG. 2002. CD38 expression and immunoglobulin variable region mutations are independent prognostic variables in chronic lymphocytic leukemia, but CD38 expression may vary during the course of the disease. *Blood* 99: 1023-1029.
59. Mehta K. 2000. Retinoid-mediated signaling in CD38 antigen expression. *Chem. Immunol.* 75: 20-38.
60. Kramer G, Steiner G, Fodinger D, Fiebiger E, Rappersberger C, Binder S, Hofbauer J and Marberger M. 1995. High expression of a CD38-like molecule in normal prostatic epithelium and its differential loss in benign and malignant disease. *J. Urol.* 154: 1636-1641.
61. Savarino A, Bottarel F, Malavasi F and Dianzani U. 2000. Role of CD38 in HIV-1 infection: an epiphenomenon of T-cell activation or an active player in virus/host interactions? *AIDS* 14: 1079-1089.
62. Ikehata F, Satoh J, Nata K, Tohgo A, Nakazawa T, Kato I, Kobayashi S, Akiyama T, Takasawa S, Toyota T and Okamoto H. 1998. Autoantibodies against CD38 (ADP-ribosyl cyclase/cyclic ADP-ribose hydrolase) that impair glucose-induced insulin secretion in noninsulin- dependent diabetes patients. *J. Clin. Invest.* 102: 395-401.
63. Pupilli C, Giannini S, Marchetti P, Lupi R, Antonelli A, Malavasi F, Takasawa S, Okamoto H and Ferrannini E. 1999. Autoantibodies to CD38 (ADP-ribosyl cyclase/cyclic ADP-ribose hydrolase) in Caucasian patients with diabetes: effects on insulin release from human islets. *Diabetes* 48: 2309-2315.
64. Antonelli A, Fallahi P, Nesti C, Pupilli C, Marchetti P, Takasawa S, Okamoto H and Ferrannini E. 2001. Anti-CD38 autoimmunity in patients with chronic autoimmune thyroiditis or Graves' disease. *Clin. Exp. Immunol.* 126: 426-431.

Enza Ferrero, MD, Dept. Genetics, Biology & Biochemistry
Via Santena, 19, 10126 Torino, Italy
Tel: +39 011 696 1734, Fax: +39 011 696 6155
e-mail: enza.ferrero@unito.it

Cyclic ADP-Ribose and NAADP. Structures, Metabolism and Functions. Lee HC (ed), pp. 81-99, Dordrecht: Kluwer.

A Yeast Expression System of High Efficiency for Producing Recombinant Enzymes

Cyrus B. Munshi and Hon Cheung Lee
Department of Pharmacology, University of Minnesota, Minneapolis, MN 55455, USA.

A YEAST EXPRESSION SYSTEM

Notwithstanding their structural and functional differences, cADPR and NAADP are synthesized by the same homologous enzymes. Three have so far been identified and they are the *Aplysia* ADP-ribosyl cyclase and the mammalian antigens, CD38 and CD157 (formerly called BST-1) [1-4]. The sequence homology among these three enzymes is shown in Figure 1. All three enzymes are widespread in occurrence and are localized in a variety of tissues [1, 5-8]. The novel multifunctionality of these enzymes calls for detailed structural studies of their catalytic properties by site-directed mutagenesis, enzyme kinetics and X-ray crystallography. Large quantities of proteins are needed for these kinds of studies, necessitating the development of an efficient expression system for producing the necessary recombinant proteins. In addition, the *Aplysia* cyclase is the most critical component of the recently developed cycling assay for cADPR [9]. CD38, on the other hand, is the only known enzyme that can hydrolyze cADPR and is used in its radioimmunoassay to verify that the signal is indeed due to cADPR [10].

For these purposes, large quantities of these enzymes are routinely needed and a yeast expression system has been developed to express the recombinant proteins [11, 12]. With this expression system, gram quantities of the cyclase and CD38 can be produced. The system is also applicable for producing recombinant nicotinamide mononucleotide adenylyl transferase (NMN-AT), a nuclear enzyme essential for the synthesis and recovery of NAD [13]. NMN-AT, as described in the chapter by Graeff and Lee in this book, is used in a cycling assay for NAADP [14].

```
CD38    1    MANCEFSPVSGDKPCCRLSRRAQLCLGVSILVLILVVVLA 40
Cyc     1    ~~~~~~~~~~~~~~~~~~~~~~~~~~~MSPVAIIACVCLA 13
CD157   1    ~~~~~~~~~~~~~~~~~~~~~MAVQGGLLSLWLWLWLSLL 19

CD38    41   VVVPRWRQTWSGPGTTKRFPETVLARC.....VKYTEIHP 75
Cyc     14   VTLTSISPS.EAIVPTRELENVFLGRCKDYEITRYLDILP 52
CD157   20   TVLLGARARWRGEGTTPHLQSIFLGRC.....AEYTTL.L 53

CD38    76   EMRHVDCQSVWDAFKGAFISKHPCNITEEDYQPLMKLGTQ 115
Cyc     53   RVRS.DCSALWKDFFKAFSFKNPCDLDLGSYKDFFTSAQQ 91
CD157   54   SLGNKNCTAIWEAFKGV.LDKDPCSVLPSDYDLFINLSRH 92

CD38    116  TVPCNKILLWSRIKDLAHQFTQVQRDMFTLEDTLLGYLAD 155
Cyc     92   QLPKNKVMFWSGVYDEAHDYANTGRKYITLEDTLPGYMLN 131
CD157   93   PIPRDKSLFWENNHLLVMSYGENTRRLVALCDVLYGKVGD 132

CD38    156  DLTWCGEFNTSKINYQSCPDWRKDCSNNPVSVFWKTVSRR 195
Cyc     132  SLVWCGQRANPGFNEKVCPDF.KTCPVQARESFWGMASSS 170
CD157   133  FLSWCRQENASGLDYQSCPT.SEDCENNAVDSYWKSASMQ 171

CD38    196  FAEAACDVVHVMLNGSRSK..IFDKNSTFGSVEVHNLQPE 233
Cyc     171  YAHSAEGEVTYMVDGSNPKVPAYRPDSFFGKYELPNL.TN 209
CD157   172  YSRDSSGVINVMLNGSEPK.GAYPTRGFFADFEIPYLQKD 210

CD38    234  KVQTLEAWVIHGGREDSRDLCQDPTIKELESIISKRNIQF 273
Cyc     210  KVTRVKVIVLHRLGEKIIEKCGAGSLLDLEKLVKAKHFAF 249
CD157   211  KVTRIEIWVMHDVGGPNVESCGEGSVKILEDRLEALGFQH 250

CD38    274  SCKNIYRPDKFLQCVKNPEDSSCTSEI~~~~~~~~~~~~~ 300
Cyc     250  DCVENPRAVLFLLCSDNPNARECRLAKRFYRIA~~~~~~~ 282
CD157   251  SCINDYRPVKFLMCVDHSTHPDCIMNSASASMRRESASLH 290

CD38     300 ~~~~~~~~~~~~~~~~~~~~~ 300
Cyc      282 ~~~~~~~~~~~~~~~~~~~~~ 282
CD157    291 AIGDASLLISLLVALASSSQA 311
```

Figure 1. **Homology profile of human CD38, *Aplysia* ADP-ribosyl cyclase (Cyc) and murine CD157.** CD38 residues in italics represent the N-terminal cytoplasmic domain. The putative hydrophobic domains of the three proteins are indicated by underlined residues. Note that CD157 has two hydrophobic regions, one each at the N- and the C-terminals. Shaded blocks highlight the homology among the proteins, with dark shading indicating sequence identity and lighter shades showing conserved substitutions. Residues doubly underlined represent potential N-linked glycosylation sites.

In this chapter the latest and optimized procedures for a yeast expression system is described. In this system, the recombinant proteins are put under the control of an alcohol oxidase promotor and are directed by the signal sequence of the α-mating factor to be secreted by the yeast into the culture media, allowing easy collection and purification of the proteins.

All the enzymes produced by the yeast expression system are catalytically active and are indistinguishable from the native proteins. They are suitable for used in crystallography as shown for the *Aplysia* cyclase [15] and are indispensable reagents for the cycling assays of cADPR and NAADP [9, 14]. Also, the expression system is particularly suitable for site-directed mutangenesis studies on the structure and function relationships of the *Aplysia* cyclase and CD38 [15-17].

MATERIALS

The yeast *Pichia pastoris* strain X-33, the expression vector pPICZαA, and zeocin are purchased from Invitrogen Corporation (Carlsbad, CA). The pYEX-S1 yeast vector is purchased from Clontech (Palo Alto, CA). Supercompetent bacterial XL1-Blue cells are obtained from Stratagene Corporation (LaJolla, CA). PCR primers are purchased from GIBCO-BRL (Gaithersburg, MD). All PCR reactions are performed on the Robocycler obtained from Stratagene (LaJolla, CA). *Taq* and *Pfu* polymerases used for PCR amplification are purchased from Roche Diagnostics/Boehringer Mannheim Corporation (Indianapolis, IN) and Stratagene Corporation, respectively. DEPC (dethylpyrocarbonate) is purchased from Sigma Chemical Co. (St. Louis, MO). All reagents and media used for yeast and bacterial cultures, including fermentation, are obtained from Fisher Scientific (Pittsburg, PA). NGD is purchased from Sigma Chemical Co. The bioreactor systems are the Bioflo III manufactured by New Brunswick Scientific Co., Inc. (Edison, NJ); and the 7.0-liter glass bioreactor from AppliKon (San Francisco, CA). The FLUOstar plate reader is purchased from BMG LabTechnologies, Inc. (Durham, NC). Electroporation chambers are purchased from GIBCO-BRL.

METHODS

cDNA's encoding for the enzymes of interest are first cloned behind the alcohol oxidase (*AOX1*) promoter of the *Pichia pastoris* expression vector pPICZαA. The *P. pastoris* yeast strain is methylotrophic, capable of growing and thriving in the presence of methanol as the sole carbon source. The *AOX1* promoter is induced by methanol, and is used to drive the expression of the gene of interest encoding the recombinant protein. Heterologous expression of proteins spliced into the pPICZαA are secreted into the medium. Secretion of the protein is facilitated by α-factor secretion peptide, Mfα1, which directs proteins to the secretory pathway. Since *P. pastoris* expresses very low levels of native proteins, the secreted recombinant protein comprises more than 90 – 95% of the total protein in the medium. The Mfα1 secretion signal consists of an 89-amino acid sequence leader with Kex2 cleavage sites occurring between arginine and glutamine in the sequence Glu-Lys-Arg-Glu-Ala-Glu-Ala. The Glu-Ala repeats are subsequently cleaved by the dipeptidyl aminopeptidase STE13. However, this second cleavage is not always complete, resulting in the production of a fusion protein with some of the residues from the Glu-Ala-Glu-Ala repeat attached to the N-terminal of the protein. The α-factor secretion peptide facilitates the high level secreted expression of soluble protein into the media, a process that greatly simplifies downstream processing of the recombinant protein.

The construct is then used to transform the methylotrophic *Pichia pastoris* X-33 yeast strain which has an intact native *AOX1* promoter allowing it to utilize methanol as a sole carbon source for both growth and expression. The presence of an intact native *AOX1* promoter allows the culture to thrive at high cell densities resulting in greater protein expression.

The pPICZαA plasmid also encodes the Zeocin (a water-soluble copper-chelated blycopeptide isolated from *Streptomyces verticillus*) resistance gene *Sh ble* which originates from the *Streptoalloteichus hindustanus* bleomycin gene. Using zeocin as a positive selection marker allows for efficient screening of the gene of interest with no detectable presence of revertants.

Construction of Human CD38, Cyclase, CD157, and NMN-AT Clones

The pPICZαA/CD38 Clone– The human CD38 was kindly provided by D.G. Jackson (John Radcliffe Hospital, Oxford, UK) [6]. In order to allow for efficient secretion of soluble protein, the intracellular and the hydrophobic motifs of CD38 must be eliminated and only the cDNA encoding for the soluble extracellular catalytically active domain be spliced into the pPICZαA expression vector [11]. The following PCR primers are used to dissociate the extracellular domain of CD38 from the rest of the protein:

(1) 5'AT**GAATTC**TGGCGCCAGACGTGGAGCGGT3'

(2) 5'AT**GCGGCCGC**TCATGTATGATGACA3'.

Bases in bold are the new restriction sites. The first primer generates an *EcoR* I restriction site at the 5' end of the cDNA encoding for the soluble domain of CD38 while the second primer introduces the *Not* I site at the 3' end. In order to minimize any errors that might evolve in the amplification process *Pfu* polymerase is used to generate the extracellular domain. Amplification is performed on the Robocycler with the following cycling program: hot start, 94° C, 2 min; denaturation, 94° C, 1min; annealing, 58° C, 1 min; extension 72° C, 1.5 min; 30 amplification cycles; final extension 72° C, 10 min. The resultant PCR product is gel purified on a 1x TAE (Tris-acetate electrophoresis buffer), 0.8 % (w/v) agarose gel, and then used during the subsequent steps in the generation of the CD38 cDNA starting with WRQTW.

The second step in the construction of this clone involves the silencing of the four putative N-linked glycosylation sites in the soluble domain [18]. Glycosylation of heterologous proteins is characteristic of the yeast expression system and adds substantial difficulty to any downstream processing of the recombinant protein. The yeast tend to hyperglycosylate the amide nitrogen of asparagine in the consensus sequence Asn-Xaa-Thr/Ser. Even though the *Pichia pastoris* yeast strain tends to glycosylate proteins to a lesser degree than other strains, removal of these sites is necessary to promote efficient secretion of the protein as well as ease the difficulty of purification from the yeast medium. The asparagines of the four N-linked glycosylation sites are changed to either aspartic acid or alanine: $Asn^{100}\rightarrow Asp^{57}$, $Asn^{164}\rightarrow Ala^{121}$, $Asn^{209}\rightarrow Asp^{166}$, and $Asn^{219}\rightarrow Asp^{176}$. Elimination of these glycosylation sites does not have any impact on the specific activity of CD38 [18]. The following PCR primers are used to mutate the glycosylation sites:

(1) $Asn^{100}\rightarrow Asp^{57}$, 5'CATCCTTG**TG**ACATTACTGA3'
(2) $Asn^{164}\rightarrow Ala^{121}$, 5'GGTGAATT**TG**CCACTTC3'
(3) $Asn^{209}\rightarrow Asp^{166}$, 5'GATGCTC**G**ATGG**C**TCCCGCAGTAAAATC3'
(4) $Asn^{219}\rightarrow Asp^{176}$, 5'AATCTTCGACAAA**G**ACAGCACTTTTGGGA3'

Bases in bold indicate points of mutations necessary to change the asparagines to either aspartic acid or alanine and thus completely eliminate the glycosylation sites.

The resultant CD38 cDNA now encodes the protein that is bereft of any potential glycosylation sites. The cDNA is gel-purified as described above and spliced into the pPICZαA vector that has been linearized by a double digestion with *EcoR* I and *Not* I restriction enzymes. Splicing is facilitated by ligation with T4 DNA ligase. Chemically competent XL1-Blue bacterial cells are then transformed by adding 3.0 μl of the ligation reaction to 50 μl of cells and heat-shocking at 42° C. The transformed bacterial cells are then plated onto LB plates (containing 1% w/v tryptone, 0.5% w/v yeast extract, 0.5 % w/v NaCl, 1.5% w/v agar and 25 μg/ml zeocin). The presence and orientation of the CD38 cDNA in the pPICZαA vector is confirmed by digestion with *BamH* I. Ten μg of a positive clone is used to transform the *P. pastoris* X-33 yeast strain. The construct is linearized with *Sac*I and purified for electrotransformation of the yeast.

Electrocompetent X-33 yeast cells are prepared as follows: Yeast cells from a frozen glycerol stock are streaked onto a YPD plate (2% w/v peptone, 2% w/v glucose, 1% w/v yeast extract, and 2% w/v agar). 10 ml of liquid YPD is inoculated with a single X-33 colony and incubated at 28° C for 8 hr. This starter culture is then used to inoculate 0.5 L of liquid YPD and grown overnight to an Abs_{600} of 1.3. The cells are washed twice in 250 ml of chilled water by centrifugation at 1500 *g* for 5 min at 4° C. The cells are then washed once with 20 ml chilled 1 M sorbitol by centrifuging at 1500 *g* for 5 min at 4° C. The supernatant is discarded and the cell pellet is resuspended in about 0.5 ml of chilled 1M sorbitol. The final Abs_{600} of the cell suspension is 225. The cells can be stored at -80° C and used up to 6 months.

Ten μg (in not more than 10 μl of H_2O) of the *Sac* I-linearized pPICZαA/CD38 clone are mixed with 20 μl of the electrocompetent X-33 cells and incubated on ice for 20 min. The DNA-cell mix is then placed between the bosses of a prechilled 0.2 cm GIBCO-BRL electroporation cuvette and electroporated at 400 V, 10 μF capacitance, and low resistance setting in a GIBCO cell porator. The electroporated cells are immediately transferred to 1 ml chilled 1M sorbitol and incubated without shaking at 28° C for 1.5 hr. One ml of YPD is then added and the incubation continued for an additional 2.0 hr at 28° C with shaking at 200 rpm. Following incubation the cells are centrifuged at 1500 *g* for 10 min and one ml of the supernatant is discarded. The cells are resuspended in the remaining 1.0 ml of the medium, and 0.3 ml of the suspension plated onto YPDS-zeocin plates (YPD media with 1 M sorbitol, and 100 μg/ml zeocin). Colonies appear in about 3 - 4 days and are screened for protein expression.

The pPICZaA/Cyclase Clone–Secretion of soluble cyclase by the yeast involves splicing of the cDNA encoding for the soluble cyclase into the pPICZαA vector [12]. The cyclase cDNA is first amplified by RT-PCR of mRNA from the ovotestes of *A. californica.* All buffers used for RNA extraction are prepared in DEPC-treated water. The ovotestes are dissected, freeze-clamped and stored overnight at -80° C. The next day the tissue is added to freshly prepared cold denaturing buffer containing 33 ml citrate-sarcosyl-β-mercaptoethanol buffer and 25 g guanidine thiocynate, and homogenized on ice with an autoclaved polytron. Total RNA is extracted with 2 M sterile sodium acetate and phenol:chloroform:isoamyl alcohol (25:24:1). The extraction mixture is placed on ice for 15 min, and then centrifuged at 10,000 *g* for 20 min at 4° C. The supernatant is then transferred into a fresh sterile tube, and the RNA precipitated by adding 2x vol of chilled (-20° C) isopropanol and centrifuging at 10,000 *g* at 4° C for 30 min. The supernatant is discarded and the pellets are resuspended again in 10 ml of ice-cold denaturing buffer, and the RNA precipitated again as outlined above. The final RNA precipitates are dissolved in DEPC-treated water and stored at -80° C.

The mRNA is then extracted from the total RNA extract by using oligo-dT columns, and mRNA dissolved in DEPC-treated water. First-strand DNA synthesis from the mRNA is prepared by using primers prepared from the published sequence of *A. californica* cyclase ([19], GenBank Accession No. M37644). The following primers are used to reverse transcribe the first-strand DNA:

(1) 5'TTTAGAAATGTCTCCCGTCGC3'

(2) 5'GAAATTGGTAAAGAGCGTACCC3'.

The first primer starts with the signal sequence of the cyclase. The second primer is used for first strand DNA synthesis with AMV (avian myeloblastosis virus) reverse transcriptase. Fifteen units of the AMV reverse transcriptase is added to the reaction mixture and the reaction mixture is incubated at room temperature for 5 min, 37° C for 60 min and then heated to 95° C for 4 min. The transcripts are then flash-frozen on a dry-ice ethanol bath and stored at -20° C. The transcripts are precipitated with 7 M ammonium acetate and ethanol, centrifuged at 12,000 *g* and redissolved in DEPC-treated water. PCR amplification of the cyclase cDNA from the transcripts is carried out with *Taq* polymerase. Thermal cycling conditions are hot start, 94° C, 2 min; denaturing, 94° C, 0.66 min; annealing, 69° C, 0.66 min; extension, 72° C; 1.5 min; no. of cycles, 30; final extension, 72° C, 5 min. The PCR reaction yields a product of the size of about 1.1 kb.

The next step involves the removal of the native secretion signal by PCR using the following primers:

(1) 5'CAT**GAATTC**ATCGTCCCCACTCGC3'

(2) 5'AAT**GCGGCCGC**GAAATTGGTAAAGAGC3'

The first primer removes the secretion signal and generates an *EcoR* I site at the 5' end site while the second primer generates a *Not* I restriction site at the 3' end of the cyclase cDNA. PCR amplification is performed using the same procedures outlined above, and *Pfu* as the DNA polymerase. The cyclase cDNA now encodes for the soluble cyclase starting with the IVPTRELEN sequence at the amino terminal. The PCR is then digested with *EcoR* I and *Not* I, purified and ligated into the *P. pastoris* expression vector pPICZαA. Supercompetent XL1-Blue bacterial cells are chemically transformed with the ligation reaction, as described above, and plated on to LB-zeocin plates. The presence and orientation of the cyclase cDNA is confirmed by digestion with *Hind* III. Ten micrograms of the positive clone are digested with *Sac* I, purified and used to transfect electrocompetent yeast X-33 cells as described in the previous section.

The pYEX/CD157 Clone–The procedure is essentially as described previously [20]. CD157 polypeptide sequence shows two putative hydrophobic regions, one each at the amino and the carboxyl terminals (Figure 1, [8]). In order to facilitate the proper expression of soluble protein, PCR is again used to remove the cDNA that encodes for the two hydrophobic domains. The resulting cDNA codes for a soluble CD157 protein starting with the 28th amino acid residue, arginine and ending with the 288th residue, serine [20]. The PCR product is inserted at the *Sac* I restriction site of the yeast expression vector pYEX-S1, in frame with the yeast secretion signal, as well as the 6xHis tag of the plasmid. The pYEX-S1 plasmid utilizes the strong phosphoglycerate kinase (PGK) promoter and the secretion signal sequence from *Kluyveromyces lactis* for expression of recombinant proteins into the yeast medium. The pYEX-S1/CD157 construct is then used to transform the *Saccharomyces cerevisiae* DY150 yeast strain. The DY150 yeast strain contains a disrupted uracil gene, making it possible to select transformants using uracil-deficient growth media. The lithium acetate method is used to transfect the DY150 strain [21]. Five ml of YPD medium is inoculated with a single colony of DY150 and grown to saturation at 30° C. The culture is then used to inoculate 300 mL of YAPD (YPD with 30 mg/l adenine hemisulfate), and grown overnight at 30° C to an OD_{600} of 0.3 – 0.5. The cells are then harvested by centrifuging at 4000 *g* for 5 min at ambient temperature. The medium is discarded and the cell pellet washed with sterile water by centrifuging at 5000 *g* for 5 min at room temperature. The water is discarded and the cell pellet resuspended in 1.5 ml TE/lithium acetate solution (10 mM Tris-Cl pH 7.5, 1 mM EDTA and 100 mM lithium acetate). The TE/lithium acetate solution is prepared fresh each time by mixing 1 vol of 1 M TE buffer, pH 7.5 with 1 vol of 1 M lithium acetate (pH 5.0) and 8 vol of sterile water. In order to achieve high transformation efficiency high-molecular weight carrier DNA is mixed with the pYEX-S1/CD157 yeast construct [21]. The high-MW DNA has to be single-stranded and consist mostly of large fragments. Five μg of pYEX-S1/CD157 are mixed with 200 μg of the carrier DNA in less than 20 μl. 200 μl of the yeast suspension is added to the DNA and mixed gently. 1.2 ml of freshly-prepared PEG solution (8 vol of 50% PEG solution, 1 vol of 1 M TE buffer pH 7.5 and 1 vol of 1 M lithium acetate stock solution pH 5.0) are added and the entire suspension heat-shocked for exactly 15 min at 42° C. The transformation mix is then centrifuged briefly, and the yeast pellet resuspended in about 0.5 ml of TE buffer and plated onto uracil-deficient minimal dextrose plates. The colonies are then screened for activity.

The pPICZαA/NMN-AT Clone–The cDNA encoding the human NMN-AT [22], Genbank Accession No. AF312734), cloned from a human placenta λgt11 cDNA library, was kindly provided to us by Dr. Giulio Magni (Istituti di Biochimica e Biologia e Genetica, Facolta di Medicina e Chirurgia, University of Ancona, via Ranieri, 60100 Ancona, Italy). The polypepetide sequence of this enzyme is shown in Figure 2.

The open reading frame encodes a 279-amino acid protein with an apparent molecular mass of 31.9 kD [22]. A Kyte and Doolittle hydropathy plot indicates that the protein is membrane-associated due the presence of several potential transmenbrane regions [22]. A single potential N-linked glycosylation (^{36}NGTG^{39}) site is present toward the N-terminal. The human NMN-AT shares a 40% sequence identity with the yeast adenylyl transferase and 33 – 34% identity with *Caenorhabditis elegans* enzymes [22].

Construction of the pPICZαA/NMN-AT clone involves three steps: (1)

removal of the putative N-linked glycosylation site by replacing Asn[36] with Asp[36], (2) removal of the internal *EcoR* I site, and (3) generation of *EcoR* I and *Not* I sites at the 5' and 3' ends, respectively, of the NMN-AT cDNA. The PCR primers used at each of these steps are listed below:

(1) Asn[36] → Asp[36]: 5'AGGACTACATG**G**ATGGAACAGG3'

(2) Removal of *EcoR* I site: 5'TGCTACCAAGAA**C**TCTAAATGG3'

(3) Generation of new *EcoR* I and *Not* I sites:

5'AT**GAATTC**ATGGAAAATTCCGAGAAGACT3' (*EcoR* I)

5'TTA**GCGGCCGC**CTATGTCTTAGCTTCTGC3' (*Not* I)

Nucleotide changes are indicated in bold. The PCR reactions are performed as described above. The resulting NMN-AT cDNA with these changes is then digested with *EcoR* I and *Not* I, purified and ligated into the linearized pPICZαA vector. The ligation mixture is used to transform supercompetent XL1-Blue bacterial cells using the protocol described in the previous section, and the resulting transformation plated onto LB-zeocin selection plates. The presence of the NMN-AT insert is confirmed by a double digest with *EcoR* I and *Not* I. Ten micrograms of the positive clone are digested with *Sac* I and the linearized construct is used to transform electrocompetent X-33 cells as described previously in this chapter.

1	MENSEKTEVV	LLACGSFNPI	TNMHLRLFEL	AKDYMNGTGR	YTVVKGIISP
51	VGDAYKKKGL	IPAYHRVIMA	ELATKNSKWV	EVDTWESLQK	EWKETLKVLR
101	HHQEKLEASD	CDHQQNSPTL	ERPGRKRKWT	ETQDSSQKKS	LEPKTKAVPK
151	VKLLCGADLL	ESFAVPNLWK	SEDITQIVAN	YGLICVTRAG	NDAQKFIYES
201	DVLWKHRSNI	HVVNEWFAND	ISSTKIRRAL	RRGQSIRYLV	PDLVQEYIEK
251	HNLYSSESED	RNAGVILAPL	QRNTAEAKT		

Figure 2. **Polypeptide sequence of human NMN-AT** [22]. The numbers to the left represent amino acid positions. Highlighted bases indicate a potential N-linked glycosylation site. This site is silenced by replacing Asn[36] by Asp[36] in order to promote efficient expression in yeast (see text). Although the hybrophobicity profile [22] indicates the presence of several membrane-associated regions, shown here by boxed residues, the protein is successfully expressed in the soluble form in yeast.

Screening of Yeast Transformants for the ADP-Ribosyl Cyclase Activity

Two ml of either YPD containing 100 μg/ml zeocin (for the CD38 and cyclase yeast clones) or uracil-deficient dextrose minimal liquid media (for CD157 clones) are inoculated with a yeast colony and incubated overnight at 28° C with vigorous shaking (250 – 300 rpm). The next day the YPD cultures are added to 10 ml of methanol induction medium consisting of 1% yeast extract, 2% peptone, 100 mM potassium phosphate pH 6.0, 1.3% yeast nitrogen base, 0.00001% biotin, and 0.5% methanol, and grown for two days at 28° C and 300 rpm. Methanol is added to the cultures to a final concentration of 0.5% every 24 hr to maintain the induction. The CD157 clones are incubated in the same uracil-deficient media for at least 2 days.

The yeast medium is screened for the ADP-ribosyl cyclase activity by the fluorimetric NGD assay [23, 24]. In this assay the NGD (a guanine analog of NAD) is cyclized to cyclic GDP-ribose (cGDPR), a fluorescent analog of cyclic ADP-ribose (cADPR). Unlike cADPR, cGDPR cyclizes at the N7 of the guanine ring instead of the N1, making it fluorescent. The hydrolyzed product, GDP-ribose, is not fluorescent either, making it convenient to monitor the cyclization reaction continuously by measuring the cGDPR fluorescence (300 nm excitation and 410 nm emission).

For the detection of the *Aplysia* cyclase and CD38, the assay is performed in 200 µl PBS pH 7.0, containing 50 µM NGD. One to 5 µl of the yeast medium can be used in this assay. Volumes greater than 5 µl are not recommended due to the high fluorescent background caused by the components in the medium. The assay is performed in a 96-well plate using the FLUOstar plate reader from BMG. Since CD157 exhibits optimum cyclizing activity at an acidic pH from 4.5 – 6.5 [4], the assay should be performed using 50 mM MES [2-(*n*-morpholino) ethanesulfonic acid] buffer pH 6.0, 100 µM NGD^+ and 10 mM $ZnCl_2$.

The concentration of the recombinant protein in the medium is calculated by comparing its activity to the specific activity of a known amount of pure protein. The yeast transformant showing the highest cyclizing activity is then selected for scaling up to the fermentation step.

Screening of X-33 Yeast Transformants for NMN-AT Activity

The spectrofluorometric assay involving the reduction of NAD by alcohol dehydrogenase (ADH) [25] is used to detect the expression of recombinant NMN-AT in the yeast cultures. The reactions leading to the formation of NADH are:

$$ATP + NMN \rightarrow NAD + PP_i$$

$$NAD + ethanol \rightarrow NADH + acetaldehyde$$

The first reaction is catalyzed by NMN-AT while the formation of NADH from NAD is catalyzed by ADH. The reaction mixture consists of 100 mM TrisHCl pH 8.0, 0.1 mM NMN, 0.5 mM ATP, 5 mM $MgCl_2$, 1.0 % ethanol, 3 U/ml ADH, and 0.01 % BSA. 200 µl of the reaction mixture is added to 5.0 µl of the yeast medium and the assay monitored by an increase in fluorescence at 355 (Ex) and 460 (Em) nm. This assay is detailed in the chapter by Graeff and Lee. The FLUOstar plate reader is used to assay multiple samples.

High Biomass Fermentation

The New Brunswick Bioflo III or the AppliKon 7-liter glass bioreactors have been successfully used. The bioreactors are equipped with microprocessor control of pH, dissolved oxygen (DO), temperature, agitation, and nutrient feed. All fermentations are performed at 28° C to minimize the impact of overheating due to inadvertent temperature fluctuations in the heating unit. The entire fermentation process consists of three phases, the starter culture phase, the biomass phase and the final induction phase.

The Starter Culture Phase–Ten ml of YPD (with 100 μg/ml zeocin) are inoculated with a single colony of the yeast transformant and the culture grown overnight at 28° C and 300 rpm to an OD_{595} of 6. The next day the entire 10-ml culture is used to inoculate 500 ml of YPD (without any antibiotic) and the culture is incubated overnight for 14 – 16 hr under the same conditions. The glass fermentation vessel is steam sterilized with the fermentation medium containing 2.67% v/v of 85% phosphoric acid, 0.0093% w/v of calcium sulfate, 1.82% w/v of potassium sulfate, 1.49% w/v of magnesium sulfate-H_2O, 0.413% w/v of potassium hydroxide, and 2.0% w/v glycerol. Following autoclaving, the pH of the fermentation medium is adjusted to 5.0 by aseptically adding 28% sodium hydroxide through the base pump attached to the fermenter. 13.1 ml of trace salts solution containing 0.6% cupric sulfate-H_2O, 0.008% sodium iodide, 0.3% manganese sulfate-H_2O, 0.02% sodium molybdate-$2H_2O$, 0.002% boric acid, 0.05% cobalt chloride, 2.0% zinc chloride, 6.5% ferrous sulfate-$7H_2O$, 0.02% biotin, and 0.5% sulfuric acid is added aseptically through the feed pump. The fermenter is then programmed to the following starting conditions: 27° C, O_2/air ratio to 1.5/1.5 vvm, agitation at 250 rpm, and the oxygen controller to maintain DO at 40%. After one hour of operation under these conditions, the fermenter is inoculated with 500 ml of the starter culture, and run until all the glycerol in the medium is exhausted indicated by a rapid increase in DO to 100% within 2 min. The DO decreases gradually as the culture grows and should not be allowed to drop below 20%. This phase takes about 16 – 24 hr (Figure 3).

The Biomass Phase–During this phase 50% glycerol containing 12 ml of trace salts solution is fed continuously to the culture. The feed rate is initially set at 9.0 ml/liter/hr of the starter culture volume and is gradually increased to a final feed rate of 18.2 ml/liter/hr over a period of one hour. This stage is marked by increases in the biomass and the oxygen requirements of the culture. The O_2/air ratio is set at 2.0/0.5 vvm, and the oxygen controller is reset to maintain DO at 60%. This phase is carried out for 4 – 8 hr until the cell biomass reaches 180 – 200 g/liter, at which point the glycerol feed is discontinued resulting in a rapid increase in the DO.

The Induction Phase–The final phase of the fermentative process is the induction of protein expression by using 100% methanol (containing 1.2% trace salts solution). Expression is initiated by supplying methanol to the culture at the rate of 3.5 ml/liter/hr for the first 3 hr. This stage of induction

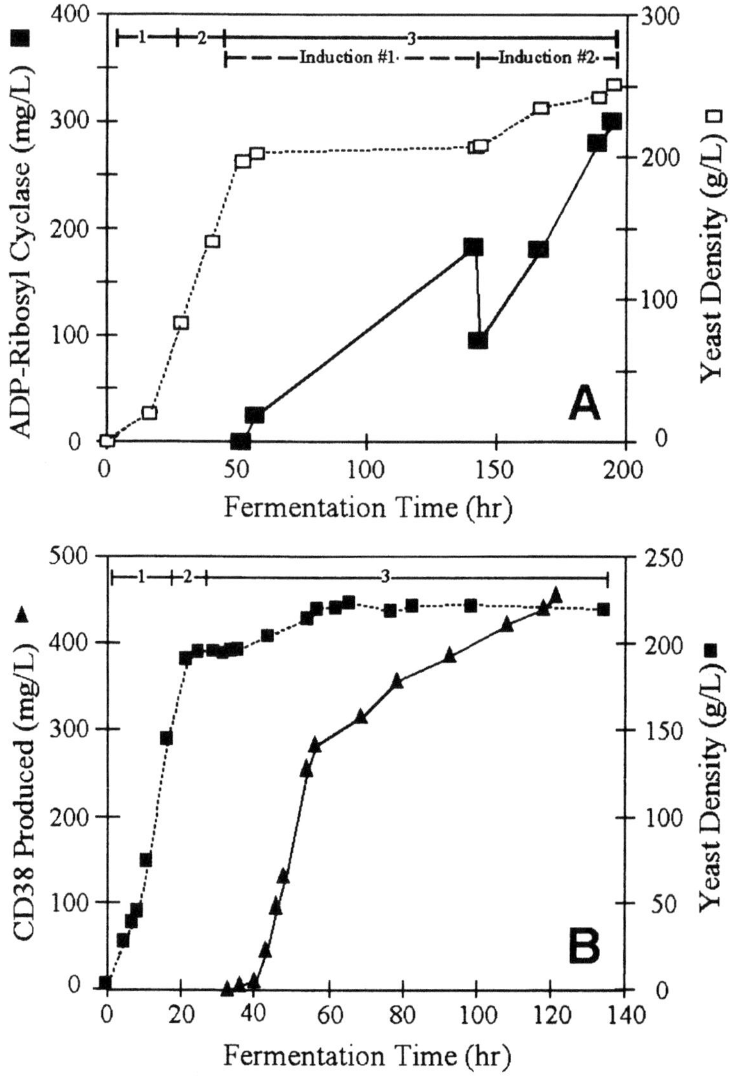

Figure 3. **High biomass fermentation of *A. californica* ADP-ribosyl cyclase and human CD38. Panel A**: Production of cyclase by two induction phases. The first phase is discontinued after 100 hours of induction, 2.0 L harvested and the cells reintroduced into the fermenter for an additional 53 hours of induction. The first induction phase produces 180 mg/L of the protein whereas the second induction phase, done at higher methanol levels, produces 300 mg/L of cyclase. **Panel B:** Expression of CD38 in a single induction phase. The cell density at the end of the CD38 fermentative process is 220 g/L and the protein concentration is 455 mg/L. For both figures, nos. 1, 2 and 3 are the Starter Culture Phase, the Biomass Phase, and the Induction Phase, respectively.

is marked by a slight decrease in the cell density and fluctuations in the DO. However, after 4 hr the DO will stabilize and methanol feed rate is increased to 8.0 ml/liter/hr and maintained for 3 hr. The methanol feed rate is then increased to 11.5 ml/liter/hr, and maintained for about 96 hr. The induction phase also produces heat, and careful attention should be paid to avoid the culture temperature from exceeding 30° C. Oxygen requirement is also greatly increased during the induction phase. A DO setting at 60% is sufficient to meet the oxygen demands throughout the induction phase. After 96 hr of induction, the fermentation is either terminated or partially harvested. If the culture is to be partially harvested, about 2.0 liter of the medium is centrifuged at 2000 *g*, the medium containing the recombinant protein flash-frozen, and the cells resuspended in freshly prepared fermentation medium (without glycerol) and reintroduced into the fermenter. The induction can be continued for an additional 50.0 hr until the final harvest. The medium is aliquoted and stored at -80° C.

The time courses of the production of *Aplysia* cyclase and CD38 are shown in Figure 3 A and B, respectively.

Expression of Recombinant Proteins in Shaker Cultures

The fermentation protocol described above is suited for large-scale production of gram quantities of the recombinant proteins. For smaller scale, as used in site-directed mutagenesis studies [15-17], the expression can be performed in shaker cultures. This alternative procedure is also good for proteins that are prone to inactivation under high-density fermentative conditions as well as for proteins that are particular sensitive to prolonged exposure to high concentrations of methenol used for induction, as is the case for NMN-AT.

A single positive colony is used to inoculate 10.0 ml of glycerol medium in a standard culture tube consisting of 1.0 % yeast extract, 2.0 % peptone, 100 mM potassium phosphate buffer pH 6.0, 1.34 % yeast nitrogen base, 0.00004 % biotin, and 1.0 % glycerol. The culture is incubated at 28° C and 275 rpm for 16 - 18 hr and then induced by 0.5% methanol. After a 5-day induction, the yeast can produce and secrete up to 40 μg/ml of the recombanant *Aplysia* cyclase in the culture medium [15]. A 10 ml-culture provides more than enough protein for enzymatic characterizations and a bio-reactor is not needed.

The procedure can also be scaled up. After incubating the initial 10-ml culture in the glycerol medium at 28° C and 275 rpm for 16 - 18 hr. The entire culture is then used to inoculate 100 ml of the glycerol medium and incubated under the same conditions for an additional 16 - 18 hr. One liter of methanol medium is then inoculated with 50 ml of the culture. The methanol medium consists of the same components as the glycerol medium, except that the glycerol is replaced by 0.5 % methanol. This is the induction phase used for expression of the recombinant protein. Induction is continued for a total of 72 hr, and 100 % methanol is added every 24 hour to a final concentration of 0.5 %. At the end of the induction phase, the yeast are centrifuged at 1,500 x *g* and the protein purified as outlined in the purification section.

Purification of the Cyclase

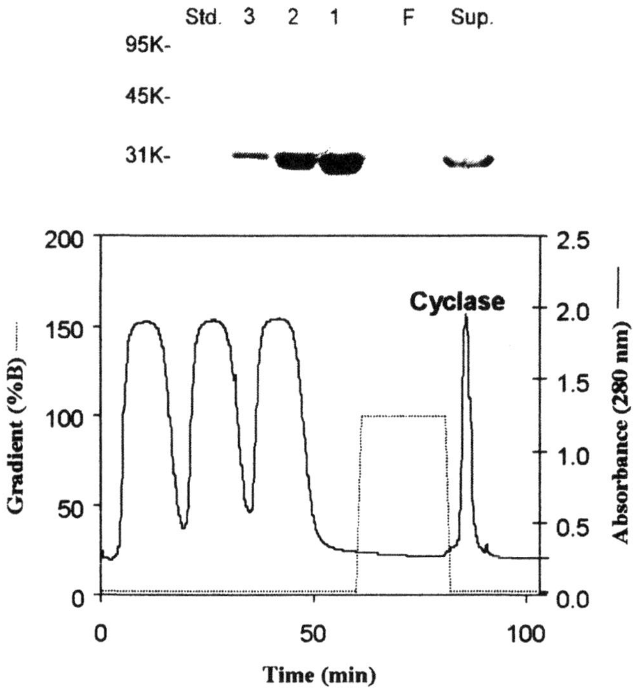

Figure 4. **Purification of ADP-ribosyl cyclase by a single step procedure using cation exchange chromatography.** The **upper panel** shows a SDS PAGE of the three fractions showing the purification of protein to homogeneity, with no detectable protein in the fall-through peaks. The **lower panel** shows the chromatographic profile of the purification. Dialyzed fermentation medium is loaded onto the column in 5-ml aliquots at 15-min intervals. The first three peaks represent the fall-through consisting of the UV-absorbing greenish material from the fermentation medium. Cyclase elutes off the cation exchange SP 5PW column as a single peak using a single NaCl step gradient, and is collected in three 1.0-ml fractions.

Fifteen ml of the fermentation medium are centrifuged at 50,000 *g* to sediment any particulate matter. The clarified medium is then dialyzed twice at 4° C with 20 mM HEPES, pH 7.5 and then filtered through a 0.4 μm filter prior to purification. The dialyzed medium is loaded on to an 8 x 75 mm Waters SP 5PW cation-exchange column. The column is equilibrated with 20 mM HEPES pH 7.5. The protein is eluted off the column at 0.5 ml/min of the elution buffer (20 mM HEPES pH 7.5 and 1.0 M NaCl.) using the following gradient: 0-40 min, 100% 20 mM HEPES pH 7.5; 40 - 41 min, linearly increased to 100% elution buffer and maintained at that level for 20 min (Figure 4). Protein elution is monitored at 280 nm. The first three peaks comprise the strongly-absorbing greenish material from the fermentation medium and usually contains very little protein and no detectable recombinant cyclase. Coomassie staining of SDS-PAGE analysis of the fractions indicate that the protein elutes off the column in three fractions and is purified to homogeneity (Figure 4). Cyclase production can reach as high as 300 mg/liter using this fermentation process. Table 1 summarizes the purification of cyclase from the fermentation medium.

Purification of CD38

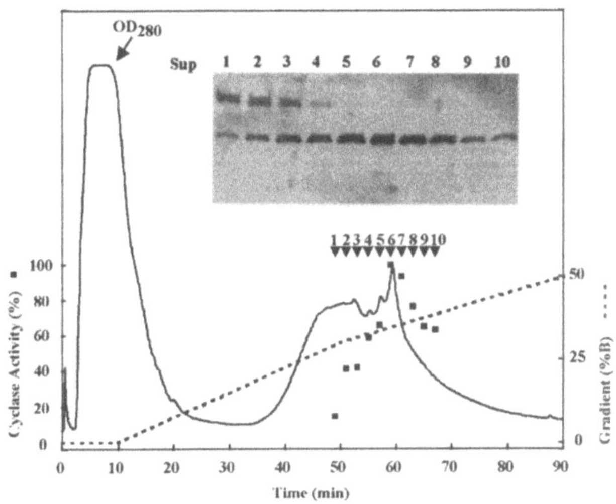

Figure 5. **Purification of recombinant CD38.** The protein is purified using an SP5 PW cation exchange column. NGD assay is used to measure the ADP ribosylcyclase activity of the eluted fractions (# 1-10). The active fractions are also analyzed by SDS PAGE to check for homogeniety of the protein.

The fermentation medium is centrifuged at 50,000 *g* and then concentrated 10-fold using a Centricon 10 concentrator. The concentrate is then dialyzed twice against 50 mM sodium acetate pH 5.0 and loaded on to a 8 x 75 mm SP 5PW column that has been equilibrated with 50 mM sodium acetate pH 5.0. The CD38 is eluted off the column at 0.5 ml/min with elution buffer (50 mM sodium acetate pH 5.0 and 1.0 M NaCl) using the following gradient profile: 0-10 min, 50 mM sodium acetate pH 5.0; 10-50 min, linearly increase to 30% elution buffer; 50-70 min, linearly increase to 40% elution buffer; 70-90 min, linearly increase to 50% elution buffer and maintain at that level for 20 min. The fall-through peak has no enzyme activity (Figure 5). The first four fractions of the elution peak contain a higher molecular mass species in addition to the 30-kDa species corresponding to the soluble CD38. The highest activity is in fractions 5-10. Using this fermentation protocol as much as 500 mg/liter can be expressed. Recoveries during the various purification steps are summarized in Table 1.

Purification of CD157

Since CD157 is expressed as fusion protein with the 6xHis-tag, purification is facilitated by utilizing Ni-NTA column chromatography [20]. The protein is first precipitated by incubation with 40 g ammonium sulfate/100 ml of the yeast medium for one hour on ice. The precipitate is sedimented by centrifugation at 22,000 *g* for 30 min at 4° C. The precipitated protein is redissolved in resuspension buffer (20 mM TRIS pH 8.0 and 1 M NaCl). The protein solution is then passed over a Ni-NTA column that has been equilibrated with the resuspension buffer and allowed to bind to the column for 60 min at 4° C. The column is then washed first with 10 vol of the resuspension buffer, then with 10 volumes of resuspension buffer containing 15 mM imidazole, and finally with 10 volumes of resuspension buffer containing 30 mM imidazole. The protein is eluted with resuspension buffer containing 250 mM imidazole. The various stages of purification are outlined in Table 1.

Table 1. **Purification of human CD38, *Aplysia* cyclase, murine CD157, and human NMN-AT from yeast cultures.**

	Culture Step	Total Protein (mg)[a]	Total Protein (mg)[b]	Total Volume (ml)	Specific Activity[c]	Purification (fold)
CD38	Medium[d]	51.8	46.3	100	6.1±1.3	1.0
	Concentrate	48.7	37.2	10	11.2±1.0	1.8
	SP5PW pool	17.6	18.0	20	17.3±1.8	2.8
Cyclase	Medium[d]	32.0	30.0	100	38.5±2.1	1.0
	Sp5PW pool	27.0	27.0	13	46.4±1.1	1.2
CD157	Medium[d]	41.86	ND	1000	ND	ND
	$(NH_4)_2SO_4$	7.6	ND	20	ND	ND
	Ni-NTA	0.032	ND	1	ND	SDS-PAGE
NMN-AT	Medium[d]	-	0.24	50	18.6	1.0
	PS[e]	-	0.21	3.5	138	1.2

a. Measured by the Bradford assay using bovine serum albumin standard.
b. Measured by the NGD$^+$ assay for CD38, Cyclase and CD157, and the fluorescence assay for NMN-AT and compared with the specific activity of a known amount of pure protein.
c. μmol cGDPR/mg/min, except for NMN-AT where it is expressed as nmol/mg/min.
d. Fermentation medium for CD38 and cyclase, and shaker-culture medium for CD157 and NMN-AT.
e. Phenylsepharose.

Purification of NMN-AT

Phenyl-sepharose chromatography is used to purify NMN-AT from the yeast medium. The phenyl sepharose column is first equilibrated with buffer A consisting of 2 M KCl and 0.1 M potassium phosphate buffer pH 7.4. Fifty ml of the yeast medium are loaded onto the column. The NMN-AT is then eluted with a linear gradient of 2.0to 0.0 M KCl in the presence of 0.1 M potassium phosphate buffer (Figure 6). Refer to chapter by Graeff and Lee for a detailed description on the chromatography of NMN-AT. Fractions are checked for activity by using the fluorescence method described earlier in this chapter. Table 1 summarizes the various purification steps involved in the isolation of the recombinant enzyme from the yeast medium.

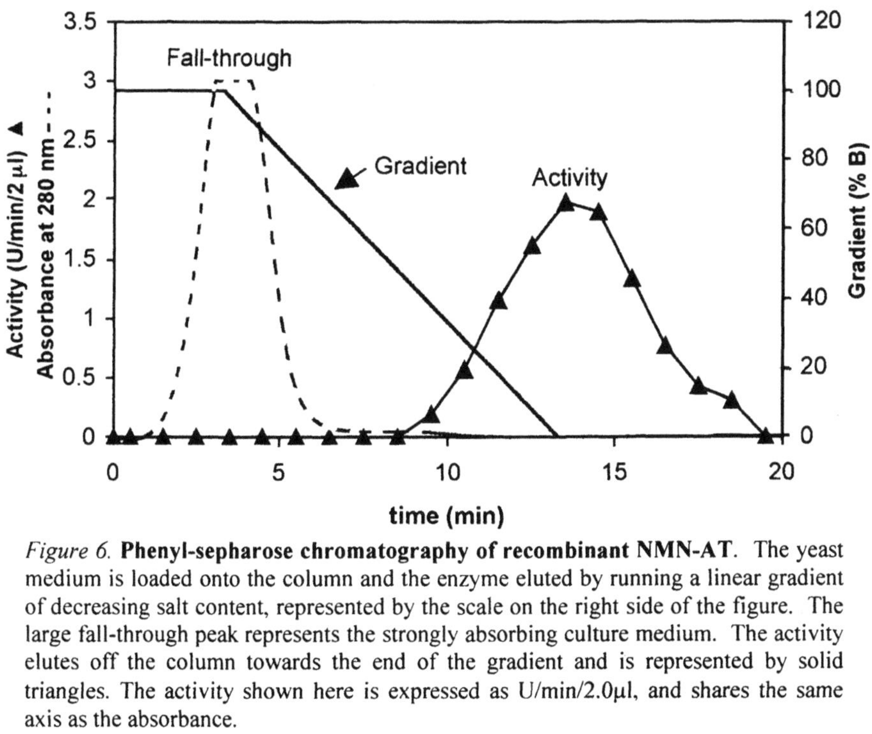

Figure 6. **Phenyl-sepharose chromatography of recombinant NMN-AT**. The yeast medium is loaded onto the column and the enzyme eluted by running a linear gradient of decreasing salt content, represented by the scale on the right side of the figure. The large fall-through peak represents the strongly absorbing culture medium. The activity elutes off the column towards the end of the gradient and is represented by solid triangles. The activity shown here is expressed as U/min/2.0µl, and shares the same axis as the absorbance.

SUMMARY

We have shown here that the yeast expression system is an invaluable tool for the production of high amounts of pure and enzymatically active protein. Unlike most other expression systems, heterologous proteins produced by the yeast systems outlined here are secreted directly into the culture media greatly simplifying what would otherwise be a very laborious extraction and purification process. The recombinant protein constitutes nearly 95% of the total protein in the medium and is purified to homogeneity by simple single-step cation exchange or phenyl-sepharose chromatography. Another advantage of this system is the suitability of the yeast to grow at high densities under appropriate fermentative conditions, an important factor contributing toward high-level production of heterologous proteins.

REFERENCES

1. Lee HC and Aarhus R. 1991. ADP-ribosyl cyclase: an enzyme that cyclizes NAD^+ into a calcium-mobilizing metabolite. *Cell Regul.* 2: 203-209.
2. Howard M, Grimaldi JC, Bazan JF, Lund FE, Santos-Argumedo L, Parkhouse RM, Walseth TF and Lee HC. 1993. Formation and hydrolysis of cyclic ADP-ribose catalyzed by lymphocyte antigen CD38. *Science* 262: 1056-1059.
3. Aarhus R, Graeff RM, Dickey DM, Walseth TF and Lee HC. 1995. ADP-ribosyl cyclase and CD38 catalyze the synthesis of a calcium-mobilizing metabolite from NADP. *J. Biol. Chem.* 270: 30327-30333.
4. Hirata Y, Kimura N, Sato K, Ohsugi Y, Takasawa S, Okamoto H, Ishikawa J, Kaisho T, Ishihara K and Hirano T. 1994. ADP ribosyl cyclase activity of a novel bone marrow stromal cell surface molecule, BST-1. *FEBS Lett.* 356: 244-248.
5. Hellmich MR and Strumwasser F. 1991. Purification and characterization of a molluscan egg-specific NADase, a second-messenger enzyme. *Cell Regul.* 2: 193-202.
6. Jackson DG and Bell JI. 1990. Isolation of a cDNA encoding the human CD38 (T10) molecule, a cell surface glycoprotein with an unusual discontinuous patern of expression during lymphocyte differentiation. *J. Immunol.* 144: 2811-2815.
7. Harada N, Santos-Argumedo L, Chang R, Grimaldi JC, Lund FE, Brannan CI, Copeland NG, Jenkins NA, Heath AW, Parkhouse RME and Howard M. 1993. Expression and cloning of a cDNA encoding a novel murine B cell activation marker. *J. Immunol.* 151: 3111-3118.
8. Itoh M, Ishihara K, Tomizawa H, Tanaka H, Kobune Y, Ishikawa J, Kaisho T and Hirano T. 1994. Molecular cloning of murine BST-1 having homology with CD38 and *Aplysia* ADP-ribosyl cyclase. *Biochem. Biophys. Res. Commun.* 203: 1309-1317.
9. Graeff R and Lee HC. 2002. A novel cycling assay for cellular cyclic ADP-ribose with nanomolar sensitivity. *Biochem. J.* 361: 379-384.
10. Graeff RM, Walseth TF and Lee HC. 1997. A radio-immunoassay for measuring endogenous levels of cyclic ADP-ribose in tissues. *Meth. Enzymol.* 280: 230-241.
11. Munshi CB, Fryxell KB, Lee HC and Branton WD. 1997. Large scale production of human CD38 in yeast by fermentation. *Meth. Enzymol.* 280: 318-330.
12. Munshi C and Lee HC. 1997. High-level expression of recombinant Aplysia ADP-ribosyl cyclase in Pichia Pastoris by fermentation. *Prot. Express. Purif.* 11: 104-110.
13. Magni G, Raffaelli N, Emanuelli M, Amici A, Natalini P and Ruggieri S. 1997. Nicotinamide-mononucleotide adenylyltransferases from yeast and other microorganism. *Meth. Enzymol.* 280: 248-255.
14. Graeff R and Lee HC. 2002. A novel cycling assay for NAADP with nanomolar sensitivity. *Biochem. J.* (in press).
15. Munshi C, Thiel DJ, Mathews II, Aarhus R, Walseth TF and Lee HC. 1999. Characterization of the active site of ADP-ribosyl cyclase. *J. Biol. Chem.* 274: 30770-30777.
16. Graeff R, Munshi C, Aarhus R, Johns M and Lee HC. 2001. A single residue at the active site of CD38 determines its NAD cyclizing and hydrolyzing activities. *J. Biol. Chem.* 276: 12169-12173.
17. Munshi C, Aarhus R, Graeff R, Walseth TF, Levitt D and Lee HC. 2000. Identification of the enzymatic active site of CD38 by site-directed mutagenesis. *J. Biol. Chem.* 275:

21566-21571.
18. Fryxell KB, O'Donoghue K, Graeff RM, Lee HC and Branton WD. 1995. Functional expression of soluble forms of human CD38 in Escherichia coli and Pichia pastoris. *Prot. Express. Purif.* 6: 329-336.
19. Glick DL, Hellmich MR, Beushausen S, Tempst P, Bayley H and Strumwasser F. 1991. Primary structure of a molluscan egg-specific NADase, a second-messenger enzyme. *Cell Regul.* 2: 211-218.
20. Hussain AMM, Lee HC and Chang CF. 1998. Functional expression of secreted mouse BST-1 in yeast. *Prot. Express. Purif.* 12: 133-137.
21. Ausubel FM, Brent R, Kingston RE, Moore DD, Seidman JG, Smith JA and Struhl K. 1992. *Short Protocols in Molecular Biology.* 2nd Edition.
22. Emanuelli M, Carnevali F, Saccucci F, Pierella F, Amici A, Raffaelli N and Magni G. 2001. Molecular cloning, chromosomal localization, tissue mRNA levels, bacterial expression, and enzymatic properties of human NMN adenylyltransferase. *J. Biol. Chem.* 276: 406-412.
23. Graeff RM, Walseth TF, Fryxell K, Branton WD and Lee HC. 1994. Enzymatic synthesis and characterizations of cyclic GDP-ribose. A procedure for distinguishing enzymes with ADP-ribosyl cyclase activity. *J. Biol. Chem.* 269: 30260-30267.
24. Graeff RM, Mehta K and Lee HC. 1994. GDP-ribosyl cyclase activity as a measure of CD38 induction by retinoic acid in HL-60 cells. *Biochem. Biophys. Res. Commun.* 205: 722-727.
25. Magni G, Emanuelli M, Amici A, Raffaelli N and Ruggieri S. 1997. Purification of human nicotinamide-mononucleotide adenylyltransferase. *Meth. Enzymol.* 280: 241-255.

Cyrus B. Munshi, Department of Pharmacology,
University of Minnesota, Minneapolis, Minnesota 55455, USA.
e-mail: munsh003@tc.umn.edu

Cyclic ADP-Ribose and NAADP. Structures, Metabolism and Functions. Lee HC (ed), pp. 101-120, Dordrecht: Kluwer.

Novel Cycling Assays for cADPR and NAADP

Richard Graeff and Hon Cheung Lee
Department of Pharmacology, University of Minnesota, Minneapolis, MN 55455, USA

INTRODUCTION

An important criterion for a second messenger is that its cellular concentrations must be responsive to the first messenger. It is thus crucial that specific and sensitive assays for cADPR and NAADP be widely available for monitoring their endogenous levels under a variety of physiological conditions. The first assay for cADPR was a bioassay based on its Ca^{2+} releasing activity in sea urchin egg homogenates [1, 2]. Using this assay, it was demonstrated that cADPR was naturally occurring in many mammalian tissues [3]. Since then, a more sensitive radioimmunoassay (RIA) for cADPR has also been developed [4, 5]. With these assays, the cellular levels of cADPR have been shown to be modulated by various surface receptor agonists, including abscisic acid [6], a plant hormone, a T-cell receptor antibody [7] and acetylcholine [8]. Intriguingly, cell permeant first messengers, such as nitric oxide [9, 10] and retinoic acid [4], metabolic factors, such as glucose [11], vitamin B12 [12], and even a physical stimulus, such as heat shock [13], can elevate cADPR levels, indicating cADPR is involved in a very broad spectrum of signaling functions.

Both the bioassay and the RIA for cADPR have several limitations preventing their wide spread adoption. In the case of the bioassay, the difficulties are numerous. First, the sample must be meticulously depleted of contaminating Ca^{2+}. Second, to minimize unwanted perturbation to the egg homogenates, samples must be added in small volumes, making it necessary to use highly concentrated samples and yet free of Ca^{2+}. Third, other agonists of Ca^{2+} release such as NAADP and IP_3 need to be removed from the samples. Finally, since the egg homogenates possess endogenous

ADP-ribosyl cyclase activity, NAD in the samples must be eliminated as well in order to avoid the conversion of NAD to cADPR.

In the case of the RIA, although it has higher sensitivity than the bioassay, it is disadvantaged by the need for specific antibodies, which are not widely available. Also, radioactive cADPR with high specific activity must be synthesized regularly, a cumbersome procedure.

To circumvent the disadvantages outlined above, an assay has been developed for measuring cADPR without the need of an antibody or radioactive label [14]. This method utilizes the reversibility of the enzyme ADP-ribosyl cyclase, which has been purified from the marine mollusk *Aplysia californica.* This enzyme normally catalyzes the formation of cADPR from NAD, with nicotinamide (Nic) as an additional product. In the presence of high concentrations of Nic, the cyclase can be reversed and catalyzes the formation of NAD from cADPR instead [15]. The resultant NAD can then be amplified and measured with a coupled-enzyme cycling assay, giving this novel assay exquisite sensitivity and convenience.

Using a similar strategy, a cycling assay for NAADP with nanomolar sensitivity has also been developed [16]. This assay makes use of alkaline phosphatase to convert NAADP to NAAD [17], which can then be assayed by a coupled enzyme cycling method using another reversible enzyme, NMN-adenylyltransferase (NMN-AT). So far, this is the only assay for NAADP that has been reported.

MATERIALS

ADP-ribosyl cyclase is prepared by a yeast expression system as described previously [18]. Native cyclase is commercially available from Sigma (St. Louis, MO) or it can be isolated from *Aplysia* as described previously [19]. Alcohol dehydrogenase from yeast (suitable for cycling), activated charcoal, diaphorase from *Clostridium kluyveri,* resazurin, Tris, NAD, nicotinamide, nucleotide pyrophosphatase from *Crotalus atrox* venom, NADase from *Neurospora crassa,* tri-n-octyl amine and 1,1,2-trichlorotrifluoroethane are from Sigma. Alkaline phosphatase from bovine intestine is from Roche (Indianapolis, IN). Centricon filters and Immobilon filter plates are from Millipore (Bedford, MA). Additional details are provided below.

A CYCLING ASSAY FOR cADPR

The assay for cADPR depends on the reverse reaction for the ADP-ribosyl cyclase, which can be written as shown in reaction 1. This reaction can be monitored by HPLC as shown in Figure 1 A.

ADP-ribosyl cyclase

1. **cADPR + nicotinamide → NAD**

Alternatively, the NAD product can be coupled to a reaction with alcohol dehydrogenase so that the fluorescent NADH product can be measured. In this case, in addition to reaction 1 above, the following reaction also applies:

alcohol dehydrogenase

2. **NAD + ethanol → NADH + H + acetaldehyde**

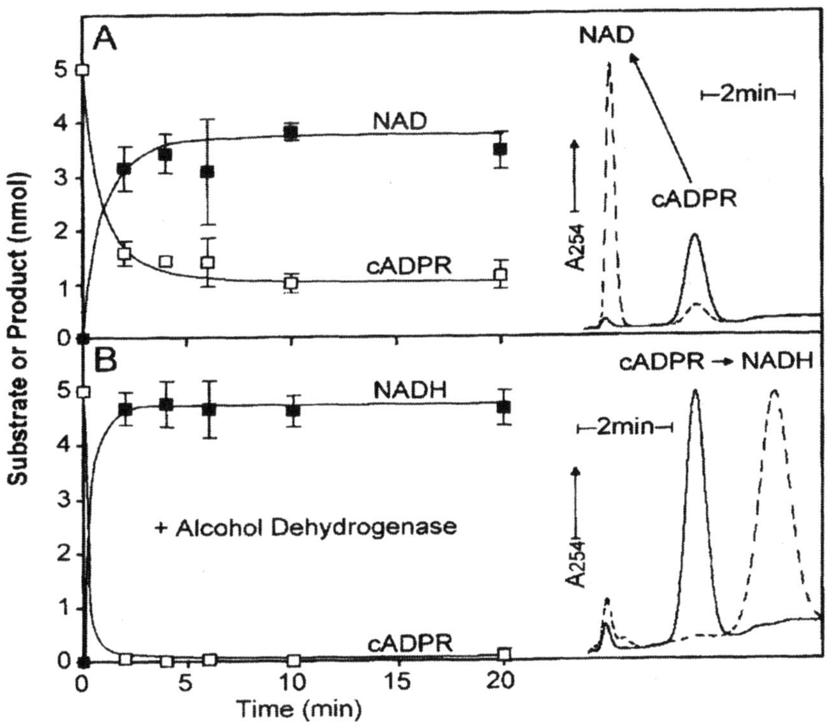

Figure 1. **Conversion of cADPR to NAD or NADH. A.** The conversion of cADPR to NAD depends on the presence of 10 mM nicotinamide and ADP-ribosyl cyclase. The reaction achieves about 80 % completion in 20 min. **B.** The same reaction with alcohol and alcohol dehydrogenase results in almost complete conversion of cADPR to NADH. To the right in each panel are chromatograms from HPLC analyses showing the peaks corresponding to NAD, cADPR, and NADH. Details of the conditions for HLPC analyses can be found in [14].

The main advantage of coupling with the alcohol dehydrogenase [20] is that the fluorescence of NADH produced can easily be monitored at 355 nm emission wavelength and 460 nm excitation wavelength using a fluorescence plate reader. Multiple samples can be assayed simultaneously, greatly improving the throughput of the assay. Also, perhaps more importantly, the coupling results in continuous conversion of NAD to NADH, which serves to pull the cyclase reaction toward completion. This is evidenced in Fig. 1B, when the substrates and products of the coupled reaction are measured directly using HPLC. After a few minutes, all the cADPR is stoichiometrically converted to NADH. This can be compared with Fig. 1A, where the conversion of cADPR to NAD is halted once a steady state is reached, which amounts to about 80% of the starting cADPR converted.

A typical reaction coupled to alcohol dehydrogenase includes the following: 1 to 20 μM cADPR, 10 mM nicotinamide, 0.1 μg/ml *Aplysia* cyclase, 1 % ethanol, and 10 μg/ml alcohol deydrogenase, in 100 mM Tris*HCl, pH 8.0. The reaction is done in a 96-well plate with each well containing typically a 200 μl reaction volume. With this method, cADPR can be measured in the micromolar range.

In order for these reactions to be useful for measuring tissue cADPR, it is necessary to have an assay that is sensitive in the nanomolar range. This is accomplished by adding one more reaction which regenerates NAD from NADH in the presence of an acceptor such as resazurin and is shown in the following reaction:

$$3. \quad \textbf{NADH + H + resazurin} \xrightarrow{\textit{diaphorase}} \textbf{NAD + resorufin}$$

The net result of the coupling of reactions 1 to 3 is a continuous fluorimetric assay for cADPR, with the resorufin fluorescence increases dependent on cADPR concentration as shown in Figure 2.

The reaction is usually conducted in 96-well plates in three steps. The first step is the addition of a 100 μl volume of standard cADPR or sample in 20 mM sodium phosphate, pH 8. The second addition is a 50 μl volume of 0.3 μg/ml *Aplysia* cyclase and 30 mM nicotinamide in 100 mM sodium phosphate, pH 8. The conversion of cADPR to NAD is allowed to proceed for 15 min. The third addition is 100 μl of a reagent that contains 2% ethanol, 100 μg/ml alcohol dehydrogenase, 10 μM resazurin, 10 μM FMN, 10 mM nicotinamide, 10 μg/ml diaphorase, and 100 mM sodium phosphate, pH 8.

The cycling assay can also be conducted in 384-well plates, and the volumes are typically 30 μl for the first addition, 15 μl for the second, and 30 μl for the third. BSA can be added to the reagent at a concentration of 0.1 mg/ml to reduce inactivation of the cyclase due to non-specific binding of the enzyme to the plate. Different sources of BSA contain various amounts of contaminating NAD, which must be determined empirically or removed by treatment with activated charcoal described below before using in the assay. The cycling reaction is allowed to proceed for 2 to 4 h and the increase in the resorufin fluorescence (excitation at 544 nm and emission at 590 nm) is measured using a fluorescence plate reader.

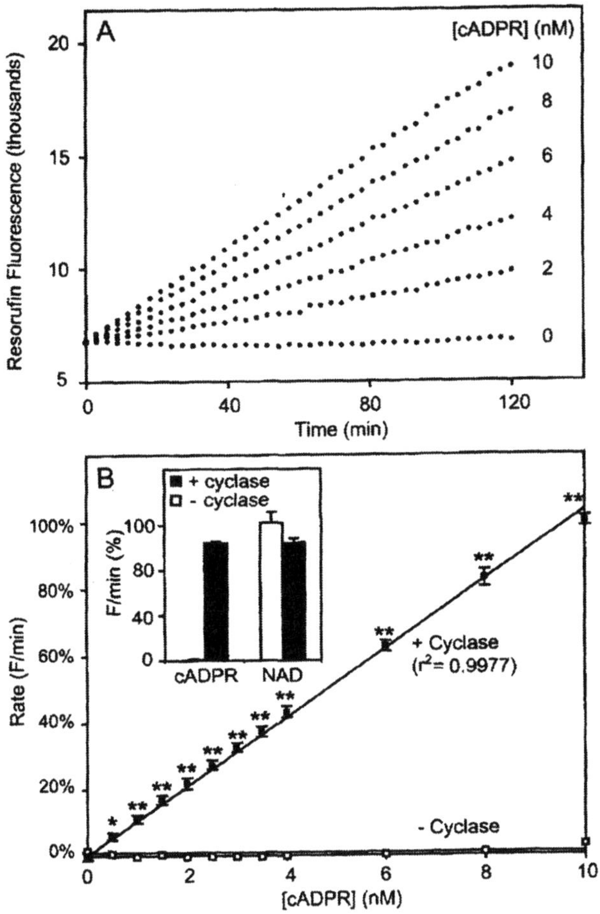

Figure 2. **A cycling assay for cADPR. A.** The rate of resorufin fluorescence increase is dependent on the concentrations of cADPR. **B.** The rates were obtained by linear regression analyses of fluorescence time courses and plotted against cADPR concentration. The inset contrasts the cycling assay for cADPR and NAD. For cADPR the assay was completely dependent on the presence of the cyclase, whereas in the case of NAD, it was totally independent of the cyclase. Values for all cADPR concentrations are significantly higher than in the absence of cADPR. * $p<10^{-5}$, ** $p<10^{-7}$ Student's t test.

The commercial diaphorase is sometimes contaminated with NAD and can be treated with charcoal to remove bound NAD. Typically, the lyophilized powder is diluted to 12 mg of protein/ml with water. Just prior to preparing the cycling reagent, 20 μl of enzyme solution is diluted with 60 μl of a 2% charcoal suspension in 20 mM phosphate, pH 7, and the enzyme/charcoal suspension is incubated for 30 min at 37°C. The charcoal is removed by centrifugation for 5 min in a microcentrifuge and the enzyme is recovered from the supernatant. Alcohol dehydrogenase is purchased as suitable for enzyme cycling, and used without charcoal treatment. It is available as a lyophilized powder and is reconstituted in water. Both alcohol dehydrogenase and diaphorase can be stored frozen in aliquots.

The cycling assay is a one-step assay, not requiring any further separation or manipulation of the samples, which makes it highly convenient and reproducible. The linearity of the assay shown in Figure 2 is quite remarkable considering that three enzyme reactions are coupled. Comparing cADPR to NAD standards (inset of Fig. 2B) the amounts of resorufin produced (resorufin fluorescence) are very similar, indicating that the conversion of cADPR into NAD is stoichiometric. As expected from the design of the assay, the resorufin produced from cADPR absolutely requires the presence of the cyclase (Figure 2B), which is in contrast to NAD (Figure 2B, inset). The difference in resorufin fluorescence produced in the presence and absence of the cyclase thus provides a convenient means to distinguish between contributions of authentic signals from that of the background, such as contaminating NAD. Even at the lowest concentration of cADPR tested (0.5 nM), the rate of resorufin production is significantly different from the control without cADPR ($P<10^{-5}$; Student's t-test). The cycling assay thus has a sensitivity better than 0.5nM cADPR. Since the sample volume is typically 0.1 ml, the sensitivity of the cyclase assay is better than 50 fmol of cADPR.

REMOVAL OF NUCLEOTIDES INTERFERING WITH THE cADPR ASSAY BY ENZYME TREATMENT

Typical samples derived from cell and tissue extracts contain various nucleotides, in addition to cADPR, which may interfere with the cycling assay. An enzyme treatment protocol is devised to remove all interfering nucleotides by treatment with a mixture of hydrolytic enzymes. The following are final concentrations of each enzyme added to the samples: 0.44 unit/ml nucleotide pyrophosphatase, 12.5 units/ml alkaline phosphatase, 0.0625 unit/ml NADase, 2.5 mM $MgCl_2$ in 20 mM sodium phosphate, pH 8.0. The incubation can proceed overnight at 37°C. The enzyme mixture hydrolyzes all nucleotides, including NAD, in the samples, except cADPR, which is resistant to these enzymes [5]. The enzymes are removed by filtration with Centricon-3 filters and samples can be recovered in the filtrate after centrifugation at 4° C and 3000g for 30 min using a Beckman JA-20 rotor.

A more convenient method for removing the enzymes before the assay is to utilize the 96-well filtering plates with an Immobilon-P membrane attached (Millipore). The best results are obtained when one row, 8 samples, is processed at a time. The wells are first treated with 100 µl of ethanol for 1 min and washed with 100 µl of water. Prior to adding sample, excess fluid is removed from the bottom of the well with a Kimwipe. Volumes of sample up to 250 μl can be filtered and collected in PCR tubes. A strip of plastic PCR tubes is aligned in parallel with the metal bars of the Millipore filtration apparatus. The tubes are placed in an empty 96-well plate underneath the filtration plate. After filtration, the samples are ready for the cycling assay. The effectiveness of the enzyme treatment is outlined in Table 1. A more detailed list has been published [14].

Table 1. **Enzyme treatment of samples removes interfering nucleotides.**

Nucleotide	Interferes with assay	Removed by
NAD	Yes	NADase, nucleotide pyrophosphatase
NADH	Yes	Acid extraction
NHD	Yes	NADase, nucleotide pyrophosphatase
NGD	Yes	NADase, nucleotide pyrophosphatase
NADP	No	NADase, nucleotide pyrophosphatase
NADPH	No	Acid extraction
NAAD	No	Nucleotide pyrophosphatase
NAADP	No	Nucleotide pyrophosphatase

PROCEDURES FOR MEASURING ENDOGENOUS LEVELS OF cADPR

Analyses of nucleotides from tissues or cells most often are done after preparing acid extracts with perchloric or trichloroacetic acid, which can effectively stop metabolism. Once the acid is added to the tissue, the sample is frozen at –80°C or maintained cold on ice. The protein is removed by centrifugation and the nucleotides are recovered from the supernatant. Removal of perchloric acid by precipitating with potassium hydroxide is not appropriate since residual potassium perchlorate can reduce the signal by as much as 50 %. Therefore, an alternative method is used by mixing the aqueous acid extracts with a solution containing 3 volume of 1,1,2-trichlorotrifluoroethane to 1 volume of tri-n-octylamine [21]. Then, 1 part of sample is added to 4 parts organic phase and vortexed for 1 min. Following centrifugation for 10 min at 1500 x g, the samples are maintained on ice until the aqueous phase easily separates from the organic phase. This aqueous layer containing the cADPR is removed and is adjusted to pH 8 with 20 mM sodium phosphate.

To quantify recovery of cADPR following the acid extraction, cADPR standards can be extracted with acid and taken through the same steps as the tissue samples. Typically, cADPR standards are prepared in 20 mM sodium phosphate, pH 8, and then treated with 0.6 M perchloric acid. The cADPR standards are neutralized along with samples, treated with enzymes, and analyzed by the cycling assay. This approach allows for a recovery to be incorporated into the experiment. The recovery of cADPR is about 80 % from the extraction step and essentially 100 % after the enzyme treatment.

To further validate the cycling assay, the standard curves of cADPR were determined in the presence and absence of extracts of HL-60 cells. Figure 3 shows that the cADPR standard curve remained perfectly linear ($r^2 = 0.997$). The slope of the regression was slightly (10%) less than that in the absence of the cell extracts. This is most likely due to a slight quenching of the resorufin fluorescence by the extracts. The y-intercept of the standard line gives a value of 1.16 ± 0.08 pmol cADPR/10^6 cells, which compares well with the value of 1.5 pmol/10^6 cells previously determined by the RIA [4, 22].

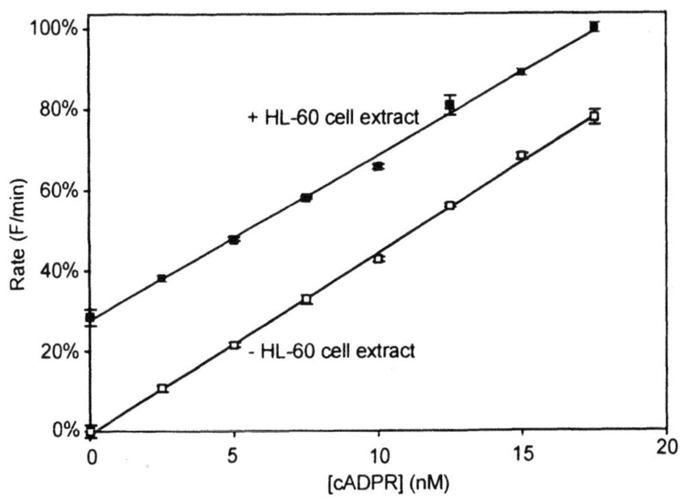

Figure 3. **Standard curve of cADPR in the presence of HL-60 cell extracts.** HL-60 cells were induced to differentiate with 1 μM retinoic acid for three days (see below). Cell extracts were prepared and treated with hydrolytic enzymes to remove contaminating nucleotides as described in the text. The standard curves for cADPR were determined in the presence and absence of the cell extracts. The slopes of the regression lines in the presence and absence of the cell extracts were 4.1%/nM and 4.5%/nM, respectively.

CHANGES IN cADPR, NAD AND NADH LEVELS IN HL-60 CELLS

HL-60 cells are derived from a human leukemia and they can be induced by retinoic acid to differentiate to a neutrophil phenotype. Coincident with differentiation, both the expression of CD38 and the cellular contents of cADPR increase [4]. Figure 4A shows the rather dramatic accumulation of cADPR for the next 96 hours after the addition of retinoic acid, during which the cells differentiated into granulocytes [23].

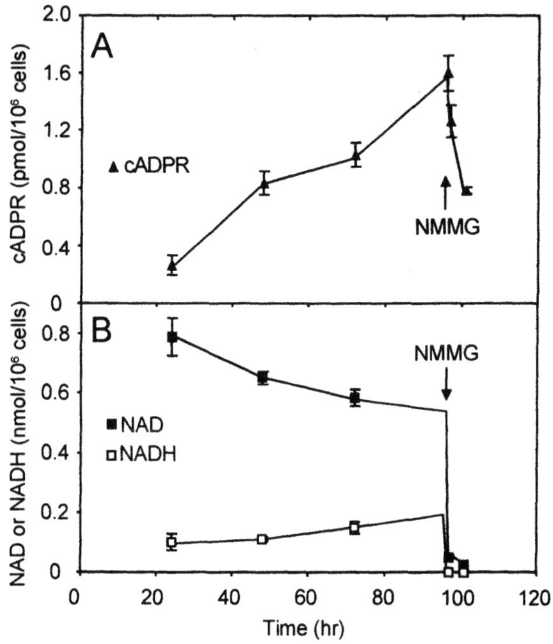

Figure 4. **Changes in endogenous levels of cADPR, NAD and NADH in human HL-60 cells during granulocytic differentiation. A.** HL60 cells were seeded at an initial density of 0.2 x 10^6 cells/ml and maintained in RPMI medium containing 10 % fetal bovine serum. The cells were induced to differentiate with 1 μM retinoic acid at time zero. At the indicated times, cells (6 x 10^6) were pelleted by centrifugation for 5 min at 700 x g and extracted with 0.5 ml of 0.6 M perchloric acid. The endogenous levels of cADPR were measured as described in the text. **B.** Samples used for the NAD measurements were extracted with perchloric acid as described for cADPR. For the NADH measurements, the samples were extracted with 0.1 N NaOH instead. At the time indicated NMMG (34 μM) was added and the change in the three nucleotide levels were measured at 1 and 5 hours afterward.

NAD is the substrate for cADPR synthesis via the ADP-ribosyl cyclase pathway. To investigate the possibility that cADPR levels in cells may be linked to the NAD levels, N-methyl-N'-nitro-N-nitrosoguanidine (MNNG) was added. MNNG is a classical reagent for activating the cellular DNA repair mechanism mediated by poly-ADP-ribose polymerase, which uses NAD as substrate for the ADP-ribosylation reaction [24]. The consequence of MNNG addition to cells was that the cellular NAD and NADH levels rapidly declined. As shown in Figure 4A, following MNNG treatment, cADPR levels decreased rapidly.

It is of interest to compare the rate of decrease of cADPR to that of the cellular NAD. This can be conveniently done using the alcohol dehydrogenase reaction, which converts the endogenous NAD in the cell extracts to NADH (reaction #2 above). NADH can then be quantified by its fluorescence. Following neutralization of the acid extracts, NAD in the samples was converted to NADH in the presence of 1 % ethanol and 9 μg/ml alcohol dehydrogenase in 100 mM Tris*HCl, pH 8. The resultant increase in NADH fluorescence was measured in a multiwell plate reader (excitation 355 nm, emission 460 nm). The fluorescence values of the extracts without the alcohol dehydrogenase treatment were used for background correction and were subtracted. Alternatively, the acid extract was diluted 400-fold in 100 mM sodium phosphate buffer, pH 8 and assayed by the complete cycling assay as described for cADPR.

The endogenous levels of NADH in cells can be readily measured using the diaphorase reaction (reaction #3 above). In this case, cells were extracted with 0.1 N NaOH instead of perchloric acid to preserve the endogenous NADH. The samples were then heated to 60 °C for 10 min and maintained at 4 °C until assayed. The heat treatment destroyed NAD in the samples. Just prior to the assay, samples were diluted 100-fold in 100 mM sodium phosphate buffer, pH 8, and assayed by the cycling assay as described for cADPR. The dilution reduced cADPR in the samples to below the detection of the cycling assay. The resultant increase in resorufin fluorescence was calibrated with NADH standards. In all cases, NAD or NADH standards were taken through the same steps as samples. Under these conditions, NAD was stable in perchloric acid at 4 °C but not in NaOH. Likewise, NADH solutions were stable in NaOH but not in perchloric acid.

As shown in Figure 4B, NADH levels remained relatively stable as the cells differentiated. During this period, the NAD levels in the HL-60 cells were also relatively stable. There may have been a slight and progressive decline in NAD concomitant with the increase in cADPR levels. Following addition of NMMG, the cellular NAD and NADH levels rapidly declined within an hour of the application. Comparing the rate of decline of cADPR (Fig. 4A) with that of NAD, it is clear that cADPR declines at a much slower rate, suggesting the two may be causally linked. These results also serve to illustrate the convenience and the utility of how the cycling assay and its two component reactions can be used for measuring cADPR, NAD and NADH levels in cells.

THE ENDOGENOUS cADPR LEVELS IN CD38 KNOCKOUT MICE

The results in HL-60 cells suggest the increased expression of CD38 induced by retinoic acid is causally related to the cADPR increase during differentiation. Conversely, it is of interest to know if CD38 expression is suppressed would the cADPR levels decrease correspondingly. The most direct way to suppress CD38 expression is to ablate its gene. CD38 knockout mice have been developed [25, 26]. The knockout mice we use were kindly supplied by Lund (Trudeau Institute, NY). Tissues from mice were frozen in

0.6 M perchloric acid and pulverized with a tissue homogenizer at 4°C. Perchloric acid was removed as described above. The results of the measurement of cADPR contents are shown in Figure 5. cADPR is endogenously present in all major organs in the wildtype mouse, with brain containing the highest level. The measured level in brain is similar to those measured in rat cerebrum and cerebellum using both the cycling assay as well as using the RIA [14], but is somewhat less than that reported by Partida-Sanchez, *et al.* [26].

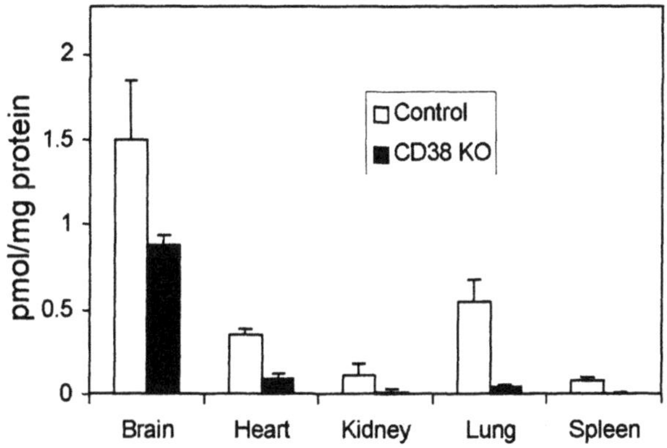

Figure 5. **cADPR content in tissues from mouse.** PCA extracts were prepared from tissues from wildtype (control) and CD38 knockout (CD38 KO) mice. cADPR was measured by the cycling assay after enzyme treatment. (n=3, average ± s.d.)

In the CD38 knockout mice, the cADPR levels are depressed in all tissues examined. In particular, the levels in kidney and spleen are below the detection limit of the cycling assay. These data show that CD38 is the major ADP-ribosyl cyclase in tissues responsible for the synthesis of cADPR. Interestingly, in agreement with a previous study [26], some tissues from CD38 knockout mice still contain cADPR. This is the case in the brain, in which the cADPR levels were more that half of the wildtype, raising the intriguing possibility of the existence of other as yet unidentified mammalian cyclases in the brain that can produce cADPR independently of CD38. If so, the brain of the CD38 knockout mice are a good source for identifying these unknown mammalian cyclases.

ADVANTAGES OF THE CYCLING ASSAY FOR cADPR

The cycling assay is highly sensitive and is capable of detecting as low as 50 fmol of cADPR, which compares favorably with the 100 fmol detection limit of the RIA [4, 5]. It is certainly better than the 10 pmol sensitivity of the HPLC assay [27] and the 4 pmol sensitivity of the bioassay [28].

In addition to sensitivity, the cycling assay has the advantage over all currently available assays in that all components of the assay, including the necessary enzymes, are available commercially, allowing this assay to be widely adopted. Indeed, the sensitivity of the cycling assay can be easily increased further to the sub-nanomolar range by either prolonging the cycling reaction or reducing the reaction volume. Unlike the RIA, the cycling assay does not require the synthesis and purification of radioactive cADPR with high specific activity, nor does it require specific antibodies against cADPR. The remarkable linearity of the cycling assay (see Figure 2) also greatly improves the accuracy and simplifies the procedure for calibration. This is in contrast to the RIA, which requires the use of curvilinear binding curves. Moreover, it is a one-step fluorimetric assay that requires no further manipulation of the samples, making it highly reproducible.

The cycling assay for cADPR can be adapted for other measurements as well. Since it is based on the assay of NAD, it can be used to measure NAD and cADPR from the same sample. This feature could be important in samples where the NAD content is expected to change, as shown in Fig. 4. Also shown in Fig. 4, the cycling assay can be used to measure NADH as well.

Another noteworthy feature of the cycling assay is its high degree of specificity. Of the more than 17 related nucleotides tested, only NAD and NADH, and to a much lesser degree, NHD, are interfering (cf. Table 1 and [14]). Potentially interfering nucleotides can be removed readily by a mixture of hydrolytic enzymes, to which cADPR is remarkably resistant. cADPR can be, however, specifically hydrolyzed by CD38, which provides a diagnostic test for the assay. In this respect, the complete dependence of the assay on the presence of ADP-ribosyl cyclase (see Figure 2) provides an additional way in which to discriminate against artifact. These advantages and safeguards make the cycling assay the best method ever devised for measuring cADPR. More importantly, the remarkable simplicity of the assay readily lends itself to automation. The possibility of high-throughput screening of a wide variety of cells and tissues under various physiological conditions is a long-sought-after goal that will lead to eventual understanding of the regulation of this novel signaling pathway mediated by cADPR.

A CYCLING ASSAY FOR NAADP

An initial attempt to develop an assay for NAADP made use of the base-exchange reaction catalyzed by the *Aplysia* ADP-ribosyl cyclase. NAADP was incubated with the cyclase in the presence of excess nicotinamide to force the reversal of the reaction and produce NADP. Unfortunately, the yield of this approach is low regardless of the pH used in the reaction (unpublished data). A much better approach [16] was found by first converting NAADP to NAAD using alkaline phosphatase [17, 29]. NAAD is a substrate for nicotinamide mononucleotide adenylyltransferase (NMN-AT), which is typically associated with the nucleus in cells and is thought to play an important role in NAD synthesis and/or NAD salvage in cells. NMN-AT normally produces NAD from nicotinamide mononucleotide (NMN) as shown in reaction 4:

$$\textbf{4.}\qquad \mathbf{NMN + ATP} \xrightarrow{\textit{NMN-AT}} \mathbf{NAD + PP_i}$$

where PP_i is inorganic pyrophosphate. Like ADP-ribosyl cyclase, NMN-AT is reversible. Another important property of NMN-AT is that it can use NAAD as a substrate to form NAMN (nicotinic acid mononucleotide) in the presence of excess PP_i as shown in reaction 5:

$$\textbf{5.}\qquad \mathbf{NAAD + PP_i} \xrightarrow{\textit{NMN-AT}} \mathbf{NAMN + ATP}$$

Incubation of NAAD with NMN-AT in the presence of excess NMN and PP_i results in production of NAD. The PP_i in the reaction helps to drive the conversion of NAAD to ATP, which can combine with NMN to form NAD in the reverse direction. NAD can then be measured with a cycling assay as described above for cADPR. In other words, NAAD can be measured by combining reactions 2 through 5.

Purification of NMN-AT–The enzyme was partially prepared from pig liver according to published methods [30]. Briefly, the liver was cut into small pieces and homogenized in buffer A containing 0.1 M potassium phosphate, pH 7.4, 1 mM dithiotreitol, 1 mM $MgCl_2$, 0.5 mM EDTA and 1 mM phenylmethylsulfonyl fluoride. After centrifugation at 13,000 x g for 20 min, the supernatant was adjusted to pH 5.0 with dropwise addition of 1 M acetic acid. The precipitated protein was collected by centrifugation and dissolved in buffer A. The protein was loaded onto a phenylsepharose column equilibrated with 3M KCl in 0.1 M potassium phosphate, pH 7.4. After washing with 2M KCl in 0.1 M potassium phosphate, pH 7.4, the NMN-AT was eluted by a linear gradient of 2 to 0 M KCL in 0.1 M potassium phosphate, pH 7.4. The enzyme could be used for the cycling assay at this step. Further purification resulted in loss of activity. NMN-AT was stable when stored at 4 °C in 1 M KCl at a protein concentration of 3.2 mg/ml.

Alternatively, NMN-AT can be produced using the yeast expression system as described in the chapter by Munshi and Lee in this book. The enzyme is purified from the yeast culture media using a similar phenylsepharose chromatographic step as described above.

The Cycling Assay Procedure–The standard assay for NAAD is done in 96-well or 384-well plates. In 384-well plates, the reactions are started by making three 30 μl additions. The first 30 μl volume contains NAAD standard or sample in 20 mM sodium phosphate pH 8. Next, 30 μl of reagent 1 is added, which contains 0.2 mM NMN, 0.2 mM PPi, 10 mM Nic, 2 mM NaF, 2 mM $MgCl_2$, 100 mM Tris-HCl, pH 8 and 0.2 mg/ml of pig NMN-AT. NMN-AT produced by the yeast expression can be purified much better and thus can be used at 0.02 mg/ml instead. This step initiates the conversion of NAAD to NAD. The third 30 μl addition is reagent 2 containing 2% ethanol, 100 μg/ml of alcohol dehydrogenase, 10 μM resazurin, 10 μg/ml diaphorase, 10 μM flavin mononucleotide, and 100 mM sodium phosphate, pH 8. The cycling reaction is allowed to proceed for 6-10 h and the increase in resorufin fluorescence (excitation 544 nm and emission 590 nm) is measured periodically using a fluorescence plate reader. A typical series of curves showing increases in fluorescence produced by NAAD standards is shown in Figure 6. One distinguishing feature of the curves is that there is an initial lag period before a steady state is achieved. The fluorescence increase becomes linear with time. A standard curve can be constructed from the linear portions of these changes in fluorescence [16].

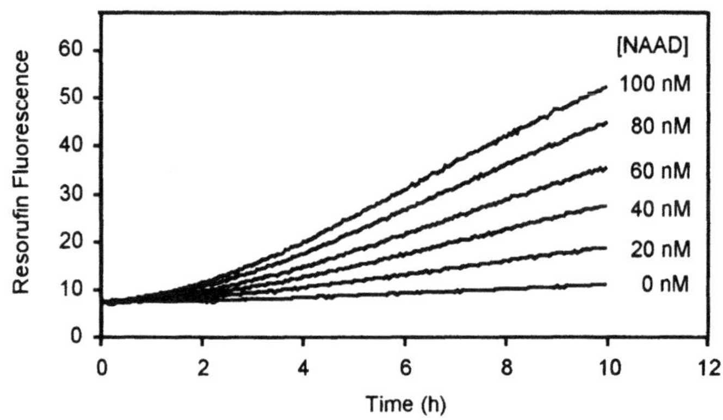

Figure 6. **Assay of NAAD by enzyme cycling.** Standard solutions of NAAD were prepared in 20 mM sodium phosphate pH 8.0. The concentrations of NAAD represent the values in the initial 30 μl of standard and are not corrected for the final dilution in the assay. After the addition of standard, reagent 1 and reagent 2 (see text), the reactions were monitored in a fluorescence plate reader for 10 h.

NAADP is not a substrate for NMN-AT and has no fluorescence increase even at concentrations as high as 100 nM as shown in Figure 7A (100 nM-AP). Treatment of NAADP with alkaline phosphatase (+AP) converts it to NAAD, which can now be detected by the cycling assay system (Fig. 7A). The conversion is performed by incubation of various concentrations of NAADP overnight at 37 °C with 10 units/ml alkaline phosphatase, 1 mM

$MgCl_2$ and 20 mM sodium phosphate, pH 8. After incubation, AP is removed by filtration with Immobilon-P plates. The conversion is stoichiometric as shown in the inset of Figure 7B. The linearity of the standard curve for NAADP obtained after the AP-treatment is close to perfect (Fig. 7B).

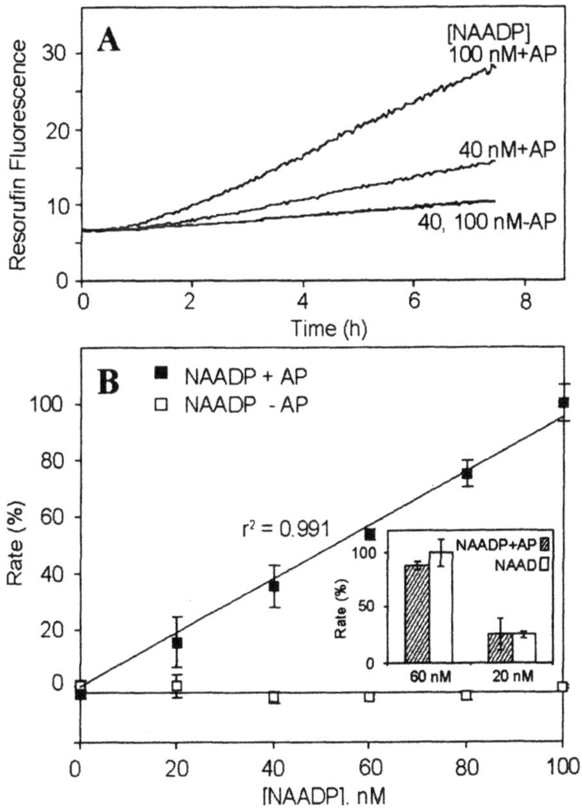

Figure 7. **The cycling assay for NAADP. A.** Conversion of NAADP to NAAD by treatment with alkaline phosphatase (+AP). NAADP itself was not a substrate for the NMN-AT and produced only background fluorescence increase without the AP treatment (-AP). **B.** A standard curve for NAADP was constructed using the rate of resorufin fluorescence increase obtained in the linear phase as shown in A. AP treatment stoichiometrically converted NAADP to NAAD. As shown in the inset, the signal produced by NAADP after the AP-treatment was essentially the same as that produced by the same concentration of NAAD.

REMOVAL OF NUCLEOTIDES INTERFERING WITH THE NAADP ASSAY BY ENZYME TREATMENT

To remove nucleotides that interfere with the assay for NAADP, a mixture of enzymes and other components is added to samples at the following final concentrations: 2.5 units/ml apyrase, 0.125 unit/ml NADase, 2 mM $MgCl_2$, 1 mM NaF, 0.1 mM PPi, 0.16 mg/ml NMN-AT and 20 mM sodium phosphate pH 8. The incubation is allowed to proceed overnight at 37 °C. This mixture hydrolyzes all nucleotides, including NAD and NAAD, but not NAADP. The enzymes are removed by filtration with Immobilon-P plates and the samples are recovered in the filtrate. The samples are next treated with AP overnight at 37 °C as described above. Prior to the cycling assay, AP is removed by filtration with Immobilon-P plates.

Table 2. **Enzyme treatment for assay of NAADP.**

Nucleotide	Interferes With Cycling Assay	Detected After First Enzyme Treatment	Detected After AP
NAADP	No	No	Yes
NAD	Yes	No	No
ATP	Yes	No	No
ADP	Yes	No	No
NAAD	Yes	No	No
GTP	Yes	No	No
NGD	Yes	No	No
NHD	Yes	No	No
NADP	No	No	No
AMP	No	No	No
GMP	No	No	No
ADPR	No	No	No
cADPR	No	No	No

As listed in Table 2, there are a number of nucleotides that can produce an interfering signal in the cycling assay before the enzyme treatment. For example, NAAD is one such nucleotide. During the first enzyme treatment, the NMN-AT in the mixture converts NAAD to NAD, which is then hydrolyzed by the NADase also present in the mixture, resulting in complete removal of the interfering NAAD. Likewise, ATP, ADP, GTP are sensitive to apyrase in the mixture, while NAD and NGD are effectively hydrolyzed by the NADase. After the first enzyme treatment, interference is removed from all of 12 nucleotides tested. NAADP is not hydrolyzed by the enzymes and can produce a signal after subsequent conversion to NAAD by a second AP-treatment. With the two-step treatment of the samples, first, the hydrolytic enzymes, and then with AP, the cycling assay can be made uniquely specific and the only nucleotide that can produce a signal afterward is NAADP. Further details are provided in [16].

A CYCLING ASSAY FOR NICOTINIC ACID

Nicotinic acid (NA) is a required substrate for the base-exchange reaction responsible for synthesizing NAADP by ADP-ribosyl cyclase and CD38

[31]. Its endogenous levels in cells have not yet been measured. In principle, changes in NA levels in cells can potentially be a factor that may regulate synthesis of NAADP. So far, no biochemical assay for NA has yet been reported. The cycling assay described above for NAAD can readily be adapted for measuring NA [16], since NA can be converted to NAAD by utilizing the base-exchange reaction catalyzed by the *Aplysia* cyclase in acidic pH [31]. Various concentrations of NAD and nicotinic acid were incubated with 0.1 μg/ml of the cyclase in 100 mM sodium acetate, pH 4. The reactions were incubated overnight at 25 °C. After adjusting the pH to 8 with Tris base, the remaining NAD in the samples was removed with NADase (0.05 unit/ml) in the presence of 1 mM $MgCl_2$ and incubated overnight at 37 °C. Following NADase treatment, the samples were filtered with Immobilon-P plates and assayed for NAAD using the cycling reaction as described above. As can be seen in Figure 8, the assay is sensitive enough for detecting NA in the 10 μM-range with the use of 100 μM NAD as co-substrate for the exchange reaction.

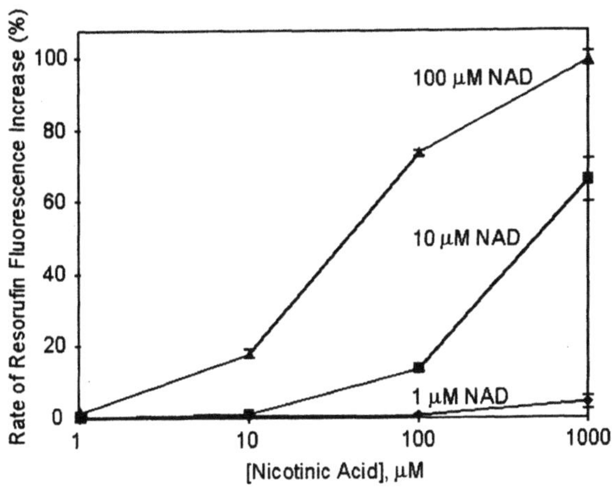

Figure 8. **A cycling assay for nicotinic acid**. NA was converted to NAAD using the base-exchange reaction catalyzed by the *Aplysia* cyclase in the presence of 1-100 μM NAD. The resulting NAAD was then measured by the cycling assay as described in Fig. 6.

SUMMARY

The basic principle behind the cycling assay for NAD is highly versatile [20]. By coupling it to ADP-ribosyl cyclase and NMN-AT, respectively, cycling assays for cADPR and NAADP have been developed. Extending the approach, we have also shown that the principle can be adapted for measuring NA and NAAD. Although not described in detail, NADP can likewise be measured by its conversion to NAD using alkaline phosphatase.

The assays are sensitive, remarkably so for cADPR, and certainly in the workable range for NAADP and NA.

The specificity of the assays comes first from the coupling enzymes used, but can be greatly improved by taking advantage of the unique properties of the nucleotides of interest. In the case of NAADP, for example, the combination of hydrolytic enzyme treatment with the alkaline phosphatase treatment renders the assay uniquely specific for NAADP.

All the assays are continuous, requiring no further separation of substrates from products, and are thus highly reproducible. Other major advantages of these assays are their remarkable linearity as well as high throughput, allowing simultaneous measurement of a large number of samples using 96-well or 384-well plate formats. These advantages will allow detailed investigation of not only the stimulus-induced changes of cADPR and NAADP levels, but also the inter-relationships with all their respective substrates, NAD, NADP and NA. With these technical advances, a more complete picture of the metabolic pathways of these two novel Ca^{2+} messengers is anticipated.

REFERENCES

1. Clapper DL, Walseth TF, Dargie PJ and Lee HC. 1987. Pyridine nucleotide metabolites stimulate calcium release from sea urchin egg microsomes desensitized to inositol trisphosphate. *J. Biol. Chem.* 262: 9561-9568.
2. Lee HC, Walseth TF, Bratt GT, Hayes RN and Clapper DL. 1989. Structural determination of a cyclic metabolite of NAD^+ with intracellular Ca^{2+}-mobilizing activity. *J. Biol. Chem.* 264: 1608-1615.
3. Walseth TF, Aarhus R, Zeleznikar RJ, Jr. and Lee HC. 1991. Determination of endogenous levels of cyclic ADP-ribose in rat tissues. *Biochim. Biophys. Acta* 1094: 113-120.
4. Takahashi K, Kukimoto I, Tokita K, Inageda K, Inoue S, Kontani K, Hoshino S, Nishina H, Kanaho Y and Katada T. 1995. Accumulation of cyclic ADP-ribose measured by a specific radioimmunoassay in differentiated human leukemic HL-60 cells with all-trans-retinoic acid. *FEBS Lett.* 371: 204-208.
5. Graeff RM, Walseth TF and Lee HC. 1997. A radio-immunoassay for measuring endogenous levels of cyclic ADP-ribose in tissues. *Meth. Enzymol.* 280: 230-241.
6. Wu Y, Kuzma J, Marechal E, Graeff R, Lee HC, Foster R and Chua NH. 1997. Abscisic acid signaling through cyclic ADP-ribose in plants. *Science* 278: 2126-2130.
7. Guse AH, da Silva CP, Berg I, Skapenko AL, Weber K, Heyer P, Hohenegger M, Ashamu GA, Schulze-Koops H, Potter BVL and Mayr GW. 1999. Regulation of calcium signalling in T lymphocytes by the second messenger cyclic ADP-ribose. *Nature* 398: 70-73.
8. Fukushi Y, Kato I, Takasawa S, Sasaki T, Ong BH, Sato M, Ohsaga A, Sato K, Shirato K, Okamoto H and Maruyama Y. 2001. Identification of cyclic ADP-ribose-dependent mechanisms in pancreatic muscarinic Ca^{2+} signaling using CD38 knockout mice. *J. Biol. Chem.* 276: 649-655.
9. Willmott N, Sethi JK, Walseth TF, Lee HC, White AM and Galione A. 1996. Nitric oxide-induced mobilization of intracellular calcium via the cyclic ADP-ribose signaling pathway. *J. Biol. Chem.* 271: 3699-3705.

10. Reyes-Harde M, Empson R, Potter BVL, Galione A and Stanton PK. 1999. Evidence of a role for cyclic ADP-ribose in long-term synaptic depression in hippocampus. *Proc. Natl. Acad. Sci. USA* 96: 4061-4066.
11. Takasawa S, Akiyama T, Nata K, Kuroki M, Tohgo A, Noguchi N, Kobayashi S, Kato I, Katada T, Okamoto H, Takasawa S, Akiyama T, Nata K, Kuroki M, Tohgo A, Noguchi N, Kobayashi S, Kato I, Katada Tet al. 1998. Cyclic ADP-ribose and inositol 1,4,5-trisphosphate as alternate second messengers for intracellular Ca^{2+} mobilization in normal and diabetic beta-cells. *J. Biol. Chem.* 273: 2497-2500.
12. Masuda W, Takenaka S, Inageda K, Nishina H, Takahashi K, Katada T, Tsuyama S, Inui H, Miyatake K and Nakano Y. 1997. Oscillation of ADP-ribosyl cyclase activity during the cell cycle and function of cyclic ADP-ribose in a unicellular organism, Euglena Gracilis. *FEBS Lett.* 405: 104-106.
13. Zocchi E, Carpaneto A, Cerrano C, Bavestrello G, Giovine M, Bruzzone S, Guida L, Franco L and Usai C. 2002. The temperature-signaling cascade in sponges involves a heat-gated cation channel, abscisic acid, and cyclic ADP-ribose. *Proc. Natl. Acad. Sci. USA* 98: 14859-14864.
14. Graeff R and Lee HC. 2002. A novel cycling assay for cellular cyclic ADP-ribose with nanomolar sensitivity. *Biochem. J.* 361: 379-384.
15. Inageda K, Takahashi K, Tokita K, Nishina H, Kanaho Y, Kukimoto I, Kontani K, Hoshino S and Katada T. 1995. Enzyme properties of Aplysia ADP-ribosyl cyclase - Comparison with NAD glycohydrolase of CD38 antigen. *J. Biochem.* 117: 125-131.
16. Graeff R and Lee HC. 2002. A novel cycling assay for NAADP with nanomolar sensitivity. *Biochem. J.* (in press).
17. Lee HC. 1997. Mechanisms of calcium signaling by cyclic ADP-ribose and NAADP. *Physiol. Rev.* 77: 1133-1164.
18. Munshi C and Lee HC. 1997. High-level expression of recombinant Aplysia ADP-ribosyl cyclase in Pichia Pastoris by fermentation. *Prot. Express. Purif.* 11: 104-110.
19. Lee HC and Aarhus R. 1991. ADP-ribosyl cyclase: an enzyme that cyclizes NAD^+ into a calcium-mobilizing metabolite. *Cell Regul.* 2: 203-209.
20. Kato T, Berger SJ, Carter JA and Lowry OH. 1973. An enzymatic cycling method for nicotinamide adenine dinucleotide with malic and alcohol dehydrogenases. *Anal. Biochem.* 53: 86-97.
21. Khym JX. 1975. An analytical system for rapid separation of tissue nucleotides at low pressures on conventional anion exchangers. *Clin. Chem.* 21: 1245-1250.
22. Graeff R, Munshi C, Aarhus R, Johns M and Lee HC. 2001. A single residue at the active site of CD38 determines its NAD cyclizing and hydrolyzing activities. *J. Biol. Chem.* 276: 12169-12173.
23. Collins SJ. 1987. The HL-60 promyelocytic leukemia cell line: Proliferation, differentiation, and cellular oncogene expression. *Blood* 70: 1233-1244.
24. Jacobson EL, Shieh WM and Huang AC. 1999. Mapping the role of NAD metabolism in prevention and treatment of carcinogenesis. *Mol. Cell. Biochem.* 193: 69-74.
25. Kato I, Yamamoto Y, Fujimura M, Noguchi N, Takasawa S and Okamoto H. 1998. CD38 disruption impairs glucose-induced increases in cyclic ADP-ribose, $[Ca^{2+}]_i$ and insulin secretion. *J. Biol. Chem.* 274: 1869-1872.
26. Partida-Sanchez S, Cockayne D, Monard S, Jacobson EL, Oppenheimer N, Garvy B, Kusser K, Goodricj S, Howard M, Harmsen A, Randall T and Lund FE. 2001. Cyclic ADP-ribose production by CD38 regulates intracellular calcium release, extracellular calcium influx and chemotaxis in neutrophils and is required for bacterial clearance in vivo. *Nature Med.* 7: 1209-1216.
27. da Silva CP, Potter BVL, Mayr GW and Guse AH. 1998. Quantification of intracellular levels of cyclic ADP-ribose by high-performance liquid chromatography. *J. Chromatogr. B* 707: 43-50.

28. Walseth TF, Wong L, Graeff RM and Lee HC. 1997. A bioassay for determining endogenous levels of cyclic ADP-ribose. *Meth. Enzymol.* 280: 287-294.
29. Lee HC, Aarhus R, Gee KR and Kestner T. 1997. Caged nicotinic acid adenine dinucleotide phosphate - Synthesis and use. *J. Biol. Chem.* 272: 4172-4178.
30. Magni G, Emanuelli M, Amici A, Raffaelli N and Ruggieri S. 1997. Purification of human nicotinamide-mononucleotide adenylyltransferase. *Meth. Enzymol.* 280: 241-255.
31. Aarhus R, Graeff RM, Dickey DM, Walseth TF and Lee HC. 1995. ADP-ribosyl cyclase and CD38 catalyze the synthesis of a calcium-mobilizing metabolite from NADP. *J. Biol. Chem.* 270: 30327-30333.

Richard Graeff, Department of Pharmacology,
University of Minnesota, Minneapolis, USA.
e-mail: graef001@maroon.tc.umn.edu

Cyclic ADP-Ribose and NAADP. Structures, Metabolism and Functions. Lee HC (ed), pp. 121-142, Dordrecht: Kluwer.

Pharmacology of Cyclic ADP-Ribose and NAADP

Synthesis and Properties of Analogs

Timothy F. Walseth and Hon Cheung Lee
Department of Pharmacology, University of Minnesota, Minneapolis, MN 55455, USA.

INTRODUCTION

Multiple Ca^{2+} stores are generally present in cells. Principal among them are the mitochondria and the endoplasmic reticulum. It is generally accepted that inositol trisphosphate mobilizes Ca^{2+} stores in the endoplasmic reticulum [1]. Also present in the organelle is another Ca^{2+} release channel, the ryanodine receptor. The discovery of two other Ca^{2+} signaling molecules, cyclic ADP-ribose (cADPR) and nicotinic acid adenine dinucleotide phosphate (NAADP) [2-4] further highlights the complexity of Ca^{2+} mobilization as a fundamental signaling mechanism [1].

Pharmacological tools are invaluable in elucidating the multiplicity of Ca^{2+} stores. In most cases, the use of specific agonists and/or antagonists provides the most straight forward approach for deciphering the intricacy of cellular mechanisms that cells use to distinguish and respond to the myriad of Ca^{2+} signaling stimuli (cf. the next chapter by Lee in this book). Indeed, pharmacological evidence is among the strongest in establishing the fact that the Ca^{2+} release mechanism activated by cADPR is independent of that mediated by inositol trisphosphate, but is related to that mediated by the ryanodine receptor [5-7]. A pharmacological approach has also been used to dissect complex Ca^{2+} signaling associated with multiple agonists in pancreatic acinar cells (see the chapter by Cancela in this book). Thus, it was found that cholecystokinin is shown to signal via the cADPR-pathway while acetylcholine signaling is dependent on inositol trisphosphate [8, 9].

The finding that ADP-ribosyl cyclase can cyclize NAD to produce cADPR [10] has since opened up the first and highly versatile way for synthesizing analogs of cADPR [11, 12]. Remarkably, the cyclase can also synthesize NAADP from NADP and nicotinic acid via a base-exchange reaction [13], which has since been utilized to synthesize a series of analogs of NAADP as well [14, 15]. These analogs not only have provided insights into structure-activity relationships of cADPR and NAADP, but have also made available many useful research tools such as antagonists, hydrolysis-resistant agonists and fluorescent analogs for investigating the physiological functions of these two novel Ca^{2+} messengers. This chapter focuses on synthesis and the properties of the analogs that have been shown to have utility in dissecting the role of cADPR or NAADP in Ca^{2+} signaling.

SYNTHESIS OF cADPR ANALOGS

cADPR is a metabolite of NAD. The structure of cADPR (see Figure 1) is confirmed by X-ray crystallographic analysis [16]. This cyclic nucleotide is unique in nature and features a N-glycosidic linkage between N^1 of adenine and the anomeric carbon of the ribosyl unit originally linked to nicotinamide in NAD.

The synthetic strategy used for most of the cADPR analogs has taken advantage of the broad substrate specificity of *Aplysia* ADP-ribosyl cyclase [10, 12, 17], the enzyme that converts NAD to cADPR. As outlined in Figure 2, the strategy relies on the synthesis of the corresponding NAD analog, followed by enzymatic conversion to the cADPR analog with ADP-ribosyl cyclase. NAD analogs have been successfully produced by one of two routes. The first route is chemical synthesis, which couples corresponding AMP analogs to β-NMN using either carbodiimide [12, 18] or diphenylphosphate [19-21]. Several AMP analogs have been produced by chemical phosphorylation of the corresponding adenosine analog [19-22]. The second route is enzymatic synthesis, which uses NAD pyrophosphorylase [11, 22] to convert the corresponding ATP analogs to the NAD analogs. Some of these ATP analogs can be enzymatically synthesized from the corresponding AMP analogs using the combined actions of adenylate kinase and creatine kinase [22]. The chemical routes of NAD synthesis are applicable to all analogs, while the enzymatic route is restricted to those analogs that can serve as substrates for either NAD-pyrophosphorylase or adenylate kinase/creatine kinase. For instance, 8-substituted AMP derivatives are not substrates for adenylate kinase, while NAD pyrophosphorylase will utilize 8-amino-NAD, but not 8-bromo- or 8-azido-NAD [11].

Recently, the total chemical synthesis of several cADPR analogs has been reported [23-28], which should allow a much wider range of analogs to be synthesized. The properties of some of the chemically synthesized analogs will be discussed below, but the details of the chemical synthesis are beyond the scope of this chapter.

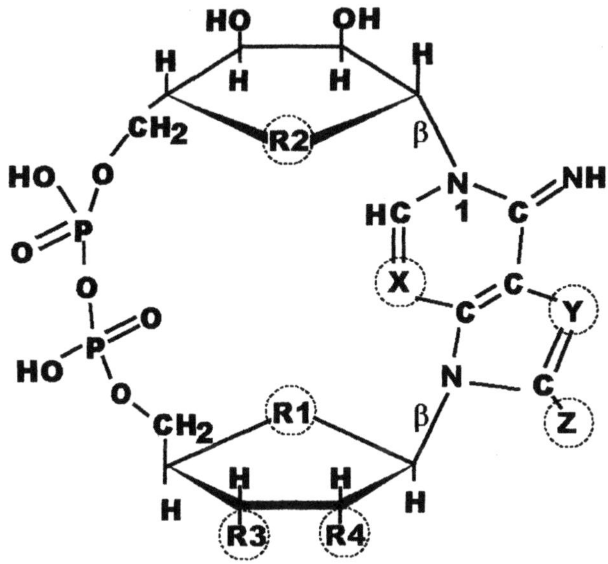

Compound	R1	R2	R3	R4	X	Y	Z	Properties
cADPR	O	O	OH	OH	N	N	H	agonist
8-NH_2-cADPR	O	O	OH	OH	N	N	NH_2	potent antagonist
8-N_3-cADPR	O	O	OH	OH	N	N	N3	antagonist, photoaffinity probe
8-Br-cADPR	O	O	OH	OH	N	N	Br	permeant antagonist
8-Br-7-deaza-cADPR	O	O	OH	OH	N	C	Br	permeant antagonist, hydrolysis-resistant
8-CH_3-cADPR	O	O	OH	OH	N	N	CH_3	antagonist
8-OCH_3-ADPR	O	O	OH	OH	N	N	OCH_3	antagonist
3'-OMe-ADPR	O	O	OCH_3	OH	N	N	H	antagonist
cArisDPR	CH_2	O	OH	OH	N	N	H	hydrolysis-resistant agonist
cADPcR	O	CH_2	OH	OH	N	N	H	hydrolysis-resistant agonist
7-deaza-cADPR	O	O	OH	OH	N	C	H	hydrolysis-resistant agonist
3-deaza-cADPR	O	O	OH	OH	C	N	H	hydrolysis-resistant agonist
cADPR-2'-P	O	O	OH	PO_4	N	N	H	agonist?
2'-deoxy-cADPR	O	O	OH	H	N	N	H	weak agonist
3'-deoxy-cADPR	O	O	H	OH	N	N	H	agonist

Figure 1. **Structure of cADPR analogs**

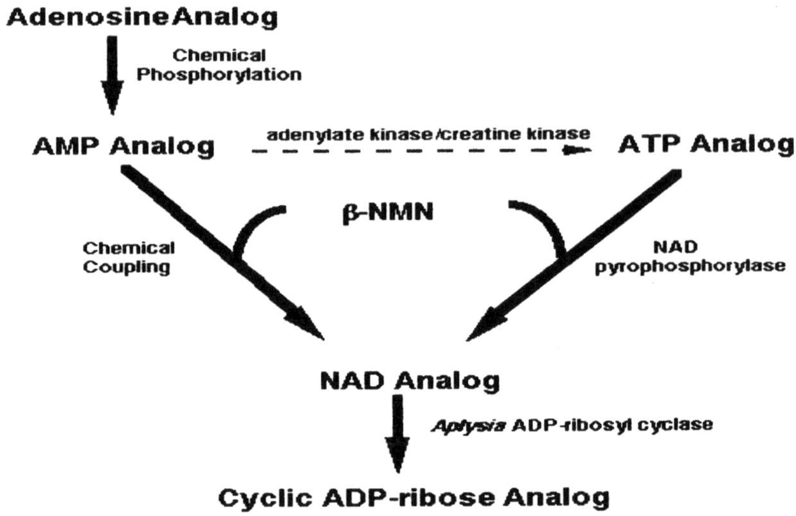

Figure 2. **A general strategy for synthesizing cADPR analogs.**

cADPR ANTAGONISTS

The first cADPR analogs synthesized are compounds with amino (8-NH_2), azido (8-N_3) or bromo (8-Br) groups substituted at the 8 position of the adenine ring [12] and their structures are shown in Figure 1. Surprisingly, all three of the analogs are found to be competitive antagonists of cADPR-induced Ca^{2+} release in a sea urchin egg homogenate system [12]. 8-NH_2-cADPR is the most potent of the three antagonists. Its effect is competitive and inhibits Ca^{2+} release induced by 135 nM cADPR with an IC_{50} of about 9 nM in sea urchin egg microsomes [12]. 8-NH_2-cADPR is also as effective as cADPR in competing for cADPR binding sites in the egg microsomes [12]. Thus, the antagonist can bind to the receptor, but is not capable of activating Ca^{2+} release mechanism. Substitution of larger sized groups at the 8-position (azido and bromo, 42 and 79 atomic units, respectively) results in a decrease in antagonistic activity. In egg homogenates, the IC_{50} values for 8-N_3-cADPR and 8-Br-cADPR inhibition of cADPR-induced Ca^{2+} release are about and 450 nM [29] and 1 μM [12], respectively.

Numerous studies have used 8-NH_2-cADPR as a specific antagonist to assess the role of cADPR in Ca^{2+} mobilization in various biological systems. These include examination of sea urchin egg fertilization [30], agonist activated Ca^{2+} release in intestinal [31] and airway smooth muscle [32], excitation contraction coupling in cardiac myocytes [33-35], T-cell activation [36], neurotransmitter release [37], neuronal Ca^{2+} regulation [38], ATP-activated potassium currents in alveolar macrophages [39], agonist-induced Ca^{2+} spiking in pancreatic acinar cells [40, 41] and abscisic acid signaling in plants [42]. In all these studies, 8-NH_2-cADPR was introduced into the cells using microinjection or detergent permeabilization.

Interestingly, as will be described in detail in a separate chapter by De Flora *et al.*, in this book, cADPR can, in some cases, evoke intracellular Ca^{2+} mobilization when applied externally [43-47]. The mechanism of how cADPR can accomplish this feat is not entirely clear at this time, although it appears that some cells may have transport mechanisms for taking up cADPR into the cells [44]. In these cases, externally applied 8-NH_2-cADPR can, likewise, inhibit the extracellular effects of cADPR [44-46].

Although the antagonistic potency of 8-Br-cADPR is not very high [12, 48], it has the remarkable advantage of being cell permeant [48]. That 8-bromo derivatives of cyclic nucleotides are cell permeant appears to be a general rule as is also documented for 8-Br-cAMP and 8-Br-cGMP. Sethi *et al.*, first demonstrated the permeability of 8-Br-cADPR in sea urchin eggs. Using heparin loaded intact *L. pictus* eggs, it was shown that 100 μM 8-Br-cADPR added to the external medium was able to inhibit the fertilization-induced Ca^{2+} transient [48]. The membrane permeant property of 8-Br-cADPR has also been used to study cADPR function in pulmonary vasoconstriction [49, 50], neutrophil chemotaxis [51] and Ca^{2+} signaling in coronary arterial smooth muscle cells [52].

A different 8-Br-derivative of cADPR, 8-Br-7-deaza-cADPR, has been synthesized by Bailey *et al.*, [20]. Similar to 8-Br-cADPR, this analog is both an antagonist and cell permeant. Moreover, it has the added advantage of being hydrolysis resistant [48]. When used in intact sea urchin eggs, 7-deaza-8-Br-cADPR was found to be about twice as effective as 8-Br-cADPR, which is most likely due to the metabolic stability of the former analog *in vivo* [48]. In addition to sea urchin eggs, 8-Br-7deaza-cADPR has also been used to antagonize cADPR actions in a T-lymphocyte cell line [53]. The availability of cell permeant cADPR antagonists, such as 8-Br-cADPR and 8-Br-7-deaza-cADPR has proved to be invaluable for studying cADPR function in intact cells.

Two other 8-substituted cADPR analogs, 8-methyl-cADPR (8-CH_3) and 8-methoxy-cADPR (8-OCH_3) have also been synthesized and shown to be antagonists of cADPR in T-lymphocytes [53, 54]. Their antagonistic potency, however, has not been directly compared with other 8-substituted analogs discussed above.

In addition to the 8-postion of the adenine ring, substitutions at the ribose can also produce antagonistic analogs. Ashamu *et al.* find that $3'_A$-O-methyl-cADPR ($3'_A$-OMe-cADPR) is an antagonist of cADPR [18], which inhibits cADPR-induced Ca^{2+} release with an IC_{50} of 5 μM in sea urchin egg homogenates. It has been proposed that the antagonistic activity of $3'_A$-OMe-cADPR may be due to reduction of the ability of the 3'-position to accept or donate a hydrogen bond when methylated [18].

Although 8-N_3-cADPR is also an antagonist, it is primarily used as a photoaffinity probe for identifying cADPR binding proteins [29]. Using ^{32}P-8-N_3-cADPR with high specific activity, Walseth *et al.*, identified two proteins of 140 kDa and 100 kDa, respectively, in *L. pictus* egg extracts as specific cADPR binding proteins [29]. Using the same probe, multiple cADPR binding proteins of masses 145, 140, 65 and 45 kDa have also been identified in the egg extracts from a different species of sea urchin, *S. purpuratus* (Gurnack and Walseth, unpublished observations).

Accumulating evidence indicates the target of cADPR is the ryanodine receptor (reviewed in [55, 56] and see also the chapter by Li *et al.*, in this book). In sea urchin eggs, the ryanodine receptor is a 597 kDa protein, based on sequencing data [57]. The fact that all the cADPR binding proteins identified by photoaffinity labelling are smaller in size, suggests that cADPR may not interact directly with the ryanodine receptor, but involve intermediates. The exact identity of these proteins and their putative roles in cADPR action remain to be elucidated.

HYDROLYSIS-RESISTANT cADPR ANALOGS

Five hydrolysis resistant cADPR agonists and one hydrolysis-resistant cADPR antagonist have been prepared and characterized. These analogs should become important tools in studying the mechanism of cADPR action, especially in systems where cADPR degradation is rapid. They are also invaluable in assessing the long-term action of cADPR in cells.

Cyclic aristeromycin diphosphoribose (cArisDPR) was the first metabolically stable cADPR analog synthesized [19]. The structure of this analog is shown in Figure 1 and it is synthesized by the chemical-enzymatic approach outlined in Figure 2, starting with aristeromycin, an adenosine analog. The notable feature of this analog is the replacement of the ribose unit normally linked to N^9 of the adenine ring in cADPR with a carbocyclic 5-membered ring. cArisDPR is an agonist with an EC_{50} of 80 nM, as compared to 30 nM for cADPR, in the *L. pictus* egg homogenates [19]. The main advantage of this analog is its resistance to enzymatic degradation. The half-lives of cArisDPR and cADPR when incubated with sea urchin egg homogenates are 170 and 15 minutes, respectively [19]. The metabolic stability of this analog is proposed to be due to a subtle difference in the binding of cArisDPR that makes the hydrolytic attack of water more difficult on the N^1-ribosyl linkage [19].

Cyclic ADP-carbocyclic-ribose (cADPcR, see Fig. 1), is another carbocyclic derivative of cADPR [23]. It is different from cArisDPR described above in that the carbocyclic ring replacing the ribose unit is linked to the N^1 of the adenine instead of N^9. This analog is prepared by an elegant total chemical strategy devised by Shuto and colleagues [23-25, 58, 59]. The chemical synthesis of cADPcR is necessary because the NAD analog containing a 2,3-dihydroxycyclopentane ring in place of the nicotinamide ribose moiety is not a substrate for ADP-ribosyl cyclase [60]. cADPcR is shown to be completely resistant to hydrolysis by recombinant CD38, a bifunctional mammalian ectoenzyme known to catalyze the synthesis and degradation of cADPR ([61] see also in the chapter by Lee *et al.*, in this book). In rat brain extracts, a more natural environment, very little degradation is seen even after two hours of incubation [23]. When injected into *A. crassispina* eggs, cADPcR induces concentration-dependent increases in intracellular Ca^{2+} which are much larger than those produced by cADPR [23].

Two other analogs have been synthesized by Fukuoka *et al.*, using similar chemical strategies as used for cADPcR. They are cyclic inosine diphosphocarbocyclic ribose (cIDPcR) and the 8-bromo derivative of cIDPcR (8-Br-cIDPcR) [26]. They differ structurally from cADPcR in that the adenine ring has been replaced by hypoxanthine or 8-Br-hypoxanthine [26]. The biological properties of these analogs have not been characterized, but judging from the similarity of their structures with cADPcR, they are expected to be hydrolysis-resistant as well.

Other alterations on the adenine ring can also result in metabolically stable analogs of cADPR. Two of these are prepared by replacing the adenine moiety of cADPR with 7-deaza- [20] or 3-deaza- adenine [22]. They differ from cADPR only by substitution of the nitrogens of the adenine at the 7- or 3-positions, respectively, with carbon (see Fig. 1). Both analogs are synthesized by the chemical-enzymatic route outlined in Fig. 2 using 7-deaza-adenosine (tubercidin) [20] or 3-deaza-adenosine [22] as starting materials.

7-deaza-cADPR is a partial agonist of Ca^{2+} release in both the *L. pictus* [20] and *S. purpuratus* [22] egg homogenates, releasing approximately two-thirds of the Ca^{2+} released by the maximal concentrations of cADPR. In the *L. pictus* system, 7-deaza-cADPR has an EC_{50} of 90 nM, which is about 3 times less potent than cADPR [20]. In the *S. purpuratus* system, the EC_{50} for 7-deaza-cADPR is 25 nM, which is about 3 times more potent than cADPR [22]. The reason for the reversal in potency between the egg systems is not known.

In contrast, 3-deaza-cADPR is a full agonist with an EC_{50} of 1 nM when evaluated in the *S. purpuratus* egg homogenates, which is about 70-times more potent than cADPR [22]. Similar remarkable effectiveness of 3-deaza-cADPR is also observed in mammalian systems. In permeabilized porcine airway smooth muscle cells, 10 nM 3-deaza-cADPR stimulates Ca^{2+} release to an extent that is equivalent to that induced by 10 μM cADPR (M.S. Kannan and T. F. Walseth, unpublished observation). In addition, equivalent increases in intracellular Ca^{2+} are observed in mononuclear cells isolated from cord blood when 10 nM 3-deaza-cADPR or 100 μM cADPR are used [44]. Interestingly, both 7-deaza- and 3-deaza- cADPR are resistant to both heat-induced and enzymatic hydrolysis [20, 22]. The metabolic stability of these analogs is most likely due to an increase in the electronegativity of the nitrogen at the N^1 position of the adenine ring, making this a poorer leaving group in a hydrolytic reaction [21]. As discussed above, the 8-bromo derivative of 7-deaza-cADPR has been prepared and shown to be a hydrolysis-resistant and cell permeant antagonist of cADPR [21, 48].

One other hydrolysis-resistant analog of cADPR that has been synthesized is cyclic adenosine triphospho-ribose (cATPR) [62]. It has three phosphates linking the two ribosyl units instead of two. cATPR is a potent stimulator of Ca^{2+} release in a rat brain microsomes, having an EC_{50} of 780 nM as compared to 22 μM for cADPR [62]. This analog is more resistant to the hydrolytic action by pig brain NADase, presumably because of the relief of ring strain by the additional phosphate unit [62]. The effectiveness of this analog has not yet been evaluated in other systems.

The five stable agonists discussed above and 8-Br-7-deaza-cADPR all have useful properties that should make these compounds important tools in studying the mechanism of cADPR action and for identifying cADPR binding proteins. This is especially true in tissues where the metabolic rate for cADPR is high. The stability of these analogs should help alleviate this problem. Analogous to 7-deaza-8-Br-cADPR, it would be of interest to brominate the 8-position of adenine of these stable agonists, which may lead to production of stable antagonists that are also cell permeant.

cADPR ANALOGS WITH N^9-RIBOSYL MODIFICATIONS

A number of analogs with modification of the ribose unit linked to N^9 of the adenine ring have been prepared and evaluated. The first, cADPR-2'-phosphate (cADPR-2'-P), is easily synthesized from NADP using ADP-ribosyl cyclase [13, 63, 64]. This compound has been found to occur naturally in mammalian tissues [65], albeit at much lower concentrations than cADPR [66]. cADPR-2'-P is active in stimulating Ca^{2+} release from rat brain microsomes [63, 64] and T-lymphocyte cell lines [54], but not in a sea urchin egg homogenates [13, 18]. These findings might reflect species differences in the cADPR-mechanism.

Ashuma *et al.*, reported an extensive structure-activity study on the role of the hydroxyl groups in the N^9-linked ribose of cADPR [18]. By replacing the 2'- and 3'-hydroxyl groups with hydrogens (2'- and 3'-deoxy-cADPR, see Fig. 1), it was found that the 3'-hydroxyl group, but not the 2'-hydroxyl group, was essential for Ca^{2+} release in *L. pictus* egg homogenates [18]. 2'-deoxy-cADPR has an EC_{50} of 5 μM, almost 100-fold higher than 3'-deoxy-cADPR [18]. Analogs with a phosphate at the 3'-position or a 2',3'-cyclic phosphate on the N^9-ribosyl unit are inactive [18, 54, 63]. In addition, as discussed above, addition of a methyl group to the 3'-hydroxyl group converts cADPR to an antagonist (3'-OMe-cADPR), suggesting that this hydroxyl group might be required for Ca^{2+} release activity, but not ligand binding [18]. Wong replaced the N^9-ribose with arabinose and found that the resulting cADPR analog had no Ca^{2+} release activity, further supporting an important role of the N^9-ribosyl unit [67].

FLUORESCENT cADPR ANALOGS

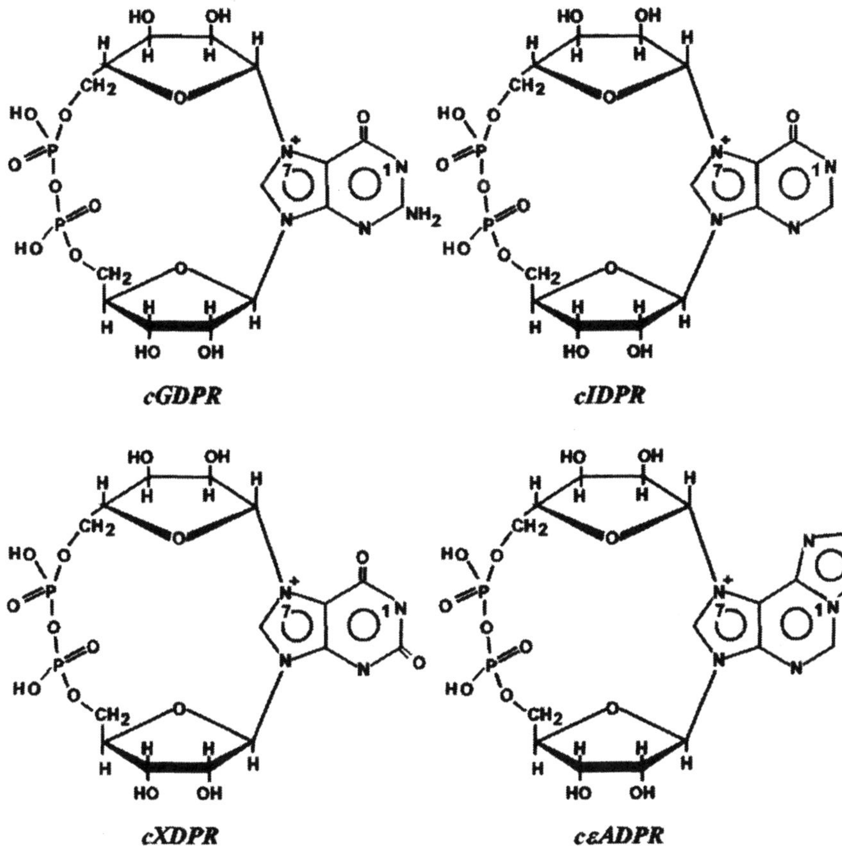

Figure 3. **Structures of fluorescent analogs of cADPR.**

NAD analogs containing guanine (NGD), hypoxanthine (NHD), xanthine (NXD) and 1, N^6-etheno-adenine (εNAD) in place of the adenine ring are substrates for ADP-ribosyl cyclase [68-72]. The resulting analogs are cyclic guanosine diphospho-ribose (cGDPR), cyclic inosine diphospho-ribose (cIDPR), cyclic xanthosine diphospho-ribose (cXDPR) and cyclic ethano-ADP-ribose (cεADPR). Their structures are shown in Figure 3. None of these analogs are active, except cεADPR, which exhibits an EC_{50} of 8μM for Ca^{2+} release in rat brain microsomes as compared to 22μM for cADPR [72].

Despite their inactivity, one important characteristic of these analogs that make them remarkably useful is that they are all fluorescent.

Spectral analyses indicate that these cADPR analogs are unique in that the site of cyclization is through the nitrogen at the 7-position of the purine ring rather than N^1 as it is in cADPR [68, 70]. This alternate site of linkage makes all the analogs fluorescent [69, 70, 72]. The reasons why ADP-ribosyl cyclase uses an alternative cyclization site with these substrates is likely to be related to the fact that the N^7 nitrogen in these NAD analogs is believed to be more nucleophilic and hence more reactive than the N^1 nitrogen [70].

The usefulness of these analogs resides in their novel fluorescent properties. For example, ADP-ribosyl cyclase activity can be continuously monitored using NGD as a substrate [69, 70, 73]. The product, cGDPR is fluorescent, while the NGD substrate is not [70]. cGDPR is also highly resistant to enzymatic hydrolysis, thus reducing the interference of product degradation during the assay [70]. Consequently, the NGD assay is widely used for measuring the cyclase activity in tissue and cell extracts [69, 74-81]. Apart from the ease and continuous nature of the NGD assay, even more important advantage, however, is that it can easily distinguish ADP-ribosyl cyclase from classical NADase, since the latter can only hydrolyze NGD to non-fluorescent GDP-ribose ([13] see also the chapter by Lee *et al.* in this book).

NHD can also be used to monitor ADP-ribosyl cyclase activity, as this substrate is cyclized to cIDPR, which is 7-times more fluorescent than cGDPR [70]. Unlike cGDPR, cIDPR is readily hydrolyzed by cADPR hydrolases [70]. This property can, however, be used as a fluorimetric assay for monitoring the cADPR hydrolase activity. Thus, hydrolysis of cIDPR to the non-fluorescent IDP-ribose catalyzed by the hydrolases results in decrease in fluorescence [70]. The continuous nature of both the NGD and the cIDPR assays provides a easy way to obtain both kinetic constants (K_m and V_{max}) of the enzymes in one measurement [69, 70, 72]. This convenience cannot be overstated.

Zhang and Sih demonstrate that the ethenocytosine derivative of NAD, nicotinamide ethenocytosine dinucleotide, can be converted to a cyclic ethenocytidine diphospho-ribose (cεCDPR) with *Aplysia* ADP-ribosyl cyclase [72]. Like cεADPR, cεCDPR is also fluorescent. Its Ca^{2+} releasing activity, however, has not been examined [72].

CAGED cADPR

Treatment of cADPR with 1-(2-nitrophenyl)diazaoethane results in the introduction of a photoactivatable 2-(nitrophenyl) ethyl caging group on one of the two phosphate moieties [11, 82]. The proposed structure of the resulting caged cADPR is shown in Figure 4. Caged cADPR is biologically inactive, but upon photolysis with UV light, can release the caging group and regenerate active cADPR [82]. The use of caged cADPR allows the precise spatial and temporal release of cADPR in intact cells and can overcome potential problems and artefacts associated with the microinjection techniques. Caged cADPR has been employed in several studies to examine the role of cADPR in Ca^{2+} homeostasis [83-88].

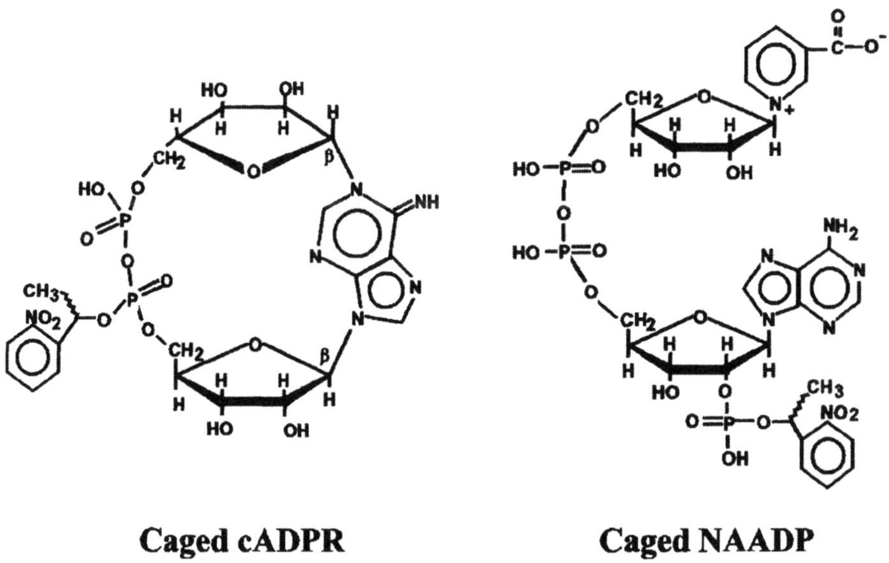

Figure 4. **Structures of caged analogs.**

OTHER cADPR ANALOGS

Huang *et al.*, recently reported the total chemical synthesis of 8-chloro-cyclic inosine diphospho-ethoxymethy ether (8-Cl-cIDP-eme), a N^1-ether strand substituted cADPR mimic [28]. The structure of 8-Cl-cIDP-eme is shown in Figure 5 and features the replacement of the adenine moiety with 8-chloro-hypoxanthine and N^1-ribose with an acyclic ethoxymethy ether. This is the first analog of cADPR with complete replacement of the terminal ribose with another group. Remarkably, the analog not only remains active but also appears to be cell permeant. A rapid and large increase in intracellular Ca^{2+} is observed when 100 μM 8-Cl-cIDP-eme is added to intact HeLa cells, while cADPR did not evoke any response over a 10 minute period, indicating it is cell permeant [28]. This data indicates that the N^1-ribosyl group of cADPR may not be totally essential for the cADPR action [28], although the acyclic ethoxymethy ether does structurally resemble a ribose. This analog is also the first 8-substituted analog of cADPR that is not antagonist. So far, 8-Cl-cIDP-eme has not yet been tested in other cell systems, but its potential is clear. A widely applicable permeant agonist of cADPR would be an invaluable tool for assessing the physiological function of cADPR.

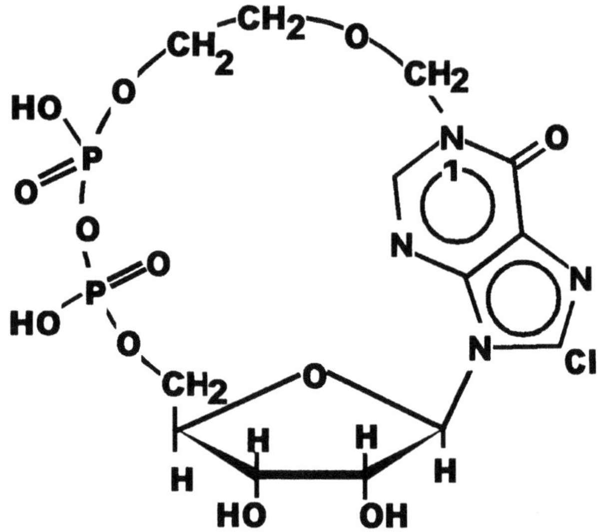

Figure 5. **Structure of 8Cl-cIDP-eme.**

Galeone *et al.*, synthesized a N[1]-carbocyclic, N[9]-butyl analog of cADPR [27]. This analog features a hypoxanthine in place of the adenine group, a carbocyclic group in place of the N[1]-ribosyl group as in cADPcR and an acyclic butyl group replacing the N[9]-ribosyl group [27]. Unfortunately, this analog is not characterized in terms of biological activity. Given the study of Ashuma *et al.* [18] on the importance of the N[9]-ribosyl unit on the biological activity of cADPR, it would appear this analog, with an acyclic butyl group replacing the N[9]-ribose, may not be biological active. However, this needs to be examined before any conclusion is drawn.

NAADP ANALOGS

NAADP is a metabolite of NADP that mobilizes Ca^{2+} via a mechanism totally distinct from cADPR or inositol 1,4,5-trisphosphate [56]. NAADP differs structurally from NADP by replacement of the nicotinamide moiety with nicotinic acid (see Figure 6). Interestingly, NAADP is synthesized by ADP-ribosyl cyclase, the same enzyme that produces cADPR [13]. NAADP synthesis occurs via a base-exchange reaction catalyzed by the cyclase, in which the nicotinamide group of NADP is exchanged for nicotinic acid [13]. The cyclase is lenient with respect to the two substrates for the base-exchange reaction, NADP and nicotinic acid. The large number of pyridine and NADP analogs available forms the basis of a flexible and general route to the synthesis of NAADP analogs. Examination of structure-activity relationships of a series NAADP analogs has revealed three structural determinants important for NAADP-induced Ca^{2+} release in sea urchin egg homogenates [14]. These three structural features are highlighted in Figure 6 and are the negative charge on the 3-carboxyl group on the nicotinic acid moiety, the amino group at the 6-position of the adenine group and the 2'-phosphate group [14]. The evidence for the importance of these moieties in NAADP action is discussed below.

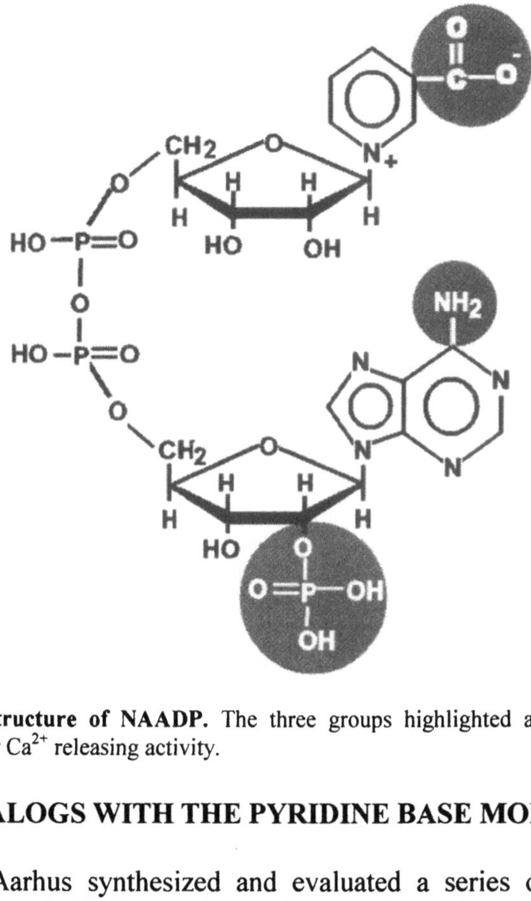

Figure 6. **Structure of NAADP.** The three groups highlighted are determinants important for Ca^{2+} releasing activity.

NAADP ANALOGS WITH THE PYRIDINE BASE MODIFIED

Lee and Aarhus synthesized and evaluated a series of pyridine base modified analogs [14]. The study provided evidence that the negative charge on the 3-position of the pyridine ring of NAADP was critical for activity [14]. The NAADP analogs prepared include compounds with pyridine 3-sulfonic acid, pyridine 3-acetic acid, pyridine 3-carbinol, quinoline 3-carboxylic acid and pyridine 4-carboxylic acid, replacing the nicotinic acid moiety (pyridine 3-carboxylic acid). Moving the carboxylic acid group from the 3-position, as it is in NAADP, to the 4-position of the pyridine ring, results in complete loss of Ca^{2+} release activity, suggesting that the presence of a substituent at the 3-position of the pyridine ring is important [14]. Analogs with substitutions of pyridine 3-sulfonic acid, pyridine 3-acetic acid or quinoline 3-carboxylic acid are able to release Ca^{2+}, albeit at much higher concentrations than NAADP. The EC_{50} values for NAADP and the pyridine 3-sulfonic acid, pyridine 3-acetic acid and quinoline 3-carboxylic acid containing NAADP analogs are about 20 nM, 3 μM, 10 μM and >10μM, respectively [14]. Replacement of the 3-carboxyl group with an uncharged carbinol group [14] or amide group [4, 89] results in the loss of biological activity, pointing to the critical importance of a negative charge at the 3-position of the pyridine ring.[14].

There appears to be some tolerance in the size of the negative group present at the 3-position of the pyridine ring. Thus, the sulfonic acid derivative remains active. Further increase in size to quinoline 3-carboxylic acid resulted in a progressive decrease in the ability of the compounds to elicit Ca^{2+} release [14]. It should be noted that pyridines substituted at the 2-position (2-carboxylic acid and 2-methyl nicotinic acid) are not substrates for the cyclase, prohibiting the synthesis of 2-substituted analogs by the base-exchange reaction [14]. Hence, the possible role of 2-substitutions on the Ca^{2+} release activity has not been investigated.

NAADP ANALOGS WITH THE ADENINE MODIFIED

Another important structural determinant of NAADP required for its biological activity is the amino group at the 6-position of the adenine ring [14, 89]. Replacement of the adenine ring with hypoxanthine (substitution of the amino group with an oxygen at the 6-position) produces deamino-NAADP [14, 89]. It is a weak agonist, only evoking Ca^{2+} release from sea urchin egg homogenates at concentrations higher than 5 μM [14].

Other modifications on the adenine ring seem to be better tolerated. Two fluorescent NAADP analogs containing 1, N^6-etheno-adenine (etheno-NAADP) [15, 89] or 1,N^6-etheno-2-aza-adenine (etheno-aza-NAADP) [15] in place of the adenine ring have been synthesized. These analogs are agonists, but are about 100-fold less effective than NAADP, with etheno-NAADP and etheno-aza-NAADP displaying EC_{50} values of 2.5 and 5 μM, respectively, in sea urchin egg homogenates [15]. Both analogs retain a nitrogen at the 6-position, which is part of the linkage via the etheno group to the N^1 nitrogen of the adenine ring. This suggests the nitrogen at the 6-position is important for activity. Etheno-NAADP and etheno-aza-NAADP are fluorescent and may prove to be useful tools for visualizing NAADP-sensitive stores in intact cells [15].

NAADP analogs containing 8-substituted adenine rings have been synthesized [90]. 8-amino- and 8-azido- NAADP both are effective agonists in the sea urchin egg Ca^{2+} release system, with EC_{50} values of 30 nM, compared to 20 nM for NAADP [90]. Attempts to use radiolabeled 8-azido-

NAADP as a photoaffinity probe for the identification of NAADP binding proteins proved unsuccessful, as only non-specific labelling was observed [90]. Another potential photoaffinity probe, a 2-azido-adenine analog of NAADP, was also prepared [90]. It is a weak agonist with an EC_{50} of 15 μM in sea urchin egg homogenates [90].

NAADP ANALOGS WITH THE ADENINE RIBOSE MODIFIED

NAADP contains a phosphate group on the 2'-position of the ribose linked to adenine (see Figure 6). Removal of the phosphate renders the resulting analog, nicotinic acid adenine dinucleotide (NAAD), completely inactive in evoking Ca^{2+} release [4, 89]. Consistently, NAAD has been shown to be unable to compete with ^{32}P-NAADP for binding to the NAADP-receptor [91]. Moving the phosphate group to the 3'-position of the ribose reduces the Ca^{2+} releasing activity by about 20-fold [14]. Replacement of the 2'-phosphate group with a cyclic phosphate linking the 2'- and 3'-positions of the of the ribose ring reduces the Ca^{2+} releasing activity by about 10-fold [14]. As discussed below, modification of the 2'-phosphate with a 2-(nitrophenyl) ethyl caging group to produce caged NAADP completely abolishes the ability to mobilize Ca^{2+} [92]. Thus, the presence of an unmodified phosphate group on the adenylyl ribose is critical for NAADP activity [14], however, there is leniency in the position to which the phosphate is linked [14].

OTHER NAADP ANALOGS

Thio-NADP, a NADP analog containing a thio-amide group at the 3-position of the pyridine ring, is reported to be an antagonist of NAADP-induced Ca^{2+} release in sea urchin egg homogenates [89]. However, Dickey *et al.*, demonstrated that this was not the case, but, in fact, the antagonistic activity was simply due to contaminating NAADP in commercial preparations of thio-NADP [93]. NAADP-induced Ca^{2+} release displays an unique self-inactivation property in which sub-threshold concentrations of NAADP desensitize the Ca^{2+} release mechanism [91, 94]. It is this self-inactivation that is responsible for the erroneous antagonistic effect reported earlier [89].

In fact, all of the synthetic NAADP analogs that are agonists (see above), also display self-inactivation of NAADP-sensitive Ca^{2+} release. Thus, while no NAADP antagonist has been identified, all NAADP agonists can be exploited as pseudo-antagonists, as sub-threshold concentrations of these analogs will self-inactivate the NAADP-sensitive Ca^{2+} release mechanism in a concentration and time-dependent manner [14, 91, 93, 94].

Caged NAADP has been synthesized by adding a photoactivatable 2-(nitrophenyl) ethyl caging group to the 2'-phosphate of NAADP [92]. Caged

NAADP is not capable of activating Ca^{2+} release in sea urchin egg homogenates unless the caging group is removed by exposure to UV light [92]. As with other caged compounds, the use of caged NAADP has provided definitive evidence that NAADP is active in releasing Ca^{2+} in live sea urchin eggs [84, 91, 92, 95, 96]. Caged NAADP has proven to be a valuable reagent in the study of the properties and localization of NAADP-sensitive Ca^{2+} release in sea urchin eggs [84, 91, 92, 95, 96] and pancreatic acinar cells [8]. Of particular interest is the demonstration that localized release of NAADP by photolysis in sea urchin eggs can produce a Ca^{2+} wave [95, 97], which can serve to encode spatial memory of prior Ca^{2+} activation ([95] see also the chapter by Churchill *et al.*, in this book).

Periodate oxidation of the cis-diol of the ribose units of nucleotides has been used as a strategy for producing chemically reactive probes for affinity labelling of nucleotide binding proteins [98-101]. Treatment of NAADP with sodium periodate results in the oxidation of the cis diol of the nicotinate ribosyl unit to a dialdehyde [102]. The adenylyl ribosyl unit is protected from oxidation due to the presence of the phosphate group on the 2'-position. The oxidized NAADP was found to bind to the NAADP receptor with low affinity, displacing ^{32}P-NAADP with an IC_{50} of about 15 μM, over 2000-fold higher than for NAADP itself [102]. This data points to the importance of an intact nicotinate ribosyl unit for NAADP action.

CONCLUDING REMARKS

The cADPR and NAADP analogs generated to date have provided a wealth of structure-activity information as well as pharmacological tools with useful properties. Most of the analogs have been synthesized using strategies relying on the broad substrate specificity of ADP-ribosyl cyclase. Total chemical synthesis of cADPR has now been achieved. In the future this approach should provide the next wave of cADPR and NAADP analogs with widely different structures that are not limited by the enzymatic approach.

ACKNOWLEDGEMENTS

We would like to acknowledge Dr. Richard Graeff and Ms. Angela Kell for many helpful discussions during the preparation of this chapter. We acknowledge grant support from the following sources: NIH DA11806, NIH GM58492 and an University of Minnesota Academic Health Center Faculty Development Grant (T.F.W) and NIH HD17484, GM61568, GM60333 (H.C.L).

REFERENCES

1. Berridge MJ. 1993. Inositol trisphosphate and calcium signaling. *Nature* 361: 315-325

2. Clapper DL, Walseth TF, Dargie PJ and Lee HC. 1987. Pyridine nucleotide metabolites stimulate calcium release from sea urchin egg microsomes desensitized to inositiol trisphosphate. *J. Biol. Chem.* 262: 9561-9568.
3. Lee HC, Walseth TF, Bratt GT, Hayes RN and Clapper DL. 1989. Structural determination of a cyclic metabolite of NAD^+ with intracellular Ca^{2+}-mobilizing activity. *J. Biol. Chem.* 264: 1608-1615.
4. Lee HC and Aarhus R. 1995. A derivative of NADP mobilizes calcium stores insensitive to inositol trisphosphate and cyclic ADP-ribose. *J. Biol. Chem.* 270: 2152-2157.
5. Dargie PJ, Agre MC and Lee HC. 1990. Comparison of Ca^{2+} mobilizing activities of cyclic ADP-ribose and inositol trisphosphate. *Cell Regul.* 1: 279-290.
6. Galione A, Lee HC and Busa WB. 1991. Ca^{2+}-induced Ca^{2+} release in sea urchin egg homogenates: Modulation by cyclic ADP-ribose. *Science* 253: 1143-1146.
7. Lee HC. 1993. Potentiation of calcium- and caffeine-induced calcium release by cyclic ADP-ribose. *J. Biol. Chem.* 268: 293-299.
8. Cancela JM, Churchill GC and Galione A. 1999. Coordination of agonist-induced Ca^{2+}-signalling patterns by NAADP in pancreatic acinar cells. *Nature* 398: 74-76.
9. Cancela JM, Gerasimenko OV, Gerasimenko JV, Tepikin AV and Petersen OH. 2000. Two different but converging messenger pathways to intracellular Ca^{2+} release: the roles of nicotinic acid adenine dinucleotide phosphate, cyclic ADP-ribose and inositol trisphosphate. *EMBO J.* 19: 2549-2557.
10. Lee HC and Aarhus R. 1991. ADP-ribosyl cyclase: an enzyme that cyclizes NAD^+ into a calcium-mobilizing metabolite. *Cell Regul.* 2: 203-209.
11. Walseth TF, Aarhus R, Gurnack ME, Wong L, Breitinger HG, *et al.* 1997. Preparation of cyclic ADP-ribose antagonists and caged cyclic ADP-ribose. *Meth. Enzymol.* 280: 294-305.
12. Walseth TF and Lee HC. 1993. Synthesis and characterization of antagonists of cyclic-ADP-ribose-induced Ca^{2+} release. *Biochim. Biophys. Acta* 1178: 235-242.
13. Aarhus R, Graeff RM, Dickey DM, Walseth TF and Lee HC. 1995. ADP-ribosyl cyclase and CD38 catalyze the synthesis of a calcium-mobilizing metabolite from NADP. *J. Biol. Chem.* 270: 30327-30333.
14. Lee HC and Aarhus R. 1997. Structural determinants of nicotinic acid adenine dinucleotide phosphate important for its calcium-mobilizing activity. *J. Biol. Chem.* 272: 20378-20383.
15. Lee HC and Aarhus R. 1998. Fluorescent analogs of NAADP with calcium mobilizing activity. *Biochim. Biophys. Acta* 1425: 263-271.
16. Lee HC, Aarhus R, Levitt D. 1994. The crystal structure of cyclic ADP-ribose. *Nature Struct. Biol.* 1: 143-144.
17. Munshi C and Lee HC. 1997. High-level expression of recombinant *Aplysia* ADP-ribosyl cyclase in *Pichia pastoris* by fermentation. *Prot. Expres. Purif.* 11: 104-110.
18. Ashamu GA, Sethi JK, Galione A and Potter BVL. 1997. Roles for adenosine ribose hydroxyl groups in cyclic adenosine 5'-diphosphate ribose-mediated Ca^{2+} release. *Biochemistry* 36: 9509-9517.
19. Bailey VC, Fortt SM, Summerhill RJ, Galione A and Potter BVL. 1996. Cyclic aristeromycin diphosphate ribose: a potent and poorly hydrolysable Ca^{2+}-mobilizing mimic of cyclic adenosine diphosphate ribose. *FEBS Lett.* 379: 227-230.
20. Bailey VC, Sethi JK, Fortt SM, Galione A and Potter B. 1997. 7-deaza cyclic adenosine 5'-diphosphate ribose - First example of a Ca^{2+}-mobilizing partial agonist related to cyclic adenosine 5'-diphosphate ribose. *Chem. Biol.* 4: 51-61.
21. Bailey VC, Sethi JK, Galione A and Potter B. 1997. Synthesis of 7-deaza-8-bromo cyclic adenosine 5'-diphosphate ribose - the first hydrolysis resistant antagonist at the cADPR receptor. *Chem. Commun.* 7: 695-696.

22. Wong L, Aarhus R, Lee HC and Walseth TF. 1999. Cyclic 3-deaza-adenosine diphosphoribose: a potent and stable analog of cyclic ADP-ribose. *Biochim. Biophys. Acta* 1472: 555-564.
23. Shuto S, Fukuoka M, Manikowsky A, Ueno Y, Nakano T, *et al.* 2001. Total synthesis of cyclic ADP-carbocyclic-ribose, a stable mimic of Ca^{2+}-mobilizing second messenger cyclic ADP-ribose. *J. Am. Chem. Soc.* 123: 8750-8759.
24. Shuto S, Fukuoka M, Abe H and Matsuda A. 2001. Intracellular Ca^{2+}-mobilizing adenine nucleotides. Synthesis and biological activity of cyclic ADP-carbocyclic-ribose and C-glycosidic analog of adenophostin A. *Nucleosides, Nucleotides Nucleic Acids* 20: 461-470.
25. Fukuoka M, Shuto S, Minakawa N, Ueno Y and Matsuda A. 2001. Synthesis and biological activities of cyclic ADP-carbocyclic-ribose and its analogs. *Nucleosides, Nucleotides Nucleic Acids* 20: 1355-1358.
26. Fukuoka M, Shuto S, Minakawa N, Ueno Y and Matsuda A. 2000. An efficient synthesis of cyclic IDP- and cyclic 8-bromo-IDP-carbocyclic-riboses using a modified hata condensation method to form an intramolecular pyrophosphate linkage as a key step. An entry to a general method for the chemical synthesis of cyclic ADP-ribose analogues. *J. Org. Chem.* 65: 5238-5248.
27. Galeone A, Mayol L, Oliviero G, Piccialli G and Varra M. 2002. Synthesis of a novel N-1 carbocyclic, N-9 butyl analogue of cyclic ADP ribose (cADPR). *Tetrahedron* 58: 363-368.
28. Huang L-J, Zhoa Y-Y, Yuan L, Min J-M and Zhang L-H. 2002. Chemical Synthesis and Calcium Release Activity of N1-Ether Strand substituted cADPR Mimic. *Bioorg. Med. Chem. Lett.* 12: 887-889.
29. Walseth TF, Aarhus R, Kerr JA and Lee HC. 1993. Identification of cyclic ADP-ribose-binding proteins by photoaffinity labeling. *J. Biol. Chem.* 268: 26686-26691.
30. Lee HC, Aarhus R and Walseth TF. 1993. Calcium mobilization by dual receptors during fertilization of sea urchin eggs. *Science* 261: 352-355.
31. Kuemmerle JF and Makhlouf GM. 1995. Agonist-stimulated cyclic ADP ribose. Endogenous modulator of Ca^{2+}-induced Ca^{2+} release in intestinal longitudinal muscle. *J. Biol. Chem.* 270: 25488-25494.
32. Prakash YS, Kannan MS, Walseth TF and Sieck GC. 1998. Role of cyclic ADP-ribose in the regulation of $[Ca^{2+}]_i$ in porcine tracheal smooth muscle. *Am. J. Physiol.* 43: C1653-C1660.
33. Rakovic S, Galione A, Ashamu GA, Potter B and Terrar DA. 1996. A specific cyclic ADP-ribose antagonist inhibits cardiac excitation-contraction coupling. *Cur. Biol.* 6: 989-996.
34. Rakovic S, Cui Y, Iino S, Galione A, Ashamu GA, *et al.* 1999. An antagonist of cADP-ribose inhibits arrhythmogenic oscillations of intracellular Ca^{2+} in heart cells. *J. Biol. Chem.* 274: 17820-17827.
35. Prakash YS, Kannan MS, Walseth TF and Sieck GC. 2000. cADP ribose and $[Ca^{2+}]_i$ regulation in rat cardiac myocytes. *Am. J. Physiol.* 279: H1482-H1489.
36. Guse AH, Berg I, Dasilva CP, Potter B and Mayr GW. 1997. Ca^{2+} entry induced by cyclic ADP-ribose in intact T-lymphocytes. *J. Biol. Chem.* 272: 8546-8550.
37. Verderio C, Bruzzone S, Zocchi E, Fedele E, Schenk U, *et al.* 2001. Evidence of a role for cyclic ADP-ribose in calcium signalling and neurotransmitter release in cultured astrocytes. *J. Neurochem.* 78: 646-657.
38. Linden DJ, Dawson TM and Dawson VL. 1995. An evaluation of the nitric oxide/cGMP/cGMP-dependent protein kinase cascade in the induction of cerebellar long-term depression in culture. *J. Neurosci.* 15: 5098-5105.
39. Ebihara S, Sasaki T, Hida W, Kikuchi Y, Oshiro T, *et al.* 1997. Role of cyclic ADP-ribose in ATP-activated potassium currents in alveolar macrophages. *J. Biol. Chem.* 272: 16023-16029.

40. Cancela JM and Petersen OH. 1998. The cyclic ADP ribose antagonist 8-NH_2-cADP-ribose blocks cholecystokinin-evoked cytosolic Ca^{2+} spiking in pancreatic acinar cells. *Pfluegers Archiv* 435: 746-748.
41. Burdakov D, Cancela JM and Petersen OH. 2001. Bombesin-induced cytosolic Ca^{2+} spiking in pancreatic acinar cells depends on cyclic ADP-ribose and ryanodine receptors. *Cell Calcium* 29: 211-216.
42. Leckie CP, McAinsh MR, Allen GJ, Sanders D and Hetherington AM. 1998. Abscisic acid-induced stomatal closure mediated by cyclic ADP-ribose. *Pro. Nat.l Acad. Sci. USA.* 95: 15837-15842.
43. De Flora A, Guida L, Franco L, Zocchi E, Pestarino M, *et al.* 1996. Ectocellular in vitro and in vivo metabolism of cADP-ribose in cerebellum. *Biochem. J.* 320: 665-672.
44. Podesta M, Zocchi E, Pitto A, Usai C, Franco L, *et al.* 2000. Extracellular cyclic ADP-ribose increases intracellular free calcium concentration and stimulates proliferation of human hemopoietic progenitors. *FASEB J.* 14: 680-690.
45. Franco L, Bruzzone S, Song PF, Guida L, Zocchi E, *et al.* 2001. Extracellular cyclic ADP-ribose potentiates ACh-induced contraction in bovine tracheal smooth muscle. *Am. J. Physiol.* 280: L98-L106.
46. Franco L, Zocchi E, Usai C, Guida L, Bruzzone S, *et al.* 2001. Paracrine roles of NAD^+ and cyclic ADP-ribose in increasing intracellular calcium and enhancing cell proliferation of 3T3 fibroblasts. *J. Biol. Chem.* 276: 21642-21648.
47. Zocchi E, Podesta M, Pitto A, Usai C, Bruzzone S, *et al.* 2001. Stroma-generated cyclic ADP-ribose stimulates the expansion of early human hemopoietic progenitors by a paracrine interaction. *FASEB J.* 15: 29.
48. Sethi JK, Empson RM, Bailey VC, Potter B and Galione A. 1997. 7-deaza-8-bromo-cyclic ADP-ribose, the first membrane-permeant, hydrolysis-resistant cyclic ADP-ribose antagonist. *J. Biol. Chem.* 272: 16358-16363.
49. Dipp M and Evans AM. 2001. Cyclic ADP-ribose is the primary trigger for hypoxic pulmonary vasoconstriction in the rat lung in situ. *Cir. Res.* 89: 77-83.
50. Wilson HL, Dipp M, Thomas JM, Lad C, Galione A and Evans AM. 2001. ADP-ribosyl cyclase and cyclic ADP-ribose hydrolase act as a redox sensor - A primary role for cyclic ADP-ribose in hypoxic pulmonary vasoconstriction. *J. Biol. Chem.* 276: 11180-11188.
51. Partida-Sanchez S, Cockayne DA, Monard S, Jacobson EL, Oppenheimer N, *et al.* 2001. Cyclic ADP-ribose production by CD38 regulates intracellular calcium release, extracellular calcium influx and chemotaxis in neutrophils and is required for bacterial clearance in vivo. *Nature Med.* 7: 1209-1216.
52. Yu JZ, Zhang DX, Zou AP, Campbell WB and Li PL. 2000. Nitric oxide inhibits Ca^{2+} mobilization through cADP-ribose signaling in coronary arterial smooth muscle cells. *Am. J. Physiol.* 279: H873-H881.
53. Guse AH, da Silva CP, Berg I, Skapenko AL, Weber K, *et al.* 1999. Regulation of calcium signalling in T lymphocytes by the second messenger cyclic ADP-ribose. *Nature* 398: 70-73.
54. Guse AH, Dasilva CP, Weber K, Armah CN, Ashamu GA, *et al.* 1997. 1-(5-phospho-beta-D-ribosyl)2'-phosphoadenosine 5'-phosphate cyclic anhydride induced Ca^{2+} release in human T-cell lines. *Eur. J. Biochem.* 245: 411-417.
55. Lee HC. 1997. Mechanisms of calcium signaling by cyclic ADP-ribose and NAADP. *Physiol. Rev.* 77: 1133-1164.
56. Lee HC. 2001. Physiological functions of cyclic ADP-ribose and NAADP as calcium messengers. *Annu. Rev. Pharmacol. Toxicol.* 41: 317-345.
57. Shiwa M, Murayama T and Ogawa Y. 2002. Molecular cloning and characterization of ryanodine receptor from unfertilized sea urchin eggs. *Am. J. Physiol.* 282: R727-R737.
58. Shuto S, Shirato M, Sumita Y, Ueno Y and Matsuda A. 1998. Synthetic studies of carbocyclic analogs of cyclic ADP-ribose - formation of a cyclic dimer, a 36-membered-ring product, in the condensation reaction of an 8-brorno-n-1-[5-(phenylthiophosphoryl)-

carbocyclic-ribosyl] inosine 5'-phosphate derivative mediated by $AgNO_3$. *Tetrahedron Lett.* 39: 7341-7344.
59. Sumita Y, Shirato M, Ueno Y, Matsuda A and Shuto S. 2000. Nucleosides and nucleotides. 192. Toward the total synthesis of cyclic ADP-carbocyclic-ribose. Formation of the intramolecular pyrophosphate linkage by a conformation-restriction strategy in a syn-form using a halogen substitution at the 8-position of the adenine ring. *Nucleosides Nucleotides Nucleic Acids* 19: 175-187.
60. Wall KA, Klis M, Kornet J, Coyle D, Ame JC, *et al.* 1998. Inhibition of the intrinsic NAD^+ glycohydrolase activity of CD38 by carbocyclic NAD analogues. *Biochem. J.* 335: 631-636.
61. Howard M, Grimaldi JC, Bazan JF, Lund FE, Santos AL, *et al.* 1993. Formation and hydrolysis of cyclic ADP-ribose catalyzed by lymphocyte antigen CD38. *Science* 262: 1056-1059.
62. Zhang FJ, Yamada S, Gu QM and Sih CJ. 1996. Synthesis and characterization of cyclic ATP-ribose - a potent mediator of calcium release. *Bioorg. Med. Chem. Lett.* 6: 1203-1208.
63. Zhang FJ, Gu QM, Jing P and Sih CJ. 1995. Enyzmatic cyclization of nicotinamide adenine dinucleotide phosphate (NADP). *Bioorg. Med. Chem. Lett.* 5: 2267-2272.
64. Vu CQ, Lu PJ, Chen CS and Jacobson MK. 1996. 2'-Phospho-cyclic ADP-ribose, a calcium-mobilizing agent derived from NADP. *J. Biol. Chem.* 271: 4747-4754.
65. Vu CQ. Coyle DL and Jacobson MK. 1997. Natural occurrence of 2'-phospho-cyclic ADP ribose in mammalian tissues. *Biochem. Biophys. Res. Commun.* 236: 723-726.
66. Walseth TF, Aarhus R, Zeleznikar RJ and Lee HC. 1991. Determination of endogenous levels of cyclic ADP-ribose in rat tissues. *Biochim. Biophys. Acta* 1094: 113-120.
67. Wong L. 1999. *Measurement of intracellular cyclic ADP-ribose and characterization of its analogs.* PH.D. thesis. University of Minnesota, Minneapolis. 176 pp.
68. Zhang FJ and Sih CJ. 1995. Novel enzymatic cyclizations of pyridine nucleotide analogs - cyclic-GDP-ribose and cyclic-HDP-ribose. *Tetrahedron Lett.* 36: 9289-9292.
69. Graeff RM, Walseth TF, Fryxell K, Branton WD and Lee HC. 1994. Enzymatic synthesis and characterizations of cyclic GDP-ribose. A procedure for distinguishing enzymes with ADP-ribosyl cyclase activity. *J. Biol. Chem.* 269: 30260-30267.
70. Graeff RM, Walseth TF, Hill HK and Lee HC. 1996. Fluorescent analogs of cyclic ADP-ribose: synthesis, spectral characterization, and use. *Biochemistry* 35: 379-386.
71. Zhang FJ and Sih CJ. 1995. Enzymatic cyclization of 1,n-6-etheno-nicotinamide adenine dinucleotide. *Bioorg. Med. Chem. Lett.* 5: 1701-1706
72. Zhang FJ and Sih CJ. 1996. Novel analogs of cyclic-ADP-ribose - 9-cyclic etheno-ADP-ribose and cyclic etheno-CDP-ribose. *Bioorg. Med. Chem. Lett.* 6: 2311-2316.
73. Graeff RM, Mehta K and Lee HC. 1994. GDP-ribosyl cyclase activity as a measure of CD38 induction by retinoic acid in HL-60 cells. *Biochem. Biophys. Res. Commun.* 205: 722-727.
74. Graeff RM, Franco L, Deflora A and Lee HC. 1998. Cyclic GMP-dependent and -independent effects on the synthesis of the calcium messengers cyclic ADP-ribose and nicotinic acid adenine dinucleotide phosphate. *J. Biol. Chem.* 273: 118-125.
75. Li NJ, Teggatz EG, Li PL, Allaire R and Zou AP. 2000. Formation and actions of cyclic ADP-ribose in renal microvessels. *Microvasc. Res.* 60: 149-159.
76. Liang M, Chini EN, Cheng JF and Dousa TP. 1999. Synthesis of NAADP and cADPR in mitochondria. *Arch. Biochem. Biophys.* 371: 317-325.
77. Looms D, Nauntofte B and Dissing S. 1998. ADP-ribosyl cyclase activity in rat parotid acinar cells. *Eur. J. Morphol.* 36: 181-185.
78. Masuda W and Noguchi T. 2000. ADP-ribosyl cyclase in rat salivary glands. *Biochem. Biophys. Res. Commun.* 270: 469-472.
79. Matsumura N and Tanuma S. 1998. Involvement of cytosolic NAD^+ glycohydrolase in cyclic ADP-ribose metabolism. *Biochem.Biophys. Res. Commun.* 253: 246-252.

80. Ziegler M, Jorcke D and Schweiger M. 1997. Identification of bovine liver mitochondrial NAD^+ glycohydrolase as ADP-ribosyl cyclase. *Biochem. J.* 326: 401-405.
81. Meszaros LG, Wrenn RW and Varadi G. 1997. Sarcoplasmic reticulum-associated and protein kinase C-regulated ADP-ribosyl cyclase in cardiac muscle. *Biochem. Biophys. Res. Commun.* 234: 252-256.
82. Aarhus R, Gee K and Lee HC. 1995. Caged cyclic ADP-ribose. Synthesis and use. *J. Biol. Chem.* 270: 7745-7749.
83. Guo XQ and Becker PL. 1997. Cyclic ADP-ribose-gated Ca^{2+} release in sea urchin eggs requires an elevated $[Ca^{2+}]$. *J. Biol. Chem.* 272: 16984-16989.
84. Lee HC and Aarhus R. 2000. Functional visualization of the separate but interacting calcium stores sensitive to NAADP and cyclic ADP-ribose. *J. Cell Sci.* 113: 4413-4420.
85. Leite MF, Burgstahler AD and Nathanson MH. 2002. Ca^{2+} waves require sequential activation of inositol trisphosphate receptors and ryanodine receptors in pancreatic acini. *Gastroenterology* 122: 415-427.
86. Lokuta AJ, Darszon A, Beltran C and Valdivia HH. 1998. Defection and functional characterization of ryanodine receptors from sea urchin eggs. *J. Physiol. London* 510: 155-164.
87. Santella L and Kyozuka K. 1997. Effects of 1-methyladenine on nuclear Ca^{2+} transients and meiosis resumption in starfish oocytes are mimicked by the nuclear injection of inositol 1,4,5-trisphosphate and cADP-ribose. *Cell Calcium* 22: 11-20.
88. Varadi A and Rutter GA. 2002. Dynamic imaging of endoplasmic reticulum Ca^{2+} concentration in insulin-secreting MIN6 cells using recombinant targeted cameleons - Roles of sarco(endo)plasmic reticulum Ca^{2+}-ATPase (SERCA)-2 and ryanodine receptors. *Diabetes* 51: S190-S201.
89. Chini EN, Beers KW and Dousa TP. 1995. Nicotinate adenine dinucleotide phosphate (NAADP) triggers a specific calcium release system in sea urchin eggs. *J. Biol. Chem.* 270: 3216-3223.
90. Dickey DM. 1999. *Characterization of the nicotinic acid adenine dinucleotide phosphate binding protein.* Ph.D. thesis. University of Minnesota, Minneapolis. 228 pp.
91. Aarhus R, Dickey DM, Graeff RM, Gee KR, Walseth TF and Lee HC. 1996. Activation and inactivation of Ca^{2+} release by $NAADP^+$. *J. Biol. Chem.* 271: 8513-8516.
92. Lee HC, Aarhus R, Gee KR and Kestner T. 1997. Caged nicotinic acid adenine dinucleotide phosphate - Synthesis and use. *J. Biol. Chem.* 272: 4172-4178.
93. Dickey DM, Aarhus R, Walseth TF and Lee HC. 1997. Thio-NADP is not an antagonist of NAADP. *Cell Biochem. Biophys.* 28: 63-73.
94. Genazzani AA, Empson R and Galione A. 1996. Unique activation properties of NAADP-induced Ca^{2+} release. *J. Biol. Chem.* 271: 11599-11602.
95. Churchill GC and Galione A. 2000. Spatial control of Ca^{2+} signaling by nicotinic acid adenine dinucleotide phosphate diffusion and gradients. *J. Biol. Chem.* 275: 38687-38692.
96. Churchill GC and Galione A. 2001. Prolonged inactivation of nicotinic acid adenine dinucleotide phosphate-induced Ca^{2+} release mediates a spatiotemporal Ca^{2+} memory. *J. Biol. Chem.* 276: 11223-11225.
97. Lee HC. 2000. NAADP: An emerging calcium signaling molecule. *J. Membr. Biol.* 173: 1-8.
98. Low A, Faulhammer HG and Sprinzl M. 1992. Affinity labeling of GTP-binding proteins in cellular extracts. *FEBS Lett.* 303: 64-68.
99. Low A, Sprinzl M and Faulhammer HG. 1993. Affinity labeling of c-H-ras p21 consensus elements with periodate-oxidized GDP and GTP. *Eur. J. Biochem.* 215: 473-479.
100. Peter ME, She J, Huber LA and Terhorst C. 1993. Labeling of adenine and guanine nucleotide-binding proteins in permeabilized cells with in situ periodate-oxidized nucleotides. *Anal. Biochem.* 210: 77-85.

101. Hohenegger M, Herrmann-Frank A, Richter M and Lehmann-Horn F. 1995. Activation and labelling of the purified skeletal muscle ryanodine receptor by an oxidized ATP analogue. *J. Biol. Chem.* 268: 293-299.
102. Yee L. 2002. *Characterization of Nicotinic Acid Adenine Dinucleotide Phosphate Signaling System*. Masters thesis. University of Minnesota, Minneapolis. 113 pp.

Timothy F. Walseth, Department of Pharmacology,
University of Minnesota, Minneapolis, MN 55455, USA.
e-mail: walseth@mail.ahc.umn.edu

Cyclic ADP-Ribose and NAADP. Structures, Metabolism and Functions. Lee HC (ed), pp. 143-166, Dordrecht: Kluwer.

Separate but Interacting Calcium Stores

Hon Cheung Lee
Department of Pharmacology, University of Minnesota, Minneapolis, MN 55455, USA.

INTRODUCTION

Mobilization of intracellular Ca^{2+} stores is a principal signaling mechanism employed by cells to respond to a wide variety of stimuli, both external and internal. One of the most dramatic examples occurs during fertilization. Immediately after sperm-egg fusion, a highly localized Ca^{2+} elevation is initiated right at the fusion site. This spark of Ca^{2+} then grows into a wave propagating across the entire egg. Intriguingly, the initiation of apoptotic cell death also involves Ca^{2+} mobilization. How cells can differentiate between Ca^{2+} signals as disparate as those governing life and death is a question of fundamental importance.

It is generally accepted that cells possess multiple Ca^{2+} stores. It is thus possible that different stimuli are linked to different types of stores. For example, mitochondria are one type of Ca^{2+} store, from which mobilization is known to be associated with apoptotic signals, while the endoplasmic reticulum is another store that serves as the source of Ca^{2+} for the fertilization wave. This possibility is now made substantive by the discoveries of at least three distinct Ca^{2+} messengers, inositol trisphosphate (IP_3) [1, 2], cyclic ADP-ribose (cADPR) [3, 4] and nicotinic acid adenine dinucleotide phosphate (NAADP) [3, 5], for mobilizing Ca^{2+} stores. In addition, Ca^{2+} stores in many cells have now been shown to be spatially segregated. Mobilization of these localized stores can thus provide a high degree of spatial specificity to the Ca^{2+} signaling process as well. In many instances, mobilization of internal Ca^{2+} stores can likewise lead to modulation of Ca^{2+} influx, adding yet another layer of complexity and versatility to Ca^{2+} mobilization as a signaling mechanism. This chapter explores the complexity and versatility of Ca^{2+} mobilization in the context of the recent advances in the understanding of Ca^{2+} signaling mediated by cADPR and NAADP.

CELLS RESPONSIVE TO cADPR AND NAADP

Cyclic ADP-ribose was first shown to be effective in mobilizing intracellular Ca^{2+} stores in sea urchin eggs [3, 4]. Since then, over fifty other cell types have been reported to be responsive as well and the number is increasing steadily. Some of these cell systems have previously been reviewed [6, 7] and other representatives are listed in Table 1. They include species from three biological kingdoms; a protozoan, *Euglena*; several species of plants; and a wide variety of animal cells, from sponges, the oldest multicellular animals, various marine invertebrates, an insect, an amphibian, to various mammals including human. It is clear that this Ca^{2+}-signaling mechanism is evolutionary conserved.

The cellular processes that cADPR is involved in are likewise diverse, from mediating thermo-sensing in sponge cells [8], cell cycle regulation in *Euglena* and cultured cells [9-11], fertilization ([3, 12, 13] and reviewed in [14]), insulin secretion (reviewed in [15]), myocyte contraction [16], neutrophil chemotaxis [17], long-term-depression and memory formation [18, 19], to clonal expansion of human hemopoietic progenitors [20, 21].

As listed in Table 1, cADPR is generally effective in the low micromolar range, although in a few cases, it is in the nanomolar range. Its action is blocked by specific antagonists such as 8-amino-cADPR, 8-Br-cADPR [22] and 7-deaza-8-Br-cADPR [23]. The last two antagonists are cell permeant and can be simply applied externally. It is also blocked by inhibitors of the ryanodine receptor including ruthenium red and ryanodine, indicating the receptor is the target of cADPR. Cells are generally impermeable to cADPR and its introduction requires either microinjection or permeabilization of the plasma membrane. However, in a few cases, including cerebellar granule cells [24], human hemopoietic progenitors [20, 21] and tracheal smooth muscles [25], cADPR appears to be permeant and can exert its Ca^{2+} mobilizing effects when high enough concentrations (~100 μM) are applied externally.

In many of the systems listed in Table 1, the endogenous levels of cADPR have also been measured. The levels have been shown to be elevated by various stimuli, which not only include agonists of surface receptors such as acetylcholine [26], abscisic acid [27] and antibodies of the T-cell receptor (OKT3) [28], but also metabolic factors such as glucose [29] and cobalamin [9]. Even physical stimuli, such as temperature [8], nuclear factors, such as retinoic acid [30] or cell-cell interaction as occurs in fertilization [13, 31], can also elevate cellular cADPR. Other signaling pathways, such as the one mediated by nitric oxide, through elevation of cGMP, can likewise stimulate

Table 1. **Cells responsive to cADPR**

Biological kingdom/ Cell/Tissue	Effective conc.	Antagonist	Endogenous level	Ref.
Protist				
Euglena gracilis	EC_{50} = 0.2 μM	RR	0-240 fmol/10^6 cells, ↑ by Cbl	[9, 32]
Plant				
red beet, vacuoles	EC_{50} = 24 nM	RR	ND	[33]
tomato *aurea* cells	~ 1 μM	$8NH_2$	+ , ↑ by ABA	[27]
C. communis micro.	100 nM	$8NH_2$	ND	[34]
Cauliflower, micro.	1 μM	$8NH_2$, RR	ND	[35]
tobacco leaves	100 nM	RR	ND	[36]
Animal				
sponge cells	10 μM	8Br	0.2-1.7 nmol/mg, ↑ by ABA, Temp.	[8]
sea urchin eggs	EC_{50} ≅ 20 nM	$8NH_2$, 8Br, RR, Ry	0-30 amol/egg, ↑ by fertilization	[3, 13, 22, 37, 38]
starfish oocytes	5-10 μM	$8NH_2$, RR	ND	[39]
Aplysia neurons	0.1-5 μM	$8NH_2$, Ry	ND	[40]
ascidian oocytes	EC_{50} = 0.1 nM	$8NH_2$, Ry	ND	[41]
insect neurons	100 nM	RR, Ry	ND	[42]
fish eggs	60 nM	ND	0.7-160 pmol/μg ↑ by fertilization	[31, 43]
frog neurons	0.3-1μM	Ry	ND	[44, 45]
rat hepatocyte nuclei	2-10 μM	$8NH_2$, Ry	~2 pmol/mg (liver)	[46, 47]
rat mesangial micro.	EC_{50} ≅ 3 μM	8Br, RR	~0.7 pmol/mg (kidney)	[46, 48]
rat pancreatic β-cell micro.	100 nM	$8NH_2$	50-700 fmol/μg, ↑ by glucose	[29, 49]
rat hippocampus	ND	7deaza8Br, 8Br, Ry	20-30 pmol/mg, ↑ by cGMP	[18]
guinea pig myocytes	EC_{50} = 2-3 μM	$8NH_2$, Ry 8Br	1 pmol/mg (rat heart)	[16, 46]
bovine arterial SR	0.01-1 μM	RR	ND	[50]
porcine tracheal smooth muscles	1-10 μM	$8NH_2$, RR	ND	[51]
rat cerebellar cells	0.1 mM (outside)	Ry	2.2 pmol/mg (brain)	[24, 46]
mouse pancreatic acinar cells	10 μM	$8NH_2$	~4 fmol//μg, ↑ by acetylcholine	[26, 52]
human hemopoietic progenitors	100 μM or 1 nM 3deaza (outside)	$8NH_2$	ND	[20]
human (Jurkat) T-lymphocytes	EC_{50} ≅ 2 μM	7deaza8Br, $8NH_2$	0.5-2 pmol/10^7 cells, ↑ by OKT3	[28, 53, 54]

Abbreviations: EC_{50}, half-maximal concentration; +, endogenous cADPR was measured qualitatively; ↑, increased; Cbl, cobalamin; RR, ruthenium red; $8NH_2$, 8-amino-cADPR; ABA, abscisic acid; 8Br, 8-bromo-cADPR; Ry, ryanodine; 7deaza8Br, 7-deaza-8-bromo-cADPR; SR, sacroplasmic reticulum; 3deaza, 3-deaza-cADPR; OKT3, specific antibody against T-cell receptor; micro., microsomes; ND, not determined.

cADPR production in cells [18, 19, 55]. The cADPR-mechanism is thus highly versatile.

The versatility and widespread occurrence of the cADPR-signaling pathway is apparently taken advantage of by bacteria to aid their cytotoxicity. Delivering effectors with important signaling functions into host cells appears to be a common strategy for bacteria to enhance their cytotoxicity. The exotoxins transferred by several pathogens, including anthrax and whooping cough, are shown to have adenylyl cyclase activity [56]. In the case of *S. pyrogenes*, the effector transferred during infection, designated SPN, acts synergistically with streptolysin O to induce cyto-toxicity [57]. Mutant bacteria lacking SPN are much less cytotoxic than the wildtype. Sequencing of the amino terminal of SPN identifies it as the ADP-ribosyl cyclase previously purified from *S. pyrogenes* cultures [58]. This bacterial cyclase catalyzes not only the synthesis of cADPR but its hydrolysis as well. Although it is a soluble protein, its enzymatic properties are similar to the mammalian CD38 [59], a membrane-bound enzyme. That bacteria have evolved to utilize ADP-ribosyl cyclase as a cytotoxic effector underscores the importance of the cADPR-signaling pathway in the normal functioning of eukaryotic cells.

Not as many cell systems have so far been reported to be responsive to NAADP. This is probably because it has only been identified since 1995 [5], and is a relatively new addition to the growing family of Ca^{2+} messengers. Nevertheless, as listed in Table 2, the responsive cells reported so far are equally widespread as those responsive to cADPR and include plants [60], invertebrates [5, 41, 61-63] and higher animals, from amphibian (frog) [64] to various mammalian systems, including human Jurkat cells [65-73].

Table 2. **Cells responsive to NAADP**

Cell or Tissue	Effective conc.	Ref.
Plants		
red beet and cauliflower microsomes	$EC_{50} \cong 96$ nM	[60]
Invertebrate animals		
sea urchin eggs	$EC_{50} \cong 30$ nM	[5]
ascidian oocytes	$EC_{50} \cong 3$ nM	[41]
starfish oocytes	$< \sim 1$ μM	[62]
Aplysia neurons	10-50 μM	[63]
Amphibian		
frog nerve-muscle preparations	$EC_{50} \cong 1$ μM	[64]
Mammals		
mouse pancreatic acinar cells	50 nM	[65]
rat brain microsomes	$EC_{50} \cong 1$ μM	[66]
human T-lymphocytes (Jurkat)	$EC_{50} \cong 50$ nM	[68]
rat kidney microsomes	$EC_{50} \cong 5$ μM	[71]
microsomes from 5 different cultured cells	$EC_{50} \cong 5$ μM	[73]
rabbit cardiac microsomes	$EC_{50} \cong 323$ nM	[69]
rat vascular smooth muscle microsomes	$EC_{50} \cong 3$ μM	[72]

In general, NAADP is more effective than cADPR and, in most systems, the half-maximal effective concentrations (EC_{50}) are in the low nanomolar range. Not much is known about the endogenous levels of NAADP in cells and tissues, due to the lack of a widely applicable assay. This is now changed with the recent introduction of a novel coupled enzyme-cycling assay [74], which is described in detail in the chapter by Graeff and Lee in this book. The sensitivity and the convenience of this assay should pave the way for detailed investigations into the regulation of NAADP metabolism.

MECHANISMS OF Ca^{2+} MOBILIZATION MEDIATED BY cADPR AND NAADP

The Ca^{2+} mobilization activated by cADPR is insensitive to heparin, indicating it is independent of the IP_3-mechanism [12]. It is, however, sensitive to inhibitors of the ryanodine receptor, such as ruthenium red, ryanodine, procaine and Mg^{2+} ([38, 75] and see Table 1). Agonists of the ryanodine receptors, caffeine and Ca^{2+}, likewise potentiate the cADPR-induced release [38, 76]. The cADPR-mechanism is also sensitized by calmodulin [77-79].

In addition to these pharmacological characteristics, accumulating evidence indicates the target of cADPR is indeed the ryanodine receptor. The cADPR-sensitive Ca^{2+} channels isolated from sea urchin eggs and reconstituted into lipid bilayers show single-channel characteristics very similar to that of the ryanodine receptor/channel [80, 81]. Conversely, reconstituted ryanodine receptors isolated from muscles also show sensitivity to cADPR [50, 82-84]. Furthermore, ryanodine binding to its receptor is directly modulated by cADPR [28, 48, 72, 85]. Introduction of cADPR into various types of muscle can likewise activate Ca^{2+} mobilization (see chapters in this book by Rakovic and Terrar, Li *et al.* and White *et al.*).

Operationally, cADPR activates Ca^{2+} mobilization by increasing the Ca^{2+} sensitivity of the Ca^{2+}-induced Ca^{2+} release mechanism, which is known to be mediated by the ryanodine receptor. This has been shown in reconstituted ryanodine receptors [50, 84], in microsomes [38, 76, 78] and in intact cells [44, 86, 87]. In this mode of operation, cADPR is a modulator of the Ca^{2+}-induced Ca^{2+} release (described in more detail in the chapter by Morgan and Galione in this book). In the presence of calmodulin, elevation of cADPR to sufficiently high levels can sensitize the release mechanism to such an extent that even basal Ca^{2+} levels in the sub-micromolar range can activate Ca^{2+}-induced Ca^{2+} release. In this manner, cADPR behaves as a regular Ca^{2+} messenger ([77, 78] and reviewed in [88]).

The Ca^{2+}-release mechanism activated by NAADP is totally independent of those activated by cADPR and IP_3. Inhibitors of the receptors of cADPR and IP_3, as well as those of the ryanodine receptor are ineffective in blocking the action of NAADP ([5] and reviewed in [89]). It is, however, sensitive to antagonists of the L-type Ca^{2+} channels, albeit with low affinity, suggesting possible similarity [62, 66, 69, 90]. The exact identity of the Ca^{2+} channel activated by NAADP, however, has yet to be determined.

Nevertheless, It is clear that a highly specific receptor is involved. This is indicated not only by the remarkable effectiveness of NAADP, being active in the low nanomolar range in many systems (cf. Table 2), but also by the fact that even small modifications of the molecule render it inactive [91]. Three critical structural determinants have been identified, including the carboxyl group of the nicotinic acid ring, the 2'-phosphate on the ribose and the 6-amino group on the adenine [91]. Specific binding of NAADP with high affinity sites has been shown in both sea urchin egg and mammalian microsomes [66, 69, 90, 92]. Interestingly, autoradiographic imaging of the distribution of the NAADP binding sites in brain shows particularly high concentrations of the receptor in the regions of the medulla, midbrain and thalamus [67]. This is in contrast to the IP_3-receptor, which is particularly concentrated in the cerebellum. This regional segregation of receptors suggests possible functional specificity.

THE cADPR- AND NAADP- SENSITIVE Ca^{2+} STORES

Organelles that are known to possess the cADPR- , NAADP- and/or IP_3 -sensitive Ca^{2+} release mechanisms are diagrammed in Figure 1. Two main points are illustrated. First, different types of Ca^{2+} stores are not necessarily distributed uniformly throughout the cell, but instead, can be spatially segregated in various regions. Despite the segregation, these stores can interact, through Ca^{2+}-induced Ca^{2+} release (CICR) and other mechanisms, to produce different forms of Ca^{2+}-signaling, including Ca^{2+} waves and Ca^{2+} oscillation. Second, although the mechanisms mediated by cADPR and NAADP are independent of each other and involve separate Ca^{2+} release channels, a single enzyme, ADP-ribosyl cyclase is responsible for producing both of them [93]. The mammalian CD38 is one such novel enzyme.

Endoplasmic Reticulum (ER)–It is generally accepted that IP_3 mobilizes Ca^{2+} from the ER. Fractionation of sea urchin egg homogenates show that the cADPR-sensitive stores co-purified with both the IP_3-sensitive stores and a marker enzyme for the ER, glucose-6-phosphatase [3, 5, 94]. Similar evidence is obtained in live eggs [95]. Centrifugation of intact eggs stratifies their organelles into defined layers. The location of the cADPR-sensitive stores can be visualised by Ca^{2+} imaging following uniform photolysis of caged cADPR preloaded into the stratified eggs by microinjection. Results indicate the cADPR-stores co-migrate with the IP_3-sensitive stores and the ER; the latter is identified by staining with fluorescent Bodipy-thapsigargin [95].

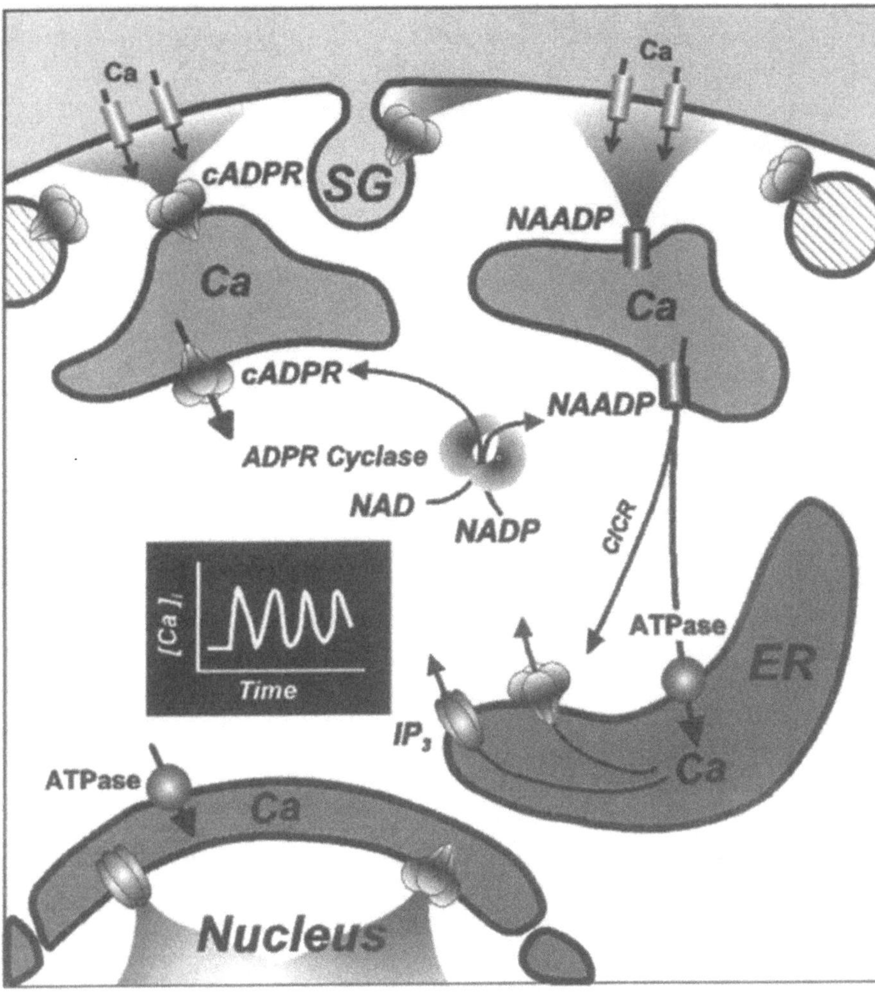

Figure 1. **Spatial segregation of separate but interacting calcium stores in cells.** A single enzyme, ADP-ribosyl (ADPR) cyclase, is responsible for synthesizing both cADPR and NAADP. Tetrameric globules and short cylinders respectively represent the cADPR- and the NAADP- mechanisms. The IP_3-receptors and the Ca^{2+} influx channels are shown as tetrameric ellipsoids and long cylinders, respectively. SG, secretory granules; ER, endoplasmic reticulum.

Evidence obtained from mammalian systems is also consistent with the ER being a cADPR-sensitive Ca^{2+} store. First, isolated cardiac sacroplasmic reticulum vesicles are found to be responsive to cADPR [69, 82]. Second, in sheep lens cells, prior emptying of the ER Ca^{2+} stores by thapsigargin, a specific inhibitor of the Ca^{2+}-ATPase in the ER, completely prevents further Ca^{2+} release by cADPR [96]. Third, direct measurements of the luminal Ca^{2+}

concentration in the ER of MIN6 cells, a pancreatic cell line, using the Ca^{2+} sensitive aequorin specifically targeted to the ER show that cADPR is indeed effective in mobilizing the ER stores [97].

Secretory Vesicles/Granules–The ER may well be the main Ca^{2+} store in cells that is sensitive to cADPR. However, other organelles have also been shown to possess the Ca^{2+}-release mechanism. Zymogen granules isolated from pancreatic acinar cells are functional Ca^{2+} stores and can be mobilized by both cADPR and IP_3. The granules are not sensitive to the ER targeting inhibitor, thapsigargin, pharmacologically distinguishing it from the ER [98]. This is also true for the dense core granules containing insulin in pancreatic β-cells (MIN6). Using aequorin specifically targeted to the secretory granules to monitor their luminal Ca^{2+}, it is demonstrated that cADPR can indeed mobilize these stores. Interestingly, in this case, IP_3 is found to be completely ineffective [97].

The cortical granules in the sea urchin eggs can also be considered secretory granules since they undergo massive exocytosis during fertilization, leading to the formation of the fertilization envelop. Vesicles isolated from the egg cortices and reconstituted into lipid bilayers reveal the presence of the cADPR-sensitive Ca^{2+} channels, suggesting the cortical granules, or other vesicles in the cortices, are cADPR-sensitive Ca^{2+} stores [81]. This appears to be true also in *Ascidian* oocytes, whose cADPR-sensitive stores are localized exclusively in the cell cortex and not in the cytoplasm [41].

Nucleus–Nuclei isolated from hepatocytes [47, 99] and a cultured cell line (MC3T3.E1) [100], can accumulate Ca^{2+} when ATP is provided. It is believed that the nuclear envelope is a functional Ca^{2+} store with the Ca^{2+}-ATPase located on the outer membrane, transporting Ca^{2+} into the envelope (cf. Fig. 1) [99]. Both cADPR and IP_3 are able to mobilize the nuclear Ca^{2+} stores. The mobilizing effect of cADPR is concentration dependent and is specifically blocked by antagonists of cADPR, 8-amino-cADPR and 8-Br-cADPR [47, 100].

Not only is the cADPR-dependent Ca^{2+} release mechanism present in the nucleus but also its metabolic enzyme, CD38. The nuclear CD38 is fully functional and can produce cADPR when NAD is provided. Electron-microscopic and biochemical studies localize the nuclear CD38 to the inner membrane of the nuclear envelope [47, 100]. It is proposed that both cADPR- and the IP_3- dependent Ca^{2+} release mechanisms are also present in the inner nuclear membrane as depicted in Fig. 1 [99]. This is supported by the observation that both cADPR and IP_3 can elevate nucleoplasmic Ca^{2+} as detected by preloading Ca^{2+}-sensitive indicators into the nucleoplasm [99, 100].

That the nucleus is indeed a functional Ca^{2+} store in live cells is shown in the starfish oocyte, which has a pronucleus large enough for microinjecting caged cADPR and a Ca^{2+} indicator directly into the nucleoplasm. Subsequent photolysis produces repetitive Ca^{2+} spiking in the nuceloplasm [39]. Although photolyzing caged IP_3 can also produce a transient increase in nuceloplasmic Ca^{2+}, no repetitive spiking is seen. It is proposed that the nuclear Ca^{2+} changes activated by cADPR and IP_3 are important for oocyte maturation in response to the hormone, 1-methyladenine [101].

Plant Vacuoles–One main structural difference between plant and animal cells is the presence of large vacuoles in the former. Patch-clamping analyses of individual vacuoles isolated from red beets reveal the presence of Ca^{2+} channels sensitive to cADPR [33]. Fractionation studies show that the cADPR-sensitive Ca^{2+} release mechanism is present not only in the vacuoles but also in the ER [35]. Mobilization of the cADPR-sensitive Ca^{2+} stores is found to be crucial in effecting closure of the stomata pores on leaves [34] and in mediating gene expression activated by the stress hormone, abscisic acid [27].

NAADP-Sensitive Ca^{2+} Stores–Fractionation studies in plants show that the NAADP-sensitive stores are exclusively associated with the ER [60]. This is in contrast to the cADPR-stores, which are present in both the vacuoles and the ER. The ER localization of the NAADP-sensitive stores does not appear to be true in animal cells. Fractionation of sea urchin egg homogenates shows a completely different distribution profile than either the ER or the cADPR-sensitive stores [5]. Stratification of live eggs by centrifugation also shows that the NAADP-sensitive Ca^{2+} stores migrate to a pole opposite to that of the ER, which contains the cADPR- and the IP_3-sensitive Ca^{2+} stores [95]. Also, the NAADP-stores in the egg are insensitive to thapsigargin, suggesting the Ca^{2+}-ATPase associated with them is different from that associated with the ER [102]. Further support comes from functional studies in *Ascidian* eggs [41] and Ca^{2+} imaging in starfish oocytes [103]; all show that the NAADP-sensitive stores are segregated exclusively in the cortex and not in cytoplasm, where the ER is located. Exactly which organelle in animal cells, if not the ER, constitutes the NAADP-sensitive Ca^{2+} store has not yet been established, however.

SPATIAL SEGREGATION OF Ca^{2+} STORES

In sea urchin eggs, all three types of Ca^{2+} stores are uniformly distributed. This is shown by preloading the eggs with caged analogs of each of the three Ca^{2+} messengers. In each case, subsequent whole-field photolysis produces uniform Ca^{2+} release throughout the eggs [104, 105]. Spatial segregation, however, can be artificially imposed by stratification, which segregates the NAADP-sensitive Ca^{2+} stores to a pole opposite to that of the cADPR- and IP_3- stores [95].

Natural segregation of stores is found in starfish oocytes. Uniform photolysis of caged NAADP preloaded into the oocytes by microinjection produces highly non-uniform Ca^{2+} mobilization, predominantly in the cortical region of the oocytes [103, 106]. Similar experiments with caged cADPR show also non-uniform Ca^{2+} mobilization, starting at a focal point in the cortex but eventually spreading throughout the entire cell [106]. In *Ascidian* oocytes, both the cADPR- and the NAADP- sensitive Ca^{2+} stores are exclusively localized in the cell cortex and when mobilized, induce rapid modulations of both membrane capacitance and Ca^{2+} currents, but produce no signal detectable by a Ca^{2+} indicator preloaded into the cytoplasm [41].

Mammalian glandular cells are highly polarized. Secretion occurs in the apical side facing the lumen of the gland, while the hormone receptors that regulate the secretion are located in the basolateral side. Using the same photolysis technique as described in sea urchin eggs, it is shown in pancreatic acinar cells that photolyzing caged IP_3 induces Ca^{2+} mobilization mainly in the apical side of the cells, while UV-regeneration of cADPR from its caged analog elevates Ca^{2+} in the basolateral side [107]. Consistently, type-2 ryanodine receptors, the type that is responsive to cADPR [82], are localized primarily in the basolateral side of the cells, while the IP_3-recptor is found only in the extreme apex of the cells [108].

The same conclusion is reached using a different method to monitor changes in the Ca^{2+} concentration of the stores directly by loading a fluorescent Ca^{2+} indicator into the lumen of the stores. The acinar cells are subsequently permeabilized. Addition of cADPR or NAADP mobilizes only the basolateral Ca^{2+} stores [109]. IP_3, on the other hand, releases Ca^{2+} from most stores throughout the cells. The latter is in variance to that described above, which could be due to the different species used or because of an alteration introduced by cell permeabilization.

Similar regional segregation is seen in parotid gland cells. Ryanodine receptors visualized by the fluorescent analog, Bodipy-ryanodine, are found mainly in the basolateral side. The Bodipy-ryanodine binding is specifically displaced by cADPR, indicating the co-localization of the cADPR-stores with the ryanodine receptor [85, 110]. Also, Ca^{2+} mobilization by cADPR is shown to be blocked by ruthenium red, an inhibitor of the ryanodine channels (cf. Table 1), and 8-Br-cADPR. The IP_3-receptors, on the other hand, like in the pancreatic acinar cells, are located mainly in the apical part of the parotid cell [85, 110].

Spatial segregation of Ca^{2+} stores is, however, not obligatory, as is found in sea urchin eggs described above. Similarly, in sheep lens cells, which are non-polarized cells, the cADPR- and IP_3 -sensitive stores are functionally overlapped and appear to be uniformly distributed throughout the cell as well [96].

Ca^{2+} MOBILIZATION AND INFLUX

Ca^{2+} influx controls a variety of important functions. In excitable cells, such as neurons and muscles, it is regulated by membrane depolarization. Mobilization of internal Ca^{2+} stores can likewise activate influx. One way is via the consequential increases in intracellular Ca^{2+}, which can directly modulate plasma membrane Ca^{2+} channels. This is the case in *Ascidian* oocytes. Infusion of cADPR or NAADP through a patched pipette mobilizes the cortical Ca^{2+} stores adjacent to the plasma membrane and effectively induces closing of the membrane Ca^{2+} channels [41]. Preloading the oocytes with a Ca^{2+} buffer prevents the Ca^{2+} conductance changes, indicating the increase in Ca^{2+} is directly responsible. This direct modulation of Ca^{2+} channels by Ca^{2+} can be remarkably store-specific because of the hierarchical organization of the Ca^{2+} stores in the cell. In the *Ascidian* oocyte, all the cADPR- and NAADP- stores are localized in the cortex. The abundance of Ca^{2+} sequestration mechanisms in these stores effectively limits Ca^{2+} diffusion between the cortex and cytoplasm. Hence, cADPR and NAADP produce no detectable Ca^{2+} signal in the cytoplasm. Conversely, mobilization of the cytoplasmic stores that are IP_3-sensitive, although effective in elevating cytosolic Ca^{2+}, is ineffective in modulating the plasma membrane conductance [41].

Another way that Ca^{2+} mobilization can modulate influx is through the capacitative entry mechanism [111], which is activated when the Ca^{2+} stores are depleted and occurs even when Ca^{2+} changes are completely buffered. A specific type of Ca^{2+} influx channel called the I_{CRAC} channel is activated by this mechanism (reviewed in [112]). It has long been known that mobilization of the IP_3-sensitive Ca^{2+} stores can activate capacitative entry in a variety of cells. That this mechanism likewise can be activated by depleting the cADPR-sensitive stores has now been shown in B-lymphocytes [113]. Gene ablation techniques are used to knockout all three types of IP_3-receptor in these cells, which renders their Ca^{2+} stores insensitive to IP_3 [114]. Passive store depletion with EGTA activates the I_{CRAC} channels after a delay. The antagonist of cADPR, 8-amino-cADPR, greatly inhibits I_{CRAC} activation, indicating that the depletion of the cADPR-sensitive stores is responsible for the activation. Conversely, infusion of cADPR, but not IP_3, facilitates I_{CRAC} activation and reduces the delay [113]. Inhibitors of the ryanodine receptor, ruthenium red and ryanodine, likewise, inhibit the capacitative entry, consistent with the idea that the target of cADPR is the ryanodine receptor.

Ca^{2+} influx is also activated by cADPR in T-lymphocytes. The Ca^{2+} stores in permeabilized T-lymphocytes are sensitive to cADPR [115]. In intact cells and in the absence of external Ca^{2+}, microinjection of cADPR produces only a small increase in Ca^{2+} due to mobilization [54]. In the presence of external Ca^{2+}, the cADPR-induced Ca^{2+} increase is greatly amplified, generating a series of Ca^{2+} oscillations as well. Imaging shows that the Ca^{2+} increase starts first at the periphery of the cells, consistent with activation of Ca^{2+} influx, and then propagates inward toward the center [54].

Similar results are seen in starfish oocytes. Photolysis of caged NAADP preloaded into the oocytes activates a localized Ca^{2+} increase that is confined to the cell cortex. The NAADP-induced cortical flash is depressed by removal of external Ca^{2+}, suggesting that mobilization of the cortical NAADP-sensitive stores is responsible for activation of the influx [103]. Whether this is due to capacitative entry or direct modulation of membrane Ca^{2-} channels remains to be determined.

Similar results are also seen in neutrophils. Permeabilized cells are responsive to cADPR, which releases Ca^{2+} from the internal stores [17]. Treatment of intact cells with fMLP, a chemotactic peptide, in the absence of external Ca^{2+}, elicits a rapid increase in internal Ca^{2+}, indicating mobilization. That the Ca^{2+} comes, at least in part, from the cADPR-sensitive stores is shown by partial suppression of the Ca^{2+} change by 8-Br-cADPR, a permeant cADPR antagonist [22]. In the presence of Ca^{2+}, fMLP induces additional Ca^{2+} influx that commences immediately after the initial mobilization [17]. The subsequent influx is specifically blocked by 8-Br-cADPR. Interestingly, another chemotactic agonist, IL-8, although capable of inducing Ca^{2+} mobilization, activates no influx. Also, the IL-8-induced mobilization is insensitive to inhibition by 8-Br-cADPR [17], indicating that Ca^{2+} influx is linked specifically to the mobilization of the cADPR-sensitive stores.

In addition to Ca^{2+} influx, another type of novel transport is mediated by surface CD38. During catalysis, CD38 can transport its enzymatic product, cADPR, into cells. This mechanism allows external NAD to elicit intracellular Ca^{2+} mobilization as has been observed in several cell types [116]. This novel influx pathway is described in the chapter by De Flora *et al.* in this book.

Ca^{2+} MOBILIZATION AND CELL FUNCTIONS

Singular Ca^{2+} stores and specialization–Cells generally possess multiple types of Ca^{2-} stores. There are, however, exceptions. In the intestine, the longitudinal muscles possess only the cADPR- but not the IP_3- sensitive stores. The opposite is true for the circular muscles [117]. This unusual and complete segregation of the two Ca^{2+}-release mechanisms in two different muscle types in the intestine is likely to be important in coordinating the peristaltic contraction executed routinely by these muscles.

Similar specialization is seen in normal pancreatic β-cells, which are found to possess only the cADPR-mechanism but not the IP_3-mechanism [118]. Internal Ca^{2+} stores isolated as microsomes are responsive to cADPR but not to IP_3 [29, 118]. These cells are sensitive to glucose, which activates CD38-mediated synthesis of cADPR, resulting in Ca^{2+} mobilization and increased secretion of insulin [29]. In contrast, β-cells from diabetic mice (*ob/ob*) produce no increase in insulin secretion when treated with glucose [49]. Microsomes isolated from them show enhanced responsiveness to IP_3 but not to cADPR [119]. These diabetic cells also lack CD38 and possess much reduced levels of the two types of ryanodine receptors that are responsive to cADPR [49]. This dichotomy between cADPR- and IP_3-responsiveness extends to cell lines derived from β-cells as well. RINm5F cells are not sensitive to cADPR and show no response to glucose [49, 119]. The glucose- responsive MIN6 cells, on the other hand, are sensitive to cADPR [97, 120]. It thus appears that there is a close correlation between the lack of response to cADPR and the insensitivity of the cells to glucose [49]. These results suggest that the specialization of Ca^{2+} stores is causally related to the physiological state of the cell. This state-dependence appears to be true also in human fibroblasts. There, sensitivity to cADPR is age-dependent, with enhanced sensitivity to cADPR seen in senescent cells as compared to younger cells that are within the first half of their mean population doubling [121].

Multiple Ca^{2+} stores and specificity–In contrast to store specialization described above, all three types of stores are generally present in a variety of animal and plant cells. In *Ascidian* oocytes, each type of store is found to subserve a separate cell function [41]. Fertilization of the oocytes is accompanied with a rapid decrease in membrane Ca^{2+} conductance, increase in membrane capacitance and activation of cytoplasmic Ca^{2+} oscillations. Both membrane events are activated by mobilizing the cADPR-sensitive Ca^{2+} stores [41]. In contrast, NAADP can trigger only the conductance decrease but not the capacitance increase, which is due to fusion of the cortical vesicles with the plasma membrane. It is likely that the cortical vesicles are cADPR-sensitive stores themselves and Ca^{2+} released from them is directly responsible for triggering membrane fusion (cf. Fig. 1). Mobilization of the IP_3-sensitive stores that are distributed in the cytosol is highly effective in producing the cytoplasmic Ca^{2+} changes, but not the two membrane events [41]. This is a clear example that each type of Ca^{2+} store can be directly and specifically linked to a particular cell function.

The subservience of one type of store to a particular function is also seen in neutrophils. The mobilization of the cADPR-sensitive stores is found to link specifically to chemotaxis [17]. The specificity is even more exquisite, since not all types of chemotaxis are linked to the cADPR-stores, only the type that is activated by the bacterial peptide, fMLP. Both fMLP and IL-8 can activate Ca^{2+} mobilization and chemotaxis, but only fMLP can induce capacitative Ca^{2+} entry induced by mobilizing the cADPR-sensitive stores. Ablation of CD38 or blockage by a cADPR antagonist inhibits the influx and eliminates the chemotactic response of neutrophils to fMLP [17], indicating the fMLP-receptor is specifically coupled to mobilization of the cADPR-sensitive stores.

Specific coupling of one receptor to a particular type of Ca^{2+} store is also seen in pancreatic acinar cells. Three different ligands, bombesin, cholecystokinin and acetylcholine, can activate Ca^{2+} mobilization in these cells. Dissection of receptor coupling is most easily done pharmacologically. Thus, 8-amino-cADPR and ryanodine are shown to block specifically the Ca^{2+} mobilization induced by bombesin, but not that induced by acetylcholine, indicating the bombesin receptor is coupled to the cADPR-sensitive stores [122]. The bombesin-receptor is not coupled to the NAADP-sensitive stores since desensitization of the NAADP-receptor has no effect on the Ca^{2+} mobilzation [123]. The cholecystokinin-receptor, on the other hand, is linked to both the NAADP- and cADPR- sensitive stores, since desensitization of the NAADP-receptor and 8-amino-cADPR both block the Ca^{2+} mobilization induced by cholecystokinin [65, 123].

Similar results are obtained in lacrimal acinar cells. Ryanodine blocks specifically the Ca^{2+} increases induced by phenylephrine but not that induced by acetylcholine [124]. Using permeabilized acinar cells, it is shown that only the cADPR-induced, but not the IP_3-induced Ca^{2+} release, is blocked by ryanodine, indicating the phenylephrine- and acetylcholine- receptors are respectively and specifically coupled to cADPR- and IP_3- sensitive stores. This store specificity extends to plant cells, where gene expression induced by the stress hormone, abscisic acid, is found to be inhibitable by 8-amino-cADPR, a cADPR antagonist, but not by heparin, a IP_3-receptor antagonist [27]. Ca^{2+} mobilization is thus a highly specific signaling mechanism even when multiple stores are present in the same cell.

Multiple Ca^{2+} stores and redundancy–The presence of multiple Ca^{2+} stores in cells can serve another important function, which is to ensure that critical Ca^{2+} signaling events will not be aborted due to unexpected errors. This is the case during fertilization, where the associated Ca^{2+} wave signals the onset of embryonic development. Without this Ca^{2+} trigger, the egg would remain dormant and eventually die instead of developing to a viable individual and achieve, in essence, immortality. In sea urchin egg, the Ca^{2+} wave that occurs during fertilization cannot be blocked by inhibiting either the cADPR- or the IP_3- sensitive mechanisms separately but requires simultaneous blockage of both [125, 126]. The employment of two separate receptors for mediating the fertilization wave provides the necessary redundancy for ensuring its successful completion.

Receptor redundancy is in fact crucial for survival. Given the dominant role of the IP_3-receptor in regulating virtually all cellular functions, it would be expected that cells lacking the receptor cannot survive. This is not the case, as shown in mice with the IP_3-receptor ablated [127]. Heterozygote mice are normal and can produce homozygote embryos lacking the IP_3-

receptor. Although most of them die in *utero* some survive till weaning. Cerebellar microsomes prepared from these mice show no response to IP_3, indicating they indeed lack functioning IP_3-sensitive Ca^{2+} stores [127]. The fact that homozygote embryos are produced indicates not only that eggs lacking functioning IP_3-stores can be fertilized but also the resulting homozygotes can, more or less, develop normally to full term fetuses.

That the cADPR-sensitive stores can indeed provide backup functions is shown in a B-lymphocyte cell line. All three types of IP_3-receptor are ablated in these cells and yet they not only survive but also can grow and proliferate [114]. They possess ryanodine-sensitive Ca^{2+} stores and depletion of these stores by cADPR can activate the Ca^{2+} entry channels [113] similar to that operated by the IP_3-sensitive stores in wildtype lymphocytes.

It is, perhaps, intuitively clear that cells must have multiple stores to provide the necessary redundancy, considering the critical importance of Ca^{2+} mobilization as a signaling mechanism.

INTERACTING Ca^{2+} STORES

The separate Ca^{2+} stores in cells do not necessarily function in isolation but can, instead, interact to produce different modes of signaling, including Ca^{2+} oscillations and waves.

In sea urchin eggs, NAADP, produced by photolyzing its caged analog preloaded into the cell, elicits a long-term Ca^{2+} oscillation [92, 105]. Removal of external Ca^{2+} does not block the oscillation, indicating it is due to mobilization of internal stores [105]. It is proposed that the cADPR-sensitive stores sequester the Ca^{2+} released from the NAADP-sensitive stores. The resultant Ca^{2+} overload sensitizes the cADPR-release mechanism and triggers spontaneous release (cf. Fig. 1) [6, 105]. This model is supported by the finding that overloading Ca^{2+} stores by repeated addition of Ca^{2+} can indeed trigger spontaneous release in egg homogenates [38]. Furthermore, blockage of the cADPR-mechanism in live eggs by 8-Br-cADPR inhibits the oscillation [61]. Complete elimination of the oscillation requires, however, blockage of both the cADPR- and IP_3- receptors, indicating both Ca^{2+} stores are responsible. The Ca^{2+} contents in these stores are assessed by discharging them with thapsigargin and measuring the resultant Ca^{2+} changes. The NAADP-sensitive stores are not discharged by thapsigargin, which is selective for the cADPR- and IP_3- sensitive stores [102]. It is found that following the NAADP-induced release, the Ca^{2+} contents of the cADPR- and IP_3- sensitive stores indeed increase [61]. This is a clear example of how the interaction between Ca^{2+} stores can lead to Ca^{2+} oscillation, a common mode of signaling seen in many cells.

Another common mode of Ca^{2+} signaling is in the form of a wave. In pancreatic acinar cells, the IP_3- and cADPR- sensitive stores are segregated, respectively, to the apical and basolateral sides [107, 109]. Acetylcholine elicits a Ca^{2+} wave starting in the apical side and propagates toward the basolateral side. The propagation of the wave can be slowed or even eliminated by 8-amino-cADPR [107]. It is proposed that acetylcholine activates production of IP_3 in the basolateral side, where the hormone receptors are located, diffuses to the apical side and triggers release from the IP_3-sensitive stores. Through amplification by the Ca^{2+}-induced Ca^{2+} release mechanism of the cADPR-sensitive stores, the Ca^{2+} signals are propagated back to the basolateral side as a wave [107]. In contrast to acetylcholine, cholecystokinin, on the other hand, acts by activating the NAADP-sensitive mechanism first, which is amplified and propagated by the cADPR- and IP_3-mechanisms via Ca^{2+}-induced Ca^{2+} release (CICR, cf. Fig. 1). Thus, either desensitizing the NAADP-receptor or preloading with 8-amino-cADPR, blocks Ca^{2+} signaling activated by cholecystokinin [65] but not that activated by acetylcholine [122]. Therefore, a specific hormone can be coupled to a specific store. The initial triggering signal is then amplified and propagated as a Ca^{2+} wave through interaction with other stores, demonstrating exquisite specificity and versatility of Ca^{2+} mobilization as a signaling mechanism.

Ca^{2+} waves are generated not only in polarized cells but also in cells where Ca^{2+} stores are distributed uniformly, as in sea urchin eggs. Whole-field photolysis of caged NAADP in the eggs produces uniform Ca^{2+} mobilization, indicating the stores are evenly distributed [92]. In contrast, local photolysis produces a Ca^{2+} wave that spreads across the entire egg [89, 128]. Analyses indicate the wave is due to a diffusion gradient of NAADP [128].

A novel possibility that this diffusion wave of NAADP can serve as spatiotemporal Ca^{2+} memory has recently been demonstrated [129]. The NAADP-receptor exhibits remarkable desensitization to even sub-threshold concentrations of NAADP and would not respond to subsequent challenge with even saturating concentrations [92, 130]. A limited diffusion gradient of NAADP can be generated in live sea urchin eggs by controlled and localized photolysis. The NAADP-stores within this spatial gradient are rendered desensitized. Subsequent whole-field photolysis can only mobilize the Ca^{2+}-stores that are outside the diffusion gradient [129]. In this manner, the spatial pattern of the first diffusion gradient is memorized by the cell. This mechanism could potentially provide positional clues in tissues. Thus, local production of sufficiently high concentrations of NAADP not only could desensitize the producing cells but also could diffuse through gap junctions to adjacent cells, forming a spatial gradient. The cells within the diffusion gradient would not produce a further Ca^{2+} response to subsequent challenge by NAADP. This positional information would be preserved as long as the NAADP-receptors in the cells remain desensitized. That Ca^{2+} messengers, such as cADPR and IP_3, can indeed diffuse through gap junctions has been demonstrated [131]. Receptor recovery could be brought about by degradation of NAADP, which is sensitive to hydrolases such as alkaline phosphatase and nucleotide pyrophosphatase [6, 105]. In this manner, positional information could be repeatedly written and erased, providing a remarkable encoding system.

A EMERGING THEME

That cells possess at least three different types of Ca^{2+} stores is now well established. A different and specific Ca^{2+} messenger mobilizes each of these stores. Specialized cells containing singular stores best suited for their functions are likely to be exceptions. More generally, cells possess multiple Ca^{2+} stores, which in some cases, can be segregated. Spatial specialization allows local Ca^{2+} changes to regulate specific functions and relieves the cell from the burden of global activation every time a local change is required, such as modulating an ion channel in the plasma membrane. Since each type of store is mobilized by a specific messenger, different stimuli can be coupled to separate stores. This specificity, together with spatial segregation of stores, allows a stimulus to activate specific local changes. As seen in pancreatic acinar cells, where Ca^{2+} mobilization can be activated in a pole different from the location of the hormone receptors. Spatial signaling is possible even in cells where the stores are evenly distributed, as in sea urchin eggs. Focal production of a Ca^{2+} messenger generates a concentration gradient, resulting in a Ca^{2+} wave that emanates from the focal point. In the case of cADPR, the wave can be regenerative due to Ca^{2+}-induced Ca^{2+} release. In contrast, a diffusive wave results from focal production of NAADP. Through receptor desensitization, this diffusive wave of NAADP can serve as Ca^{2+} memory, encoding spatial-temporal information. Specific interaction between different types of Ca^{2+} stores, on the other hand, can produce Ca^{2+} oscillations, a common mode of signaling observed in a wide variety of cells. It is thus clear that Ca^{2+} mobilization as a signaling mechanism is intricate and complex, providing sufficient versatility and specificity for cells to respond to a myriad of stimuli impinging on them constantly.

REFERENCES

1. Streb H, Irvine RF, Berridge MJ and Schulz I. 1983. Release of Ca^{2+} from a nonmitochondrial intracellular store in pancreatic acinar cells by inositol-1,4,5-trisphosphate. *Nature* 306: 67-69.
2. Berridge MJ. 1983. Rapid accumulation of inositol trisphosphate reveals that agonists hydrolyse polyphosphoinositides instead of phosphatidylinositol. *Biochem. J.* 212: 849-858.
3. Clapper DL, Walseth TF, Dargie PJ and Lee HC. 1987. Pyridine nucleotide metabolites stimulate calcium release from sea urchin egg microsomes desensitized to inositol trisphosphate. *J. Biol. Chem.* 262: 9561-9568.

4. Lee HC, Walseth TF, Bratt GT, Hayes RN and Clapper DL. 1989. Structural determination of a cyclic metabolite of NAD^+ with intracellular Ca^{2+}-mobilizing activity. *J. Biol. Chem.* 264: 1608-1615.
5. Lee HC and Aarhus R. 1995. A derivative of NADP mobilizes calcium stores insensitive to inositol trisphosphate and cyclic ADP-ribose. *J. Biol. Chem.* 270: 2152-2157.
6. Lee HC. 1997. Mechanisms of calcium signaling by cyclic ADP-ribose and NAADP. *Physiol. Rev.* 77: 1133-1164.
7. Lee HC. 2001. Physiological functions of cyclic ADP-ribose and NAADP as calcium messengers. *Ann. Rev. Pharmacol. Toxicol.* 41: 317-345.
8. Zocchi E, Carpaneto A, Cerrano C, Bavestrello G, Giovine M, Bruzzone S, Guida L, Franco L and Usai C. 2002. The temperature-signaling cascade in sponges involves a heat-gated cation channel, abscisic acid, and cyclic ADP-ribose. *Proc. Natl. Acad. Sci. USA* 98: 14859-14864.
9. Masuda W, Takenaka S, Inageda K, Nishina H, Takahashi K, Katada T, Tsuyama S, Inui H, Miyatake K and Nakano Y. 1997. Oscillation of ADP-ribosyl cyclase activity during the cell cycle and function of cyclic ADP-ribose in a unicellular organism, *Euglena Gracilis*. *FEBS Lett.* 405: 104-106.
10. Zocchi E, Daga A, Usai C, Franco L, Guida L, Bruzzone S, Costa A, Marchetti C and Deflora A. 1998. Expression of CD38 increases intracellular calcium concentration and reduces doubling time in HeLa and 3T3 cells. *J. Biol. Chem.* 273: 8017-8024.
11. Franco L, Zocchi E, Usai C, Guida L, Bruzzone S, Costa A and De Flora A. 2001. Paracrine roles of NAD^+ and cyclic ADP-ribose in increasing intracellular calcium and enhancing cell proliferation of 3T3 fibroblasts. *J. Biol. Chem.* 276: 21642-21648.
12. Dargie PJ, Agre MC and Lee HC. 1990. Comparison of Ca^{2+} mobilizing activities of cyclic ADP-ribose and inositol trisphosphate. *Cell Regul.* 1: 279-290.
13. Kuroda R, Kontani K, Kanda Y, Katada T, Nakano T, Satoh Y-I, Suzuki N and Kuroda H. 2001. Increase of cGMP, cADP-ribose and inositol 1,4,5-trisphosphate preceding Ca^{2+} transients in fertilization of sea urchin eggs. *Development* 128: 4405-4414.
14. Lee HC. 1996. Cyclic ADP-ribose and calcium signaling in eggs. *Biol. Signals* 5: 101-110.
15. Okamoto H. 1999. The CD38-cyclic ADP-ribose signaling system in insulin secretion. *Mol. Cell. Biochem.* 193: 115-118.
16. Iino S, Cui Y, Galione A and Terrar DA. 1997. Actions of cADP-ribose and its antagonists on contraction in guinea pig isolated ventricular myocytes - Influence of temperature. *Circ. Res.* 81: 879-884.
17. Partida-Sanchez S, Cockayne D, Monard S, Jacobson EL, Oppenheimer N, Garvy B, Kusser K, Goodricj S, Howard M, Harmsen A, Randall T and Lund FE. 2001. Cyclic ADP-ribose production by CD38 regulates intracellular calcium release, extracellular calcium influx and chemotaxis in neutrophils and is required for bacterial clearance in vivo. *Nature Med.* 7: 1209-1216.
18. Reyes-Harde M, Empson R, Potter BVL, Galione A and Stanton PK. 1999. Evidence of a role for cyclic ADP-ribose in long-term synaptic depression in hippocampus. *Proc. Natl. Acad. Sci. USA* 96: 4061-4066.
19. Reyes-Harde M, Potter BVL, Galione A and Stanton PK. 1999. Induction of hippocampal LTD requires nitric-oxide-stimulated PKG activity and Ca^{2+} release from cyclic ADP-ribose-sensitive stores. *J. Neurophysiol.* 82: 1569-1576.
20. Podesta M, Zocchi E, Pitto A, Usai C, Franco L, Bruzzone S, Guida L, Bacigalupo A, Scadden DT, Walseth TF, De Flora A and Daga A. 2000. Extracellular cyclic ADP-ribose increases intracellular free calcium concentration and stimulates proliferation of human hemopoietic progenitors. *FASEB J.* 14: 680-690.
21. Zocchi E, Podesta M, Pitto A, Usai C, Bruzzone S, Franco L, Guida L, Bacigalupo A and De Flora A. 2001. Paracrinally stimulated expansion of early human hemopoietic progenitors by stroma-generated cyclic ADP-ribose. *FASEB J.* 15: 1610-1612.

22. Walseth TF and Lee HC. 1993. Synthesis and characterization of antagonists of cyclic-ADP-ribose-induced Ca^{2+} release. *Biochim. Biophys. Acta* 1178: 235-242.
23. Sethi JK, Empson RM, Bailey VC, Potter BVL and Galione A. 1997. 7-Deaza-8-bromo-cyclic ADP-ribose, the first membrane-permeant, hydrolysis-resistant cyclic ADP-ribose antagonist. *J. Biol. Chem.* 272: 16358-16363.
24. Deflora A, Guida L, Franco L, Zocchi E, Pestarino M, Usai C, Marchetti C, Fedele E, Fontana G and Raiteri M. 1996. Ectocellular in vitro and in vivo metabolism of cADP-ribose in cerebellum. *Biochem. J.* 320: 665-671.
25. Franco L, Bruzzone S, Song P, Guida L, Zocchi E, Walseth TF, Crimi E, Usai C, De Flora A and Brusasco V. 2001. Extracellular cyclic ADP-ribose potentiates ACh-induced contraction in bovine tracheal smooth muscle. *Am. J. Physiol.* 280: L98-L106.
26. Fukushi Y, Kato I, Takasawa S, Sasaki T, Ong BH, Sato M, Ohsaga A, Sato K, Shirato K, Okamoto H and Maruyama Y. 2001. Identification of cyclic ADP-ribose-dependent mechanisms in pancreatic muscarinic Ca^{2+} signaling using CD38 knockout mice. *J. Biol. Chem.* 276: 649-655.
27. Wu Y, Kuzma J, Marechal E, Graeff R, Lee HC, Foster R and Chua NH. 1997. Abscisic acid signaling through cyclic ADP-ribose in plants. *Science* 278: 2126-2130.
28. Guse AH, da Silva CP, Berg I, Skapenko AL, Weber K, Heyer P, Hohenegger M, Ashamu GA, Schulze-Koops H, Potter BVL and Mayr GW. 1999. Regulation of calcium signalling in T lymphocytes by the second messenger cyclic ADP-ribose. *Nature* 398: 70-73.
29. Kato I, Yamamoto Y, Fujimura M, Noguchi N, Takasawa S and Okamoto H. 1998. CD38 disruption impairs glucose-induced increases in cyclic ADP-ribose, $[Ca^{2+}]_i$ and insulin secretion. *J. Biol. Chem.* 274: 1869-1872.
30. Takahashi K, Kukimoto I, Tokita K, Inageda K, Inoue S, Kontani K, Hoshino S, Nishina H, Kanaho Y and Katada T. 1995. Accumulation of cyclic ADP-ribose measured by a specific radioimmunoassay in differentiated human leukemic HL-60 cells with all-trans-retinoic acid. *FEBS Lett.* 371: 204-208.
31. Polzonetti V, Cardinali M, Mosconi G, Natalini P, Meiri I and Carnevali O. 2002. Cyclic ADPR and calcium signaling in sea bream (Sparus aurata) egg fertilization. *Mol. Reprod. Dev.* 61: 213-217.
32. Masuda W, Takenaka S, Tsuyama S, Tokunaga M, Yamaji R, Inui H, Miyatake K and Nakano Y. 1997. Inositol 1,4,5-trisphosphate and cyclic ADP-ribose mobilize Ca^{2+} in a protist, *Euglena Gracilis*. *Comp. Biochem. Physiol.* 118: 279-283.
33. Allen GJ, Muir SR and Sanders D. 1995. Release of Ca^{2+} from individual plant vacuoles by both InsP3 and cyclic ADP-ribose. *Science* 268: 735-737.
34. Leckie CP, McAinsh MR, Allen GJ, Sanders D and Hetherington AM. 1998. Abscisic acid-induced stomatal closure mediated by cyclic ADP-ribose. *Proc. Natl. Acad. Sci. USA* 95: 15837-15842.
35. Navazio L, Mariani P and Sanders D. 2001. Mobilization of Ca^{2+} by cyclic ADP-ribose from the endoplasmic reticulum of cauliflower florets. *Plant Physiol.* 125: 2129-2138.
36. Durner J, Wendehenne D and Klessig DF. 1998. Defense gene induction in tobacco by nitric oxide, cyclic GMP, and cyclic ADP-ribose. *Proc. Natl. Acad. Sci. USA* 95: 10328-10333.
37. Rusinko N and Lee HC. 1989. Widespread occurrence in animal tissues of an enzyme catalyzing the conversion of NAD^+ into a cyclic metabolite with intracellular Ca^{2+}-mobilizing activity. *J. Biol. Chem.* 264: 11725-11731.
38. Galione A, Lee HC and Busa WB. 1991. Ca^{2+}-induced Ca^{2+} release in sea urchin egg homogenates: modulation by cyclic ADP-ribose. *Science* 253: 1143-1146.
39. Santella L and Kyozuka K. 1997. Effects of 1-methyladenine on nuclear Ca^{2+} transients and meiosis resumption in starfish oocytes are mimicked by the nuclear injection of inositol 1,4,5-trisphosphate and cADP-ribose. *Cell Calium* 22: 11-20.

40. Mothet JP, Fossier P, Meunier FM, Stinnakre J, Tauc L and Baux G. 1998. Cyclic ADP-ribose and calcium-induced calcium release regulate neurotransmitter release at a cholinergic synapse of Aplysia. *J. Physiol.* 507.2: 405-414.
41. Albrieux M, Lee HC and Villaz M. 1998. Calcium signaling by cyclic ADP-ribose, NAADP, and inositol trisphosphate are involved in distinct functions in *Ascidian* oocytes. *J. Biol. Chem.* 273: 14566-14574.
42. Messutat S, Heine M and Wicher D. 2001. Calcium-induced calcium release in neurosecretory insect neurons:fast and slow responses. *Cell Calcium* 30: 199-211.
43. Fluck R, Abraham V, Miller A and Galione A. 1999. Microinjection of cyclic ADP-ribose triggers a regenerative wave of Ca^{2+} release and exocytosis of cortical alveoli in medaka eggs. *Zygote* 7: 285-292.
44. Hua SY, Tokimasa T, Takasawa S, Furuya Y, Nohmi M, Okamoto H and Kuba K. 1994. Cyclic ADP-ribose modulates Ca^{2+} release channels for activation by physiological Ca^{2+} entry in bullfrog sympathetic neurons. *Neuron* 12: 1073-1079.
45. Brailoiu E and Miyamoto D. 2000. Inositol trisphosphate and cyclic adenosine diphosphate-ribose increase quantal transmitter release at frog motor nerve terminals: Possible involvement of smooth endoplasmic reticulum. *Neuroscience* 95: 927-931.
46. Graeff R and Lee HC. 2002. A novel cycling assay for cellular cyclic ADP-ribose with nanomolar sensitivity. *Biochem. J.* 361: 379-384.
47. Khoo KM, Han M-K, Park JB, Chae SW, Kim U-H, Lee HC, Bay BH and Chang CF. 2000. Localization of the cyclic ADP-ribose-dependent calcium signaling pathway in hepatocyte nucleus. *J. Biol. Chem.* 275: 24807-24817.
48. Yusufi ANK, Cheng J, Thompson MA, Dousa TP, Warner GM, Walker HJ and Grande JP. 2001. cADP-ribose/ryanodine channel/Ca^{2+}-release signal transduction pathway in mesangial cells. *Am. J. Physiol.* 281: F91-F102.
49. Takasawa S, Akiyama T, Nata K, Kuroki M, Tohgo A, Noguchi N, Kobayashi S, Kato I, Katada T, Okamoto H, Takasawa S, Akiyama T, Nata K, Kuroki M, Tohgo A, Noguchi N, Kobayashi S, Kato I, Katada T, *et al.* 1998. Cyclic ADP-ribose and inositol 1,4,5-trisphosphate as alternate second messengers for intracellular Ca^{2+} mobilization in normal and diabetic beta-cells. *J. Biol. Chem.* 273: 2497-2500.
50. Li P-L, Tang W-X, Valdivia HH, Zou A-P and Campbell WB. 2001. cADP-ribose activates reconstituted ryanodine receptors from coronary arterial smooth muscle. *Am. J. Physiol.* 280: H208-H215.
51. Prakash YS, Kannan MS, Walseth TF and Sieck GC. 1998. Role of cyclic ADP-ribose in the regulation of $[Ca^{2+}]_i$ in porcine tracheal smooth muscle. *Am. J. Physiol.* 43: C1653-C1660.
52. Cancela JM and Petersen OH. 1998. The cyclic ADP ribose antagonist 8-NH2-cADP-ribose blocks cholecystokinin-evoked cytosolic Ca^{2+} spiking in pancreatic acinar cells. *Pflug. Arch.* 435: 746-748.
53. Guse AH, Dasilva CP, Emmrich F, Ashamu GA, Potter BVL and Mayr GW. 1995. Characterization of cyclic adenosine diphosphate-ribose-induced Ca^{2+} release in T lymphocyte cell lines. *J. Immunol.* 155: 3353-3359.
54. Guse AH, Berg I, Dasilva CP, Potter BVL and Mayr GW. 1997. Ca^{2+} entry induced by cyclic ADP-ribose in intact T-lymphocytes. *J. Biol. Chem.* 272: 8546-8550.
55. Galione A, White A, Willmott N, Turner M, Potter BV and Watson SP. 1993. cGMP mobilizes intracellular Ca^{2+} in sea urchin eggs by stimulating cyclic ADP-ribose synthesis. *Nature* 365: 456-459.
56. Drum CL, Yan SZ, Bard J, Shen YQ, Lu D, Soelaiman S, Grabarek Z, Bohm A and Tang WJ. 2002. Structural basis for the activation of anthrax adenylyl cyclase exotoxin by calmodulin. *Nature* 415: 396-402.
57. Madden JC, Ruiz N and Caparon M. 2001. Cytolysin-mediated translocation (CMT): A functional equivalent of type III secretion in Gram-positive bacteria. *Cell* 104: 143-152.

58. Karasawa T, Takasawa S, Yamakawa K, Yonekura H, Okamoto H and Nakamura S. 1995. NAD^+-glycohydrolase from Streptococcus pyogenes shows cyclic ADP-ribose forming activity. *FEMS Microbiol. Lett.* 130: 201-204.
59. Howard M, Grimaldi JC, Bazan JF, Lund FE, Santos-Argumedo L, Parkhouse RM, Walseth TF and Lee HC. 1993. Formation and hydrolysis of cyclic ADP-ribose catalyzed by lymphocyte antigen CD38. *Science* 262: 1056-1059.
60. Navazio L, Bewell MA, Siddiqua A, Dickinson GD, Galione A and Sanders D. 2000. Calcium release from the endoplasmic reticulum of higher plants elicited by the NADP metabolite nicotinic acid adenine dinucleotide phosphate. *Proc. Natl. Acad. Sci. USA* 97: 8693-8698.
61. Churchill GC and Galione A. 2001. NAADP induces Ca^{2+} oscillations via a two-pool mechanism by priming IP_3- and cADPR-sensitive Ca^{2+} stores. *EMBO J.* 20: 2666-2671.
62. Santella L, Kyozuka K, Genazzani AA, De Riso L and Carafoli E. 2000. Nicotinic acid adenine dinucleotide phosphate-induced Ca^{2+} release. Interactions among distinct Ca^{2+} mobilizing mechanisms in starfish oocytes. *J. Biol. Chem.* 275: 8301-8306.
63. Chameau P, Van De Vrede Y, Fossier P and Baux G. 2001. Ryanodine-, IP_3- and NAADP-dependent calcium stores control acetylcholine release. *Pflugers Arch.* 443: 289-296.
64. Brailoiu E, Miyamoto MD and Dun NJ. 2001. Nicotinic acid adenine dinucleotide phosphate enhances quantal neurosecretion at the frog neuromuscular junction: possible action on synaptic vesicles in the releasable pool. *Mol. Pharmacol.* 60: 718-724.
65. Cancela JM, Churchill GC and Galione A. 1999. Coordination of agonist-induced Ca^{2+}-signalling patterns by NAADP in pancreatic acinar cells. *Nature* 398: 74-76.
66. Bak J, White P, Timár G, Missiaen L, Genazzani AA and Galione A. 1999. Nicotinic acid adenine dinucleotide phosphate triggers Ca^{2+} release from brain microsomes. *Curr. Biol.* 9: 751–754.
67. Patel S, Churchill GC, Sharp T and Galione A. 2000. Widespread distribution of binding sites for the novel Ca^{2+}-mobilizing messenger, nicotinic acid adenine dinucleotide phosphate, in the brain. *J. Biol. Chem.* 275: 36495-36497.
68. Berg I, Potter BVL, Mayr GW and Guse AH. 2000. Nicotinic acid adenine dinucleotide phosphate ($NAADP^+$) is an essential regulator of T-lymphocyte Ca^{2+}-signaling. *J. Cell Biol.* 150: 581-588.
69. Bak J, Billington RA, Timar G, Dutton AC and Genazzani AA. 2001. NAADP receptors are present and functional in the heart. *Curr. Biol.* 11: 987-990.
70. Mojzisova A, Krizanova O, Zacikova L, Kominkova V and Ondrias K. 2001. Effect of nicotinic acid adenine dinucleotide phosphate on ryanodine calcium release channel in heart. *Eur. J. Physiol.* 441: 674-677.
71. Cheng J, Yusufi AN, Thompson MA, Chini EN and Grande JP. 2001. Nicotinic acid adenine dinucleotide phosphate: A new Ca^{2+} releasing agent in kidney. *J. Am. Soc. Nephrol.* 12: 54-60.
72. Yusufi AN, Cheng J, Thompson MA, Burnett JC and Grande JP. 2002. Differential mechanisms of Ca^{2+} release from vascular smooth muscle cell microsomes. *Exp. Biol. Med.* 227: 36-44.
73. Yusufi AN, Cheng J, Thompson MA, Chini EN and Grande JP. 2001. Nicotinic acid-adenine dinucleotide phosphate (NAADP) elicits specific microsomal Ca^{2+} release from mammalian cells. *Biochem. J.* 353: 531-536.
74. Graeff R and Lee HC. 2002. A novel cycling assay for NAADP with nanomolar sensitivity. *Biochem. J.* (in press).
75. Graeff RM, Podein RJ, Aarhus R and Lee HC. 1995. Magnesium ions but not ATP inhibit cyclic ADP-ribose-induced calcium release. *Biochem. Biophys. Res. Commun.* 206: 786-791.
76. Lee HC. 1993. Potentiation of calcium- and caffeine-induced calcium release by cyclic ADP-ribose. *J. Biol. Chem.* 268: 293-299.

77. Lee HC, Aarhus R, Graeff R, Gurnack ME and Walseth TF. 1994. Cyclic ADP ribose activation of the ryanodine receptor is mediated by calmodulin. *Nature* 370: 307-309.
78. Lee HC, Aarhus R and Graeff RM. 1995. Sensitization of calcium-induced calcium release by cyclic ADP-ribose and calmodulin. *J. Biol. Chem.* 270: 9060-9066.
79. Tanaka Y and Tashjian AH, Jr. 1995. Calmodulin is a selective mediator of Ca^{2+}-induced Ca^{2+} release via the ryanodine receptor-like Ca^{2+} channel triggered by cyclic ADP-ribose. *Proc. Natl. Acad. Sci. USA* 92: 3244-3248.
80. Perez CF, Marengo JJ, Bull R and Hidalgo C. 1998. Cyclic ADP-ribose activates caffeine-sensitive calcium channels from sea urchin egg microsomes. *Am. J. Physiol.* 274: C430-C439.
81. Lokuta AJ, Darszon A, Beltran C and Valdivia HH. 1998. Detection and functional characterization of ryanodine receptors from sea urchin eggs. *J. Physiol.* 510.1: 155-164.
82. Meszaros LG, Bak J and Chu A. 1993. Cyclic ADP-ribose as an endogenous regulator of the non-skeletal type ryanodine receptor Ca^{2+} channel. *Nature* 364: 76-79.
83. Sitsapesan R and Williams AJ. 1995. Cyclic ADP-ribose and related compounds activate sheep skeletal sarcoplasmic reticulum Ca^{2+} release channel. *Am. J. Physiol.* 268: C1235-C1240.
84. Sonnleitner A, Conti A, Bertocchini F, Schindler H and Sorrentino V. 1998. Functional properties of the ryanodine receptor type 3 (Ryr3) Ca^{2+} release channel. *EMBO J.* 17: 2790-2798.
85. Zhang X, Wen J, Bidasee KR, Besch Jr HR, Wojcikiewicz RJ, Lee B and Rubin RP. 1999. Ryanodine and inositol trisphosphate receptors are differentially distributed and expressed in rat parotid gland. *Biochem. J.* 340: 519-527.
86. Empson RM and Galione A. 1997. Cyclic ADP-ribose enhances coupling between voltage-gated Ca^{2+} entry and intracellular Ca^{2+} release. *J. Biol. Chem.* 272: 20967-20970.
87. Hashii M, Minabe Y and Higashida H. 2000. cADP-ribose potentiates cytosolic Ca2+ elevation and Ca^{2+} entry via L-type voltage-activated Ca^{2+} channels in NG108-15 neuronal cells. *Biochem. J.* 345: 207-215.
88. Lee HC. 1996. Modulator and messenger functions of cyclic ADP-ribose in calcium signaling. *Re. Prog. Horm. Res.* 51: 355-388.
89. Lee HC. 2000. NAADP: An emerging calcium signaling molecule. *J. Memb. Biol.* 173: 1-8.
90. Genazzani AA, Mezna M, Dickey DM, Michelangeli F, Walseth TF and Galione A. 1997. Pharmacological properties of the Ca^{2+}-release mechanism sensitive to NAADP in the sea urchin egg. *Brit. J. Pharm.* 121: 1489-1495.
91. Lee HC and Aarhus R. 1997. Structural determinants of nicotinic acid adenine dinucleotide phosphate important for its calcium-mobilizing activity. *J. Biol. Chem.* 272: 20378-20383.
92. Aarhus R, Dickey DM, Graeff RM, Gee KR, Walseth TF and Lee HC. 1996. Activation and inactivation of Ca^{2+} release by $NAADP^{+}$. *J. Biol. Chem.* 271: 8513-8516.
93. Aarhus R, Graeff RM, Dickey DM, Walseth TF and Lee HC. 1995. ADP-ribosyl cyclase and CD38 catalyze the synthesis of a calcium-mobilizing metabolite from NADP. *J. Biol. Chem.* 270: 30327-30333.
94. Lee HC. 1991. Specific binding of cyclic ADP-ribose to calcium-storing microsomes from sea urchin eggs. *J. Biol. Chem.* 266: 2276-2281.
95. Lee HC and Aarhus R. 2000. Functional visualization of the separate but interacting calcium stores sensitive to NAADP and cyclic ADP-ribose. *J. Cell Sci.* 113: 4413-4420.
96. Churchill GC and Louis CF. 1999. Imaging of intracellular calcium stores in single permeabilized lens cells. *Am. J. Physiol.* 276: C426-C434.
97. Mitchell KJ, Pinton P, Varadi A, Tacchetti C, Ainscow EK, Pozzan T, Rizzuto R and Rutter GA. 2001. Dense core secretory vesicles revealed as a dynamic Ca^{2+} store in neuroendocrine cells with a vesicle-associated membrane protein aequorin chimaera. *J. Cell Biol.* 155: 41-51.

98. Gerasimenko OV, Gerasimenko JV, Belan PV and Petersen OH. 1996. Inositol trisphosphate and cyclic ADP-ribose-mediated release of Ca^{2+} from single isolated pancreatic zymogen granules. *Cell* 84: 473-480.
99. Gerasimenko OV, Gerasimenko JV, Tepikin AV and Petersen OH. 1995. ATP-dependent accumulation and inositol trisphosphate- or cyclic ADP-ribose-mediated release of Ca^{2+} from the nuclear envelope. *Cell* 80: 439-444.
100. Adebanjo OA, Anandatheerthavarada HK, Koval AP, Moonga BS, Biswas G, Sun L, Sodam BR, Bevis PJR, Huang CLH, Epstein S, Lai FA, Avadhani NG and Zaidi M. 1999. A new function for CD38/ADP-ribosyl cyclase in nuclear Ca^{2+} homeostasis. *Nature Cell Biol.* 1: 409-414.
101. Santella L, Deriso L, Gragnaniello G and Kyozuka E. 1998. Separate activation of the cytoplasmic and nuclear calcium pools in maturing starfish oocytes. *Biochem. Biophys. Res. Commun.* 252: 1-4.
102. Genazzani AA and Galione A. 1996. Nicotinic acid-adenine dinucleotide phosphate mobilizes Ca^{2+} from a thapsigargin-insensitive pool. *Biochem. J.* 315: 721-725.
103. Lim D, Kyozuka K, Gragnaniello G, Carafoli E and Santella L. 2001. $NAADP^+$ initiates the Ca^{2+} response during fertilization of starfish oocytes. *FASEB J.* 15: 2257-2267.
104. Aarhus R, Gee K and Lee HC. 1995. Caged cyclic ADP-ribose - synthesis and use. *J. Biol. Chem.* 270: 7745-7749.
105. Lee HC, Aarhus R, Gee KR and Kestner T. 1997. Caged nicotinic acid adenine dinucleotide phosphate - Synthesis and use. *J. Biol. Chem.* 272: 4172-4178.
106. Nusco GA, Lim D, Sabala P and Santella L. 2002. Ca^{2+} response to cADPr during maturation and fertilization of starfish oocytes. *Biochem. Biophys. Res. Commun.* 290: 1015-1021.
107. Leite MF, Burgstahler AD and Nathanson MH. 2002. Ca^{2+} waves require sequential activation of inositol trisphosphate receptors and ryanodine receptors in pancreatic acini. *Gastroenterology* 122: 415-427.
108. Leite MF, Dranofff JA, Gao L and Nathanson MH. 1999. Expression and subcellular localization of the ryanodine receptor in rat pancreatic acinar cells. *Biochem. J.* 337: 305-309.
109. Krause E, Göbel A and Schulz I. 2002. Cell side-specific sensitivities of intracellular Ca^{2+} stores for inositol 1,4,5-trisphosphate,cyclic ADP-ribose and Nicotinic adenine dinucleotide phosphate in permeabilized pancreatic acinar cells from mouse. *J. Biol. Chem.* 11696-11702:
110. Zhang XJ, Wen JY, Bidasee KR, Besch HR and Rubin RP. 1997. Ryanodine receptor expression is associated with intracellular Ca^{2+} release in rat parotid acinar cells. *Am. J. Physiol.* 42: C1306-C1314.
111. Putney JW, Jr. 1986. A model for receptor-regulated calcium entry. *Cell Calcium* 7: 1-12.
112. Parekh AB and Penner R. 1997. Store depletion and calcium influx. *Physiol. Rev.* 77: 901-930.
113. Kiselyov K, Shin DM, Shcheynikov N, Kurosaki T and Muallem S. 2001. Regulation of Ca^{2+}-release-activated Ca^{2+} current (I_{CRAC}) by ryanodine receptors in inositol 1,4,5-trisphosphate-receptor-deficient DT40 cells. *Biochem. J.* 360: 17-22.
114. Sugawara H, Kurosaki M, Takata M and Kurosaki T. 1997. Genetic evidence for involvement of type 1, type 2 and type 3 inositol 1,4,5-trisphosphate receptors in signal transduction through the B-cell antigen receptor. *EMBO J.* 16: 3078-3088.
115. Guse AH, Dasilva CP, Weber K, Ashamu GA, Potter BVL and Mayr GW. 1996. Regulation of cADP-ribose-induced Ca^{2+} release by Mg^{2+} and inorganic phosphate. *J. Biol. Chem.* 271: 23946-23953.
116. De Flora A, Franco L, Guida L, Bruzzone S, Usai C and Zocchi E. 2000. Topology of CD38. *Chem. Immunol.* 75: 79-98.

117. Kuemmerle JF and Makhlouf GM. 1995. Agonist-stimulated cyclic ADP ribose. Endogenous modulator of Ca^{2+}-induced Ca^{2+} release in intestinal longitudinal muscle. *J. Biol. Chem.* 270: 25488-25494.
118. Takasawa S, Nata K, Yonekura H and Okamoto H. 1993. Cyclic ADP-ribose in insulin secretion from pancreatic beta cells. *Science* 259: 370-373.
119. Islam MS, Larsson O and Berggren PO. 1993. Cyclic ADP-ribose in beta cells. *Science* 262: 584-586.
120. Varadi A and Rutter GA. 2002. Dynamic imaging of endoplasmic reticulum Ca^{2+} concentration in insulin-secreting MIN6 cells using recombinant targeted cameleons: Roles of sarco(endo)plasmic reticulum Ca^{2+}-ATPase (SERCA)-2 and ryanodine receptors. *Diabetes* 51, Suppl 1: S190-S201.
121. Huang MS, Adebanjo O, Moonga BS, Goldstein S, Lai FA, Lipschitz DA and Zaidi M. 1998. Upregulation of functional ryanodine receptors during in vitro aging of human diploid fibroblasts. *Biochem. Biophys. Res. Commun.* 245: 50-52.
122. Burdakov D, Cancela JM and Petersen OH. 2001. Bombesin-induced cytosolic Ca^{2+} spiking in pancreatic acinar cells depends on cyclic ADP-ribose and ryanodine receptors. *Cell Calcium* 29: 211-216.
123. Burdakov D and Galione A. 2000. Two neuropeptides recruit different messenger pathways to evoke Ca^{2+} signals in the same cell. *Curr. Biol.* 10: 993-996.
124. Gromada J, Jorgensen TD and Dissing S. 1995. The release of intracellular Ca^{2+} in lacrimal acinar cells by alpha-, beta-adrenergic and muscarinic cholinergic stimulation: the roles of inositol triphosphate and cyclic ADP-ribose. *Pflug. Arch.* 429: 751-761.
125. Lee HC, Aarhus R and Walseth TF. 1993. Calcium mobilization by dual receptors during fertilization of sea urchin eggs. *Science* 261: 352-355.
126. Galione A, McDougall A, Busa WB, Willmott N, Gillot I and M. W. 1993. Redundant mechanisms of calcium-induced calcium release underlying calcium waves during fertilization of sea urchin eggs. *Science* 261: 348-352.
127. Matsumoto M, Nakagawa T, Inoue T, Nagata E, Tanaka K, Takano H, Minowa O, Kuno J, Sakakibara S, Yamada M, Yoneshima H, Miyawaki A, Fukuuchi Y, Furuichi T, Okano H, Mikoshiba K and Noda T. 1996. Ataxia and epileptic seizures in mice lacking type 1 inositol 1,4,5-trisphosphate receptor. *Nature* 379: 168-171.
128. Churchill GC and Galione A. 2000. Spatial control of Ca^{2+} signaling by nicotinic acid adenine dinucleotide phosphate diffusion and gradients. *J. Biol. Chem.* 275: 38687-38692.
129. Churchill GC and Galione A. 2001. Prolonged inactivation of nicotinic acid adenine dinucleotide phosphate-induced Ca^{2+} Release mediates a spatiotemporal Ca^{2+} memory. *J. Biol. Chem.* 276: 11223-11225.
130. Genazzani AA, Empson RM and Galione A. 1996. Unique inactivation properties of NAADP-sensitive Ca^{2+} release. *J. Biol. Chem.* 271: 11599-11602.
131. Churchill GC and Louis CF. 1998. Roles of Ca^{2+}, inositol trisphosphate and cyclic ADP-ribose in mediating intercellular Ca^{2+} signaling in sheep lens cells. *J. Cell Sci.* 111: 1217-1225.

Hon Cheung Lee, Department of Pharmacology, University of Minnesota,
321 Church St., SE, 4-145 Jackson Hall, Minneapolis, MN 55455, USA.
Tel: 612-625-7120 (office), 612-625-4641 (lab)
Fax: 612-625-0991
e-mail: leehc@tc.umn.edu
Web site: http://www.tc.umn.edu/~leehc

Cyclic ADP-Ribose and NAADP. Structures, Metabolism and Functions. Lee HC (ed), pp. 167-197, Dordrecht: Kluwer.

Sensitizing Calcium-Induced Calcium Release

Role of cADPR as an Endogenous Modulator

Anthony J. Morgan and Antony Galione

Department of Pharmacology, Oxford University, Mansfield Road, Oxford OX1 3QT, UK.

REGULATION OF Ca^{2+} RELEASE

The primary mechanism by which cells effect rapid increases in their intracellular Ca^{2+} concentration ($[Ca^{2+}]_i$) is by the opening (gating) of specific channels which facilitate the passage of Ca^{2+} across a membrane down the electrochemical gradient generated by various pumps and/or exchangers. Whether at the plasma membrane or on intracellular organelles (*e.g.* endoplasmic reticulum (ER), mitochondria, Golgi), such channels have to be tightly controlled to prevent deleterious or inappropriate elevations of the $[Ca^{2+}]_i$. Therefore, cells have developed a seemingly varied repertoire for their discrete regulation. However, this apparent complexity belies the fact that activation, in essence, encompasses only three modes of operation: (a) ligand-gated ion channels (b) voltage-gated ion channels (c) channels regulated by protein-protein interactions.

Let us concentrate upon the first paradigm, that is, the regulation of channels by specific ligands (either from within or without the cell, depending upon membrane topology). The first example that probably leaps to mind is the ubiquitous activation of intracellular channels following binding of organic second messenger molecules *e.g.* inositol 1,4,5-trisphosphate (IP_3) to the IP_3 receptor. But this is to overlook the role that smaller species (such as ions) can have upon channel function. Indeed, it is contingent upon any biological system to incorporate points of feedback (positive or negative) so that a given pathway functions effectively and in a manner coordinated with the cellular needs. Thus, hardly surprisingly, evolution has provided a means by which Ca^{2+} itself can feedback upon these channels and thereby influence its own homeostasis.

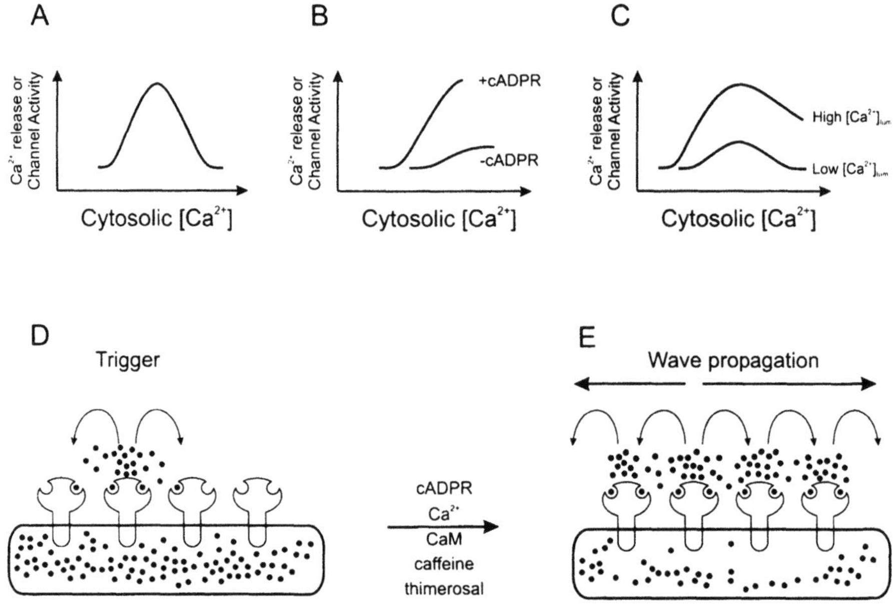

Figure 1. **Ca^{2+}-induced Ca^{2+} release.** (**A**) Biphasic effect of cytosolic Ca^{2+} upon RYR function, either measured as Ca^{2+} release or channel activity. (**B**) The stimulation by Ca^{2+} is shifted and enhanced by cADPR (on an arbitrary scale compared to (A), so not directly comparable. (**C**) Effect of luminal Ca^{2+} upon sensitivity of RYR to cytosolic Ca^{2+} (based upon [114]). (**D**) and (**E**) show the effect of CICR-sensitizing agents upon Ca^{2+} stores and excitability of the cytosol. In the absence of agents, Trigger Ca^{2+} does not recruit neighbouring RYR as the $[Ca^{2+}]$ is below their threshold. The addition of CICR agents renders the neighbouring RYRs sensitive to the local Ca^{2+} and a progressive recruitment away from the trigger site is manifest as a regenerative wave via CICR.

Such has been the benefit of this feedback, that channels of both the plasma membrane as well as of intracellular organelles have adopted regulation by Ca^{2+} (other factors may also come into play, as we shall discuss later): I_{CRAC} currents, representing one possible route of Ca^{2+} entry into the cell, show Ca^{2+} sensitivity (reviewed in [1]), as do the IP_3 receptors that mediate Ca^{2+} release from intracellular stores (summarized in [2]). Moreover, the similarity does not stop there, since the Ca^{2+} effects upon channel gating critically depend upon the Ca^{2+} concentration range – displaying a bell-shaped dependency, both channel classes are *stimulated* by modest increases in the local $[Ca^{2+}]$, whereas they are subsequently *inhibited* beyond a certain $[Ca^{2+}]$, a safeguard if the $[Ca^{2+}]_i$ goes too high (Figure 1A). In the case of intracellular channels, the promotion of Ca^{2+} release by Ca^{2+} positive feedback has been generically termed **Ca^{2+}-induced Ca^{2+} release** (CICR). To date, only two families of intracellular release channel have been shown to exhibit CICR *viz* IP_3 receptors and the ryanodine receptors (RYRs). The novel NAADP-sensitive Ca^{2+} release pathway is unique in being entirely insensitive to cytoplasmic Ca^{2+} and as such is incapable of supporting CICR *per se* (reviewed in [3]). For the rest of this discussion, however, we will focus our attention upon RYRs.

Ca^{2+}-Induced Ca^{2+} Release

Historically, Ca^{2+}-induced Ca^{2+} release (CICR) was a term originally used to describe the activation of Ca^{2+} release from skeletal muscle sarcoplasmic reticulum (SR) in response to an increase in the $[Ca^{2+}]$ in the myoplasmic environment surrounding this organelle *in situ* [4, 5]. These studies were extended to cardiac muscle preparations where a similar phenomenon was observed [6-8], and CICR is now considered to be the principal mechanism in cardiac excitation-contraction coupling. By contrast, in *skeletal* muscle, a rather specialized direct protein-protein interaction of sarcolemmal voltage-sensitive proteins with SR Ca^{2+} release channels is thought to be the primary mechanism of activation under physiological conditions (voltage-induced Ca^{2+} release - VICR), but this is beyond the scope of this review.

This is all very well, but why develop CICR at all, what is its *raison d'être*? Put simply, it affords the cell a means to amplify small, highly localized Ca^{2+} signals into larger Ca^{2+} release events. Firstly, this can be considered as a highly localized autocatalytic, "intramolecular" positive feedback that ensures a robust release from weakly stimulated Ca^{2+} release channels (or cluster of channels) – this phenomenon is thought to underlie the small elementary Ca^{2+} release events such as quarks and sparks [9]. Secondly, the Ca^{2+} signal can be propagated in space away from an initiating site by the progressive recruitment (activation) of neighbouring channel complexes ("intermolecular" positive feedback), the 'runner' between the complexes being Ca^{2+} itself (Figure 1D & E) – this potentially amplifies the localized elementary events into a regenerative wave of Ca^{2+} release that spreads across the entire cell (once a threshold of stimulus intensity is breached, as was first suggested in Medaka eggs [10]). This latter, "propagative" role of CICR is a highly important consideration given that the considerable Ca^{2+} buffering capacity of the cytosol limits Ca^{2+} diffusion to only a few microns beyond a given point source of entry/release [11].

Returning to the above example in the light of this, we can envisage that in cardiac tissue after a depolarizing action potential, small 'trigger' increases in $[Ca^{2+}]$ due to voltage-dependent Ca^{2+} entry across the sarcolemma can be amplified into a disproportionately large global Ca^{2+} release from the SR via CICR which result in a coordinated activation of the contractile apparatus. We shall now turn to the elements and characteristics of the the CICR machinery.

MOLECULAR ELEMENTS

Pharmacological characterization of CICR well-preceded the molecular identification of the components of the SR Ca^{2+} release pathway. The plant alkaloid ryanodine, extracted from the South American plant named after the Scottish botanist John Ryan, was found to potently inhibit CICR, while the methylxanthine, caffeine, could potentiate the effects [12]. The former effect was the basis of use of ryanodine as drug for euthanasia and alligator poison by South American Indians, and latterly for use as a somewhat dangerous insecticide.

The high selectivity and affinity of ryanodine was just one method exploited to purify its molecular target by the use of affinity chromatography [13]; along with immunoaffinity chromatography [14] and other conventional biochemical techniques [15, 16] this resulted in the identification of a large membrane protein complex termed the ryanodine receptor (RYR) from mammalian skeletal muscle. Four identical 565 kDa subunits assembled to form large quatrefoil tetrameric structures as observed by electron microscopy [13, 15], similar to the foot structures earlier observed between sarcolemma and SR in skeletal muscle cells by the same technique [17-20]. Importantly, these proteins when incorporated into artificial lipid bilayers were able to function as a large conductance Ca^{2+} channel which could be activated by increasing *cis* ("cytoplasmic") Ca^{2+} concentrations, and modulated by drugs such as ryanodine and caffeine in a similar fashion to the way in which these drugs affected CICR [13, 14].

A number of RYRs have now been cloned and sequenced (reviewed [21-24]): three mammalian forms have been identified, type 1 being predominant in skeletal muscle, type 2 in heart and brain, and type 3 expressed at low levels in a wide range of tissues. In addition, RYRs have been cloned and sequenced from invertebrates including *Drosophila* [25], *C. elegans* [26] and sea urchin eggs where a single species-specific isoform appears to be expressed [27]. Molecular studies and immunolabelling clearly demonstrated the widespread distribution of RYRs not only in a wide range of organisms but also in an extensive range of cell types. However, it is only in the last few years that their functional roles and regulation have been studied in detail. One important area of research has been the finding that in many cell types, cyclic ADP-ribose is an important regulator of RYR-mediated Ca^{2+} signaling and hence CICR.

Cyclic ADP-Ribose-Induced Ca^{2+} Release

Background

As is discussed in more detail in other chapters of this book, cyclic ADP-ribose (cADPR) is a novel cyclic nucleotide which is endogenously synthesized from the common pyridine nucleotide β-NAD by the action of an ADP-ribosyl cyclase (of which there are a variety of forms, including the T cell antigen CD38 [28]). First discovered by Lee and colleagues to release Ca^{2+} from sea urchin microsomes [29]), cADPR has now been described as an activator of Ca^{2+} release throughout the animal and plant kingdom. It soon became apparent from pharmacological criteria as well as heterologous desensitization experiments that cADPR activated an intracellular Ca^{2+} channel distinct from that utilized by the ubiquitous IP_3 (or NAADP for that matter) and which appeared to be a RYR. Accordingly, activators of RYRs (*e.g.* low concentrations of ryanodine, caffeine) potentiate cADPR-induced Ca^{2+} release [30-32], while classical inhibitors of RYR (high concentrations of ryanodine, Mg^{2+}, procaine, ruthenium red) abrogate cADPR-induced Ca^{2+} release [30, 31, 33]. By contrast, neither IP_3 nor IP_3 receptor antagonists (*e.g.* heparin) affected the response [29, 34, 35].

Experimental Approaches for Assessing cADPR-Induced Ca^{2+} Release

The hydrophilic nature of cADPR renders it cell-impermeant to all intents and purposes (but see below). Therefore, this limitation to experimental design has to be overcome by the usual measures *i.e.* microinjection, introduction via a patch pipette, the use of permeabilised cells, and/or broken cell preparations. However crude the measurement of Ca^{2+} release fluorimetrically in a broken cell preparation may first appear, it has nevertheless proved decisive in pinpointing cADPR as a Ca^{2+}-mobilizing messenger in the first instance, and in addition, has been instrumental in defining the pharmacology of the Ca^{2+} release process. It is still a valuable technique that is universally employed to probe RYR regulation by cADPR and CICR [36].

An additional useful weapon in the experimenter's armoury has been the synthesis of an inactive precursor, caged cADPR, which photolyses to yield active cADPR upon exposure to UV light [36]. Although this impermeant compound requires access to the cell interior via microinjection or patching, this drawback is more than offset by the ability to precisely control cADPR delivery in a single intact cell in both time and space [37].

The ability to experimentally modify cADPR metabolism (and hence cADPR levels – see below) is not well developed pharmacologically. One cell-permeant agent that has been used to block cADPR synthesis is nicotinamide (IC_{50} ~1.5mM) which is not a true inhibitor but merely drives the equilibrium reaction in reverse [38]. By contrast, potent ADP-ribosyl cyclase or hydrolase inhibitors based on NAD analogues have demonstrable, appropriate effects upon Ca^{2+} release in sea urchin homogenate [39]. Another, more potent inhibitor is arabinosyl 2'-fluoro-2'-deoxynicotinamide mononucleotide (~145nM affinity) [40] though this hasn't yet been tested biologically.

As with studies of all receptors, a great advancement for the investigator is the development of a range of potent and selective analogues with either agonist or antagonistic properties at the appropriate binding site. The cADPR field has proven no exception and, despite difficult synthetic chemistry, we are fortunate to have a few useful molecular tools at our disposal (see the Chapter by Walseth and Lee of this book for details). Amongst the more common antagonists are derivatives modified on the 8-position such as 8-amino-cADPR and 8-Br-cADPR, each of which has its own advantages (*i.e.* the former is more effective, but the latter is found to be weakly cell-permeant [41, 42]). More metabolically stable agents are the 7-deaza analogues with both agonist (7-deaza-cADPR) [43] and cell-permeant antagonist properties (7-deaza-8-bromo-cADPR) [42], and both having reasonable potencies. These have proven invaluable in highlighting the importance of cADPR in Ca^{2+} signaling.

Effect of cADPR Upon Intracellular Ca^{2+} Signals

At the intact and single cell level, the organization of Ca^{2+} signals attains a complexity that cannot be appreciated in either broken cell preparations, or in cell populations. Ranging from the subcellular, highly localized elemental Ca^{2+} release events (*e.g.* Ca^{2+} sparks) to Ca^{2+} oscillations (and their spatial counterpart, global Ca^{2+} waves), this hierarchy of Ca^{2+} signals is profoundly sensitive to cADPR, underscoring its importance in a more physiologically relevant context. In cardiac cells, for example, each Ca^{2+} spark is thought to represent the brief opening of a small cluster of RYR channels [9]. Again, consistent with cADPR-induced gating of the RYR, delivery of cADPR by flash photolysis in intact cells [44] or bath application in permeabilized cells [45] enhanced the frequency of these local Ca^{2+} release events. At higher stimulus intensities, elementary local events may coalesce and, under appropriate conditions, transform into full-blown global Ca^{2+} signals such as repetitive waves. Indeed, recent work in exocrine pancreas suggests that cADPR may even be crucial for orchestrating this local to global transition [46], a property undoubtedly connected with its ability to sensitize the cytosol to CICR (see below).

A natural corollary of the regulation of local events is the propensity of cADPR to modulate full-blown Ca^{2+} responses: in the temporal domain, cADPR influences Ca^{2+} oscillations in pancreatic acinar cells, [47-49] as indeed it can in sea urchin eggs [50], dorsal root ganglion neurons [51], and porcine tracheal smooth muscle [52]. Even potentially arrythmogenic Ca^{2+} oscillations of the heart seem to involve cADPR [53] which has implications for cADPR in pathophysiology.

In the light of these data, it is to be expected that cADPR modulates Ca^{2+} in the spatial domain as well. Cyclic ADPR seems to both affect and effect intracellular Ca^{2+} waves, as exemplified by observations in cardiomyocytes [54], pancreatic acinar cells [55], and eggs [56, 57]. The downstream ramifications of these phenomena are slowly being unraveled, but to date cADPR has been implicated in scenarios as diverse as fertilization [58, 59], neurotransmitter release [60], plant gene expression [61], mammalian gene expression [62], hypoxic pulmonary vasoconstriction [63, 64], long-term synaptic depression [65] and as a link between intermediary metabolism and Ca^{2+} signals [66]. Undoubtedly more will follow as we begin to understand more about the whys and wherefores of cADPR.

Intercellular Messages

In addition to its generally well-accepted role in regulating the intricacies of intracellular signals, a picture is emerging of cADPR as an intercellular messenger (see the Chapter by De Flora *et al.* in this book). For example, the propagation of intercellular Ca^{2+} waves in sheep lens cells seem to rely at least in part upon cADPR passing through gap junctions to stimulate Ca^{2+} release in neighbouring cells [67]. Its actions may not only be restricted to the intracellular face: notwithstanding the majority of cell types, recent experiments have highlighted the unexpected ability of some cell types to transport cADPR across their plasma membranes, probably facilitated by CD38 itself [68]. Consequently, cADPR (and perhaps its precursor β-NAD) has been put forward as an extracellular paracrine factor that regulates, amongst things, haematopoetic progenitor expansion [69] or tracheal smooth muscle contraction [70]. Such evidence expands the repertoire for cADPR to include intercellular as well as an intracellular Ca^{2+} signaling role.

cADPR AND Ca^{2+}-INDUCED Ca^{2+} RELEASE

Thus far, we have seen that cADPR is a potent Ca^{2+} mobilizing agent that acts (somehow) upon the RYR family in a host of species and circumstances. Our presentation of its properties has been somewhat descriptive inasmuch that we have merely highlighted some of its basic features and the repercussions of its activity. However, we have not yet addressed the underlying mechanisms of how cADPR works at a detailed molecular level. We shall now turn our attention to what is currently known of its locus of action and its relation to CICR.

RyR and Ca^{2+}-induced Ca^{2+} Release

It has already been alluded to, the phenomenon we now know as CICR was first described in muscle preparations where modest-high concentrations of Ca^{2+} could stimulate Ca^{2+} release from Ca^{2+} stores of the SR. Subsequently, it was shown that caffeine mimicked this release response (even in the absence of elevated [Ca^{2+}]) by shifting the Ca^{2+} sensitivity of the RYR channel such that even 'basal' [Ca^{2+}] is sufficient to promote release [5] (but in addition to sensitizing CICR, caffeine also enhances the rate of release [71]).

But there is other compelling evidence that caffeine utilizes the CICR mode of RYR activation. Like caffeine, ryanodine itself is now thought to also activate RYR by sensitizing them to Ca^{2+} (rather than just locking the channel in a sub-conductance state) [72, 73]. This corroborates former proposals that [^{3}H]ryanodine binding is a reliable index of the CICR activation state (*i.e.* of the open channel conformation) of RYRs [23, 24]. Accordingly, an excellent correlation exists between ryanodine binding and Ca^{2+} release: both display a bell-shaped dependency upon [Ca^{2+}] [71, 74], both are *inhibited* by Mg^{2+}, ruthenium red, procaine, and tetracaine [19, 71, 75-77] whereas both are *enhanced* by 4-Chloro-3-ethylphenol [76], and the closely related 4-chloro-*m*-cresol [78].

Just as Ca^{2+} (*i.e.* CICR) promotes [^{3}H]ryanodine binding, so caffeine has similar effects upon binding in DRG neurons [79], purified frog RYR [71], cardiac and skeletal RYRs [80] and recombinant RyR3 [19] (but see reports of marginal caffeine effects [76, 78]). Overall, this further strengthens the idea that caffeine sensitizes CICR by inducing a similar conformational change in the RYR structure as does Ca^{2+}.

If ryanodine, caffeine and Ca^{2+} induce CICR, this begs the question how do the effects of cADPR compare? Does cADPR also activate by sensitizing RYRs to Ca^{2+}? Taking [^{3}H]ryanodine binding as an index of CICR, there are certainly some confirmatory reports that cADPR enhances [^{3}H]ryanodine binding to cardiac SR vesicles [81], rat brain microsomes [82], and in T-cells [83, 84] while cADPR competes with [^{3}H]ryanodine for its binding sites in parotid cells [85] and mesangial cells [86]. Either way, this would tend to suggest that cADPR indeed recruits RYRs via a CICR mechanism.

This notion is borne out directly by Ca^{2+} release measurements. In sea urchin homogenates, the mobilization of Ca^{2+} by cADPR is greatly potentiated by caffeine, or ryanodine, evidence that they act via a similar mechanism [30, 31, 87, 88]. Furthermore, both Ca^{2+} [87, 89] and its surrogate, Sr^{2+} [31] greatly enhanced cADPR-induced Ca^{2+} release in these vesicular subcellular preparations. Even in intact cells, the latency of the Ca^{2+} release response to uncaging of cADPR was profoundly dependent upon the resting [Ca^{2+}] (or Ca^{2+} buffering capacity) suggesting that Ca^{2+}

positive feedback is a critical prerequisite [90]. Conversely, the CICR induced by Ca^{2+} alone was potentiated by the addition of cADPR [87].

Confirmation that the above effects seen with Ca^{2+} release mirror alterations in channel activity (rather than pumps) comes from electrophysiological recordings: the open probability (NP_o) of RYRs reconstituted into lipid membranes was dose-dependently increased by application of Ca^{2+} to the *cis*- (cytoplasmic) chamber and again, cADPR enhanced such "CICR" in a ryanodine-sensitive manner in smooth muscle [91] and sea urchin preparations [74]. In another report, cADPR unequivocally sensitized a RyR3 to CICR in smooth muscle CICR as evidenced by a leftward shift in the relationship between NP_o and $[Ca^{2+}]$ [92]. Taken together, the data overwhelmingly reinforce cADPR as an activator, or sensitizer, of CICR at the RYR (Figure 1B).

cADPR, CICR and Calmodulin

That Ca^{2+} stimulates the RYR has long been known, but the molecular identity of the Ca^{2+} sensor(s) responsible has remained obscure; whether stimulation is primarily a consequence of direct binding of Ca^{2+} to the RYR or indirectly via an accessory protein still remains to be established, and, of course, they are not mutually exclusive. Analysis of the primary sequence of RYR isoforms suggests that there are both inhibitory and stimulatory sites on the receptor itself [24], as exemplified by the detection of Ca^{2+}-binding EF hand motifs in lobster RYRs [93] although their relevance to other homologues has been questioned by some [94]. Whether these sites alone subserve the major stimulatory role *in situ* is a matter for debate, because of the profound effects of accessory proteins. Nonetheless, potentiation of cADPR-mediated Ca^{2+} release by Sr^{2+} [32] certainly provides some experimental support for the idea that there are multiple sites of cation action (see below for details).

In the absence of a better understanding of the primary Ca^{2+} sensor sites, it is germane to discuss the involvement of the ubiquitous small (16.7kDa) Ca^{2+}-binding protein calmodulin (CaM) in regulating RYR function which has by far the most overwhelming support as one mediator of Ca^{2+} stimulation/inhibition (recently reviewed in [95]). Although a role for CaM in inhibiting mammalian RYR has been known for some 15 years or so [96], its role in *stimulating* Ca^{2+} release evolved from studies with sea urchin microsomes which were rendered insensitive to cADPR when purified on Percoll gradients [97]. It transpired that CaM was the essential factor that had been removed during the isolation procedure, and when reconstituted, restored a robust cADPR-induced Ca^{2+} release [97]. Follow up work by the same group and by others established that CaM acts by sensitizing the CICR mechanism to Ca^{2+}, irrespective of whether this is initiated by Ca^{2+}, caffeine, Sr^{2+}, ryanodine or cADPR [32, 80, 87, 98, 99]. As with cADPR and caffeine,

CaM enhances [^{3}H]ryanodine binding [80]. Therefore, CaM is yet another factor that stimulates RYR by shifting their Ca^{2+} sensitivity and initiating CICR.

Interestingly, the potentiating effect of CaM is consistently greatest with the endogenous ligand, cADPR, rather than caffeine [32, 87, 98] which may indicate subtle differences in their activation mechanisms. It is also noteworthy that Sr^{2+} alleviated the cADPR requirement for CaM, which may well substantiate multiple sites of action for stimulatory cations as already alluded to [32]. In light of this, it is tempting to speculate on the role of RYR EF hand motifs [93] and to query the absolute requirement for CaM (although in a physiological context, it is open to question whether CaM will be limiting or not).

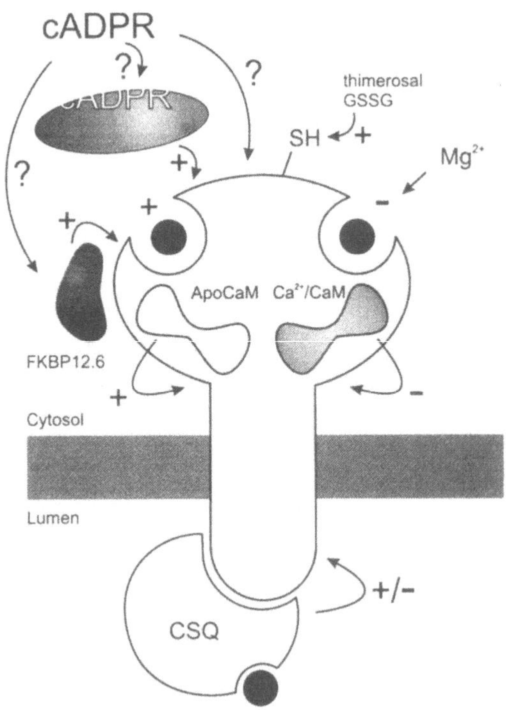

Figure 2. **Regulation of CICR at the ryanodine receptor stimulatory factors** are arbitrarily depicted on the left hand side. inhibitory factors on the right. cADPR is proposed to activate the receptor either directly. or via accessory proteins such as FKBP12.6 or a putative cADPR receptor. Ca^{2+} ions bind to both stimulatory (+) or inhibitory cytosolic sites (-) on the RYR itself, as well as to the luminal proteins such as calsequestrin (CSQ) (which regulates the RYR depending upon the luminal [Ca^{2+}] (+/-)). Mg^{2+} inhibits the RYR via the Ca^{2+} (-) site. Calmodulin either stimulates RYR in the Ca^{2+}-free form (ApoCaM) or inhibits at another distinct but neighbouring site when Ca^{2+} bound (Ca^{2+}/CaM). Disulfide reagents such as oxidized glutathione (GSSG) or thimerosal also sensitize RYR to CICR.

Despite this ongoing interest in CaM, it has yet to yield a definitive explanation of how precisely the protein is involved in RYR activation by cADPR. The simplest model would involve the allosteric modulation of the RYR complex by direct binding of CaM. Circumstantial evidence that CaM remains bound to RYR came from work in broken cell preparations: whereas addition of exogenous CaM has no further effect in some systems [74, 89, 100], CaM antagonists typically inhibit channel activity [74], or Ca^{2+} release [32, 98, 99], and even [^{32}P]cADPR binding [101]. One explanation for such a discrepancy between the effects of CaM supplementation and inhibition is that binding sites are already occupied by endogenous CaM [100] which needs to be stripped away [102].

A more direct investigation of the physical association between RYR and CaM has been undertaken over the years [24, 95]. For example, [^{35}S]CaM binding to both the skeletal and cardiac isoforms of the RYR has been demonstrated directly, albeit with different affinities [80]. Our current understanding has culminated in an interesting model whereby both apocalmodulin (the Ca^{2+}-free form) and the Ca^{2+}/CaM complex actually bind to distinct but adjacent sites on the RyR1, each with a 1:1 stoichiometry per RYR subunit (*i.e.* four per tetrameric complex) [103]. Crucially, this has a functional correlate in that apoCaM activates the RYR, whereas the Ca^{2+}/CaM complex inhibits [103], with the opposing roles of CaM predominantly depending upon the occupation of its lower affinity C terminus Ca^{2+}-binding sites [104]. The fascinating aspect of this scheme is that (a) the biphasic Ca^{2+} dependency mirrors that for CICR (low [Ca^{2+}] stimulating, high [Ca^{2+}] inhibiting RYR) and (b) the proximity of the CaM-binding sites means that CaM need never dissociate from the RYR (*i.e.* no Ca^{2+}-dependent loss), but is rather trapped and flips from one site to another during $[Ca^{2+}]_i$ oscillations, as the authors speculate [103] (Figure 2).

As attractive a molecular model as the above is, its relevance to cADPR and CaM in CICR is unproven. Certainly, there are key differences between the skeletal muscle RyR1 activation and cardiac RyR2 which should engender a cautionary view: the RyR1 physically couples to sarcolemmal channels, whereas the RyR2 is more likely to be modulated by cADPR (but see [105, 106] for cADPR effects upon skeletal muscle RYRs). That said, some optimism for the significance of CaM is that the RyR2 does indeed bind and respond to CaM, albeit in an inhibitory manner [102]. We can use this as a starting model which can be adapted as we move forward in understanding the interplay between CICR and cADPR and CaM.

As well as the consequences of direct CaM binding, indirect effects of its accessory proteins have also been proposed. Other proteins containing a CaM subunit may have profound consequences for RYR function as exemplified by CaM kinase II [107, 108], and calcineurin [109] which may regulate the phosphorylation state and thereby the activity of the RYR channel complex. With the exception of one report with CaM kinase II however [107], very little evidence points to this having any physiological bearing upon cADPR-stimulated Ca^{2+} release, rather reminiscent of the CaM story above; indeed, the entire issue of RYR phosphorylation and cADPR is barely understood and demands further investigation.

Overall, the precise loci of action of Ca^{2+} are unclear, and while CaM mediates some of the potentiating effects, this is more likely to be an additional "fine tuner" as opposed to the primary allosteric CICR site.

Luminal Ca^{2+} and CICR

Ca^{2+} effects upon the cytosolic face of the RYR have dominated discussion up to this point, but this is to overlook the impact of Ca^{2+} within the Ca^{2+} store itself (i.e. of the ER/SR lumen). More than just the reservoir and hence the driving force for release, Ca^{2+} within the store functionally interacts with the RYR to affect its gating properties. As with IP_3 receptors [110], the more replete the store with Ca^{2+}, the more sensitive the channel to CICR: for skeletal or cardiac muscle RYR incorporated into lipid bilayers, *trans* (luminal) Ca^{2+} stimulated NP_o in a concentration-dependent manner from μM to mM, an appropriate range for the $[Ca^{2+}]$ within the stores [111-113]. More importantly in terms of our discussion, the higher the luminal (*trans*) $[Ca^{2+}]$, the more potent and efficacious is cytosolic (*cis*) Ca^{2+} at stimulating channel activity [114] and Ca^{2+}- or ryanodine-stimulated Ca^{2+} release [115] *i.e.* verification that store loading modulates CICR (Figure 1C).

Mechanistically speaking, how luminal Ca^{2+} affects CICR is somewhat controversial. While in part it has been attributed to luminal Ca^{2+} permeating the channel pore and activating RYR from the cytosolic face [111, 116] there is good evidence that Ca^{2-} truly modulates RYR function by interacting with luminal sites [112, 114]. The precise location of these sites is unknown as yet, but they must either be on the RYR itself [112, 117], or on some accessory protein such as calsequestrin (CSQ) whose inhibition of the RYR is inversely correlated with the total luminal $[Ca^{2+}]$ (Figure 2) [117] (note: the issue may be a little more complex given that, upon stimulation of Ca^{2+} release, the Ca^{2+} bound to CSQ might be displaced by K^+ in an electrically compensating exchange pathway [118]. This would serve to increase the free $[Ca^{2+}]$ and hence the driving force for release in the lumen [118, 119], but at the same time inhibit the RYR when CSQ is in the Ca^{2+}-free form).

As for its relevance to cADPR regulation of CICR, the role of luminal Ca^{2+} has hardly been directly investigated, probably since it has been tacitly assumed (quite understandably) that it will mirror Ca^{2+}-stimulation of RYRs. To date, a luminal Ca^{2+} dependence of cADPR-stimulated RYR activity has only been demonstrated in sheep skeletal muscle [120] and sea urchin egg homogenate [121]. Somewhat at odds with the general concensus of its activation mechanism, there is also an isolated report that the cADPR augmentation of Ca^{2+} sparks may be due to an enhancement of store loading, and hence a potentiation of RYR activity via luminal Ca^{2+} [122]. A more rigorous analysis of the relationship between cADPR and luminal Ca^{2+} is clearly required.

SULFHYDRYL AGENTS AND CICR

Besides modulators such as cADPR, RYRs have also been sensitized to CICR by chemical means, both *in situ* and *in vitro*. The effects of agents which modify –SH groups on proteins have proven useful experimental tool in mimicking CICR and, in so doing, has thrown light upon the intricacies of channel regulation. The RYR contains 80-100 cysteine moieties per subunit, a quarter of which are available for covalent modification [123], and which can be subdivided based on functional criteria [123-125]. Most of the residues crucial for channel activity seem to reside in the cytoplasmic and membrane-spanning regions of the RYR judging by the effect of oxidants added to either the cytosolic or luminal face of the receptor [124], but oxidation has different effects depending upon RYR isoform, the agent used as well as the degree of modifation [123]. Amongst the many possible oxidants used over the years, thimerosal, 4,4'-dithiodipyridine (4,4' DTDP) and N-ethylmaleimide (NEM) have been the most common, and, of these, the mercuric compound thimerosal has probably received the greatest attention due to its potent effects on $[Ca^{2+}]_i$ signals in intact cells (the activities of thimerosal are reviewed in [126]).

As might be anticipated from the mode of action, their effects are pleiotropic, and different mechanisms have been proposed to underlie the stimulation of RYR, including relief from Ca^{2+} or Mg^{2+} inhibition [127, 128] as well as a sensitization of RYR to the stimulatory effects of Ca^{2+} (assessed either functionally or inferred from [^{3}H]ryanodine binding) [128-132]. However, despite a clear leftward shift in the Ca^{2+} concentration-response curve in some RYR preparations [128, 131], others have shown no alteration of the Ca^{2+} dependence or the ryanodine binding affinity (*i.e.* stimulation is not due to sensitized CICR) [133]. In the latter case, it would appear that thimerosal actually overrides the requirement for Ca^{2+} (indeed, these conditions may have 'locked' the receptor conformation since thimerosal also rendered the channel insensitive to other classical modulators including ruthenium red and Mg^{2+} [133]). The most likely explanation of this discrepancy between workers is the *extent* of oxidation, which is known to have a whole spectrum of consequences for channel function, ranging from activation through to inhibition [127, 131, 132].

The upshot of all this is that the usefulness of disulfide-reactive agents is twofold: (a) their judicious use allows CICR to be sensitized independently of changes in either trigger Ca^{2+} or cADPR levels (b) consequently, they have strengthened the case for the role of CICR via RYR in several systems.

The earliest reports that thimerosal could induce or potentiate RYR-mediated Ca^{2+} release originated from sea urchin and mouse egg preparations [121, 129, 130]. In the intact cell scenario, CICR sensitization can be manifest as the generation of simple [134] or complex Ca^{2+} signals, such as oscillations [130, 135] and an enhancement of the Ca^{2+} wave induced by trigger Ca^{2+} [136].

In broken cell preparations, thiol oxidation bears all the hallmarks of a CICR activator: it promotes [^{3}H]ryanodine binding [125, 128, 137], enhances RYR channel NP_o in lipid bilayers [124, 125, 128, 137, 138], and stimulates Ca^{2+} release via the RYR *per se* [121, 139, 140]. Of particular interest in the context of this review, is the fact that sulfhydryl modification also potentiates cADPR-induced Ca^{2+} release [121, 140, 141], but whether this reflects an enhancement of cADPR binding or not remains to be determined. Taken together, the data indicate that myriad thiol groups are of crucial importance for different aspects of RYR function, the most notable being its sensitivity to stimulation by Ca^{2+} (CICR). That cADPR-induced activation is also affected is in accord with its role as a CICR modulator.

Although it is beyond the scope of this chapter, it is worth pointing out that reactive thiols on RYRs are not merely an experimental curiosity but are real candidates for physiological sensors of the cellular environment, whether via their S-nitrosylation by the messenger nitric oxide (in a CaM-dependent manner) [142] or as redox and pO_2 sensors [143, 144]. Hence, Ca^{2+} signals are sensitive to redox status at the level of the RYR as well as the alteration of cADPR levels as mentioned earlier [64].

The reader should also note that while we have confined our discussion to effects upon the RYR, –SH modification will naturally affect other Ca^{2+} homeostatic mechanisms such as IP_3 receptors and Ca^{2+} pumps, as well as channels and exchangers for other ions [126]. Hence, agents such as thimerosal should in no way be considered specific activators of RYRs, but they can provide useful information when combined with appropriate controls.

RYRs AS TARGETS FOR cADPR

It is a truth universally acknowledged that a messenger in possession of good activity is in need of a receptor. In that regard, cADPR must be considered somewhat orphaned, because the unequivocal identification of its protein target has proven frustratingly problematic. Clearly, specific intracellular targets must be present on intracellular membranes, and indeed early on, high affinity specific binding of [^{32}P]cADPR was demonstrated in sea urchin egg microsomes, with a K_d ~17 nM and B_{max}~ 25 fmol / mg protein [145]. However, there is little direct data on cADPR binding elsewhere. An exception is in rabbit longitudinal smooth muscle, where specific binding to dispersed permeabilized cells was reported (IC_{50} ~ 1.9 nM) [146].

More recently, our laboratory extensively characterized sea urchin cADPR binding sites, in particular with respect to the effects of classical CICR modulators such as caffeine, ruthenium red, and ryanodine. We demonstrated that there was little correlation between their activity upon function (Ca^{2+} release) and [^{32}P]cADPR binding *i.e.* these agents do not affect cADPR-induced Ca^{2+} release by altering the cADPR binding to its receptor, rather they acted downstream, by somehow modifying the 'gain' of the CICR system [101]. In view of these data (together with previous [^{3}H]ryanodine binding studies), one can speculate that these agents modulate RYR by exposing or occluding either activating or inhibitory sites which in turn influences the endogenous modulator, cADPR.

Candidate cADPR Receptors

Whether cADPR directly interacts with RyRs or does so via any of the many accessory proteins that interact with the huge cytoplasmic domains of RyRs [147] has been the subject of much speculation over the last few years. Overall, it would be fair to say that the weight of evidence discounts a direct binding of cADPR to the RYR, and favours some intermediary protein, a cADPR receptor (Figure 2) [28, 148].

The most damning evidence against direct cADPR binding comes from photoaffinity labelling studies of membrane proteins with [^{32}P]azido-cADPR which implicated polypeptides too small to be a RYR (*viz* 100 and 140kDa) [149] (although these do have a similar MW to N-terminus proteolytic fragments [150]). Of the accessory proteins that interact with RYR, one family that has attracted most attention as a potential cADPR receptor has been that which includes the immunophilins FKBP12 and FKBP12.6 (for **FK**-506 **B**inding **P**rotein, FK-506 being an immunosuppressant drug). These small proteins (M_r ~ 12kDa) are prolyl isomerases that bind to and stabilize RYR channel activity [151] and which may even be important for propagating channel gating by coordinating adjacent RYR complexes in either skeletal or cardiac muscle [152, 153].

Their potential for RYR regulation entered a new realm when cADPR was reported to not only bind to FKBP12.6 with high affinity (K_d ~35nM), but to cause a translocation of the immunophilin away from the RYRs to the cytosol of pancreatic islet microsomes – according to the authors, the consequent stripping of this stabilizing, inhibitory factor could feasibly account for the stimulation of Ca^{2+} release by cADPR [154]. Surprisingly, little confirmatory evidence has followed this report (and even FK-506/FKBP affecting cADPR responses *e.g.* [155, 156] does not exclude an independent cADPR receptor). On the contrary, the most convincing evidence currently available is inconsistent with FKBPs as cADPR receptors: for example, the association of recombinant FBKP isoforms with RyR3 was unaffected by cADPR [157]. Furthermore, the immunosuppressants FK-506 or rapamycin which would release FKBPs from RYR did not modulate cADPR binding to sea urchin membranes [101]. Taken together, the paucity of data recommends that, for the moment at least, we treat the idea of the FKBPs as universal cADPR receptors with some caution. All in all, the identity of the cADPR receptor is still to be unequivocally resolved, and it remains to be seen whether FKBPs, CaM or some other hitherto unknown protein fulfil this function.

cADPR: CICR Activator or Modulator?

Throughout the early years of cADPR research, doubt surrounded this novel nucleotide as a true second messenger, in the sense that it was questionable whether it was really rapidly synthesized upon demand in response to hormonal stimulation (*cf.* IP_3) to evoke an explosive Ca^{2+} response. An alternative school of thought proposed that dynamic, acute changes in cADPR did not occur, but rather it was the ambient, 'basal' levels of cADPR that set the threshold for CICR; or put another way, cADPR was the "modulator" that regulated the excitability of the medium, the degree to which depended upon the *chronically* controlled [cADPR] (Figure 1D & E).

Of course, the truth may actually lie somewhere between the two extremes as they are not incompatible concepts. But since those early days, we have made great inroads into measuring dynamic changes in cADPR levels thanks to the development of, first, a bioassay [158], then a radioreceptor assay [64, 65, 159], radioimmunoassay [160] as well as new fluorescent techniques for the determination of nanomolar levels of cADPR in cell extracts [161]. Using these protocols, investigators have observed that cADPR does indeed fulfil the criterion for a dynamic second messenger, being produced rapidly on demand during egg fertilization [162], glucose stimulation of pancreatic β cells [163] (albeit controversially [164]), activation of the T cell receptor [84], cholecystokinin stimulation of smooth muscle [146], and in adrenal chromaffin cells challenged with acetylcholine [165].

The more recent exploitation of powerful genetic approaches using CD38-deficient mice has provided stronger, more direct evidence of a role for acutely-stimulated cADPR production: it has been confirmed that acetylcholine evokes rapid increases in cADPR involving CD38 in pancreatic acinar cells [166], as does glucose in pancreatic β cells [167], while in neutrophils the sustained Ca^{2+} response to fMLP is also dependent upon cADPR synthesis and which proves to be crucial for mounting a robust inflammatory chemotactic response [168].

Chronically, slow-acting agonists that work on a transcriptional time scale have also been noted to enhance cADPR levels, as exemplified by retinoic acid (RA)-induced differentiation of progenitor HL-60 cells [160]. This probably reflects an enhancement of ADP-ribosyl cyclase activity and hence cADPR synthesis, if we extrapolate from similar studies in estrogen-stimulated rat uterus [169], or smooth muscle incubated with RA, calcitriol, or the thyroid hormone, triiodothyronine (T_3) [170].

The ramifications of these increases in cADPR levels for CICR are readily apparent in the acute studies inasmuch that they were accompanied by a rapid increase in $[Ca^{2+}]_i$ [84, 146, 162, 163, 165]. However, it is not yet clear how an increase in cADPR functionally manifests itself in the chronically-treated cells [160, 169, 170], an altogether more complex scenario; one might anticipate a cADPR-dependent sensitization of the CICR mechanism and an enhanced sensitivity of Ca^{2+} stores to a subsequent acute challenge, but this awaits vital experimental confirmation.

Taken together, the data indicate that CICR and cADPR can be regulated from outside the cell via cell-surface receptors (or intracellular transcription factors). We anticipate that future work will demonstrate agonist-specific changes in cADPR, and hence agonist-specific modulation of cytosolic excitability via CICR.

INTERACTION OF cADPR WITH OTHER Ca^{2+} HOMEOSTATIC MECHANISMS

Thus far, we have considered the cADPR/RYR system in isolation for the sake of clarity, but seldom, if ever, would this couple be shielded from other elements of the Ca^{2+} homeostatic machinery in an intact cell. It is inconceivable that RYR-mediated Ca^{2+} release would *not* affect other Ca^{2+} channels (and extrusion processes), or, conversely, that local fluctuations in $[Ca^{2+}]_i$ mediated by these other proteins would not affect RYR, in view of their very nature as CICR amplifiers.

cADPR as an Amplifier of Ca^{2+} Signals

The role of Ca^{2+} pumps in shaping CICR is often overlooked, but they have more than a passive role of merely mopping up $[Ca^{2+}]_i$ "after the event". As discussed and modeled previously [171], repression of local Ca^{2+} domains around CICR channels by SERCA (sarco-endoplasmic reticulum Ca^{2+}-ATPase) greatly influences the Ca^{2+} release profile in time and space, which has some experimental support [172, 173]. Indeed, the pancreatic acinar cell, often used as a model for CICR and cADPR, displays Ca^{2+} oscillations which are exquisitely sensitive to SERCA activity [174]. Whilst pending experimental verification, it would be expected that Ca^{2+} pumps suppress cADPR-mediated CICR, especially considering the substantive role of Ca^{2+} buffering in shaping the kinetics of cADPR responses [90].

Besides the possible inhibition meted out by pumps/exchangers/buffers, *increases* in local Ca^{2+} as a result of neighbouring Ca^{2+} channel activity have very definite *stimulatory* effects upon RYR via CICR. Whether Ca^{2+} is mobilized from intracellular stores or from the extracellular reservoir matters not to the RYRs which cannot distinguish the Ca^{2+} source (cellular architecture considerations aside). In terms of amplifying Ca^{2+} influx across the plasmalemma, the seminal example is that of excitation-contraction coupling in cardiac muscle where Ca^{2+} entry via VOCCs is amplified by CICR from internal stores [9]. But it turns out that this is not a unique paradigm, for in neuronal-derived NG108-15 cells cADPR converted what would have otherwise been a halo of L-channel mediated Ca^{2+} influx into a global cytosolic Ca^{2+} response, via CICR at the RYR [155, 175].

Turning to the interaction between cADPR and other *intracellular* Ca^{2+} release channel families, we now know that the vast majority of cell types express not one but multiple types of intracellular channels, often with differences in their relative abundance and spatial distribution [21, 22, 176]. Because of its inherent CICR properties, the RYR is aptly poised to amplify trigger Ca^{2+} released from other channels such as the IP_3 receptor or the NAADP-stimulated channel. To date, the most extensively studied model systems investigating such channel "cross-talk" (for which "channel-chatter" has specifically been coined [3]) have been the sea urchin egg and the pancreatic acinar cell.

In the first example, a primary Ca^{2+} trigger in eggs can be supplied by NAADP-mediated (or possibly IP_3-mediated [58]) Ca^{2+} release which is then in turn amplified, at least in part, by a cADPR-dependent CICR mechanism – this conclusion was corroborated by the inhibition of NAADP-induced rapid Ca^{2+} waves by the antagonist, 8-amino-cADPR [177]. Extending these studies to cover a longer time period, NAADP oscillatory responses were also facilitated via cADPR-dependent Ca^{2+} release, [50]. However, in this instance, the Ca^{2+} released by NAADP did not only act on the cytosolic face of adjacent RYRs, but also had to be sequestered into the ER to sensitize them via luminal sites (see above) – such shuttling of Ca^{2+} between stores to prime a second store validates the so-called two pool model of oscillations [50].

For mammalian cells, the situation may be a similar one in cholecystokinin (CCK)-stimulated pancreatic acinar cells. It transpires that this particular agonist recruits two putative second messenger pathways, NAADP and cADPR: again, NAADP seems to provide trigger Ca^{2+} to stimulate regenerative cADPR-dependent CICR during oscillations and waves [178].

However, when acinar cells are stimulated with another agonist, acetylcholine (ACh), the story is not so clear cut. Based on some reports, its sole messenger appears to be IP_3 judging by the effects of heparin and the insensitivity of ACh-induced Ca^{2+} oscillations to the cADPR antagonist 8-amino-cADPR [47, 179] or ruthenium red [47]. On the other hand, both ACh- and pure IP_3-evoked oscillations are blocked by high, inhibitory concentrations of ryanodine [179] (*cf.* earlier contradictory reports with lower ryanodine levels [47]). One interpretation would be that IP_3-dependent responses are indeed facilitated by RYR, but without the necessity for cADPR.

However, other reports do in fact implicate cADPR in the ACh-stimulated response. Recent studies using CD38-deficient mice indicate that (in addition to IP_3) cADPR levels are acutely modulated by ACh via the T cell antigen and which contribute to $[Ca^{2+}]_i$ oscillations at the RYR [166]. Similarly, ACh-induced $[Ca^{2+}]_i$ waves were facilitated by cADPR through the basolateral region (and IP_3 in the apical region) [55].

The reason underlying the apparently differential contribution of cADPR in ACh responses in different workers' hands is unclear, but may relate to several issues: (a) the use of single cells *versus* intact acini [55] (b) confounding side-effects of ruthenium red (*e.g.* inhibition of mitochondria, important in these cells [180]) as well as (c) the recruitment of different second messenger complements when the concentration of agonist is increased over only a narrow range [46]. The bottom line, though, is that both IP_3 and NAADP may provide 'trigger' Ca^{2+} for subsequent amplification by cADPR and CICR.

Taken together, a precedence exists for the downstream actions of cADPR/RYR as amplifiers of Ca^{2+} release in either invertebrate or mammalian tissues. It is hoped that analysis of other systems will unravel an analogous cross-talk between Ca^{2+}-release messengers to reveal subtle differences in Ca^{2+} regulation by different agonists.

cADPR-Induced Ca^{2+} Release Upstream of Other Channels

The other side of the coin puts cADPR *upstream* of other Ca^{2+} homeostatic processes, which has not been as readily detected. If processes are positively regulated by Ca^{2+} to enhance $[Ca^{2+}]_i$, then RYRs this time find themselves acting to supply the trigger Ca^{2+} pulse which is amplified elsewhere (an indirect, 'heterologous' variety of CICR at a molecularly distinct channel, if you will). However, whilst this is not strictly-speaking CICR, it has been briefly included for the sake of completeness.

Channels such as I_{CRAC} provide our first candidates where plasma membrane Ca^{2+} channels are stimulated by RYR and cADPR, perhaps in subplasmalemmal restricted domains [181, 182]. As already alluded to, these channels display a similar bell-shaped dependency on the $[Ca^{2+}]$ as do intracellular release channels [1], but in these circumstances the predominant effect of RYR activation is unlikely to be merely a consequence of elevated cytosolic Ca^{2+} – it is more likely to be due to the emptying of stores *per se* activating the capacitative Ca^{2+} entry pathway [1, 183].

Returning to the notion of intracellular "channel chatter", cADPR-induced Ca^{2+} release is demonstrably amplified by channels downstream from RYRs in our prototypical mammalian model, the pancreatic acinar cell: the effects of heparin upon cADPR-induced Ca^{2+} oscillations are consistent with an ancillary contribution from CICR at the IP_3 receptor [47]. A less well-understood positive feedback operates in Jurkat T cells where cADPR-mediated responses are in turn amplified by NAADP [184], the opposite situation to that commented upon above in sea urchin and pancreas. To date, assessment of Ca^{2+} feedback upon NAADP-induced Ca^{2+} release has yielded negative results in systems as diverse as sea urchin, brain microsomes and plants e.g. [185], reviewed in [3]. If this also holds true for the T cell system, we are left with the conundrum of how a non-CICR channel such as the NAADP receptor operates and senses upstream events such as IP_3- and cADPR-mediated Ca^{2+} release. Clearly, these are early days awaiting clarification of the precise mechanisms of action.

POTENTIAL NOVEL ASPECTS OF cADPR SIGNALING

On the issue of CICR, a new archetype was revealed some years ago which does not use the classical RYR as the amplifier and generator of Ca^{2+} signals, but rather utilized a rather novel mode of operation of mitochondria. There has been a resurgence of interest in these organelles over the last ten years owing to their propensity to take up Ca^{2+} from intracellular stores due in part to their close abuttal to intracellular release channels (excellently reviewed in [186]). Uptake is driven by the mitochondrial membrane potential and facilitated by a Ca^{2+} uniporter, but they are not an infinite reservoir, and eventually release Ca^{2+} back to the cytosol such that Ca^{2+} dynamically changes in the mitochondrial matrix to faithfully mirror fluctuations in the cytosol.

However, it has been demonstrated that, under appropriate conditions, mitochondria may discretely accumulate large amounts of Ca^{2+} until it is rapidly released in a coordinated manner. This explosive release into the cytosol can be mediated either by a channel termed the permeability transition pore [187] or via a reversal of the Ca^{2+} uniporter [188] and which may contribute to agonist-evoked Ca^{2+} spiking [187]. Mitochondria may even undergo successive rounds of uptake and release, even in space which propagates across a mitochondrial network, and this has been summarily termed mitochondrial CICR (mCICR) [187].

What has recently made this relevant to the cADPR field is that the mitochondrial Ca^{2+} uniporter may in fact be molecularly related to the RYR [189] – in addition to immunological criteria, it is ruthenium red-sensitive, and can even be stimulated by cytosolic Ca^{2+} to release Ca^{2+} from primed mitochondria [188], a mode of mCICR more closely related to classical CICR. It should be obvious that this raises the intriguing possibility that cADPR modulates mitochondrial CICR, and indeed cADPR metabolism has been reported in (or on) these pyridine nucleotide-packed organelles [190] as evidenced by an NAD glycohydrolase activity that can both synthesise and hydrolyse cADPR [191].

The breakdown product of cyclic ADP-ribose, ADP-ribose (ADPR), was recently appended to the list of candidate second messengers when it was discovered to profoundly stimulate plasmalemmal Ca^{2+} influx channels [192], including those of the TRP family to elevate intracellular Ca^{2+} [193, 194]. The channel itself also may serve a role to self-terminate the putative signal for it contains an ADPR pyrophosphatase motif [194]. A scheme whereby a cascade of Ca^{2+}-elevating messengers is recruited from cADPR through to ADPR is currently a highly speculative one for there are other cellular sources of ADPR, but nonetheless, we eagerly await future work to clarify the role cADPR as a precursor in this novel pathway.

SUMMARY

All in all, we have seen that RYRs are a major effector of the phenomenon of Ca^{2+}-induced Ca^{2+} release (CICR), which allows local elevations of Ca^{2+} to be amplified in time and space through positive feedback. The "trigger Ca^{2+}" which recruits these RYR can be from sources including Ca^{2+} influx (*e.g.* VOCCs) or release from stores (IP_3, or NAADP receptors). These interactions are crucial for coordinating Ca^{2+} sparks, oscillations and waves as well as the transition along this hierarchy. The nucleotide cyclic ADP-ribose is the physiological ligand which sensitizes the RYR to CICR in a manner similar to classical pharmacological activators, although which protein cADPR binds to remains uncertain. How Ca^{2+} itself stimulates RYR is also unclear, for there is evidence for multiple sites of action, including both the RYR itself as well as the accessory protein calmodulin, which greatly facilitates cADPR effects. There are still many years of fruitful research ahead of us before we fully understand the precise nature of how cADPR controls CICR.

REFERENCES

1. Lewis RS. 2001. Calcium Signaling Mechanisms in T Lymphocytes. *Annu. Rev. Immunol.* 19: 497-521.
2. Taylor CW. 1998. Inositol trisphosphate receptors: Ca^{2+}-modulated intracellular Ca^{2+} channels. *Biochim. Biophys. Acta* 1436: 19-33.
3. Patel S, Churchill GC and Galione A. 2001. Coordination of Ca^{2+} signalling by NAADP. *Trends Biochem. Sci.* 26: 482-489.
4. Ford LE and Podolsky RJ. 1970. Regenerative calcium release within muscle cells. *Science* 167: 58-59.

5. Endo M, Tanaka M and Ogawa Y. 1970. Calcium induced release of calcium from the sarcoplasmic reticulum of skinned skeletal muscle fibres. *Nature* 228: 34-36.
6. Fabiato A and Fabiato F. 1975. Contractions induced by a calcium-triggered release of calcium from the sarcoplasmic reticulum of single skinned cardiac cells. *J. Physiol.* 249: 469-495.
7. Fabiato A. 1981. Myoplasmic free calcium concentration reached during the twitch of an intact isolated cardiac cell and during calcium-induced release of calcium from the sarcoplasmic reticulum of a skinned cardiac cell from the adult rat or rabbit ventricle. *J. Gen. Physiol.* 78: 457-497.
8. Fabiato A. 1983. Calcium-induced release of calcium from the cardiac sarcoplasmic reticulum. *Am. J. Physiol.* 245: C1-C14.
9. Niggli E. 1999. Localized intracellular calcium signaling in muscle: calcium sparks and calcium quarks. *Annu. Rev. Physiol.* 61: 311-335.
10. Ridgway EB, Gilkey JC and Jaffe LF. 1977. Free calcium increases explosively in activating medaka eggs. *Proc. Natl. Acad. Sci. USA* 74: 623-627.
11. Allbritton NL, Meyer T and Stryer L. 1992. Range of messenger action of calcium ion and inositol 1,4,5-trisphosphate. *Science* 258: 1812-1815.
12. Sutko JL, Airey JA, Welch W and Ruest L. 1997. The pharmacology of ryanodine and related compounds. *Pharmacol. Rev.* 49: 53-98.
13. Lai FA, Erickson HP, Rousseau E, Liu QY and Meissner G. 1988. Purification and reconstitution of the calcium release channel from skeletal muscle. *Nature* 331: 315-319.
14. Imagawa T, Smith JS, Coronado R and Campbell KP. 1987. Purified ryanodine receptor from skeletal muscle sarcoplasmic reticulum is the Ca^{2+}-permeable pore of the calcium release channel. *J. Biol. Chem.* 262: 16636-16643.
15. Inui M, Saito A and Fleischer S. 1987. Purification of the ryanodine receptor and identity with feet structures of junctional terminal cisternae of sarcoplasmic reticulum from fast skeletal muscle. *J. Biol. Chem.* 262: 1740-1747.
16. Inui M, Saito A and Fleischer S. 1987. Isolation of the ryanodine receptor from cardiac sarcoplasmic reticulum and identity with the feet structures. *J. Biol. Chem.* 262: 15637-15642.
17. Wagenknecht T and Radermacher M. 1997. Ryanodine receptors: structure and macromolecular interactions. *Curr. Opin. Struct. Biol.* 7: 258-265
18. Sharma MR, Penczek P, Grassucci R, Xin HB, Fleischer S and Wagenknecht T. 1998. Cryoelectron microscopy and image analysis of the cardiac ryanodine receptor. *J. Biol. Chem.* 273: 18429-18434.
19. Liu Z, Zhang J, Sharma MR, Li P, Chen SR and Wagenknecht T. 2001. Three-dimensional reconstruction of the recombinant type 3 ryanodine receptor and localization of its amino terminus. *Proc. Natl. Acad. Sci. USA* 98: 6104-6109.
20. Felder E and Franzini-Armstrong C. 2002. Type 3 ryanodine receptors of skeletal muscle are segregated in a parajunctional position. *Proc. Natl. Acad. Sci. USA* 99: 1695-1700.
21. Sutko JL and Airey JA. 1996. Ryanodine receptor Ca^{2+} release channels: does diversity in form equal diversity in function? *Physiol. Rev.* 76: 1027-1071.
22. Shoshan-Barmatz V and Ashley RH. 1998. The structure, function, and cellular regulation of ryanodine-sensitive Ca^{2+} release channels. *Int. Rev. Cytol.* 183: 185-270.
23. Ogawa Y, Kurebayashi N and Murayama T. 1999. Ryanodine receptor isoforms in excitation-contraction coupling. *Adv. Biophys.* 36: 27-64.
24. Meissner G. 1994. Ryanodine receptor/Ca^{2+} release channels and their regulation by endogenous effectors. *Annu. Rev. Physiol.* 56: 485-508.
25. Hasan G and Rosbash M. 1992. Drosophila homologs of two mammalian intracellular Ca^{2+}-release channels: identification and expression patterns of the inositol 1,4,5-triphosphate and the ryanodine receptor genes. *Development* 116: 967-975.
26. Sakube Y, Ando H and Kagawa H. 1993. Cloning and mapping of a ryanodine receptor homolog gene of *Caenorhabditis elegans*. *Ann. NY Acad. Sci.* 707: 540-545.

27. Shiwa M, Murayama T and Ogawa Y. 2002. Molecular cloning and characterization of ryanodine receptor from unfertilized sea urchin eggs. *Am. J. Physiol. Regul. Integr. Comp. Physiol.* 282: R727-R737.
28. Lee HC. 2001. Physiological Functions of Cyclic ADP-Ribose and NAADP as calcium messengers. *Annu. Rev. Pharmacol. Toxicol.* 41: 317-345.
29. Clapper DL, Walseth TF, Dargie PJ and Lee HC. 1987. Pyridine nucleotide metabolites stimulate calcium release from sea urchin egg microsomes desensitized to inositol trisphosphate. *J. Biol. Chem.* 262: 9561-9568.
30. Galione A, Lee HC and Busa WB. 1991. Ca^{2+}-induced Ca^{2+} release in sea urchin egg homogenates: modulation by cyclic ADP-ribose. *Science* 253: 1143-1146.
31. Lee HC. 1993. Potentiation of calcium- and caffeine-induced calcium release by cyclic ADP-ribose. *J. Biol. Chem.* 268: 293-299.
32. Lee HC, Aarhus R and Graeff RM. 1995. Sensitization of calcium-induced calcium release by cyclic ADP-ribose and calmodulin. *J. Biol. Chem.* 270: 9060-9066.
33. Graeff RM, Podein RJ, Aarhus R and Lee HC. 1995. Magnesium ions but not ATP inhibit cyclic ADP-ribose-induced calcium release. *Biochem. Biophys. Res. Commun.* 206: 786-791.
34. Koshiyama H, Lee HC and Tashjian AH. 1991. Novel mechanism of intracellular calcium release in pituitary cells. *J. Biol. Chem.* 266: 16985-16988.
35. White AM, Watson SP and Galione A. 1993. Cyclic ADP-ribose-induced Ca^{2+} release from rat brain microsomes. *FEBS Lett* 318: 259-263.
36. Galione A, Cancela J-M, Churchill G, Genazzani A, Lad C, *et al.* 2000. Methods in cADPR and NAADP research. In *Methods in Calcium Signalling*, ed. JJ Putney, pp. 249-296. Boca Raton: CRC Press
37. Churchill GC and Galione A. 2001. Prolonged inactivation of NAADP-induced Ca^{2+} release mediates a spatiotemporal Ca^{2+} memory. *J. Biol. Chem.* 276: 11223-11225.
38. Kim H, Jacobson EL and Jacobson MK. 1993. Synthesis and degradation of cyclic ADP-ribose by NAD glycohydrolases. *Science* 261: 1330-1333.
39. Migaud ME, Pederick RL, Bailey VC and Potter BV. 1999. Probing *Aplysia californica* adenosine 5'-diphosphate ribosyl cyclase for substrate binding requirements: design of potent inhibitors. *Biochemistry* 38: 9105-9114.
40. Sauve AA, Deng H-T, Angeletti RH and Schramm VL. 2000. A Covalent Intermediate in CD38 Is Responsible for ADP-ribosylation and Cyclization Reactions. *J. Am. Chem. Soc.* 122: 7855-7859.
41. Walseth TF and Lee HC. 1993. Synthesis and characterization of antagonists of cyclic-ADP-ribose-induced Ca^{2+} release. *Biochim. Biophys. Acta* 1178: 235-242.
42. Sethi JK, Empson RM, Bailey VC, Potter BV and Galione A. 1997. 7-Deaza-8-bromo-cyclic ADP-ribose, the first membrane-permeant, hydrolysis-resistant cyclic ADP-ribose antagonist. *J. Biol. Chem.* 272: 16358-16363.
43. Bailey VC, Sethi JK, Fortt SM, Galione A and Potter BVL. 1997. 7-Deaza cyclic adenosine 5'-diphosphate ribose: First example of a Ca^{2+}-mobilizing partial agonist related to cyclic adenosine 5'- diphosphate ribose. *Chem. Biol.* 4: 51-61.
44. Cui Y, Galione A and Terrar DA. 1999. Effects of photoreleased cADP-ribose on calcium transients and calcium sparks in myocytes isolated from guinea-pig and rat ventricle. *Biochem. J.* 342: 269-273.
45. Lukyanenko V and Gyorke S. 1999. Ca^{2+} sparks and Ca^{2+} waves in saponin-permeabilized rat ventricular myocytes. *J. Physiol. (Lond)* 521: 575-585.
46. Cancela JM, Van Coppenolle F, Galione A, Tepikin AV and Petersen OH. 2002. Transformation of local Ca^{2+} spikes to global Ca^{2+} transients: the combinatorial roles of multiple Ca^{2+} releasing messengers. *EMBO J.* 21: 909-919.
47. Thorn P, Gerasimenko O and Petersen OH. 1994. Cyclic ADP-ribose regulation of ryanodine receptors involved in agonist evoked cytosolic Ca^{2+} oscillations in pancreatic acinar cells. *EMBO J.* 13: 2038-2043.

48. Burdakov D, Cancela JM and Petersen OH. 2001. Bombesin-induced cytosolic Ca^{2+}spiking in pancreatic acinar cells depends on cyclic ADP-ribose and ryanodine receptors. *Cell Calcium* 29: 211-216.
49. Cancela JM. 2001. Specific Ca^{2+} signaling evoked by cholecystokinin and acetylcholine: the roles of NAADP, cADPR, and IP_3. *Annu. Rev. Physiol.* 63: 99-117.
50. Churchill GC and Galione A. 2001. NAADP induces Ca^{2+} oscillations via a two-pool mechanism by priming IP_3- and cADPR-sensitive Ca^{2+} stores. *EMBO J.* 20: 2666-2671.
51. Currie KP, Swann K, Galione A and Scott RH. 1992. Activation of Ca^{2+}-dependent currents in cultured rat dorsal root ganglion neurones by a sperm factor and cyclic ADP-ribose. *Mol. Biol. Cell* 3: 1415-1425.
52. Prakash YS, Kannan MS, Walseth TF and Sieck GC. 1998. Role of cyclic ADP-ribose in the regulation of $[Ca^{2+}]i$ in porcine tracheal smooth muscle. *Am. J. Physiol.* 274: C1653-C1660.
53. Rakovic S, Cui Y, Iino S, Galione A, Ashamu GA, *et al.* 1999. An antagonist of cADP-ribose inhibits arrhythmogenic oscillations of intracellular Ca^{2+} in heart cells. *J. Biol. Chem.* 274: 17820-17827.
54. Prakash YS, Kannan MS, Walseth TF and Sieck GC. 2000. cADP ribose and $[Ca^{2+}]_i$ regulation in rat cardiac myocytes. *Am. J. Physiol. Heart Circ. Physiol.* 279: H1482-H1489.
55. Leite MF, Burgstahler AD and Nathanson MH. 2002. Ca^{2+} Waves require sequential activation of inositol trisphosphate receptors and ryanodine receptors in pancreatic acini. *Gastroenterology* 122: 415-427.
56. Willmott N, Sethi J, Walseth TF, Lee HC, White AM and Galione A. 1996. Nitric oxide induced mobilization of intracellular calcium via the cyclic ADP-ribose signalling pathway. *J. Biol. Chem.* 271: 3699-3705.
57. Fluck R, Abraham V, Miller A and Galione A. 1999. Microinjection of cyclic ADP-ribose triggers a regenerative wave of Ca^{2+} release and exocytosis of cortical alveoli in medaka eggs. *Zygote* 7: 285-292.
58. Lee HC, Aarhus R and Walseth TF. 1993. Calcium mobilization by dual receptors during fertilization of sea urchin eggs. *Science* 261: 352-355.
59. Polzonetti V, Cardinali M, Mosconi G, Natalini P, Meiri I and Carnevali O. 2002. Cyclic ADPR and calcium signaling in sea bream (Sparus aurata) egg fertilization. *Mol. Reprod. Dev.* 61: 213-217.
60. Mothet JP, Fossier P, Meunier FM, Stinnakre J, Tauc L and Baux G. 1998. Cyclic ADP-ribose and calcium-induced calcium release regulate neurotransmitter release at a cholinergic synapse of *Aplysia*. *J. Physiol. (Lond)* 507: 405-414.
61. Wu Y, Kuzma J, Marechal E, Graeff R, Lee HC, *et al.* 1997. Abscisic acid signaling through cyclic ADP-ribose in plants. *Science* 278: 2126-2130.
62. Adebanjo OA, Anandatheerthavarada HK, Koval AP, Moonga BS, Biswas G, *et al.* 1999. A new function for CD38/ADP-ribosyl cyclase in nuclear Ca^{2+} homeostasis. *Nature Cell Biol.* 1: 409-414.
63. Dipp M and Evans AM. 2001. Cyclic ADP-ribose is the primary trigger for hypoxic pulmonary vasoconstriction in the rat lung in situ. *Circ. Res.* 89: 77-83.
64. Wilson HL, Dipp M, Thomas JM, Lad C, Galione A and Evans AM. 2001. ADP-ribosyl cyclase and cyclic ADP-ribose hydrolase act as a redox sensor: a primary role for cADPR in hypoxic pulmonary vasoconstriction. *J. Biol. Chem.* 276: 11180-11188.
65. Reyes-Harde M, Empson R, Potter BV, Galione A and Stanton PK. 1999. Evidence of a role for cyclic ADP-ribose in long-term synaptic depression in hippocampus. *Proc. Natl. Acad. Sci. USA* 96: 4061-4066.
66. Sun L, Adebanjo OA, Koval A, Anandatheerthavarada HK, Iqbal J, *et al.* 2002. A novel mechanism for coupling cellular intermediary metabolism to cytosolic Ca^{2+} signaling via CD38/ADP-ribosyl cyclase, a putative intracellular NAD^+ sensor. *FASEB J.* 16: 302-314.

67. Churchill G and Louis C. 1998. Roles of Ca^{2+}, inositol trisphosphate and cyclic ADP-ribose in mediating intercellular Ca^{2+} signaling in sheep lens cells. *J. Cell Sci.* 111: 1217-1225.
68. Franco L, Zocchi E, Usai C, Guida L, Bruzzone S, *et al.* 2001. Paracrine roles of NAD^+ and cyclic ADP-ribose in increasing intracellular calcium and enhancing cell proliferation of 3T3 fibroblasts. *J. Biol. Chem.* 276: 21642-21648.
69. Zocchi E, Podesta M, Pitto A, Usai C, Bruzzone S, *et al.* 2001. Paracrinally stimulated expansion of early human hemopoietic progenitors by stroma-generated cyclic ADP-ribose. *FASEB J.* 15: 1610-1612.
70. Franco L, Bruzzone S, Song P, Guida L, Zocchi E, *et al.* 2001. Extracellular cyclic ADP-ribose potentiates ACh-induced contraction in bovine tracheal smooth muscle. *Am. J. Physiol. Lung Cell Mol. Physiol.* 280: L98-L106.
71. Murayama T, Kurebayashi N and Ogawa Y. 2000. Role of Mg^{2+} in Ca^{2+}-induced Ca^{2+} release through ryanodine receptors of frog skeletal muscle: modulations by adenine nucleotides and caffeine. *Biophys. J.* 78: 1810-1824.
72. Masumiya H, Li P, Zhang L and Chen SR. 2001. Ryanodine sensitizes the Ca^{2+} release channel (ryanodine receptor) to Ca^{2+} activation. *J. Biol. Chem.* 276: 39727-39735.
73. Du GG, Guo X, Khanna VK and MacLennan DH. 2001. Ryanodine sensitizes the cardiac Ca^{2+} release channel (ryanodine receptor isoform 2) to Ca^{2+} activation and dissociates as the channel is closed by Ca^{2+} depletion. *Proc. Natl. Acad. Sci. USA* 98: 13625-13630.
74. Lokuta AJ, Darszon A, Beltran C and Valdivia HH. 1998. Detection and functional characterization of ryanodine receptors from sea urchin eggs. *J. Physiol. (Lond)* 510: 155-164.
75. Bodding M. 2001. Histamine-induced Ca^{2+} release in bovine adrenal chromaffin cells. *Naunyn Schmiedebergs Arch. Pharmacol.* 364: 508-515.
76. DiJulio DH, Watson EL, Pessah IN, Jacobson KL, Ott SM, *et al.* 1997. Ryanodine receptor type III (Ry3R) identification in mouse parotid acini. Properties and modulation of [3H]ryanodine-binding sites. *J. Biol. Chem.* 272: 15687-15696.
77. Rosa R, Sanfeliu C, Rodriguez-Farre E, Frandsen A, Schousboe A and Sunol C. 1997. Properties of ryanodine receptors in cultured cerebellar granule neurons: effects of hexachlorocyclohexane isomers and calcium. *J. Neurosci. Res.* 47: 27-33.
78. Fessenden JD, Wang Y, Moore RA, Chen SR, Allen PD and Pessah IN. 2000. Divergent functional properties of ryanodine receptor types 1 and 3 expressed in a myogenic cell line. *Biophys. J.* 79: 2509-2525.
79. Lokuta AJ, Komai H, McDowell TS and Valdivia HH. 2002. Functional properties of ryanodine receptors from rat dorsal root ganglia. *FEBS Lett* 511: 90-96.
80. Fruen BR, Bardy JM, Byrem TM, Strasburg GM and Louis CF. 2000. Differential Ca^{2+} sensitivity of skeletal and cardiac muscle ryanodine receptors in the presence of calmodulin. *Am. J. Physiol. Cell Physiol.* 279: C724-C733.
81. Meszaros LG, Bak J and Chu A. 1993. Cyclic ADP-ribose as an endogenous regulator of the non-skeletal type ryanodine receptor Ca^{2+} channel. *Nature* 364: 76-79.
82. Singh AK. 1999. Early developmental changes in intracellular Ca^{2+} stores in rat brain. *Comp. Biochem. Physiol. A Mol. Integr. Physiol.* 123: 163-172.
83. Bourguignon LY, Chu A, Jin H and Brandt NR. 1995. Ryanodine receptor-ankyrin interaction regulates internal Ca^{2+} release in mouse T-lymphoma cells. *J. Biol. Chem.* 270: 17917-17922.
84. Guse AH, da Silva CP, Berg I, Skapenko AL, Weber K, *et al.* 1999. Regulation of calcium signalling in T lymphocytes by the second messenger cyclic ADP-ribose. *Nature* 398: 70-73.
85. Zhang X, Wen J, Bidasee KR, Besch Jr HR, Wojcikiewicz RJ, *et al.* 1999. Ryanodine and inositol trisphosphate receptors are differentially distributed and expressed in rat parotid gland. *Biochem. J.* 340: 519-527.

86. Yusufi AN, Cheng J, Thompson MA, Dousa TP, Warner GM, *et al.* 2001. cADP-ribose/ryanodine channel/Ca^{2+}-release signal transduction pathway in mesangial cells. *Am. J. Physiol. Renal Physiol.* 281: F91-F102.
87. Tanaka Y and Tashjian AH. 1995. Calmodulin is a selective mediator of Ca^{2+}-induced Ca^{2+} release via the ryanodine receptor-like Ca^{2+} channel triggered by cyclic ADP-ribose. *Proc. Natl. Acad. Sci. USA* 92: 3244-3248.
88. Buck WR, Hoffmann EE, Rakow TL and Shen SS. 1994. Synergistic calcium release in the sea urchin egg by ryanodine and cyclic ADP ribose. *Dev. Biol.* 163: 1-10.
89. Panfoli I, Burlando B and Viarengo A. 1999. Cyclic ADP-ribose-dependent Ca^{2+} release is modulated by free [Ca^{2+}] in the scallop sarcoplasmic reticulum. *Biochem. Biophys. Res. Commun.* 257: 57-62.
90. Guo X and Becker PL. 1997. Cyclic ADP-ribose-gated Ca^{2+} release in sea urchin eggs requires an elevated [Ca^{2+}]. *J. Biol. Chem.* 272: 16984-16989.
91. Li PL, Tang WX, Valdivia HH, Zou AP and Campbell WB. 2001. cADP-ribose activates reconstituted ryanodine receptors from coronary arterial smooth muscle . *Am. J. Physiol. Heart Circ. Physiol.* 280: H208-H215.
92. Sonnleitner A, Conti A, Bertocchini F, Schindler H and Sorrentino V. 1998. Functional properties of the ryanodine receptor type 3 (RyR3) Ca^{2+} release channel. *EMBO J.* 17: 2790-2798.
93. Xiong H, Feng X, Gao L, Xu L, Pasek DA, *et al.* 1998. Identification of a two EF-hand Ca^{2+} binding domain in lobster skeletal muscle ryanodine receptor/Ca^{2+} release channel. *Biochemistry* 37: 4804-4814.
94. Zhang JJ, Williams AJ and Sitsapesan R. 1999. Evidence for novel caffeine and Ca^{2+} binding sites on the lobster skeletal ryanodine receptor. *Br. J. Pharmacol.* 126: 1066-1074.
95. Balshaw DM, Yamaguchi N and Meissner G. 2002. Modulation of intracellular calcium-release channels by calmodulin. *J. Membr. Biol.* 185: 1-8.
96. Meissner G and Henderson JS. 1987. Rapid calcium release from cardiac sarcoplasmic reticulum vesicles is dependent on Ca^{2+} and is modulated by Mg^{2+}, adenine nucleotide, and calmodulin. *J. Biol. Chem.* 262: 3065-3073.
97. Lee HC, Aarhus R, Graeff R, Gurnack ME and Walseth TF. 1994. Cyclic ADP ribose activation of the ryanodine receptor is mediated by calmodulin. *Nature* 370: 307-309.
98. Ozawa T. 1999. Ryanodine-sensitive Ca^{2+} release mechanism of rat pancreatic acinar cells is modulated by calmodulin. *Biochim. Biophys. Acta* 1452: 254-262.
99. Zhang X, Wen J, Bidasee KR, Besch HR, Jr. and Rubin RP. 1997. Ryanodine receptor expression is associated with intracellular Ca^{2+} release in rat parotid acinar cells. *Am. J. Physiol.* 273: C1306-C1314.
100. Perez CF, Marengo JJ, Bull R and Hidalgo C. 1998. Cyclic ADP-ribose activates caffeine-sensitive calcium channels from sea urchin egg microsomes. *Am. J. Physiol. Cell Physiol.* 43: C430-C439.
101. Thomas JM, Masgrau R, Churchill GC and Galione A. 2001. Pharmacological characterization of the putative cADP-ribose receptor. *Biochem. J.* 359: 451-457.
102. Balshaw DM, Xu L, Yamaguchi N, Pasek DA and Meissner G. 2001. Calmodulin binding and inhibition of cardiac muscle calcium release channel (ryanodine receptor). *J. Biol. Chem.* 276: 20144-20153.
103. Samso M and Wagenknecht T. 2002. Apocalmodulin and Ca^{2+}-calmodulin bind to neighboring locations on the ryanodine receptor. *J. Biol. Chem.* 277: 1349-1353.
104. Rodney GG, Krol J, Williams B, Beckingham K and Hamilton SL. 2001. The carboxy-terminal calcium binding sites of calmodulin control calmodulin's switch from an activator to an inhibitor of RYR1. *Biochemistry* 40: 12430-12435.
105. Yamaguchi N and Kasai M. 1997. Potentiation of depolarization-induced calcium release from skeletal muscle triads by cyclic ADP-ribose and inositol 1,4,5-trisphosphate. *Biochem. Biophys. Res. Commun.* 240: 772-777.

106. Fulceri R, Rossi R, Bottinelli R, Conti A, Intravaia E, *et al.* 2001. Ca^{2+} release induced by cyclic adp ribose in mice lacking type 3 ryanodine receptor. *Biochem. Biophys. Res. Commun.* 288: 697-702.
107. Takasawa S, Ishida A, Nata K, Nakagawa K, Noguchi N, *et al.* 1995. Requirement of calmodulin-dependent protein kinase II in cyclic ADP- ribose-mediated intracellular Ca^{2+} mobilization. *J. Biol. Chem.* 270: 30257-30259.
108. Dulhunty AF, Laver D, Curtis SM, Pace S, Haarmann C and Gallant EM. 2001. Characteristics of irreversible ATP activation suggest that native skeletal ryanodine receptors can be phosphorylated via an endogenous CaMKII. *Biophys. J.* 81: 3240-3252.
109. Bandyopadhyay A, Shin DW, Ahn JO and Kim DH. 2000. Calcineurin regulates ryanodine receptor/Ca^{2+}-release channels in rat heart. *Biochem. J.* 352 Pt 1: 61-70.
110. Berridge M, Lipp P and Bootman M. 2000. The versatility and universality of calcium signalling. *Nature Mol. Cell Biol. Rev.* 1: 11-21.
111. Tripathy A and Meissner G. 1996. Sarcoplasmic reticulum luminal Ca^{2+} has access to cytosolic activation and inactivation sites of skeletal muscle Ca^{2+} release channel. *Biophys. J.* 70: 2600-2615.
112. Ching LL, Williams AJ and Sitsapesan R. 2000. Evidence for Ca^{2+} activation and inactivation sites on the luminal side of the cardiac ryanodine receptor complex. *Circ. Res.* 87: 201-206.
113. Lukyanenko V, Gyorke I and Gyorke S. 1996. Regulation of calcium release by calcium inside the sarcoplasmic reticulum in ventricular myocytes. *Pflugers Arch.* 432: 1047-1054.
114. Gyorke I and Gyorke S. 1998. Regulation of the cardiac ryanodine receptor channel by luminal Ca^{2+} involves luminal Ca^{2+} sensing sites. *Biophys. J.* 75: 2801-2810.
115. Gilchrist JS, Belcastro AN and Katz S. 1992. Intraluminal Ca^{2+} dependence of Ca^{2+} and ryanodine-mediated regulation of skeletal muscle sarcoplasmic reticulum Ca^{2+} release. *J. Biol. Chem.* 267: 20850-20856.
116. Herrmann-Frank A and Lehmann-Horn F. 1996. Regulation of the purified Ca^{2+} release channel/ryanodine receptor complex of skeletal muscle sarcoplasmic reticulum by luminal calcium. *Pflugers Arch.* 432: 155-157.
117. Beard NA, Sakowska MM, Dulhunty AF and Laver DR. 2002. Calsequestrin is an inhibitor of skeletal muscle ryanodine receptor calcium release channels. *Biophys. J.* 82: 310-320.
118. Nguyen T, Chin WC and Verdugo P. 1998. Role of Ca^{2+}/K^{+} ion exchange in intracellular storage and release of Ca^{2+}. *Nature* 395: 908-912.
119. Ikemoto N, Antoniu B, Kang JJ, Meszaros LG and Ronjat M. 1991. Intravesicular calcium transient during calcium release from sarcoplasmic reticulum. *Biochemistry* 30: 5230-5237.
120. Sitsapesan R and Williams AJ. 1995. Cyclic ADP-ribose and related compounds activate sheep skeletal sarcoplasmic reticulum Ca^{2+} release channel. *Am. J. Physiol.* 268: C1235-C1240.
121. Galione A, McDougall A, Busa WB, Willmott N, Gillot I and Whitaker M. 1993. Redundant mechanisms of calcium-induced calcium-release underlying calcium waves during fertilization of sea-urchin eggs. *Science* 261: 348-352.
122. Lukyanenko V, Gyorke I, Wiesner TF and Gyorke S. 2001. Potentiation of Ca^{2+} release by cADP-Ribose in the heart is mediated by enhanced SR Ca^{2+} uptake into the sarcoplasmic reticulum. *Circ. Res.* 89: 614-622.
123. Dulhunty A, Haarmann C, Green D and Hart J. 2000. How many cysteine residues regulate ryanodine receptor channel activity? *Antioxid. Redox. Signal.* 2: 27-34.
124. Eager KR and Dulhunty AF. 1999. Cardiac ryanodine receptor activity is altered by oxidizing reagents in either the luminal or cytoplasmic solution. *J. Membr. Biol.* 167: 205-214.

125. Sun J, Xu L, Eu JP, Stamler JS and Meissner G. 2001. Classes of thiols that influence the activity of the skeletal muscle calcium release channel. *J. Biol. Chem.* 276: 15625-15630.
126. Elferink JG. 1999. Thimerosal: a versatile sulfhydryl reagent, calcium mobilizer, and cell function-modulating agent. *Gen. Pharmacol.* 33: 1-6.
127. Donoso P, Aracena P and Hidalgo C. 2000. Sulfhydryl oxidation overrides Mg^{2+} inhibition of calcium-induced calcium release in skeletal muscle triads. *Biophys. J.* 79: 279-286.
128. Suko J and Hellmann G. 1998. Modification of sulfhydryls of the skeletal muscle calcium release channel by organic mercurial compounds alters Ca^{2+} affinity of regulatory Ca^{2+} sites in single channel recordings and [^{3}H]ryanodine binding. *Biochim. Biophys. Acta* 1404: 435-450.
129. Swann K. 1991. Thimerosal causes calcium oscillations and sensitizes calcium-induced calcium release in unfertilized hamster eggs. *FEBS Lett* 278: 175-178.
130. Swann K. 1992. Different triggers for calcium oscillations in mouse eggs involve a ryanodine-sensitive calcium store. *Biochem. J.* 287: 79-84.
131. Marengo JJ, Hidalgo C and Bull R. 1998. Sulfhydryl oxidation modifies the calcium dependence of ryanodine-sensitive calcium channels of excitable cells. *Biophys. J.* 74: 1263-1277.
132. Aghdasi B, Zhang JZ, Wu Y, Reid MB and Hamilton SL. 1997. Multiple classes of sulfhydryls modulate the skeletal muscle Ca^{2+} release channel. *J. Biol. Chem.* 272: 3739-3748.
133. Abramson JJ, Zable AC, Favero TG and Salama G. 1995. Thimerosal interacts with the Ca^{2+} release channel ryanodine receptor from skeletal muscle sarcoplasmic reticulum. *J. Biol. Chem.* 270: 29644-29647.
134. Suzuki YJ, Cleemann L, Abernethy DR and Morad M. 1998. Glutathione is a cofactor for H_2O_2-mediated stimulation of Ca^{2+}-induced Ca^{2+} release in cardiac myocytes. *Free Radic. Biol. Med.* 24: 318-325.
135. Machaty Z, Wang WH, Day BN and Prather RS. 1999. Calcium release and subsequent development induced by modification of sulfhydryl groups in porcine oocytes. *Biol. Reprod.* 60: 1384-1391.
136. McDougall A, Gillot I and Whitaker M. 1993. Thimerosal reveals calcium-induced calcium release in unfertilised sea urchin eggs. *Zygote* 1: 35-42.
137. Zable AC, Favero TG and Abramson JJ. 1997. Glutathione modulates ryanodine receptor from skeletal muscle sarcoplasmic reticulum. Evidence for redox regulation of the Ca^{2+} release mechanism. *J. Biol. Chem.* 272: 7069-7077.
138. Feng W, Liu G, Allen PD and Pessah IN. 2000. Transmembrane redox sensor of ryanodine receptor complex. *J. Biol. Chem.* 275: 35902-35907.
139. Masuda W, Takenaka S, Tsuyama S, Tokunaga M, Yamaji R, *et al.* 1997. Inositol 1,4,5-trisphosphate and cyclic ADP-ribose mobilize Ca^{2+} in a protist, Euglena gracilis. *Comp. Biochem. Physiol. C-Pharmacol. Toxicol. Endocr.* 118: 279-283.
140. Tanaka Y and Tashjian AH, Jr. 1994. Thimerosal potentiates Ca^{2+} release mediated by both the inositol 1,4,5-trisphosphate and the ryanodine receptors in sea urchin eggs. Implications for mechanistic studies on Ca^{2+} signaling. *J. Biol. Chem.* 269: 11247-11253.
141. Chini EN, Liang M and Dousa TP. 1998. Differential effect of pH upon cyclic-ADP-ribose and nicotinate-adenine dinucleotide phosphate-induced Ca^{2+} release systems. *Biochem. J.* 335: 499-504.
142. Sun J, Xin C, Eu JP, Stamler JS and Meissner G. 2001. Cysteine-3635 is responsible for skeletal muscle ryanodine receptor modulation by NO. *Proc. Natl. Acad. Sci. USA* 98: 11158-11162.
143. Feng W, Liu G, Allen PD and Pessah IN. 2000. Transmembrane redox sensor of ryanodine receptor complex. *J. Biol. Chem.* 275: 35902-35907.
144. Eu JP, Sun J, Xu L, Stamler JS and Meissner G. 2000. The skeletal muscle calcium release channel: coupled O_2 sensor and NO signaling functions. *Cell* 102: 499-509.

145. Lee HC. 1991. Specific binding of cyclic ADP-ribose to calcium-storing microsomes from sea urchin eggs. *J. Biol. Chem.* 266: 2276-2281.
146. Kuemmerle JF and Makhlouf GM. 1995. Agonist-stimulated cyclic ADP ribose. Endogenous modulator of Ca^{2+}- induced Ca^{2+} release in intestinal longitudinal muscle. *J. Biol. Chem.* 270: 25488-25494.
147. MacKrill JJ. 1999. Protein-protein interactions in intracellular Ca^{2+}-release channel function. *Biochem. J.* 337: 345-361.
148. Sitsapesan R, McGarry SJ and Williams AJ. 1995. Cyclic ADP-ribose, the ryanodine receptor and Ca^{2+} release. *Trends Pharmacol. Sci.* 16: 386-391.
149. Walseth TF, Aarhus R, Kerr JA and Lee HC. 1993. Identification of cyclic ADP-ribose-binding proteins by photoaffinity labeling. *J. Biol. Chem.* 268: 26686-26691.
150. Paul-Pletzer K, Palnitkar SS, Jimenez LS, Morimoto H and Parness J. 2001. The skeletal muscle ryanodine receptor identified as a molecular target of [^{3}H]azidodantrolene by photoaffinity labeling. *Biochemistry* 40: 531-542.
151. Brillantes AB, Ondrias K, Scott A, Kobrinsky E, Ondriasova E, *et al.* 1994. Stabilization of calcium release channel (ryanodine receptor) function by FK506-binding protein. *Cell* 77: 513-523.
152. Marx SO, Ondrias K and Marks AR. 1998. Coupled gating between individual skeletal muscle Ca^{2+} release channels (ryanodine receptors). *Science* 281: 818-821.
153. Marx SO, Gaburjakova J, Gaburjakova M, Henrikson C, Ondrias K and Marks AR. 2001. Coupled gating between cardiac calcium release channels (ryanodine receptors). *Circ. Res.* 88: 1151-1158.
154. Noguchi N, Takasawa S, Nata K, Tohgo A, Kato I, *et al.* 1997. Cyclic ADP-ribose binds to FK506-binding protein 12.6 to release Ca^{2+} from islet microsomes. *J. Biol. Chem.* 272: 3133-3136.
155. Hashii M, Minabe Y and Higashida H. 2000. cADP-ribose potentiates cytosolic Ca^{2+} elevation and Ca^{2+} entry via L- type voltage-activated Ca^{2+} channels in NG108-15 neuronal cells. *Biochem. J.* 345: 207-215.
156. Tang WX, Chen YF, Zou AP, Campbell WB and Li PL. 2002. Role of FKBP12.6 in cADPR-induced activation of reconstituted ryanodine receptors from arterial smooth muscle. *Am. J. Physiol. Heart Circ. Physiol.* 282: H1304-H1310.
157. Bultynck G, Rossi D, Callewaert G, Missiaen L, Sorrentino V, *et al.* 2001. The conserved sites for the FK506-binding proteins in ryanodine receptors and inositol 1,4,5-trisphosphate receptors are structurally and functionally different. *J. Biol. Chem.* 276: 47715-47724.
158. Walseth TF, Wong L, Graeff RM and Lee HC. 1997. Bioassay for determining endogenous levels of cyclic ADP-ribose. *Meth. Enzymol.* 280: 287-294.
159. Lee HC, Galione A and Walseth TF. 1994. Cyclic ADP-ribose: metabolism and calcium mobilizing function. *Vitam. Horm.* 48: 199-257.
160. Takahashi K, Kukimoto I, Tokita K, Inageda K, Inoue S, *et al.* 1995. Accumulation of cyclic ADP-ribose measured by a specific radioimmunoassay in differentiated human leukemic HL-60 cells with all- trans-retinoic acid. *FEBS Lett* 371: 204-208.
161. Graeff R and Lee HC. 2002. A novel cycling assay for cellular cADP-ribose with nanomolar sensitivity. *Biochem. J.* 361: 379-384.
162. Kuroda R, Kontani K, Kanda Y, Katada T, Nakano T, *et al.* 2001. Increase of cGMP, cADP-ribose and inositol 1,4,5-trisphosphate preceding Ca^{2+} transients in fertilization of sea urchin eggs. *Development* 128: 4405-4414.
163. Takasawa S, Akiyama T, Nata K, Kuroki M, Tohgo A, *et al.* 1998. Cyclic ADP-ribose and inositol 1,4,5-trisphosphate as alternate second messengers for intracellular Ca^{2+} mobilization in normal and diabetic beta-cells. *J. Biol. Chem.* 273: 2497-2500.
164. Webb DL, Islam MS, Efanov AM, Brown G, Kohler M, *et al.* 1996. Insulin exocytosis and glucose-mediated increase in cytoplasmic free Ca^{2+} concentration in the pancreatic beta-cell are independent of cyclic ADP-ribose. *J. Biol. Chem.* 271: 19074-19079.

165. Morita K, Kitayama S and Dohi T. 1997. Stimulation of cyclic ADP-ribose synthesis by acetylcholine and its role in catecholamine release in bovine adrenal chromaffin cells. *J. Biol. Chem.* 272: 21002-21009.
166. Fukushi Y, Kato I, Takasawa S, Sasaki T, Ong BH, *et al.* 2001. Identification of cyclic ADP-ribose-dependent mechanisms in pancreatic muscarinic Ca^{2+} signaling using CD38 knockout mice. *J Biol Chem* 276: 649-655.
167. Kato I, Yamamoto Y, Fujimura M, Noguchi N, Takasawa S and Okamoto H. 1999. CD38 disruption impairs glucose-induced increases in cyclic ADP-ribose, $[Ca^{2+}]_i$, and insulin secretion. *J. Biol. Chem.* 274: 1869-1872.
168. Partida-Sanchez S, Cockayne DA, Monard S, Jacobson EL, Oppenheimer N, *et al.* 2001. Cyclic ADP-ribose production by CD38 regulates intracellular calcium release, extracellular calcium influx and chemotaxis in neutrophils and is required for bacterial clearance in vivo. *Nature Med.* 7: 1209-1216.
169. Chini EN, de Toledo FG, Thompson MA and Dousa TP. 1997. Effect of estrogen upon cyclic ADP ribose metabolism: beta-estradiol stimulates ADP ribosyl cyclase in rat uterus. *Proc. Natl. Acad. Sci. USA* 94: 5872-5876.
170. de Toledo FG, Cheng J, Liang M, Chini EN and Dousa TP. 2000. ADP-Ribosyl cyclase in rat vascular smooth muscle cells: properties and regulation. *Circ. Res.* 86: 1153-1159.
171. Dupont G and Swillens S. 1996. Quantal release, incremental detection, and long-period Ca^{2+} oscillations in a model based on regulatory Ca^{2+}-binding sites along the permeation pathway. *Biophys. J.* 71: 1714-1722.
172. Morgan AJ and Jacob R. 1998. Differential modulation of the phases of a Ca^{2+} spike by the store Ca^{2+}-ATPase in human umbilical vein endothelial cells. *J Physiol* 513: 83-101.
173. Brini M, Bano D, Manni S, Rizzuto R and Carafoli E. 2000. Effects of PMCA and SERCA pump overexpression on the kinetics of cell Ca^{2+} signalling. *EMBO J.* 19: 4926-4935.
174. Petersen CC, Petersen OH and Berridge MJ. 1993. The role of endoplasmic reticulum calcium pumps during cytosolic calcium spiking in pancreatic acinar cells. *J. Biol. Chem.* 268: 22262-22264.
175. Empson RM and Galione A. 1997. Cyclic ADP-ribose enhances coupling between voltage-gated Ca^{2+} entry and intracellular Ca^{2+} release. *J. Biol. Chem.* 272: 20967-20970.
176. Taylor CW, Genazzani AA and Morris SA. 1999. Expression of inositol trisphosphate receptors. *Cell Calcium* 26: 237-251.
177. Churchill GC and Galione A. 2000. Spatial control of Ca^{2+} signaling by nicotinic acid adenine dinucleotide phosphate diffusion and gradients. *J. Biol. Chem.* 275: 38687-38692.
178. Cancela JM, Churchill GC and Galione A. 1999. Coordination of agonist-induced Ca^{2+}-signalling patterns by NAADP in pancreatic acinar cells. *Nature* 398: 74-76.
179. Cancela JM, Gerasimenko OV, Gerasimenko JV, Tepikin AV and Petersen OH. 2000. Two different but converging messenger pathways to intracellular Ca^{2+} release: the roles of nicotinic acid adenine dinucleotide phosphate, cyclic ADP-ribose and inositol trisphosphate. *EMBO J.* 19: 2549-2557.
180. Tinel H, Cancela JM, Mogami H, Gerasimenko JV, Gerasimenko OV, *et al.* 1999. Active mitochondria surrounding the pancreatic acinar granule region prevent spreading of inositol trisphosphate-evoked local cytosolic Ca^{2+} signals. *EMBO J.* 18: 4999-5008.
181. Kiselyov K, Shin DM, Shcheynikov N, Kurosaki T and Muallem S. 2001. Regulation of Ca^{2+}-release-activated Ca^{2+} current (I_{CRAC}) by ryanodine receptors in inositol 1,4,5-trisphosphate-receptor-deficient DT40 cells. *Biochem. J.* 360: 17-22.
182. Paltauf-Doburzynska J, Posch K, Paltauf G and Graier WF. 1998. Stealth ryanodine-sensitive Ca^{2+} release contributes to activity of capacitative Ca^{2+} entry and nitric oxide synthase in bovine endothelial cells. *J. Physiol.* 513: 369-379.
183. Guse AH, Berg I, da Silva CP, Potter BV and Mayr GW. 1997. Ca^{2+} entry induced by cyclic ADP-ribose in intact T-lymphocytes. *J. Biol. Chem.* 272: 8546-8550.

184. Berg I, Potter BV, Mayr GW and Guse AH. 2000. Nicotinic acid adenine dinucleotide phosphate (NAADP$^+$) is an essential regulator of T-lymphocyte Ca^{2+}-signaling. *J. Cell Biol.* 150: 581-588.
185. Chini EN and Dousa TP. 1996. Nicotinate-adenine dinucleotide phosphate-induced Ca^{2+}-release does not behave as a Ca^{2+}-induced Ca^{2+}-release system. *Biochem. J.* 316: 709-711.
186. Duchen MR. 1999. Contributions of mitochondria to animal physiology: from homeostatic sensor to calcium signalling and cell death. *J. Physiol.* 516: 1-17.
187. Ichas F, Jouaville LS and Mazat JP. 1997. Mitochondria are excitable organelles capable of generating and conveying electrical and calcium signals. *Cell* 89: 1145-1153.
188. Montero M, Alonso MT, Albillos A, Garcia-Sancho J and Alvarez J. 2001. Mitochondrial Ca^{2+}-induced Ca^{2+} release mediated by the Ca^{2+} uniporter. *Mol. Biol. Cell* 12: 63-71.
189. Beutner G, Sharma VK, Giovannucci DR, Yule DI and Sheu SS. 2001. Identification of a ryanodine receptor in rat heart mitochondria. *J. Biol. Chem.* 276: 21482-21488.
190. Liang M, Chini EN, Cheng J and Dousa TP. 1999. Synthesis of NAADP and cADPR in mitochondria. *Arch. Biochem. Biophys.* 371: 317-325.
191. Ziegler M, Jorcke D and Schweiger M. 1997. Identification of bovine liver mitochondrial NAD$^+$ glycohydrolase as ADP-ribosyl cyclase. *Biochem. J.* 326: 401-405.
192. Wilding M, Russo GL, Galione A, Marino M and Dale B. 1998. ADP-ribose gates the fertilization channel in ascidian oocytes. *Am. J. Physiol.* 275: C1277-C1283.
193. Sano Y, Inamura K, Miyake A, Mochizuki S, Yokoi H, *et al.* 2001. Immunocyte Ca^{2+} influx system mediated by LTRPC2. *Science* 293: 1327-1330.
194. Perraud AL, Fleig A, Dunn CA, Bagley LA, Launay P, *et al.* 2001. ADP-ribose gating of the calcium-permeable LTRPC2 channel revealed by Nudix motif homology. *Nature* 411: 595-599.

Antony Galione, Department of Pharmacology,
Oxford University, Mansfield Road,
Oxford OX1 3QT, UK
e-mail: antony.galione@pharm.ox.ac.uk

Cyclic ADP-Ribose and NAADP. Structures, Metabolism and Functions. Lee HC (ed), pp. 199-215, Dordrecht: Kluwer.

Spatial and Temporal Control of Calcium Signaling by NAADP

Grant C. Churchill, Sandip Patel, Justyn M. Thomas and Antony Galione
Department of Pharmacology, Oxford University, Mansfield Road, Oxford OX1 3QT, UK.

IMPORTANCE OF Ca^{2+} SIGNALING IN TIME AND SPACE

Within an individual cell Ca^{2+} plays many roles, some of which are seemingly contradictory. For example, in a smooth muscle cell, Ca^{2+} controls growth, differentiation, gene expression, and paradoxically both contraction and relaxation [1]. How can Ca^{2+} be linked to such a multitude of processes and maintain any specificity? For many of these processes, if not all, the answer is that Ca^{2+} increases are under tight control both spatially and temporally [2, 3]. The importance of the spatiotemporal aspect of a Ca^{2+} signal arises from necessity; Ca^{2+} is an atom and unlike other intracellular messengers, it cannot be created or destroyed in cells, only moved around. Spatially, localized Ca^{2+} increases control secretion in pancreatic acinar cells [4, 5], growth cone migration in neurones [6], and relaxation in smooth muscle cells [1], whereas global Ca^{2+} signals control contraction of all types of muscle cells and fertilization of eggs from marine invertebrates to mammals [7-9]. Temporally, the frequency of repetitive increases in Ca^{2+} (oscillations) control cell division [10] and gene expression [11, 12]. Cells decode the Ca^{2+} oscillations via enzymes, which exhibit different kinetics of activation and inactivation by Ca^{2+}, effectively tuning their activity to certain Ca^{2+} oscillation frequencies [3, 12].

Ca^{2+}-Mobilizing Messengers

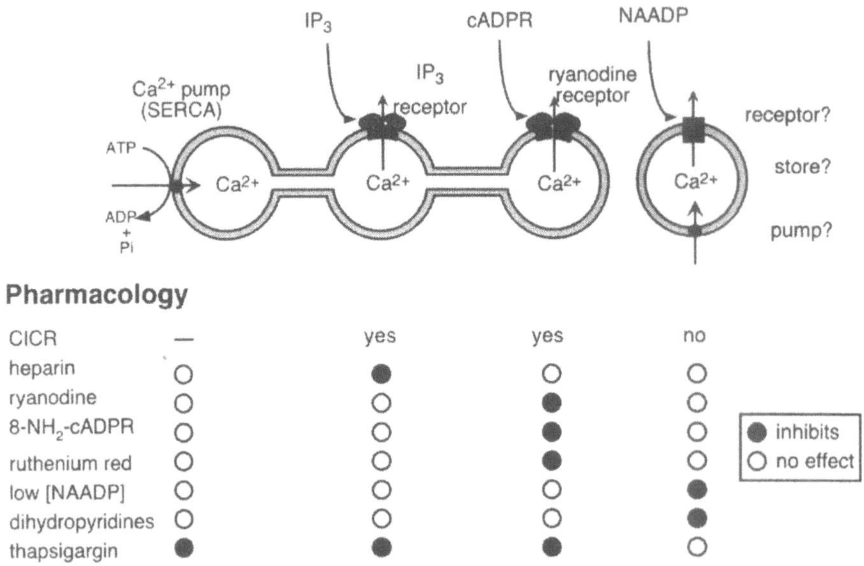

Figure 1. **Mechanisms and pharmacology of Ca^{2+} regulation by intracellular stores in the sea urchin egg.** Data are summarized from [3, 5. 14, 17, 18, 21, 23, 56, 67. 71. 80].

Ca^{2+}-mobilizing intracellular compounds in the sea urchin currently include inositol 1,4,5-trisphosphate (IP_3) [3], cyclic ADP-ribose (cADPR) [13-15] and nicotinic acid adenine dinucleotide phosphate (NAADP) [16-19] (certain other cell types also respond to sphingosine 1-phosphate [20]). The common property of these messengers is that they release Ca^{2+} from intracellular stores (Fig. 1). The molecular details of activation and Ca^{2+} release varies with messenger, and the amount known is strongly related to the length of time the messenger has been studied. IP_3 was discovered first and therefore best characterized [3, 21]. IP_3 is produced by the action of phospholipase C, which hydrolyzes the head group (IP_3) off phosphatidyl inositol 4,5-bisphosphate. IP_3 diffuses through the cytosol and activates its receptor-channel on the endoplasmic reticulum, thereby releasing Ca^{2+} (Fig. 1.). As the archetypal messenger, IP_3 has served as a paradigm for how other Ca^{2+}-releasing messengers are expected to function. This paradigm holds well for cADPR [14, 17], which acts on ryanodine receptors [22], but does not hold for NAADP. The properties and function of NAADP are markedly different from IP_3 and cADPR [17, 18]. Indeed, it has been the unexpected properties of NAADP that has spurred the emerging concept that spatiotemporal Ca^{2+} signals are due to both multiple intracellular Ca^{2+} stores, multiple Ca^{2+}-mobilizing messengers (Fig. 1) and interactions between these [3, 23, 24].

Unique Properties of NAADP

NAADP operates through a pathway that is pharmacologically and physically distinct from IP_3- and cADPR-sensitive Ca^{2+} release pathways [17, 18, 25] (Fig. 1). In the sea urchin egg, NAADP is unique among the Ca^{2+}-releasing messengers in at least four respects. First, NAADP-mediated Ca^{2+} release is neither potentiated nor inhibited by divalent cations [26, 27]. Second, NAADP potently self-inactivates even when present at concentrations less than those required to activate detectable Ca^{2+} release [17, 28, 29]. Third, once bound to its receptor, radiolabelled NAADP is occluded and cannot be displaced by nonlabelled NAADP, which is thought to be related to its self-inactivation mechanism [28, 30, 31]. Fourth, NAADP mobilizes Ca^{2+} from a store that is distinct pharmacologically and physically from the IP_3- and cADPR-sensitive Ca^{2+} stores [16, 29, 32].

TRADITIONAL MECHANISMS UNDERLYING Ca^{2+} OSCILLATIONS AND WAVES

Hierarchical Structure Supports Spatiotemporal Ca^{2+} Signals

Traditionally, spatiotemporal Ca^{2+} signaling has meant that mediated by IP_3 receptors and/or ryanodine receptors [3]. Such Ca^{2+} signals have a unifying underlying mechanism based on the principle of positive and negative feedback by Ca^{2+} itself on the channels. Both the IP_3 receptor and the ryanodine receptor exhibit regulation by Ca^{2+} whereby increases in Ca^{2+} several-fold over resting activate the channel, whereas increases 5-100 fold over resting Ca^{2+} inhibit the channel [33, 34]. The regenerative autocatalytic positive feedback by Ca^{2+} is termed Ca^{2+}-induced Ca^{2+} release (CICR) (Fig. 2). The sensitivity to CICR is increased by IP_3 acting on the IP_3 receptor [21, 35, 36] and cADPR acting on the ryanodine receptor [22] (Fig. 1 and 2).

Spatiotemporal Ca^{2+} signaling is controlled by a hierarchy of events [2, 3, 35, 36]. The most fundamental events are the opening of single channels which gives rise to what are termed blips when mediated by IP_3 receptors [37] and quarks when mediated by ryanodine receptors [38] (Fig. 2). Channels are arranged in clusters [35, 36, 39]. Thus, when one channel opens, it is likely that the others in the cluster will also open, giving rise to an elementary event called a puff when mediated by IP_3 receptors [36] or spark when mediated by ryanodine receptors [40] (Fig. 2).

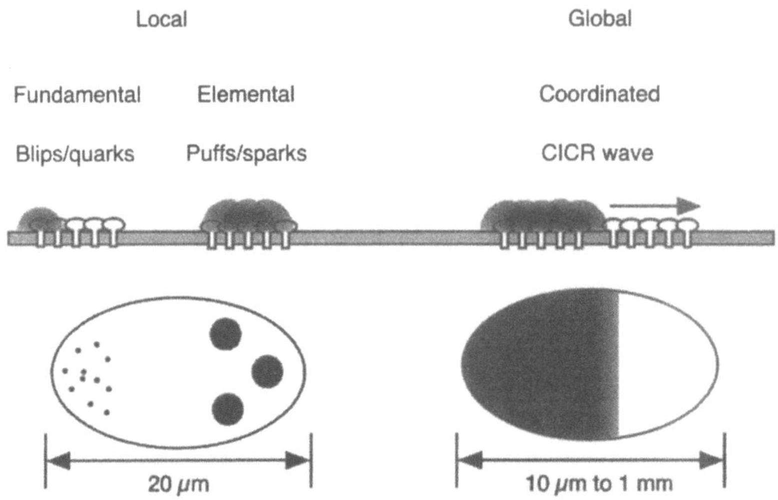

Figure 2. **Spatiotemporal Ca^{2+} signaling mediated by controlling the sensitivity of Ca^{2+}-induced Ca^{2+} release.** Ca^{2+} signals are organized into a hierarchy of events. Single channels and clusters of channels give rise to localized responses, whereas coupling between these clusters gives rise to global Ca^{2+} waves.

The Ca^{2+} emanating from the clusters of CICR channels remains confined to a small volume of space, at least initially. This is due to the strong buffering of Ca^{2+} by large, relatively immobile buffers in the cytoplasm making it poorly diffusable [41] as well as sequestration by organelles [3, 35, 36]. However, as the Ca^{2+} concentration increases, the buffers become saturated, the free Ca^{2+} concentration increases and Ca^{2+} diffuses farther and farther from the cluster of channels as a function of concentration [41]. The response of the neighbouring channels to this Ca^{2+} depends largely on their sensitivity to Ca^{2+}. When either IP_3 [35, 36] or cADPR [42] is elevated sufficiently, the cytoplasm becomes an excitable medium in which any increase in Ca^{2+} will propagate through CICR. This is the regenerative mechanism underlying Ca^{2+} waves [35, 36]. Depending on the concentrations of IP_3 and cADPR, a Ca^{2+} increase can remain localized, spread abortively, or propagate with constant amplitude and velocity (5-100 μm/s) throughout an entire cell [35, 36, 39, 43], group of cells [44] or entire regions of organ such as the brain [45] and liver [46].

Mechanisms of Ca^{2+} Oscillations

Ca^{2+} oscillations are the temporal analogue of Ca^{2+} waves [3, 47, 48]. Several theoretical models have been proposed to explain Ca^{2+} oscillations, and can be classified as to the number of Ca^{2+} stores required (one or two) and whether the level of IP_3 must also oscillate [49-51]. Depending on the cell type and experimental conditions, evidence exists for mechanisms requiring oscillating levels of IP_3 [52-54] and control by positive and subsequently negative feedback of Ca^{2+} on Ca^{2+} release channels with a steady level of IP_3 [54-56]. Additionally, Berridge and colleagues [49, 50] proposed a two-pool mechanism in which Ca^{2+} was transferred from IP_3-sensitive stores into other CICR stores, which overload and spontaneously release Ca^{2+}.

NAADP-MEDIATED Ca^{2+} WAVES

When NAADP was injected into a sea urchin egg, it gave rise to an apparently regenerative Ca^{2+} wave [57]. Given that NAADP does not exhibit CICR [26] we investigated how it could mediate a Ca^{2+} wave. We tested three possibilities: (1) NAADP released trigger Ca^{2+} and this trigger Ca^{2+} ignited a CICR wave mediated by CICR channels (i.e., the IP_3 receptor and ryanodine receptor); (2) NAADP released Ca^{2+} locally that stimulated the production of either IP_3 or cADPR, which could then diffuse globally, making the entire egg an excitable medium; (3) NAADP diffusion underlied the spread of the Ca^{2+} increase.

Propagation by NAADP Diffusion

These experiments required releasing controlled amounts of NAADP with great spatial precision. This was achieved with the use of caged NAADP [58] and a confocal microscope modified to enable spatially localized photolysis [59]. When we photoreleased NAADP locally at low concentrations, even though the local increase in Ca^{2+} was very high, the resulting Ca^{2+} wave was not global and appeared diffusive (Fig. 3). A plot comparing the distance at which the Ca^{2+} increase reached half its maximum versus time revealed that velocity (the slope of this plot) decreased with time (and distance) (Fig. 3). Processes governed by diffusion do not propagate directly with time but rather directly with the square root of time [60]. When the data are transformed so that distance squared (the algebraic equivalent of the square root of time for this purpose) is plotted against time the values fit a straight line (Fig. 3), consistent with diffusion. The slope of the line has the units of distance squared divided by time, the dimensions of a diffusion coefficient. The diffusion coefficient can be calculated from this slope by dividing it by the dimension term for diffusion. In one dimension a particle can go two ways so the coefficient is 2, for 2 dimensional diffusion a particle can go in four directions, and in 3 dimensions, 6 directions [60]. In this specific case, the number is, arguably, 6. This yields a diffusion coefficient

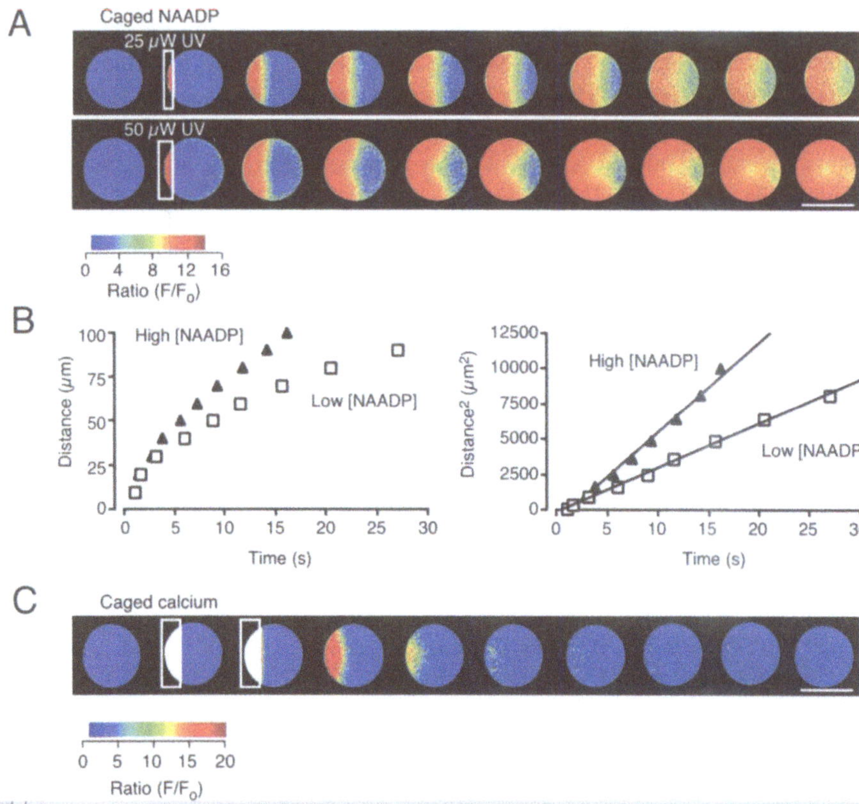

Figure 3. **NAADP mediates the spread of Ca^{2+} increase by the diffusion of NAADP itself. A.** The extent of the Ca^{2+} wave depends on the concentration of NAADP. NAADP was photoreleased (boxed region) to a relatively high or low concentration by varying the power of the UV laser (indicated in μW). Images are false colored to indicate the ratio as indicated by the calibration scale. Scale bar, 100 μm. **B**. Plots of distance and distance squared by time demonstrating the decrease in wave velocity (slope) with distance. Distance was taken as the point at which the ratio reached half of its maximum. **C.** Ca^{2+} diffuses poorly in the sea urchin egg cytosol and fails to trigger a regenerative Ca^{2+} wave. Ca^{2+} was photreleased from caged Ca^{2+} in the boxed area. Sea urchin eggs were microinjected with fluo-3 (50 μM) and either caged NAADP (0.5 μM) or caged Ca^{2+} (NP-EGTA, 500 μM) as indicated. Modified from Figure 1 of Galione and Churchill [42] by permission of the American Society for Biochemistry and Molecular Biology.

of 50 to 150 $\mu m^2/s$ depending on the concentration of NAADP. This range for the diffusion coefficient overlaps with that for Ca^{2+} in a cell [41]. Upon photorelease, Ca^{2+} did not spread significantly (Fig 3), however, suggesting that the NAADP-mediated Ca^{2+} increase is spread by diffusion of NAADP itself. The concentration dependence of the diffusion, like Ca^{2+} [41], suggests that NAADP diffusion is slowed by binding to immobile buffers, likely the NAADP receptors. Diffusion control of the spreading of a Ca^{2+} increase contrasts markedly to the regenerative Ca^{2+} waves propagated by CICR via either IP_3 receptors or ryanodine receptors.

NAADP Responses are Amplified Locally by CICR

Although diffusion controls the spatial domain of the NAADP-mediated Ca^{2+} increase (Fig. 3), we found there is local amplification by CICR [42]. Blocking IP_3 receptors with heparin [61] or cADPR receptors with 8-amino-cADPR [62] alone had minor effects on the NAADP-mediated Ca^{2+} wave [42]. However, when both inhibitors were included together there was a dramatic inhibition. The lag time between the start of photolysis and the beginning of the Ca^{2+} increase was lengthened to the extent that the equivalent amount of NAADP that gave a maximum response in the control did not elicit any response in the doubly inhibited egg. Only after much higher amounts of NAADP were liberated was there a Ca^{2+} increase, which was characterized by a slow rise time, and slow velocity. Our interpretation is that when either CICR amplification system is functioning, small amounts of NAADP are sufficient for a maximal response (Fig. 4). CICR amplification of the NAADP response in sea urchin is consistent with that reported in pancreatic acinar cells [63] and starfish oocytes [64].

Model for the Control of Spatial Ca^{2+} Signaling by NAADP

NAADP controls Ca^{2+} signaling by diffusing, forming gradients and triggering amplification via CICR (Fig. 4). A gradient in the concentration of NAADP defines the domain over which the wave propagates. Functional coupling to either IP_3 receptors or ryanodine receptors contributes to the amplitude of the Ca^{2+} increase, but the effect is spatially localized. Both Ca^{2+} and NAADP could activate Ca^{2+} release from the clusters of channels, but unlike Ca^{2+}, NAADP can diffuse between the clusters to coordinate a diffusive Ca^{2+} wave.

Given the mechanism by which NAADP mediates spatial Ca^{2+} signaling, NAADP can be tuned to be either a local or global messenger. In contrast, IP_3 and cADPR are always global messengers [41, 65], and Ca^{2+} is always a local messenger [41]. A potentially important implication of diffusive NAADP-mediated Ca^{2+} waves is that they encode information such as the intensity of the originating signal, distance from the originating signal and the direction of the signal.

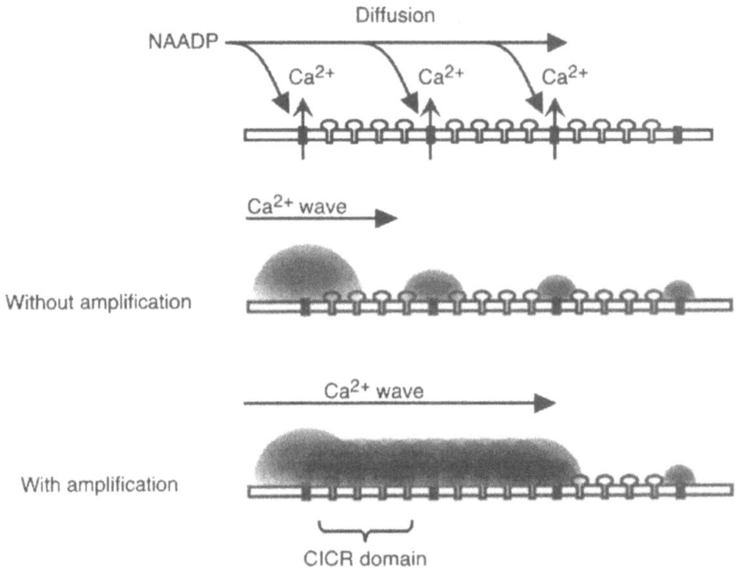

Figure 4. **Mechanism by which NAADP mediates a Ca^{2+} wave.** NAADP diffusion defines the domain of the Ca^{2+} increase. The Ca^{2+} released by NAADP is amplified by IP_3 and ryanodine receptors, which exhibit Ca^{2+}-induced Ca^{2+} release (CICR). This amplification is local and therefore does not propagate. Black rectangles depict putative NAADP-sensitive Ca^{2+} channels. Mushroom shapes depict both IP_3 and ryanodine receptors. Hemispherical clouds depict Ca^{2+} with shade corresponding to concentration.

NAADP MEDIATES A SPATIOTEMPORAL Ca^{2+} MEMORY

The unique ability of NAADP to self-desensitize at concentrations that do not release Ca^{2+} [17, 28, 29], combined with its concentration-dependent diffusion [42], suggested another intriguing possibility: that an initial spatially restricted NAADP signal affects later signals. Indeed, we found that this effect can serve as a memory of the NAADP gradient such that past responses affect subsequent responses [66].

NAADP Forms a Spatial Memory

A local release of a low concentration of NAADP resulted in a Ca^{2+} increase that spread only part way across an egg, thus forming a Ca^{2+} gradient (Fig. 5A). When Ca^{2+} returned to near resting levels, a global release of NAADP resulted in a Ca^{2+} increase that was an inversion of the first Ca^{2+} gradient (Fig. 5A). This second Ca^{2+} gradient is a direct consequence of the first NAADP gradient that formed subcellular zones where the NAADP response was completely inhibited, partially inhibited and not inhibited. In contrast to NAADP, a local release of either cADPR or IP_3 did not affect the pattern of Ca^{2+} increase elicited by a later global release of messenger (Fig. 5A) [66]. These experiments illustrate a key difference between Ca^{2+} signals generated by IP_3 and cADPR and those generated by NAADP. Namely, that NAADP responses exhibit a spatial memory of the previous response that shapes the spatial pattern of subsequent Ca^{2+} increases.

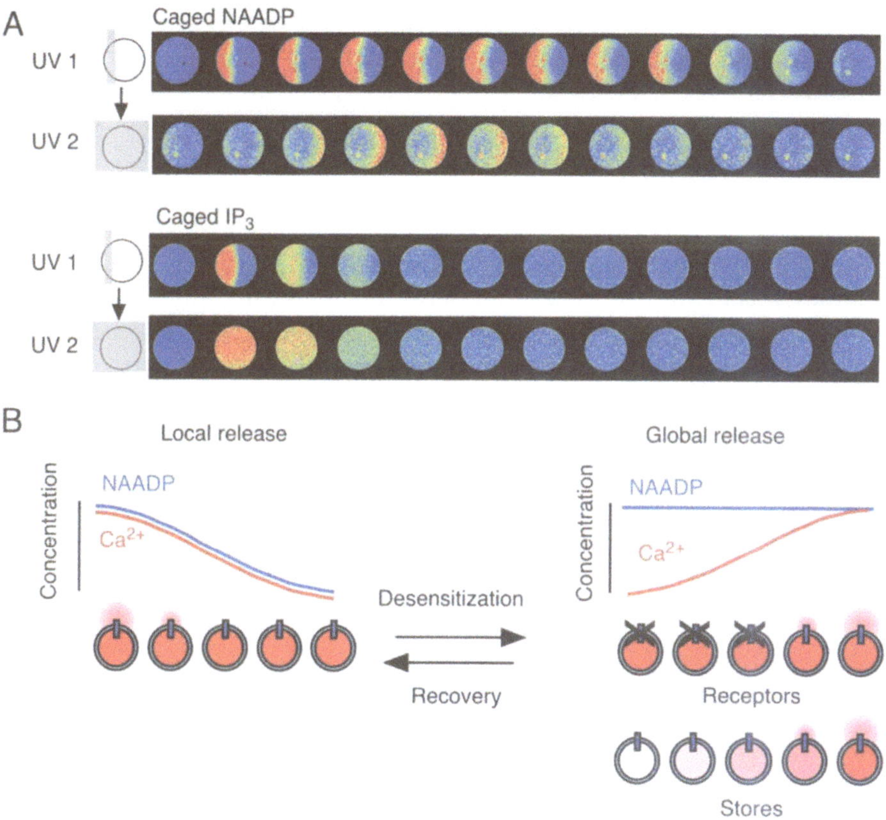

Figure 5. **NAADP mediates a spatiotemporal Ca^{2+} memory. A.** Confocal images showing the response in sea urchin eggs to a local initial release of messenger followed by global release of the messenger approximately 60 s later. Ca^{2+} concentration is mapped by color with warmer colors representing higher concentrations. The diagrams to the left of the images show where UV light was used to photorelease the indicated caged compound. Sea urchin eggs were microinjected with Oregon Green 488 BAPTA Dextran (10 μM) and either caged NAADP (0.5 μM) or caged IP_3 (5 μM) as indicated. **B.** Possible origin of the spatiotemporal Ca^{2+} memory. The local release and diffusion of NAADP forms a Ca^{2+} gradient. Prolonged desensitization of the NAADP response results in a "memory" of the initial gradient. The memory results from either receptor inactivation or Ca^{2+} store depletion. Re-sensitization provides a temporal component to the memory. Modified from Figures 2 and 3 of Galione and Churchill [66] by permission of the American Society for Biochemistry and Molecular Biology.

Recovery of NAADP Sensitivity Provides a Temporal Memory

To determine the reversibility of spatially localized NAADP inactivation, eggs were locally desensitized to NAADP and then given various periods of recovery before a global release of NAADP. Surprisingly, the response recovered progressively with time until full recovery after about 20 minutes [66]. This reversibility of the inactivation to NAADP in intact eggs means that the NAADP-mediated spatiotemporal Ca^{2+} memory is short-term rather than permanent. We suggest that recovery occurs by either receptor resensitization, store refilling or both (Fig. 5B). The mechanism of recovery may involve NAADP degradation as reported by Lee [67].

Possible Physiological Implications

The implications for NAADP forming a subcellular spatiotemporal Ca^{2+} memory are manifold. An obvious extension of this property would be in controlling embryo polarity and developmental patterning [68-70]. NAADP could easily set up a gradient (of both NAADP and Ca^{2+}). Recent data suggests that such a mechanism may exist during fertilization of the sea urchin egg. Sea urchin sperm not only contain endogenous NAADP (Billington and Genazzani, personal communication) but synthesize sufficient NAADP to activate an egg, at least locally (Churchill and Galione, unpublished observations). By delivering NAADP from a point source, the sperm would impose an NAADP gradient, which would serve as a marker for where the sperm bound, and how far every point within the egg was from this location.

NAADP-INDUCED Ca^{2+} OSCILLATIONS

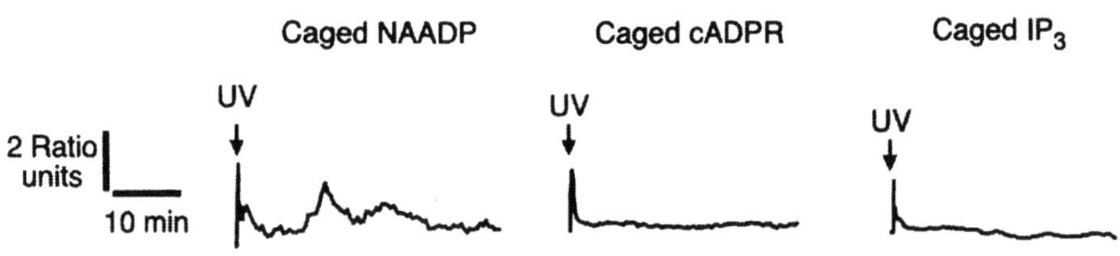

Figure 6. **The photorelease of NAADP but not cADPR or IP_3 induces Ca^{2+} oscillations in sea urchin eggs.** Sea urchin eggs were microinjected with Oregon Green 488 BAPTA Dextran (10 μM) and either caged NAADP (0.5 μM), caged cADPR (5 μM) or caged IP_3 (5 μM) as indicated.

NAADP but not IP_3 or cADPR induces long-lasting (at least 80 min) Ca^{2+} oscillations when photoreleased in sea urchin eggs [17, 28, 58, 71]. The response to NAADP is comprised of an initial Ca^{2+} increase mediated by NAADP photorelease, a first oscillation which is readily separated or appears as only a shoulder on the primary Ca^{2+} increase, and finally long-lasting Ca^{2+} spikes of irregular pattern (Fig. 6). The first Ca^{2+} oscillation can also take on a spatial form. When NAADP was released in only a region of the egg a Ca^{2+} wave propagated away from this site, but several tens of seconds later an 'echo' Ca^{2+} wave back-propagated to the origin of the first wave [72]. The time of onset, duration, amplitude and frequency of the Ca^{2+} increases is decidedly variable. Often there are small Ca^{2+} transients of short duration (~1 min) superimposed on larger prolonged Ca^{2+} transients (~5-15 min). This suggests that there are at least two independent oscillatory mechanisms operating simultaneously.

A clue to the mechanism came from the fact that only NAADP and not IP_3 or cADPR could induce these oscillations. Lee *et al.* [17, 28, 58] suggested a mechanism whereby Ca^{2+} was transferred from the NAADP-sensitive Ca^{2+} stores into the CICR stores overloading them and causing spontaneous release. Spontaneous release results from overloading of a Ca^{2+} store and is well-characterised for both IP_3 and ryanodine receptors [73]. Lee's model is essentially a modification of the original two-pool model where it was envisaged that IP_3-released Ca^{2+} was taken up into a pool containing ryanodine receptors - which were considered the only CICR channel at the time – thereby setting up cycles of Ca^{2+} uptake and release [49, 50]. Lee *et al.*'s [17, 28, 58] model suggesting cooperativity between Ca^{2+} release pathways accounted for all the properties of the NAADP-induced Ca^{2+} oscillations. We recently tested this model and found it to be correct.

Oscillations Require Functional CICR Pathways

The two-pool model makes several predictions that, importantly, are testable. As expected for such a mechanism, the oscillations were affected by interfering with either the IP_3 system with heparin [61] or the cADPR system with 8-amino-cADPR [74]. Neither of these agents fully inhibited the oscillations when applied alone, but did when co-applied [71]. The requirement for dual block is similar to that reported for inhibition of fertilization-induced Ca^{2+} waves in sea urchin eggs [75, 76], NAADP-induced Ca^{2+} release in starfish eggs [77] and for amplification of NAADP-mediated Ca^{2+} waves in sea urchin eggs [42].

Both IP_3 and cADPR release Ca^{2+} from stores dependent on sarco(endo)plasmic recticulum Ca^{2+}-ATPases (SERCA), whereas NAADP does not [16, 29]. Thapsigargin is an inhibitor of SERCAs [61]. Therefore, treating cells with thapsigargin selectively drains their serca-dependent stores, which are generally taken to be synonomous with the endoplasmic rectiulum [3, 23]. Following exposure to thapsigargin, NAADP still mobilized Ca^{2+} but failed to induce long-term Ca^{2+} oscillations [71]. This demonstrated that the thapsigargin-sensitive Ca^{2+} stores are physically distinct from the NAADP-sensitive stores, consistent with the demonstration of spatial separation of the NAADP-sensitive Ca^{2+} stores from the IP_3- and cADPR-sensitive Ca^{2+} stores in sea urchin eggs whose organelles were stratified by centrifugation [32]. Additionally, the Ca^{2+} transient elicited by thapsigargin is proportional to the Ca^{2+} load of the CICR stores. The size of this Ca^{2+} transient was increased after NAADP-mediated Ca^{2+} mobilization [71], directly demonstrating transfer of Ca^{2+} from one store to the other, consistent with the prediction of the two-pool model.

NAADP Operates via a Two-Pool Mechanism

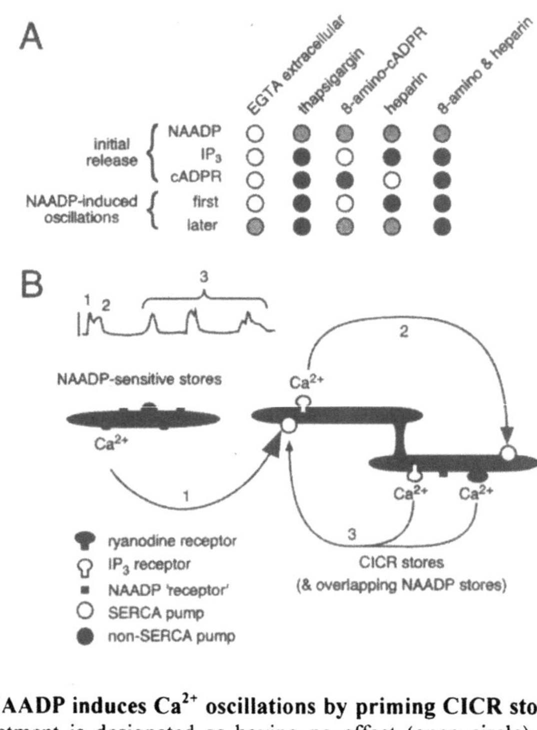

Figure 7. **NAADP induces Ca^{2+} oscillations by priming CICR stores. A,** The effect of each treatment is designated as having no effect (open circle), partial inhibition (gray circle) and complete inhibition (black circle). **B.** Cartoon of the proposed mechanism. NAADP releases Ca^{2+} from a functionally separate Ca^{2+} store (designated 1). This Ca^{2+} is taken up into the CICR stores, thereby priming them. This results in overloading and spontaneous Ca^{2+} release initially from IP_3-sensitive Ca^{2+} stores (designated 2). Subsequently, Ca^{2+} is taken up into and released from both IP_3- and cADPR-sensitive Ca^{2+} stores (designated 3). Figure reproduced from Churchill and Galione [71] with permission from Oxford University Press.

The mechanism by which NAADP induces Ca^{2+} oscillations and a summary of the evidence for this mechanism are presented in Fig. 7. The initial Ca^{2+} release mediated by NAADP is largely independent of any additional processes (pathway 1, Fig. 7) except for an amplification by CICR pathways, as described previously [42]. The Ca^{2+} released by NAADP is taken up into CICR stores until they overload and spontaneously release Ca^{2+} as demonstrated previously [22, 46, 73]. A necessary feature of this mechanism is that the released Ca^{2+} is taken up into thapsigargin-sensitive stores faster than into thapsigargin-insensitive (NAADP-sensitive) stores, but how this is accomplished is unknown. The first Ca^{2+} pulse after the NAADP-induced transient requires IP_3-dependent Ca^{2+} release (pathway 2, Fig. 7). Both IP_3- and cADPR-sensitive Ca^{2+} stores participate in the later cycles of Ca^{2+} uptake, overload, and release (pathway 3, Fig. 7). This mechanism meets the criteria of a two-pool model for Ca^{2+} oscillations [49, 51] because one Ca^{2+} store is the source of priming Ca^{2+} (NAADP sensitive) and a separate Ca^{2+} store mediates the oscillations (CICR), and oscillations can be triggered by any increase in Ca^{2+} that primes the CICR stores. As such, this mechanism was the first direct demonstration of the two-pool model for Ca^{2+} oscillations.

Possible Physiological Implications

Ca^{2+} oscillations with a similar 'signature' to the NAADP-induced oscillations occur after fertilization of sea urchin eggs with increases occurring before nuclear envelope breakdown, metaphase-anaphase transition and cytokinesis [10, 78]. This does not necessarily imply that these events are triggered by an initial surge in NAADP, but, rather, suggests that the underlying mechanism responsible for endogenous Ca^{2+} oscillations during development, which is unknown [3], is similar in both processes. Because these Ca^{2+} oscillations do not require continued presence of an extracellular agonist, like most Ca^{2+} oscillations in mammalian cells [3, 48, 49, 51], the mechanisms may have direct relevance to internally controlled Ca^{2+} oscillation that underlie processes such as circadian rhythms [79] and cell division [10, 80].

The relatively small thapsigargin-induced Ca^{2+} transient demonstrates that the CICR stores are relatively unfilled in the egg. An intriguing possibility is that Ca^{2+} store loading occurs in sea urchin fertilization during the latent period – the time between sperm-egg fusion and the initiation of the Ca^{2+} wave [8, 81]. Priming of the intracellular Ca^{2+} stores has been suggested before in conjunction with the sperm serving as a Ca^{2+} conduit [82]. One of the roles of NAADP may be to release Ca^{2+} required for loading of the CICR stores, which can then propagate a regenerative Ca^{2+} wave.

GENERALIZED ROLES FOR NAADP IN Ca^{2+} SIGNALING

The characteristics of NAADP-mediated Ca^{2+} signaling differ markedly from the other Ca^{2+} release pathways. Because of this, NAADP plays several unique and critical roles in Ca^{2+} signaling. First, NAADP is an initiator of

Ca^{2+} signals. NAADP releases a small amount of Ca^{2+} which then stimulates neighbouring CICR channels, resulting in amplification of the signal [5, 63]. In resting cells, the NAADP-induced CICR is localized to only channels in close proximity to the site of NAADP-mediated Ca^{2+} release [42]. In cells whose cytosol is an excitable medium, NAADP would trigger a global Ca^{2+} wave. NAADP may be the answer to a long-standing problem of how Ca^{2+} waves are initiated from a specific subcellular region if the whole egg is excitable. Second, NAADP can be tuned to be a local or global messenger depending on its concentration [42]. At low concentrations, NAADP induces localized Ca^{2+} increases, whereas higher concentrations mediate global Ca^{2+} waves (by NAADP diffusion).

Lastly, although the mechanisms by which NAADP controls Ca^{2+} signaling are fundamentally different from the CICR-based mechanisms of IP_3 and cADPR, all three have features that are conserved at the functional rather than mechanistic level. For spatial Ca^{2+} signaling, NAADP relies on simple diffusion rather than reaction-diffusion. Also, to turn off the Ca^{2+} increase, NAADP relies on self-inactivation rather than Ca^{2+}-mediated inhibition. These observations suggest that evolution selects function not mechanism for controlling Ca^{2+} signaling. How the function is reached varies with the individual messenger. IP_3 and cADPR followed the same road, whereas NAADP took the one less travelled by.

REFERENCES

1. Bolton TB. Prestwich SA, Zholos AV and Gordienko DV. 1999. Excitation-contraction coupling in gastrointestinal and other smooth muscles. *Annu. Rev. Physiol.* 61: 85-115.
2. Berridge MJ. 1997. Elementary and global aspects of calcium signalling. *J. Physiol.* 499: 291-306.
3. Berridge M, Lipp P and Bootman M. 2000. The versatility and universality of calcium signalling. *Nature Mol. Cell Biol. Rev.* 1: 11-21.
4. Petersen OH, Burdakov D and Tepikin AV. 1999. Polarity in intracellular calcium signaling. *Bioessays* 21: 851-860.
5. Cancela JM. 2001. Specific Ca^{2+} signaling evoked by cholecystokinin and acetylcholine: the roles of NAADP, cADPR, and IP_3. *Annu. Rev. Physiol.* 63: 99-117.
6. Spitzer NC, Lautermilch NJ, Smith RD and Gomez TM. 2000. Coding of neuronal differentiation by calcium transients. *Bioessays* 22: 811-817.
7. Gilkey JC, Jaffe LF, Ridgway EB and Reynolds GT. 1978. A free calcium wave traverses the activating egg of the Medaka, *Oryzias latipes*. *J. Cell Biol.* 76: 448-466.
8. Stricker SA. 1999. Comparative biology of calcium signaling during fertilization and egg activation in animals. *Dev. Biol.* 211: 157-176.
9. Ciapa B and Chiri S. 2000. Egg activation: Upstream of the fertilization calcium signal. *Biol Cell* 92: 215-233.
10. Whitaker M and Larman MG. 2001. Calcium and mitosis. *Semin. Cell Dev. Biol.* 12: 53-58.
11. Dolmetsch RE, Lewis RS, Goodnow CC and Healy JI. 1997. Differential activation of transcription factors induced by Ca^{2+} response amplitude and duration. *Nature* 386: 855-858.
12. Berridge MJ. 1997. The AM and FM of calcium signalling. *Nature* 386: 759-760.

13. Lee HC, Walseth TF, Bratt GT, Hayes RN and Clapper DL. 1989. Structural determination of a cyclic metabolite of NAD with intracellular calcium-mobilizing activity. *J Biol Chem* 264: 1608-1615.
14. Galione A and Churchill G. 2000. Cyclic ADP-ribose as a calcium mobilizing messenger. *Science STKE.* www.stke.org/cgi/content/full/OC_sigtrans;2000/41/pe1.
15. Lee HC. 2001. Physiological functions of cyclic ADP-ribose and NAADP as calcium messengers. *Annu. Rev. Pharmacol. Toxcicol.* 41: 317-345.
16. Lee HC and Aarhus R. 1995. A derivative of NADP mobilizes calcium stores insensitive to inositol trisphosphate and cyclic ADP-Ribose. *J. Biol. Chem.* 270: 2152-2157.
17. Lee HC. 1997. Mechanisms of calcium signaling by cyclic ADP-ribose and NAADP. *Physiol. Rev.* 77: 1133-1164.
18. Patel S, Churchill GC and Galione A. 2001. Coordination of Ca^{2+} signalling by NAADP. *Trend. Biochem Sci.* 26: 482-489.
19. Chini EN, Beers KW and Dousa TP. 1995. Nicotinate adenine dinucleotide phosphate (NAADP) triggers a specific calcium release system in sea urchin eggs. *J. Biol. Chem.* 270: 3216-3223.
20. Young KW and Nahorski SR. 2001. Intracellular sphingosine 1-phosphate production: A novel pathway for Ca^{2+} release. *Semin. Cell Dev. Biol.* 12: 19-25.
21. Patel S, Joseph SK and Thomas AP. 1999. Molecular properties of inositol 1,4,5-trisphosphate receptors. *Cell Calcium* 25: 247-264.
22. Galione A, Lee HC and Busa WB. 1991. Ca^{2+}-Induced Ca^{2+} release in sea-urchin egg homogenates - Modulation by cyclic ADP-ribose. *Science* 253: 1143-1146.
23. Pozzan T, Rizzuto R, Volpe P and Meldolesi J. 1994. Molecular and cellular physiology of intracellular calcium stores. *Physiol. Rev.* 74: 595-636.
24. Lee HC. 2000. Multiple calcium stores: Separate but interacting. *Science STKE.* www.stke.org/cgi/content/full/OC_sigtrans;2000/40/pe1.
25. Genazzani AA and Galione A. 1997. A Ca^{2+} release mechanism gated by the novel pyridine nucleotide, NAADP. *Trends Pharmacol. Sci.* 18: 108-110.
26. Chini EN and Dousa TP. 1996. Nicotinate-adenine dinucleotide phosphate-induced Ca^{2+}-release does not behave as a Ca^{2+}-induced Ca^{2+}-release system. *Biochem. J.* 316: 709-711.
27. Bak J, White P, Timar G, Missiaen L, Genazzani AA and Galione A. 1999. Nicotinic acid adenine dinucleotide phosphate triggers Ca^{2+} release from brain microsomes. *Curr. Biol.* 9: 751-754.
28. Aarhus R, Dickey DM, Graeff RM, Gee KR, Walseth TF and Lee HC. 1996. Activation and inactivation of Ca^{2+} release by $NAADP^+$. *J. Biol. Chem.* 271: 8513-8516.
29. Genazzani AA and Galione A. 1996. Nicotinic acid-adenine dinucleotide phosphate mobilizes Ca^{2+} from a thapsigargin-insensitive pool. *Biochem. J.* 315: 721-725.
30. Billington RA and Genazzani AA. 2000. Characterization of $NAADP^+$ binding in sea urchin eggs. *Biochem. Biophys. Res. Commun.* 276: 112-116.
31. Patel S, Churchill GC and Galione A. 2000. Unique kinetics of nicotinic acid-adenine dinucleotide phosphate (NAADP) binding enhance the sensitivity of NAADP receptors for their ligand. *Biochem. J.* 352: 725-729.
32. Lee HC and Aarhus R. 2000. Functional visualization of the separate but interacting calcium stores sensitive to NAADP and cyclic ADP-ribose. *J. Cell Sci.* 113: 4413-4420.
33. Bezprozvanny I, Watras J and Ehrlich BE. 1991. Bell-shaped calcium-response curves of Ins(1,4,5)P_3- and calcium-gated channels from endoplasmic reticulum of cerebellum. *Nature* 351: 751-754.
34. Meissner G. 1994. Ryanodine receptor/Ca^{2+} release channels and their regulation by endogenous effectors. *Annu. Rev. Physiol.* 56: 485-508.
35. Lechleiter JD and Clapham DE. 1992. Molecular mechanisms of intracellular calcium excitability in X. laevis oocytes. *Cell* 69: 283-294.
36. Parker I, Choi J and Yao Y. 1996. Elementary events of $InsP_3$-induced Ca^{2+} liberation in Xenopus oocytes: Hot spots, puffs and blips. *Cell Calcium* 20: 105-121.

37. Parker I and Yao Y. 1996. Ca^{2+} transients associated with openings of inositol trisphosphate- gated channels in Xenopus oocytes. *J. Physiol.* 491: 663-668.
38. Lipp P and Niggli E. 1996. Submicroscopic calcium signals as fundamental events of excitation–contraction coupling in guinea-pig cardiac myocytes. *J. Physiol.* 492: 31-38.
39. Bootman MD and Berridge MJ. 1996. Subcellular Ca^{2+} signals underlying waves and graded responses in HeLa cells. *Curr. Biol.* 6: 855-865.
40. Cheng H, Lederer WJ and Cannell MB. 1993. Calcium sparks: elementary events underlying excitation-contraction coupling in heart muscle. *Science* 262: 740-744.
41. Allbritton NL, Meyer T and Stryer L. 1992. Range of messenger action of calcium ion and inositol 1,4,5-trisphosphate. *Science* 258: 1812-1815.
42. Churchill GC and Galione A. 2000. Spatial control of Ca^{2+} signaling by nicotinic acid adenine dinucleotide phosphate diffusion and gradients. *J. Biol. Chem.* 275: 38687-38692.
43. Gilkey JC, Jaffe LF, Ridgway EB and Reynolds GT. 1978. A free calcium wave traverses the activating egg of the medaka, *Oryzias latipes*. *J. Cell Biol.* 76: 448-466.
44. Sanderson MJ, Charles AC, Boitano S and Dirksen ER. 1994. Mechanisms and function of intercellular calcium signaling. *Mol. Cell Endocrinol.* 98: 173-187.
45. Nedergaard M, Cooper AJ and Goldman SA. 1995. Gap junctions are required for the propagation of spreading depression. *J. Neurobiol.* 28: 433-444.
46. Robb-Gaspers LD and Thomas AP. 1995. Coordination of Ca^{2+} signaling by intercellular propagation of Ca^{2+} waves in the intact liver. *J. Biol. Chem.* 270: 8102-8107.
47. Dupont G and Goldbeter A. 1992. Oscillations and waves of cytosolic calcium: insights from theoretical models. *Bioessays* 14: 485-493.
48. Thomas AP, Renard DC and Rooney TA. 1991. Spatial and temporal organization of calcium signalling in hepatocytes. *Cell Calcium* 12: 111-126.
49. Berridge MJ and Galione A. 1988. Cytosolic Calcium Oscillators. *FASEB J.* 2: 3074-3082.
50. Berridge MJ. 1991. Cytoplasmic calcium oscillations: a two pool model. *Cell Calcium* 12: 63-72.
51. Tsien RW and Tsien RY. 1990. Calcium channels, stores, and oscillations. *Annu. Rev. Cell Biol.* 6: 715-760.
52. Harootunian AT, Kao JP, Paranjape S and Tsien RY. 1991. Generation of calcium oscillations in fibroblasts by positive feedback between calcium and IP_3. *Science* 251: 75-78.
53. Hirose K, Kadowaki S, Tanabe M, Takeshima H and Iino M. 1999. Spatiotemporal dynamics of inositol 1,4,5-trisphosphate that underlies complex Ca^{2+} mobilization patterns. *Science* 284: 1527-1530.
54. Nash MS, Young KW, Challiss RA and Nahorski SR. 2001. Intracellular signalling. Receptor-specific messenger oscillations. *Nature* 413: 381-382.
55. Hajnoczky G and Thomas AP. 1997. Minimal requirements for calcium oscillations driven by the IP_3 receptor. *EMBO J.* 16: 3533-3543.
56. Wakui M, Potter BV and Petersen OH. 1989. Pulsatile intracellular calcium release does not depend on fluctuations in inositol trisphosphate concentration. *Nature* 339: 317-320.
57. Perez Terzic CM, Chini EN, Shen SS, Dousa TP and Clapham DE. 1995. Ca^{2+} release triggered by nicotinate adenine dinucleotide phosphate in intact sea urchin eggs. *Biochem. J.* 312: 955-959.
58. Lee HC, Aarhus R, Gee KR and Kestner T. 1997. Caged nicotinic acid adenine dinucleotide phosphate. Synthesis and use. *J. Biol. Chem.* 272: 4172-4178.
59. Galione A, Cancela J-M, Churchill G, Genazzani A, Lad C, *et al.* 2000. Methods in cADPR and NAADP research. In *Methods in Calcium Signalling*, ed. JJ Putney, pp. 249-296. Boca Raton: CRC Press
60. Hille B. 2001. *Ion channels of excitable membranes*. Sunderland, Mass.: Sinauer. xviii, 814 pp.

61. Taylor CW and Broad LM. 1998. Pharmacological analysis of intracellular Ca^{2+} signalling: problems and pitfalls. *Trends Pharmacol. Sci.* 19: 370-375.
62. Walseth TF and Lee HC. 1993. Synthesis and characterization of antagonists of cyclic-ADP-ribose-induced calcium release. *Biochim. Biophys. Acta* 1178: 235-242.
63. Cancela JM, Churchill GC and Galione A. 1999. Coordination of agonist-induced Ca^{2+}-signalling patterns by NAADP in pancreatic acinar cells. *Nature* 398: 74-76.
64. Santella L, Kyozuka K, Genazzani AA, De Riso L and Carafoli E. 2000. Nicotinic acid adenine dinucleotide phosphate-induced Ca^{2+} release. Interactions among distinct Ca^{2+} mobilizing mechanisms in starfish oocytes. *J. Biol. Chem.* 275: 8301-8306.
65. Petersen OH. 1995. Inositol trisphosphate and cyclic ADP-ribose as long range messengers generating local subcellular calcium signals. *J. Physiol.* 89: 125-127.
66. Churchill GC and Galione A. 2001. Prolonged inactivation of NAADP-induced Ca^{2+} release mediates a spatiotemporal Ca^{2+} memory. *J. Biol. Chem.* 276: 11223-11225.
67. Lee HC. 2000. NAADP: An emerging calcium signaling molecule. *J. Membr. Biol.* 173: 1-8.
68. Jaffe LF. 1999. Organization of early development by calcium patterns. *Bioessays* 21: 657-667.
69. Webb SE and Miller AL. 2000. Calcium signalling during zebrafish embryonic development. *Bioessays* 22: 113-23.
70. Creton R, Speksnijder JE and Jaffe LF. 1998. Patterns of free calcium in zebrafish embryos. *J. Cell Sci.* 111: 1613-1622.
71. Churchill GC and Galione A. 2001. NAADP induces Ca^{2+} oscillations via a two-pool mechanism by priming IP_3- and cADPR-sensitive Ca^{2+} stores. *EMBO J.* 20: 2666-2671.
72. Galione A, Patel S and Churchill GC. 2000. NAADP-induced calcium release in sea urchin eggs. *Biol. Cell* 92: 197-204.
73. Missiaen L, Taylor CW and Berridge MJ. 1991. Spontaneous calcium release from inositol trisphosphate-sensitive calcium stores. *Nature* 352: 241-244.
74. Walseth TF and Lee HC. 1993. Synthesis and characterization of antagonists of cyclic-ADP-ribose-induced calcium release. *Biochim. Biophys. Acta* 1178: 235-242.
75. Lee HC, Aarhus R and Walseth TF. 1993. Calcium mobilization by dual receptors during fertilization of sea urchin eggs. *Science* 261: 352-325.
76. Galione A, McDougall A, Busa WB, Willmott N, Gillot I and Whitaker M. 1993. Redundant mechanisms of calcium-induced calcium release underlying calcium waves during fertilization of sea-urchin eggs. *Science* 261: 348-352
77. Santella L, Kyozuka K, Genazzani AA, De Riso L and Carafoli E. 2000. Nicotinic acid adenine dinucleotide phosphate-induced Ca^{2+} release. Interactions among distinct Ca^{2+} mobilizing mechanisms in starfish oocytes. *J. Biol. Chem.* 275: 8301-8306.
78. Poenie M, Alderton J, Tsien RY and Steinhardt RA. 1985. Changes of free calcium levels with stages of the cell division cycle. *Nature* 315: 147-149.
79. Ding JM, Buchanan GF, Tischkau SA, Chen D, Kuriashkina L, *et al.* 1998. A neuronal ryanodine receptor mediates light-induced phase delays of the circadian clock. *Nature* 394: 381-384.
80. Twigg J, Patel R and Whitaker M. 1988. Translational control of $InsP_3$-induced chromatin condensation during the early cell cycles of sea urchin embryos. *Nature* 332: 366-369.
81. McDougall A, Shearer J and Whitaker M. 2000. The initiation and propagation of the fertilization wave in sea urchin eggs. *Biol. Cell* 92: 205-214.
82. Jaffe LF. 1997. On the validity of the sperm conduit model of fertilization. *Bioessays* 19: 835-836.

Grant C. Churchill, Department of Pharmacology,
Oxford University, Mansfield Road, Oxford OX1 3QT, UK.
e-mail: grant.churchill@pharmacology.oxford.ac.uk

Cyclic ADP-Ribose and NAADP. Structures, Metabolism and Functions. Lee HC (ed), pp. 217-240, Dordrecht: Kluwer.

Regulation of Immune Responses by CD38 and cADPR

Frances E. Lund, Troy D. Randall and Santiago Partida-Sánchez
Trudeau Institute, PO Box 59 Saranac Lake NY 12983, USA.

INTRODUCTION

CD38 was identified more than 20 years ago as a cell surface protein expressed on leukocyte subsets [1, 2]. Over the last decade, antibodies to human CD38 have been used to classify tumors from Multiple Myeloma [3, 4] and Chronic Lymphocytic Leukemia patients [5, 6], to identify which HIV^{+} patients will develop AIDS [7] and to subset a variety of leukocyte populations including hematopoietic progenitors [8, 9], germinal center B cells [10, 11] and regulatory T cells [12, 13]. However, despite the clear utility of using anti-CD38 antibodies as diagnostic and prognostic tools, it was unclear whether CD38 itself played an important functional role on normal lymphocytes. Since CD38 is expressed on the plasma membrane, it was postulated that CD38 was involved in leukocyte signaling or adhesion [14-16]. Although numerous experiments demonstrated that antibody mediated crosslinking of CD38 activated a variety of different signaling pathways in lymphocytes (reviewed in ref. [17, 18]), it was unclear how CD38 initiated signal transduction since the cytoplasmic tail of CD38 does not contain any known signaling motifs [2, 19]. The demonstration that CD38 was an ecto-enzyme [20-24] capable of catalyzing the production of several calcium mobilizing molecules, including cyclic adenosine diphosphate ribose (cADPR), nicotinic acid-adenine dinucleotide phosphate (NAADP) and adenosine diphosphate ribose (ADPR) opened up the exciting new possibility that CD38 might mediate its signaling functions on lymphocytes through the production of these calcium-mobilizing metabolites. However, multiple labs have now shown that anti-CD38 mediated signaling appears to be controlled independently of CD38 enzyme activity [25-27]. In contrast, the enzymatic activity of CD38 is essential for immune responses *in vivo*, as demonstrated using CD38 deficient mice [28]. For example, neutrophils lacking either CD38 or cADPR made a defective calcium response upon engagement of the G-protein coupled formylated peptide receptor (FPR) [28]. This functional defect resulted in inefficient recruitment of CD38 KO neutrophils to sites of bacterial infection *in vivo* [28]. Taken together, these data suggest that a primary function of CD38 *in vivo* may be to regulate intracellular calcium levels in cells responding to engagement of cell surface receptors. This review will focus on the recent efforts of our laboratory and others to dissect the role of CD38 in cells of the immune system.

ROLE OF CD38 IN INNATE AND ADAPTIVE IMMUNE RESPONSES

Crosslinking of CD38 Induces Signal Transduction in Lymphocytes Independently of Enzyme Activity

Numerous studies have shown that crosslinking of CD38 on primary leukocytes or cell lines *in vitro* has profound effects on cell activation, proliferation, differentiation and death. There are previous reviews that discuss these experiments in detail [17, 18, 30], therefore, in this section we will outline those studies that have suggested potential functional roles for CD38 in immune responses *in vivo*. Our laboratory has assessed the signal transduction pathway induced by CD38 crosslinking in murine B lymphocytes. We found that triggering of CD38 on mature B cells induces activation, proliferation, and cytokine production [31-34]. Work from Takatsu's group further showed that anti-CD38 stimulation of primary B cells in combination with the cytokine IL-5 could induce isotype switching to IgG1 and could induce the differentiation of B cells into antibody secreting plasma cells [35-37]. In contrast, experiments performed by the Campana laboratory showed that crosslinking of CD38 on progenitor B cells did not induce activation, but rather, induced apoptosis [25, 27, 38-40]. Although these results initially suggested that mouse and human CD38 were functionally distinct, more recent experiments have indicated that crosslinking of CD38 on mature B lymphocytes, whether mouse or human in origin, induces activation [41, 42], while crosslinking of CD38 on mouse or human B cell progenitors induces cell death [27]. Thus, crosslinking of CD38 can induce very different biological outcomes even within cells of the same lineage, depending on the state of maturation of the cells.

Crosslinking of CD38 has also been shown to induce signaling in T lymphocytes [43-45] and myeloid lineage cells [46, 47]. In general, stimulation of primary human peripheral blood T cells with anti-CD38 antibodies induces activation of T cells and results in upregulation of activation markers, proliferation and cytokine production [14, 43, 48]. The published experiments are less clear about the role for CD38 on myeloid lineage cells. The Campana laboratory has shown that anti-CD38 stimulation of myeloid progenitors induces programmed cell death [47], while Mehta's group has shown that anti-CD38 stimulation of myeloid leukemic cells can induce activation [46].

Regardless of the experimental details, it was clear that crosslinking of CD38 on B cells, T cells and myeloid cells triggers biologically important responses. Thus, a number of studies were performed to identify the relevant signaling pathways that induced signaling when CD38 was crosslinked. It was shown by several groups that CD38 was not competent to induce signaling on its own [27, 33, 34, 44]. Instead, CD38 appeared to "borrow" the signal transduction apparatus associated with other cellular receptors. For example, we showed that CD38 signaling was dependent on expression of a functional B cell receptor (BCR) complex in mature B cells [33, 34], while Malavasi and Sancho demonstrated that CD38 mediated signal transduction is dependent on the T cell receptor (TCR) complex in T cells [44]. We then showed that the cytoplasmic tail of murine CD38 could be entirely replaced without affecting anti-CD38 mediated cytokine production in mature B cells [26]. Likewise, Campana's group demonstrated that the cytoplasmic tail of human CD38 could be truncated down to a single amino acid without altering anti-CD38 mediated apoptosis of progenitor B cells [27]. These results indicated that CD38 must somehow tap into the intracellular signaling scaffold associated with other receptors, such as the BCR and TCR. Consistent with this idea, CD38 crosslinking activated signaling molecules that are physically associated with either the BCR [32, 36, 37] or TCR complex [45, 49, 50].

Since many of the molecules involved in anti-CD38 mediated signaling regulate or are responsive to cytosolic free calcium, we predicted that anti-CD38 stimulation might modulate CD38 enzyme activity and augment production of calcium mobilizing second messengers like cADPR. However, using both mouse and human B cells, we and the Campana group have shown that anti-CD38 antibodies do not increase the total enzymatic activity of CD38 or preferentially induce cADPR production by CD38 [25, 26, 34]. Furthermore, we found that the enzymatic activity of CD38 is not required for anti-CD38 mediated signal transduction [26]. For example, B cells expressing a mutant form of murine CD38 (E150L-D151V) signaled normally when triggered with anti-CD38 antibodies, despite the fact that the enzyme activity of this mutant was less than 1% of the wild-type CD38 [26]. In more recent experiments, we have found that precursor B cells expressing enzyme inactive mutant forms of CD38 undergo anti-CD38 mediated apoptosis in an equivalent manner to cells expressing the non-mutated form of CD38 (manuscript in preparation). Finally, we have shown that inhibition of CD38 catalytic activity using modified enzyme substrates does not alter anti-CD38 mediated signaling (manuscript in preparation). Thus, despite the fact that CD38 can catalyze the production of calcium regulating metabolites, anti-CD38 mediated signaling in B lymphocytes appears to be controlled independently of enzyme activity.

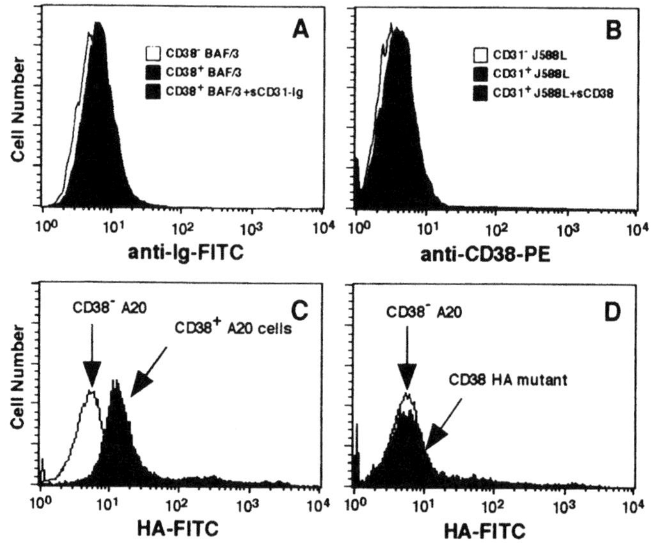

Figure 1. **Murine CD38 specifically binds hyaluronic acid but does not interact with CD31.** (**A**) To determine whether murine CD38 and CD31 represent a receptor/ligand pair, soluble recombinant murine CD31-Ig was incubated with parental BAF/3 cells that do not express CD38 or with BAF/3 cells that were stably transfected with murine CD38. Fluorochrome labeled anti-Ig antibodies were used to detect bound CD31. No specific binding was observed between murine CD38 and CD31. (**B**) Likewise, soluble recombinant CD38 was used to stain parental J558L that do not express murine CD31 or J558L cells stably transfected with murine CD31. Fluorochrome-labeled anti-CD38 antibodies were then used to detect bound soluble CD38. Again, no specific binding between CD31 and CD38 was observed. (**C**) In contrast, when FITC coupled hyaluronic acid (HA) was incubated with CD38 negative A20 cells or with A20 cells that had been stably transfected with murine CD38, HA specifically bound to the CD38 expressing A20 cells. (**D**) Furthermore, binding of HA to CD38 could be abolished when HA-binding sites in CD38 were mutated (HA mutant, see ref [26]).

The experimental findings described above have left us with the difficulty of trying to explain how the crosslinking of CD38 results in the activation of signaling complexes associated with other receptors like the BCR and TCR. However, a recent series of experiments from the Sancho laboratory may point the way. They demonstrated that a substantial fraction of membrane CD38 is located in the lipid rafts of T cells [50]. Upon anti-CD38 stimulation, an even larger fraction of CD38 was found in the rafts. Therefore, Sancho and colleagues have proposed that crosslinking cell surface CD38 with antibodies results in the passive formation of larger and more stable rafts [50]. In turn, this leads to the formation of supramolecular assemblies of TCR signaling components, which activates the TCR signaling complex. This model fits the available data for B cells as well.

If this model is correct, we must evaluate whether larger, more stable raft complexes form when CD38 is engaged by its physiologic ligand(s). To date, two different ligands for human CD38 have been identified, hyaluronic acid (HA) [51] and CD31 [52, 53]. In the mouse, however, only HA appears to bind CD38 (Fig. 1, bottom panel) as we have found no evidence that murine CD38 and murine CD31 bind to each other with measurable affinity (Fig. 1, top panel). Unfortunately, there is no convincing published data showing that direct addition of purified HA or CD31 to CD38 expressing mouse or human lymphocytes can induce the formation of competent signaling complexes. Thus, it is possible that the use of anti-CD38 antibodies to stimulate lymphocytes *in vitro* may not be representative of what occurs when CD38 is engaged by its ligands *in vivo*.

CD38 is Required for Optimal Adaptive and Innate Immune Responses

Despite the questions remaining about the *in vivo* relevance of the *in vitro* experimental models that use anti-CD38 antibodies to mediate signaling, the *in vitro* data suggested that CD38 might modulate BCR and TCR signaling by stabilizing the formation of rafts. Therefore, we generated CD38 deficient mice [29] to test whether CD38 regulated B or T lymphocyte development, activation or effector function. We found that the CD38 KO mice were viable and fertile and had apparently normal hematopoiesis despite the loss of CD38 on hematopoietic stem cells [29]. Additionally, we found that B and T lymphocytes develop normally and are found at the expected frequencies in almost all cellular compartments. Furthermore, the serum levels of all immunoglobulin isotypes were normal in unimmunized animals. Moreover, CD38 deficient B and T cells responded normally in a number of *ex vivo* activation, proliferation and functional assays [18, 29]. Although disappointing, these data suggested that CD38 was not required for B or T cell development nor was it absolutely required for BCR or TCR dependent responses.

Despite the fact that isolated lymphocytes from the CD38 KO mice were functionally competent in *in vitro* assays, it became obvious that there were a number of functional abnormalities in the immune responses of the CD38 KO mice. For example, upon immunization with low doses of antigen using the weak adjuvant, alum, we found that CD38 KO mice made poor antibody responses relative to WT mice [29]. This effect was even more pronounced upon secondary challenge. In light of Takatsu's data showing that CD38 triggering could induce isotype switching to IgG1 [35, 36], we also examined the ability of B cells to produce antibody of appropriate isotypes after antigen challenge and found that the production of all antibody isotypes were reduced in CD38 KO mice [29]. Interestingly, if mice were immunized with a large dose of antigen or with a more potent adjuvant, such as CFA, the difference in antibody production between CD38KO and WT mice was minimized [29]. Thus, if CD38 KO mice were exposed to antigen in the context of a weak inflammatory challenge, they were incapable of responding to the antigen, yet if the inflammatory challenge was robust, the immune response was normal. While these data fit with our original model that CD38 modulated BCR signals, it could equally suggest a problem with the initiation of the inflammatory response. In fact, we have recently found that bone marrow-derived myeloid cells are responsible for the reduced antibody response of the CD38 KO mice (manuscript in preparation), suggesting that the defective humoral immune response in the CD38 KO mice may be due to an inability of the CD38 KO myeloid cells to respond to weak inflammatory signals.

CD38 Deficient Neutrophils are Unable to Migrate to Sites of Infection and are Responsible for the Defective Innate Immune Response of CD38 KO Mice

Since myeloid cells, like macrophages, neutrophils, eosinophils and dendritic cells, respond rapidly to cell damage and infection, they provide an important early line of defense against bacteria and viruses. Thus, we predicted that if these cells were functionally defective in CD38 KO mice, the mice would be more susceptible to bacterial infections. To test this hypothesis, CD38 KO mice were evaluated for their ability to resist various experimental bacterial infections, including a pulmonary infection with *Streptococcus pneumoniae*. Interestingly, we found that *S. pneumoniae* spread rapidly from the local site of infection (lung) to systemic sites (blood) in CD38 KO mice (ref. [28], and Fig. 2). This rapid bacterial dissemination rendered the CD38 KO mice at least 10-fold more susceptible to *S. pneumoniae* infection than their wild-type counterparts [28]. In addition, the CD38 KO mice died with accelerated kinetics compared to control animals, with most CD38 KO mice dying within the first 24-36 hours of infection (Fig. 2). Thus, in addition to their defects in antibody production, CD38 KO mice also were defective in their ability to mount a rapid innate immune response to infection.

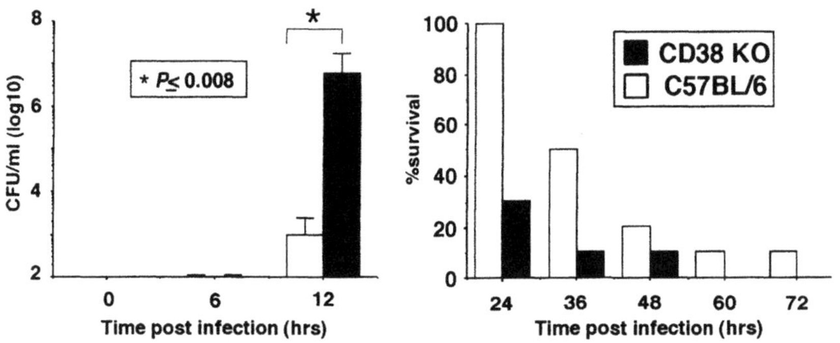

Figure 2. ***S. pneumoniae*** **infection rapidly disseminates in CD38 deficient mice.** CD38 KO and C57BL/6 mice were infected intratracheally with *S. pneumoniae*. The number of bacteria in the blood increased rapidly in the CD38 KO mice (left panel), resulting in death within the first 24-36 hours of infection (right panel).

Since the defect in the innate immune response was so striking in the CD38 KO mice, we set out to identify the cell types or tissues that were responsible for the increased susceptibility of the mice to pulmonary infection with *S. pneumoniae*. Experiments using various chimeric and knockout animals demonstrated that CD38 expression on bone marrow-derived myeloid cells is required for these cells to control bacterial infections [28]. Furthermore, we demonstrated that CD38 KO neutrophils were unable to efficiently migrate to the site of infection in the lung. For example, five-fold fewer neutrophils accumulated in the lungs of the CD38 KO mice relative to WT mice in the first 12 hours after infection [28]. Since neutrophils are responsible for controlling bacterial growth in the first hours to days after infection [54, 55], we believe that the failure of the CD38 KO neutrophils to accumulate at the infection site allowed for rapid systemic dissemination of the bacteria, resulting in sepsis and death.

To determine why the CD38 KO neutrophils were reduced in number at the infection/inflammatory site, we measured whether the CD38 KO neutrophils could be activated by inflammatory products such as the chemokine Interleukin-8 (IL-8) or the n-formyl-methionyl-leucyl-phenylalanine peptide (fMLP). While both IL-8 and fMLP are potent activators of neutrophils [56-58], IL-8 is a chemokine that is produced by cells at the site of infection [59, 60] and fMLP is a chemoattractant that is produced as a degradation product from bacteria [61, 62]. To test whether these molecules could induce the activation of CD38 KO neutrophils, we performed migration assays (ref. [28] and Fig. 3). We first placed CD38 KO or WT neutrophils in the top chamber of a transwell plate that contained either fMLP or IL-8 at equal concentration in the top and bottom chambers of the transwell (chemokinesis assay). Alternatively, we placed neutrophils in the top chamber of a transwell plate that contained IL-8 or fMLP in the bottom chamber only (chemotaxis assay). We then determined the number of neutrophils that migrated to the bottom chamber within 30 minutes.

We found that CD38 KO neutrophils could be activated by both fMLP and IL-8 and were equivalent to WT cells in their ability to rapidly migrate non-directionally (chemokinesis). However, when we tested whether the neutrophils from CD38 KO mice were capable of migrating to directionally toward gradients of fMLP or IL-8 (chemotaxis), we found that CD38KO neutrophils were unable to efficiently migrate in the direction of the fMLP gradient (Fig. 3). This was a selective defect in migration, because the CD38 KO neutrophils were perfectly competent to migrate toward the IL-8 gradient [28]. Since the receptors for IL-8 (CXCR1 and CXCR2) and fMLP (FPR1, FPR2 and FPRL1) are 7 transmembrane G-protein coupled receptors (GPCRs), this suggested that CD38 was required for optimal signal transduction through some, but not all, GPCRs. Additionally, since the CD38 KO neutrophils could be activated by fMLP, this suggested that only some of the signals delivered through the FPR are dependent on CD38.

Taken together, this work indicates that CD38 is necessary for optimal innate and adaptive humoral immune responses. Although purified lymphocytes appear to function normally, particularly in response to antigen receptor signaling, we have found that CD38 expression on at least one cell type, the neutrophil, is absolutely required to respond to triggering through the FPR receptor. In the next sections we focus on describing experiments that were designed to determine the mechanism by which CD38 regulates GPCR signaling in neutrophils.

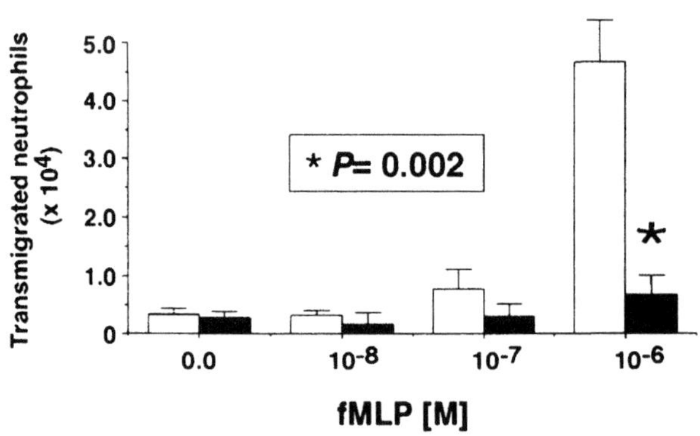

Figure 3. **CD38 KO neutrophils are unable to follow a chemotactic gradient of fMLP.** Neutrophils were purified from the bone marrow of CD38 KO and C57BL/6 mice and placed in the top chamber of a transwell containing various concentrations of fMLP in the bottom chamber. The number of neutrophils that migrated to the bottom chamber in response to the chemotactic gradient was determined 30 minutes later.

CD38 and cADPR Regulate Intracellular Free Calcium Levels in Response to FPR Ligation and Control Neutrophil Chemotaxis to Bacterially-Derived Peptides

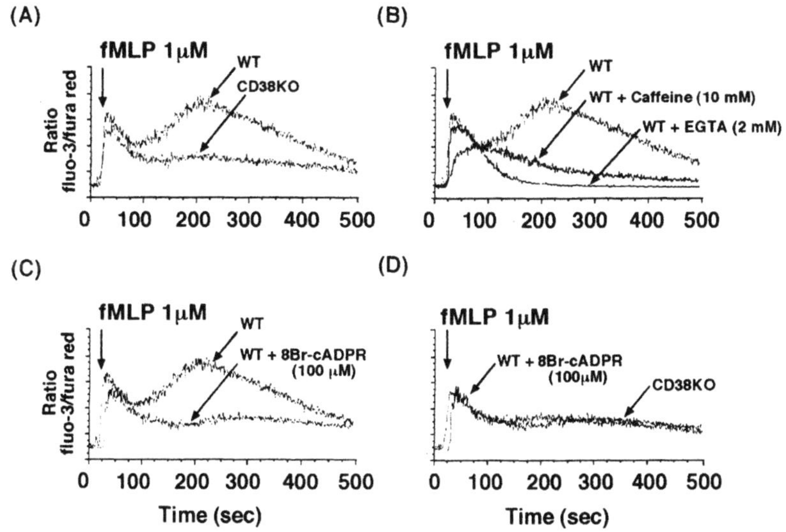

Figure 4. **CD38 and cADPR regulate intracellular calcium release and extracellular calcium influx in neutrophils.** CD38 KO or C57BL/6 (WT) neutrophils were stimulated with fMLP and intracellular free calcium levels were measured by FACS. (**A**) In WT neutrophils, fMLP induced an initial rapid release of intracellular calcium followed by an extended influx of extracellular calcium. In contrast, intracellular calcium release was reduced and extracellular calcium influx was abolished in fMLP-stimulated CD38 KO neutrophils. (**B**) Pretreatment of WT neutrophils with caffeine dramatically reduced fMLP-induced intracellular calcium release and extracellular calcium influx, while addition of EGTA to the media blocked extracellular calcium influx. (**C**) Pretreatment with 8-Br-cADPR reduced intracellular calcium release of fMLP-stimulated WT neutrophils and largely eliminated extracellular calcium influx. (**D**) The calcium mobilization of 8-Br-cADPR treated, fMLP-stimulated WT neutrophils was indistinguishable from the response of fMLP-stimulated CD38 KO neutrophils.

Since CD38 KO neutrophils were unable to migrate toward the bacterial chemoattractant, fMLP, we suspected that CD38 was necessary for proper signaling through the FPR. Since CD38 can catalyze the production of calcium mobilizing metabolites such as cADPR [20], and since ligation of the FPR has been shown to induce intracellular calcium accumulation through a combination of intracellular calcium release and extracellular calcium influx [63-65], we postulated that CD38 might regulate cytosolic free calcium levels in FPR activated neutrophils. We found that stimulation of WT neutrophils with fMLP resulted in an initial rapid release of intracellular calcium and a more extended influx of extracellular calcium

(Fig. 4A-B). However, when we stimulated CD38 KO neutrophils with fMLP, we found that the initial release of intracellular calcium was reduced by approximately 20% compared to fMLP-stimulated WT neutrophils (Fig. 4A). Furthermore, we saw that the calcium influx phase of the response was essentially eliminated in the fMLP-stimulated CD38 KO neutrophils (Fig. 4A). These data suggested that CD38 regulated FPR signaling by modulating the intracellular calcium release response and controlling the calcium influx response.

To test whether the changes in intracellular calcium levels in response to FPR triggering in neutrophils were regulated by cADPR, we stimulated WT neutrophils with fMLP in the presence or absence of the cADPR antagonist, 8-Br-cADPR [66, 67], and then measured changes in intracellular free calcium levels [28]. Interestingly, we found that stimulation of WT neutrophils with fMLP in the presence of 8-Br-cADPR reduced the initial intracellular calcium release and blocked the extracellular calcium influx (Fig. 4C). Indeed, treatment of the WT cells with the cADPR antagonist rendered them indistinguishable from the CD38 KO cells (Fig. 4D). These results suggested that CD38, through its production of cADPR, regulated calcium mobilization in fMLP-stimulated cells.

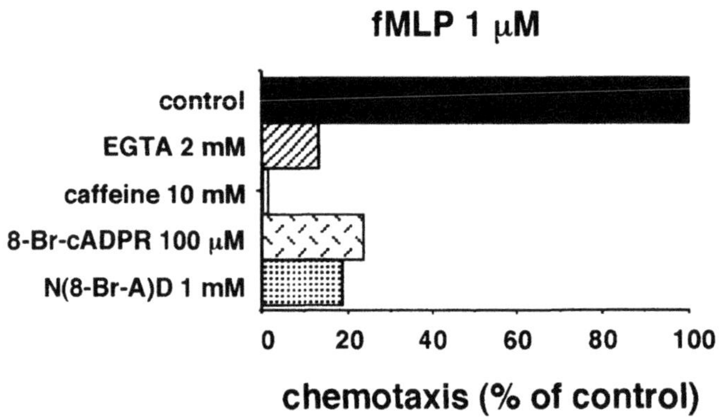

Figure 5. **Neutrophil chemotaxis to fMLP is dependent on cADPR production by CD38.** C57BL/6 (WT) neutrophils were incubated for 20 minutes in media alone (control) or in media containing EGTA. 8-Br-cADPR. N(8-Br-A)D$^+$ or caffeine. The cells were then placed in the top chamber of a transwell containing 1 μM fMLP in the bottom chamber. Neutrophil chemotaxis to fMLP was measured after 30 minutes.

The defective calcium mobilization observed after stimulation of CD38 KO neutrophils with fMLP correlated with the defective chemotactic response of these cells to fMLP. Therefore, it was possible that neutrophil chemotaxis to fMLP required cADPR-dependent calcium mobilization. To address this possibility, we tested 8-Br-cADPR treated neutrophils in a chemotaxis assay to see if blockade of cADPR-dependent calcium mobilization would alter the ability of neutrophils to chemotax toward an fMLP gradient [28]. Strikingly, we found that 8-Br-cADPR blocked neutrophil chemotaxis to fMLP (Fig. 5). Thus, these data indicate that CD38-generated cADPR modulates intracellular calcium release, controls extracellular calcium influx, and is required for chemotaxis of neutrophils in response to FPR ligation.

Modified Substrates of CD38 can Modulate Neutrophil Chemotaxis

Since we knew that CD38-generated cADPR appeared to regulate neutrophil chemotaxis to fMLP, we postulated that we could modulate neutrophil migration simply by modifying the CD38 substrate, NAD^+. To test this hypothesis, we incubated CD38KO and WT neutrophils with an altered substrate for CD38, N(8-Br-A)D^+ [68], and then assayed their ability to migrate toward fMLP [28]. We reasoned that N(8-Br-A)D^+ would be utilized as a substrate by CD38 and would be converted at a low frequency to 8-Br-cADPR, the cADPR antagonist we had used in our earlier experiments. We found that inclusion of N(8-Br-A)D^+ in the assay was sufficient to block the chemotaxis of WT neutrophils to fMLP (Fig. 5). This inhibition was CD38 dependent, because the N(8-Br-A)D^+ had no effect on CD38 KO neutrophils [28]. Thus, by providing CD38 with an altered substrate, we could efficiently block neutrophil chemotaxis. If these data can be extrapolated to *in vivo* models, we would predict that substrate analogs of CD38 might represent potential drugs that could be used to modulate inflammatory responses *in vivo*. However, it is clear that it is not desirable to eliminate neutrophil chemotaxis to sites of infection. Rather, we propose that these analogs would be therapeutically useful in situations like organ transplantation, ischemia/reperfusion injuries, burn, allergy/asthma and autoimmune disease where neutrophils contribute to inflammation-mediated tissue damage [69-73].

REMAINING QUESTIONS

The results described in the previous sections begin to discern how CD38 can regulate innate immune responses at a cellular and molecular level. If CD38 regulates intracellular calcium levels in granulocytes that have been stimulated with inflammatory cytokines or chemokines, then one can easily envision that CD38 would be necessary for optimal innate, cellular and humoral immune responses, all of which are exquisitely regulated by calcium-dependent signal transduction molecules. However, we are still left with a number of unanswered mechanistic questions that our data does not address. For example, there are a number of features regarding the enzyme activity of CD38 that complicate our understanding of how CD38-generated cADPR could regulate signaling through the FPR on neutrophils. First, unlike the *Aplysia* cyclase [74], CD38 is primarily localized to the plasma membrane of leukocytes [2]. Thus, the active site of CD38 is located outside the cell, rather than inside the cell. Second, the known physiologic substrates for CD38, NAD^+ and NADP, are predominantly localized inside the cell and are not found at high extracellular concentrations in normal serum [75]. This implies that the active site of CD38 may not have access to substrate under normal physiologic conditions and may be effectively inert unless substrate is made available to the extracellular milieu. Third, unlike the *Aplysia* enzyme, CD38 is predominantly an NAD^+ glycohydrolase, and under V_{max} conditions, approximately 97-99% of the product produced by CD38 is ADPR and not cADPR or NAADP [20, 76]. Fourth, when substrate is present, CD38 is a constitutively active enzyme and does not need to be phosphorylated or otherwise structurally modified to efficiently catalyze enzyme reactions [20]. Finally, CD38 does not appear to be physically associated with other cell surface proteins expressed on neutrophils. Thus, we are left with the conundrum of how a constitutively active ecto-enzyme that has limited access to substrate is able to produce membrane impermeant metabolites that regulate intracellular calcium release upon triggering of other physically unassociated receptors. In the next sections we outline some of the preliminary experiments that have been done to answer these questions and then present our newest working model.

Where is CD38 Localized in Neutrophils?

We have measured abundant CD38 enzyme activity on total cell membrane fractions and on the plasma membrane of intact neutrophils [28]. This indicates that enzymatically active CD38 is present on the plasma membrane, but does not rule out the possibility that CD38 is also localized to intracellular membranes or found free within the cytosol. Experiments are underway to assess these possibilities, but based on analysis of the CD38 gene [77, 78], it seems unlikely that a soluble cytoplasmic form of CD38 is encoded within the gene. Therefore, our current working model depicts CD38 producing cADPR and regulating calcium responses from its plasma membrane location.

How Does CD38 Access Substrate?

If we presume that CD38 functions as a constitutively active ecto-enzyme, then we must identify the source of substrate for extracellular CD38. As mentioned earlier, NAD(P) is present only at very low concentrations in serum from normal animals [75]. This would predict that CD38 does not produce ADPR or cADPR most of the time simply because substrate is unavailable. Furthermore, it would suggest that substrate availability may be a major way in which CD38 enzyme activity is regulated *in vivo*. So, when might substrate be available? We hypothesize that substrate is made available to ecto-CD38 every time a cell undergoes a non-apoptotic cell death and releases its cytosolic contents. Under conditions of infection or inflammation, when cells are rapidly undergoing necrosis, the local or systemic concentrations of NAD^+ and NADP might become quite high. Then CD38 would have access to an abundant source of substrate that would disappear when the injury or infection was resolved. One prediction from this model is that lack of CD38 would result in higher extracellular concentrations of NAD^+, particularly at sites of infection and inflammation. This hypothesis is currently being tested in the CD38 KO mice.

How Does cADPR Arrive at Its Intracellular Location?

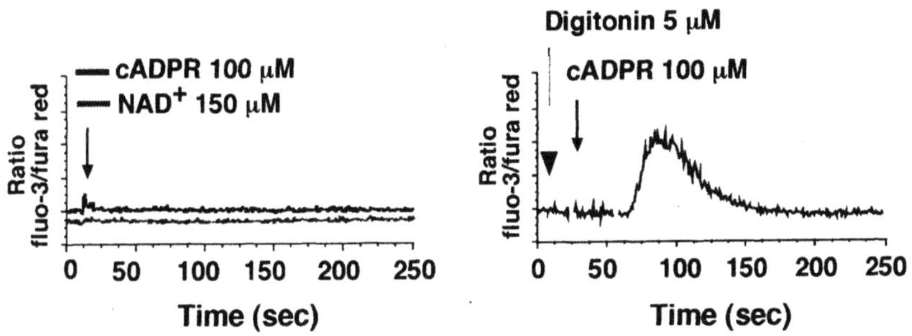

Figure 6. **Purified NAD^+ and cADPR do not induce intracellular calcium release in intact neutrophils.** (**Left panel**) The addition of either cADPR or NAD^+ directly to neutrophils does not induce intracellular calcium release. (**Right panel**) However, cADPR added to neutrophils permeabilized with digitonin induces intracellular calcium release.

If CD38 accesses substrate and generates products outside the cell, how do the membrane impermeant products arrive at their presumed site of action inside the cell? This question is one that has long perplexed those who have studied CD38 enzymology and a number of models have been proposed. Recently it has been shown that cADPR is first produced, and then directly transported by CD38 across the plasma membrane [79]. This model for cADPR internalization has not been directly tested in neutrophils, however, we know that direct addition of cADPR to neutrophils does not induce intracellular calcium release (Fig. 6). Instead, we find that the plasma membrane of the neutrophils must be permeabilized before addition of cADPR induces intracellular calcium release (Fig. 6). This suggests that under normal physiologic conditions cADPR must be transported across the plasma membrane to the cytosol to mediate intracellular calcium release.

However, whether CD38 or another protein is required for the transport of cADPR in neutrophils is completely unknown.

Is cADPR a Calcium Mobilizing Second Messenger in Neutrophils?

We know that CD38-generated cADPR is necessary for calcium signaling after ligation of the FPR in neutrophils. However, there is no evidence that CD38 and the FPR receptor are physically associated nor is there any direct evidence that cADPR is produced by CD38 in response to fMLP stimulation. For example, we have found that both intracellular calcium release and extracellular calcium influx is normal in WT neutrophils stimulated by fMLP in NAD-free media (unpublished data). This suggests that substrate may not have to be present at the time of calcium mobilization and that cADPR may be preformed and present in the cytosol. Based on our preliminary data, we postulate that cADPR is generated in very small quantities by CD38 whenever substrate is present (ie during an inflammatory response) and is then transported to an intracellular location where it is "stored" until it is either utilized or degraded. Therefore, we favor the model proposed by Galione and colleagues that cADPR, at least in neutrophils, acts as a modulator of calcium-induced calcium release (CICR) from Ryanodine Receptor (RyR) gated stores [80]. If this model is correct, both free cytosolic calcium and cADPR will be needed to induce release of calcium from the RyR-gated stores in neutrophils. It is known that neutrophils activated by ligation of the FPR release intracellular calcium from inositol trisphosphate receptor (IP_3R) and sphingosine-1 phosphate (SPP) receptor gated stores [81, 82]. If cADPR is also present at sufficient concentration, we propose that calcium will also be released from cADPR-dependent RyR-gated stores, leading to efficient extracellular calcium influx and eventually chemotaxis. If no cADPR is present, then only IP_3 and SPP mediated intracellular calcium release will occur and extracellular calcium influx will be greatly reduced, resulting in impaired chemotaxis.

How Does cADPR Regulate Intracellular Calcium Release and Extracellular Calcium Influx in fMLP Stimulated Neutrophils?

In both smooth and skeletal muscle it is believed that cADPR induces intracellular calcium release by binding, either directly or indirectly, to RyRs localized in the sarcoplasmic reticulum [83, 84]. RyRs can release calcium not only in response to cADPR but also to a number of other stimuli including cytosolic free calcium (CICR), ryanodine and caffeine [85]. It has been shown that cADPR can directly induce calcium release from RyR-gated stores and can further sensitize the RyR to respond to intracellular free calcium by releasing additional calcium [86]. This cADPR dependent amplification of the CICR process allows initiation of calcium release from RyR gated stores at a lower threshold and can prolong the RyR-mediated calcium signal [86]. Thus, cADPR appears to serve as a rheostat to regulate CICR [80, 87]. To test whether cADPR can modulate RyR-mediated calcium release in neutrophils, we first determined whether RyRs were present and functional in normal neutrophils. We found that neutrophils express mRNA for RyR3 [88] and, when permeabilized, could respond directly to ryanodine by releasing intracellular calcium [28]. Furthermore, the ryanodine dependent calcium release could be blocked by the addition of ruthenium red, a ryanodine receptor antagonist [89]. We also found that direct addition of cADPR to permeabilized neutrophils could induce a small, but reproducible, increase in intracellular calcium levels (Fig. 6) due to intracellular calcium release [28]. This cADPR induced calcium response could be completely blocked by the addition of 8-Br-cADPR [28]. Taken together, these data indicated that neutrophils respond to intracellular cADPR by releasing intracellular calcium. Although we hypothesize that cADPR directly regulates RyR-gated stores, the ryanodine antagonist, ruthenium red, did not antagonize cADPR-mediated calcium release and the cADPR antagonist, 8-Br-cADPR, did not antagonize ryanodine-mediated calcium release (unpublished data). This result may be due to the fact that cADPR and ryanodine appear to interact with the RyRs at different sites [90]. Although this experiment did not provide evidence for a direct relationship between cADPR and RyRs in neutrophils, we have recently found that preincubation of neutrophils with caffeine, a known RyR agonist, desensitized the calcium influx response of fMLP-stimulated neutrophils (Fig. 4B) and blocked chemotaxis to fMLP (Fig. 5). These data suggest that release of the RyR-gated calcium pool must be required for the chemotaxis of neutrophils to fMLP.

Another important issue that remains unresolved is how cADPR selectively induces extracellular calcium influx. It is known that in several cell types [91-93], including T lymphocytes [94], cADPR and RyRs appear to regulate calcium influx through store operated channels (SOCs). It is known that neutrophils, like other bone marrow derived cells, express SOCs that can be activated by intracellular calcium store depletion and/or increased intracellular free calcium levels [63, 64, 95, 96]. SOCs are activated in neutrophils upon triggering of some, but not all, membrane chemokine receptors. Interestingly, SOCs are known to be activated in fMLP stimulated neutrophils [97]. In addition to SOCs, it has been suggested that neutrophils might also express other plasma membrane calcium channels such as voltage-gated L-type channels. Although there is no direct evidence that neutrophils actually express conventional L-type channels, treatment of neutrophils with voltage-gated L channel inhibitors, such as verapamil, has been shown to alter neutrophil responses to fMLP [98].

If we assume that SOCs are the predominant plasma membrane calcium channels expressed by neutrophils, we are still left with the question of how they are opened specifically in response to cADPR-mediated intracellular calcium release. One possibility is that all intracellular calcium stores must be depleted before the SOC is activated and the cADPR-dependent store is simply the last store to be released. This seems unlikely given that the amount of calcium released by cADPR is actually quite small in comparison to that released by IP_3 and/or SPP [28]. Alternatively, perhaps the SOCs can specifically "sense" depletion of the cADPR-gated calcium stores via a conformational change in the RyR induced by cADPR and calcium binding. Finally, it is possible that cADPR and IP_3R gated calcium stores are localized in distinct sites within the cell, allowing SOCs to be activated by one or the other depending on the type of stimulus delivered to the cell. Interestingly, neutrophils have been reported to sequester calcium in cytoplasmic vesicles referred to as calciosomes [99] and to have spatially distinct stores of calcium within the cell [100].

Does CD38 Catalyzed ADPR or NAADP Influence Calcium Responses in Neutrophils?

In the experiments that have been done to date, only the role of CD38 generated cADPR in neutrophils has been examined. However, it is certainly possible that NAADP and/or ADPR might also play a role in CD38 mediated regulation of calcium signaling in neutrophils. It was recently published that ADPR can open the plasma membrane LTRPC2 cation channel and allow the influx of extracellular calcium [101, 102]. Given that the most striking feature of the fMLP stimulated CD38 KO neutrophils is the loss of calcium influx [28] and that ADPR is the most predominant metabolite produced from the CD38 enzyme reaction [20], it is tempting to speculate that CD38 generated ADPR might also be involved regulating calcium influx in the neutrophils. However, direct addition of NAD^+ to CD38 expressing neutrophils did not induce calcium influx (Fig. 6), despite the fact that abundant ADPR was generated. This suggests that either neutrophils don't express the LTRPC2 channel proteins or that the production of ADPR by CD38 is not sufficient to induce activation of these channels in neutrophils.

It is also possible that NAADP could play a role in calcium signaling in neutrophils. Indeed experiments from the Guse laboratory showed that NAADP could induce intracellular calcium mobilization in Jurkat T cells that were triggered through the TCR [103]. They demonstrated that all intracellular calcium release was dependent on NAADP and suggested that NAADP was necessary to induce sufficient intracellular free calcium to sensitize the IP_3R and RyRs to IP_3 and cADPR [103]. If this is true for neutrophils, we would need to demonstrate that CD38 is expressed in intracellular acidic vesicles as NAADP is generated at low pH values [24].

Working Model for CD38-Mediated Regulation of Innate Immune Responses.

Our new working model depicting the mechanism by which CD38 regulates the innate immune responses to bacterial pathogens is shown in Figure 7.

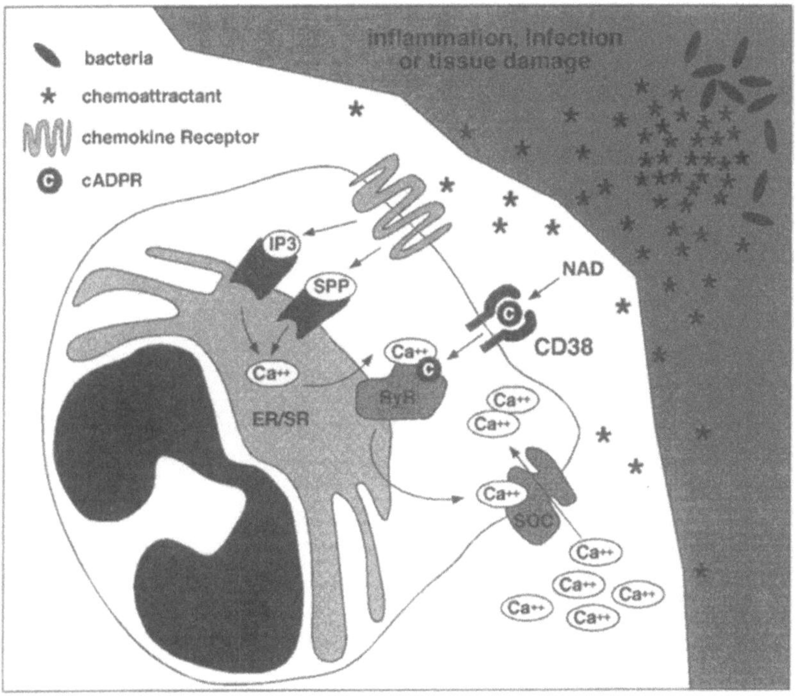

Figure 7. **A proposed model to explain how CD38 regulates G protein coupled receptor signaling in neutrophils.**

We hypothesize that during the acute phase of a bacterial infection, the cells within the infected tissue are damaged and die. The damaged cells produce inflammatory mediators and the dying cells release NAD^+, resulting in transient local and systemic concentrations of these metabolites. CD38 expressing neutrophils encounter NAD^+ and produce cADPR that is transported inside the cell where it can then interact with RyRs. The bacteria, meanwhile replicate and release proteins that are degraded into chemotactic fMLP peptides. CD38-expressing neutrophils in the periphery are activated by the inflammatory mediators and the fMLP peptides. Engagement of the

FPR on neutrophils results in the release of intracellular calcium from IP_3 and/or SPP controlled stores. The free cytosolic calcium binds to RyRs cooperatively with cADPR and induces calcium release from the RyRs. This results in activation and opening of the SOCs and sustained calcium influx. This extended calcium influx potentiates neutrophil chemotaxis to the fMLP peptide and allows for efficient migration of the neutrophil to the site of infection. Once at the site, the neutrophil can rapidly phagocytose and kill the bacteria, thus preventing bacterial dissemination.

If this model is correct, we would predict that compounds that alter CD38's enzyme activity may be useful in modulating the migration of neutrophils to sites of infection or damage. Excitingly, in preliminary experiments we have found that the loss of CD38 provides protection from immune-mediated tissue damage in at least one *in vivo* model of inflammation induced damage. In the future we hope to test CD38 inhibitors in this type of *in vivo* model. If these experiments are successful, it would significantly raise the promise of using CD38 enzyme analogues to treat human disease. Thus, someday we could move from using antibodies to CD38 as prognostic markers of disease to using CD38 enzyme modulators to treat disease.

ACKNOWLEDGEMENTS

The authors gratefully acknowledge Dr. D. Cockayne and M. Howard for originally providing us with the CD38 KO mice. This work was supported by NIH grant AI-43629 and the Trudeau Institute.

REFERENCES

1. Reinherz EL, Kung PC, Goldstein G, Levey RH and Schlossman SF. 1980. Discrete stages of intrathymic differentiation: analysis of normal thymocytes and leukemic lymphoblasts of the T lineage. *Proc. Natl. Acad. Sci. USA.* 77: 1588-1592.
2. Jackson DG and Bell JI. 1990. Isolation of a cDNA encoding the human CD38 (T10) molecule, a cell surface glycoprotein with an unusual discontinuous pattern of expression during lymphocyte differentiation. *J. Immunol.* 144: 2811-2815.
3. Kawano MM, Mihara K, Tsujimoto T, Huang N and Kuramoto A. 1995. A new phenotypic classification of bone marrow plasmacytosis. *Int. J. Hematol.* 61: 179-188.
4. San Miguel JF, Garcia-Sanz R, Gonzalez M, Moro MJ, Hernandez JM, *et al.* 1995. A new staging system for multiple myeloma based on the number of S-phase plasma cells. *Blood* 85: 448-455.
5. Damle RN, Wasil T, Fais F, Ghiotto F, Valetto A, *et al.* 1999. Ig V gene mutation status and CD38 expression as novel prognostic indicators in chronic lymphocytic leukemia. *Blood* 94: 1840-1847.
6. Ibrahim S, Keating M, Do KA, O'Brien S, Huh YO, *et al.* 2001. CD38 expression as an important prognostic factor in B-cell derived lymphocytic leukemia. *Blood* 98: 181-186.
7. Liu Z, Cumberland WG, Hultin LE, Prince HE, Detels R and Giorgi JV. 1997. Elevated CD38 antigen expression on $CD8^+$ T cells is a stronger marker for the risk of chronic HIV disease progression to AIDS and death in the multicenter AIDS cohort study than

CD4⁻ cell count, soluble immune activation markers, or combinations of HLA-DR and CD38 expression. *J. Acquir. Immune Defic. Syndr. Hum. Retrovirol.* 16: 83-92.
8. Terstappen LWMM, Huang S, Safford M, Landsorp PM and Loken MR. 1991. Sequential generations of hematopoietic colonies derived from single nonlineage-committed $CD34^+$ $CD38^-$ progenitor cells. *Blood* 77: 1218-1227.
9. Randall TD, Lund FE, Howard MC and Weissman IL. 1996. Expression of murine CD38 defines a population of long-term reconstituting hematopoietic stem cells. *Blood* 87: 4057-4067.
10. Liu YJ, Malisan F, de Bouteiller O, Guret C, Lebecque S, *et al.* 1996. Within germinal centers, isotype switching of immunoglobulin genes occurs after the onset of somatic mutation. *Immunity* 4: 241-250.
11. Oliver AM, Martin F and Kearney JF. 1997. Mouse CD38 is down-regulated on germinal center B cells and mature plasma cells. *J. Immunol.* 158: 1108-1115.
12. Bean AGD, Godfrey DI, Ferlin WG, Santos-Argumedo L, Parkhouse RME, *et al.* 1995. CD38 expression on mouse T cells: CD38 defines functionally distinct subsets of αβ $TCR^+CD4^-CD8^-$ thymocytes. *Int. Immunol.* 7: 213-221.
13. Read S, Mauze S, Asseman C, Bean A, Coffman R and Powrie F. 1998. $CD38^+$ $CD45RB^{low}$ $CD4^+$ T cells: a population of T cells with immune regulatory activities *in vitro*. *Eur. J. Immunol.* 28: 3435-3447.
14. Funaro A, Spagnoli GC, Ausiello CM, Alessio M, Roggero S, *et al.* 1990. Involvement of the multilineage CD38 molecule in a unique pathway of cell activation and proliferation. *J. Immunol.* 145: 2390-2396.
15. Malavasi F, Funaro A, Alessio M, De Monte LB, Ausiello CM, *et al.* 1992. CD38: A multi-lineage cell activation molecule with a split personality. *Int. J. Clin. Lab. Res.* 22: 73-80.
16. Malavasi F, Funaro A, Roggero S, Horenstein A, Calosso L and Mehta K. 1994. Human CD38: a glycoprotein in search of a function. *Immunol. Today* 15: 95-97.
17. Mehta K, Shahid U and Malavasi F. 1996. Human CD38, a cell-surface protein with multiple functions. *FASEB J.* 10: 1408-1417.
18. Lund FE, Cockayne DA, Randall TD, Solvason N, Schuber F and Howard MC. 1998. CD38: A new paradigm in lymphocyte activation and signal transduction. *Immunol. Rev.* 161: 79-93.
19. Harada N, Santos-Argumedo L, Chang R, Grimaldi JC, Lund FE, *et al.* 1993. Expression cloning of a cDNA encoding a novel murine B cell activation marker. *J. Immunol.* 151: 3111-3118.
20. Howard M, Grimaldi JC, Bazan JF, Lund FE, Santos-Argumedo L, *et al.* 1993. Formation and hydrolysis of cyclic ADP-ribose catalyzed by lymphocyte antigen CD38. *Science* 262: 1056-1059.
21. Takasawa S, Togho A, Noguchi N, Koguma T, Nata K, *et al.* 1993. Synthesis and hydrolysis of cyclic ADP-ribose by human leukocyte antigen CD38 and inhibition of the hydrolysis by ATP. *J. Biol. Chem.* 268: 26052-4.
22. Summerhill RJ, Jackson DG and Galione A. 1993. Human lymphocyte antigen CD38 catalyzes the production of cyclic ADP-ribose. *FEBS Lett.* 335: 231-233.
23. Zocchi E, Franco L, Guida L, Benatti U, Bargellesi A, *et al.* 1993. A single protein immunologically identified as CD38 displays NAD^+ glycohydrolase, ADP-ribosyl cyclase and cyclic ADP-ribosyl hydrolase activities at the outer surface of human erythrocytes. *Biochem. Biophys. Res. Comm.* 196: 1459-1465.
24. Aarhus R, Graeff RM, Dickey DM, Walseth TF and Lee HC. 1995. ADP-ribosyl cyclase and CD38 catalyze the synthesis of a calcium mobilizing metabolite from NADP. *J. Biol. Chem.* 270: 30327-30333.
25. Kumagai M, Coustan-Smith E, Murray DJ, Silvennoinen O, Murti KG, *et al.* 1995. Ligation of CD38 suppresses human B lymphopoiesis. *J. Exp. Med.* 181: 1101-1110.

26. Lund FE, Muller-Steffner HM, Yu N, Stout CD, Schuber F and Howard M. 1999. CD38 signaling is controlled by its ectodomain but occurs independently of enzymatically generated ADP-ribose or cyclic ADP-ribose. *J. Immunol.* 162: 2693-2702.
27. Kitanaka A, Suzuki T, Ito C, Nishigaki H, Coustan-Smith E, *et al.* 1999. CD38-mediated signaling events in murine pro-B cells expressing human CD38 with or without its cytoplasmic domain. *J. Immunol.* 162: 1952-1958.
28. Partida-Sanchez S, Cockayne DA, Monard S, Jacobson EL, Oppenheimer N, *et al.* 2001. Cyclic ADP-ribose production by CD38 regulates intracellular calcium release, extracellular calcium influx and chemotaxis in neutrophils and is required for bacterial clearance *in vivo*. *Nature Med.* 7: 1209-1216.
29. Cockayne D, Muchamuel T, Grimaldi JC, Muller-Steffner H, Randall TD, *et al.* 1998. Mice deficient for the ecto-NAD$^-$ glycohydrolase CD38 exhibit altered humoral immune responses. *Blood* 92: 1324-1333.
30. Shubinsky G and Schlesinger M. 1997. The CD38 lymphocyte differentiation marker- New insight into its ectoenzymatic activity and its role as a signal transducer. *Immunity* 7: 315-324.
31. Santos-Argumedo L, Teixeira C, Preece G, Kirkham PA and Parkhouse RME. 1993. A B lymphocyte surface molecule mediating activation and protection from apoptosis via calcium channels. *J. Immunol.* 151: 3119-3130.
32. Santos-Argumedo L, Lund FE, Heath AW, Solvason N, Wu WW, *et al.* 1994. CD38 unresponsiveness of *xid* B cells implicates Bruton's tyrosine kinase (*btk*) as a regulator of CD38 induced signal transduction. *Int. Immunol.* 7: 163-170.
33. Lund FE, Solvason NW, Cooke MP, Heath AW, Grimaldi JC, *et al.* 1995. Signaling through murine CD38 is impaired in antigen receptor unresponsive B cells. *Eur. J. Immunol.* 25: 1338-1345.
34. Lund FE, Yu N, Kim K-M, Reth M and Howard MC. 1996. Signaling through CD38 augments B cell antigen receptor (BCR) responses and is dependent on BCR expression. *J. Immunol.* 157: 1455-1467.
35. Yamashita Y, Miyake K, Kikuchi Y, Takatsu K, Noda S and Kosugi A. 1995. A monoclonal antibody against a murine CD38 homologue delivers a signal to B cells for prolongation of survival and protection against apoptosis *in vitro*: unresponsiveness of X-linked immunodeficient B cells. *Immunology* 85: 248-255.
36. Kikuchi Y, Yasue T, Miyake K, Kimoto M and Takatsu K. 1995. CD38 ligation induces tyrosine phosphorylation of Bruton tyrosine kinase and enhanced expression of interleukin 5-receptor α chain: Synergistic effects with interleukin 5. *Proc. Natl. Acad. Sci. USA.* 92: 11814-11818.
37. Yasue T, Nishizumi H, Aizawa S, Yamamoto T, Miyake K, *et al.* 1997. A critical role of Lyn and Fyn for B cell responses to CD38 ligation and interleukin 5. *Proc. Natl. Acad. Sci. USA.* 94: 10307-10312.
38. Kitanaka A, Ito C, Nishigaki H and Campana D. 1996. CD38 mediated growth suppression of B-cell progenitors requires activation of phosphatidylinositol 3-kinase and involves its association with the protein product of the *c-cbl* proto-oncogene. *Blood* 88: 590-598.
39. Silvennoinen O, Nishigaki H, Kitanaka A, Kumagai M, Ito C, *et al.* 1996. CD38 signal transduction in human B cell precursors. *J. Immunol.* 156: 100-107.
40. Kitanaka A, Ito C, Coustan-Smith E and Campana D. 1997. CD38 ligation in human B cell progenitors triggers tyrosine phosphorylation of CD19 and association of CD19 with Lyn and phosphatidylinositol 3-kinase. *J. Immunol.* 159: 184-192.
41. Zupo S, Rugari E, Dono M, Taborelli G, Malavasi F and Ferrarini M. 1994. CD38 signaling by agonistic monoclonal antibody prevents apoptosis of human germinal center B cells. *Eur. J. Immunol.* 24: 1218-1222.

42. Funaro A, Morra M, Calosso L, Zini MG, Ausiello CM and Malavasi F. 1997. Role of the human CD38 molecule in B cell activation and proliferation. *Tissue Antigens* 49: 7-15.
43. Ausiello CM, Urbani F, la Sala F, Funaro A and Malavasi F. 1995. CD38 ligation induces discrete cytokine mRNA expression in human cultured lymphocytes. *Eur. J. Immunol.* 25: 1477-1480.
44. Morra M, Zubiaur M, Terhorst C, Sancho J and Malavasi F. 1998. CD38 is functionally dependent on the TCR/CD3 complex in human T cells. *FASEB J.* 12: 581-592.
45. Zubiaur M, Guirado M, Terhorst C, Malavasi F and Sancho J. 1999. The CD3-gamma delta epsilon transducing module mediates CD38-induced protein-tyrosine kinase and mitogen-activated protein kinase activation in Jurkat T cells. *J. Biol. Chem.* 274: 20633-20642.
46. Konopleva M, Estrov Z, Zhao S, Andreeff M and Mehta K. 1998. Ligation of cell surface CD38 protein with agonistic monoclonal antibody induces a cell growth signal in myeloid leukemia cells. *J. Immunol.* 161: 4702-4708.
47. Todisco E, Suzuki T, Srivannaboon K, Coustan-Smith E, Raimondi SC, *et al.* 1999. CD38 ligation inhibits normal and leukemic myelopoiesis. *Blood* 95: 535-542.
48. Ausiello CM, la Sala A, Ramoni C, Urbani F, Funaro A and Malavasi F. 1996. Secretion of IFN-γ, IL-6, granulocyte-macrophage colony-stimulating factor and IL-10 cytokines after activation of human purified T lymphocytes upon CD38 ligation. *Cell. Immunol.* 173: 192-197.
49. Zubiaur M, Izquierdo M, Terhorst C, Malavasi F and Sancho J. 1997. CD38 ligation results in activation of the Raf-1/mitogen-activated protein kinase and the CD3-ζ/ζ-associated protein-70 signaling pathways in Jurkat T lymphocytes. *J. Immunol.* 159: 193-205.
50. Zubiaur M, Fernandez O, Ferrero E, Salmeron J, Malissen B, *et al.* 2002. CD38 is associated with lipid rafts and upon receptor stimulation leads to Akt/Protein Kinase B and erk activation in the absence of the CD3-ζ immune receptor tyrosine-based activation motifs. *J. Biol. Chem.* 277: 13-22.
51. Nishina H, Inageda K, Takahashi K, Hoshino S, Ikeda K and Katada T. 1994. Cell surface antigen CD38 identified as ecto-enzyme of NAD glycohydrolase has hyaluronate-binding activity. *Biochem. Biophys. Res. Comm.* 203: 1318-1323.
52. Deaglio S, Dianzani U, Horenstein AL, Fernandez JE, van Kooten C, *et al.* 1996. Human CD38 ligand: A 120-KDA protein predominantly expressed on endothelial cells. *J. Immunol.* 156: 727-734.
53. Deaglio S, Morra M, Mallone R, Ausiello CM, Prager E, *et al.* 1998. Human CD38 (ADP-Ribosyl Cyclase) is a counter-receptor of CD31, an Ig superfamily member. *J. Immunol.* 160: 395-402.
54. Garvy BA and Harmsen AG. 1996. The importance of neutrophils in resistance to pneumococcal pneumonia in adult and neonatal mice. *Inflammation* 20: 499-512.
55. Bergeron Y, Ouellet N, Deslauriers A, Simard M, Olivier M and Bergeron MG. 1998. Cytokine kinetics and other host factors in response to pneumococcal pulmonary infection in mice. *Infect. Immun.* 66: 912-922.
56. Le Y, Oppenheim J and Wang JM. 2001. Pleiotropic roles of formyl peptide receptors. *Cytokine and Growth Factor Rev.* 12: 91-105.
57. Mukaida N. 2000. Interleukin-8: an expanding universe beyond neutrophil chemotaxis and activation. *Int. J. Hematol.* 72: 391-398.
58. Zeilhofer HU and Schorr W. 2000. Role of interleukin-8 in neutrophil signaling. *Curr. Opin. Hematol.* 7: 178-182.
59. Mukaida N, Harada A and Matsushima K. 1998. Interleukin-8 (IL-8) and monocyte chemotactic and activating factor (MCAF/MCP-1) chemokines essentially involved in inflammatory and immune reactions. *Cytokine Growth Factor. Rev.* 9: 9-23.

60. Khair OA, Davies RJ and Devalia JL. 1996. Bacterial-induced release of inflammatory mediators by bronchial epithelial cells. *Eur. Respir. J.* 9: 1913-1922.
61. Schiffmann E, Corcoran BA and Wahl SM. 1975. N-formylmethionyl peptides as chemoattractants for leucocytes. *Proc. Natl. Acad. Sci. USA.* 72: 1059-1062.
62. Marasco WA, Phan SH, Krutzsch H. Showell HJ, Feltner DE, *et al.* 1984. Purification and identification of formyl-methionyl-leucyl-phenylalanine as the major peptide neutrophil chemotactic factor produced by *Escherichia coli*. *J. Biol. Chem.* 259: 5430-5409.
63. von Tscharner V, Prod'hom B. Baggiolini M and Reuter H. 1986. Ion channels in human neutrophils activated by a rise in free cytosolic calcium concentration. *Nature* 324: 369-372.
64. Demaurex N. Monod A. Lew DP and Krause K-H. 1994. Characterization of receptor-mediated and store-regulated Ca^{2+} influx in human neutrophils. *Biochem. J.* 297: 595-601.
65. Gao J-L, Lee EJ and Murphy PM. 1999. Impaired antibacterial host defense in mice lacking the N-formylpeptide receptor. *J. Exp. Med.* 189: 657-662.
66. Lee HC. 2001. Physiological functions of cyclic ADP-ribose and NAADP as calcium messengers. *Annu. Rev. Pharmacol. Toxicol.* 41: 317-345.
67. Sethi JK. Empson RM. Bailey VC. Potter BVL and Galione A. 1997. 7-Deaza-8-bromo-cyclic ADP-ribose, the first membrane-permeant, hydrolysis-resistant cyclic ADP-ribose antagonist. *J. Biol. Chem.* 272: 16358-16563.
68. Abdallah MA, Biellmann JF, Nordström B and Brändén CI. 1975. The conformation of adenosine diphosphoribose and 8-bromoadenosine diphosphoribose when bound to liver alcohol dehydrogenase. *Euro. J. Biochem.* 50: 475-481.
69. Lentsch AB and Ward PA. 2000. Regulation of inflammatory vascular damage. *J. Pathol.* 190: 343-348.
70. Pillinger MH and Abramson SB. 1995. The neutrophil in rheumatoid arthritis. *Rheumatoid Arthritis* 21: 691-714.
71. Hanson PR. 1995. Role of neutrophils in myocardial ischemia and reperfusion. *Circulation* 91: 1872-1885.
72. McColl SR, Staykova MA. Wozniak A. Fordham S. Bruce J and Willenborg DO. 1998. Treatment with anti-granulocyte antibodies inhibits the effector phase of experimental autoimmune encephalomyelitis. *J. Immunol.* 161: 6421-6426.
73. Linden A. 2001. Role of interleukin-17 and the neutrophil in asthma. *Int. Arch. Allergy Immunol.* 126: 179-184.
74. Lee HC and Aarhus R. 1991. ADP-ribosyl cyclase: an enzyme that cyclizes NAD^+ into a calcium mobilizing metabolite. *Cell Regul.* 2: 203-209.
75. Kim U-H, Kim J-S, Han MK, Park B-H and Kim H-R. 1993. Purification and characterization of NAD glycohydrolase from rabbit erythrocytes. *Arch. Biochem. Biophys.* 305: 147-152.
76. Kim H, Jacobson EL and Jacobson MK. 1993. Synthesis and degradation of cyclic ADP-ribose by NAD glycohydrolases. *Science* 261: 1330-1333.
77. Nata K. Takamura T. Karasawa T. Kumagai T. Hashioka W, *et al.* 1997. Human gene encoding CD38 (ADP-ribosyl cyclase/cyclic ADP-ribose hydrolase): organization, nucleotide sequence and alternative splicing. *Gene* 186: 285-292.
78. Ferrero E and Malavasi F. 1997. Human CD38, a leukocyte receptor and ectoenzyme, is a member of a novel eukaryotic gene family of nicotinamide adenine dinucleotide$^+$-converting enzymes. *J. Immunol.* 159: 3858-3865.
79. Franco L. Guida L. Bruzzone S. Zocchi E, Usai C and De Flora A. 1998. The transmembrane glycoprotein CD38 is a catalytically active transporter responsible for the generation and influx of the second messenger cyclic ADP-ribose across membranes. *FASEB J.* 12: 1507-1520.

80. Galione A and White A. 1994. Ca^{2+} release induced by cyclic ADP-ribose. *Trends in Cell Biol.* 4: 431-436.
81. Prentki M, Wollheim CB and Lew PD. 1984. Ca^{2+} homeostasis in permeabilized human neutrophils:characterization of Ca^{2+} sequestering pools and the action of inositol 1, 4, 5-trisphosphate. *J. Biol. Chem.* 259: 13777-13782.
82. Alemany R, Meyer zu Herigndorf D, van Koppen CJ and Jakobs KH. 1999. Formyl peptide receptor signaling in HL-60 cells through sphingosine kinase. *J. Biol. Chem.* 274: 3994-3999.
83. Meszaros L, Bak J and Chu A. 1993. Cyclic ADP-ribose as an endogenous regulator of the non-skeletal type ryanodine receptor Ca^{2+} channel. *Nature* 354: 76-78.
84. Sitsapesan R, McGarry SJ and Williams AJ. 1995. Cyclic ADP-ribose, the ryanodine receptor and Ca^{2+} release. *Trends Pharmacol. Sci.* 16: 386-391.
85. Meissner G. 1994. Ryanodine receptor/Ca^{2+} release channels and their regulation by endogenous effectors. *Ann. Rev. Physiol.* 56: 485-508.
86. Lee HC. 1993. Potentiation of calcium- and caffeine-induced calcium release by cyclic ADP-ribose. *J. Biol. Chem.* 268: 293-299.
87. Peterson OH and Cancela JM. 1999. New Ca^{2+}-releasing messengers: are they important in the nervous system? *Trends Neurosci.* 22: 488-494.
88. Giannini G, Conti A, Mammarella S, Scrobogna M and Sorrentino V. 1995. The ryanodine receptor/calcium channel genes are widely and differentially expressed in murine brain and peripheral tissues. *J. Cell Biol.* 128: 893-904.
89. Galione A, Lee HC and Busa WB. 1991. Ca^{2+}-induced Ca^{2+}release in sea urchin egg homogenates: modulation by cyclic ADP-ribose. *Science* 253: 1143-6.
90. Thomas JM, Masgrau R, Churchill GC and Galione A. 2001. Pharmacological characterization of the putative cADP-ribose receptor. *Biochem. J.* 359: 451-7.
91. Bennett DL, Bootman MD, Berridge MJ and Cheek TR. 1998. Ca^{2+} entry into PC12 cells initiated by ryanodine receptors or inositol 1,4,5-trisphosphate receptors. *Biochem. J.* 329: 349-357.
92. Kiselyov KI, Shin DM, Wang Y, Pessah IN, Allen PD and Muallem S. 2000. Gating of store-operated channels by conformational coupling to ryanodine receptors. *Mol. Cell* 6: 421-431.
93. Kiselyov K, Shin DM, Shcheynikov N, Kurosaki T and Muallem S. 2001. Regulation of Ca^{2+} -release activated Ca^{2+} current (Icrac) by ryanodine receptors in inositol 1,4,5-trisphosphate-receptor-deficient DT40 cells. *Biochem. J.* 360: 17-22.
94. Guse AH, da Silva CP, Berg I, Skapenko AL, Weber K, *et al.* 1999. Regulation of calcium signaling in T lymphocytes by the second messenger cyclic ADP-ribose. *Nature* 398: 70-73
95. Montero M, Garcia-Sancho J and Alvarez J. 1994. Phosphorylation down-regulates the store-operated Ca^{2+} entry pathway of human neutrophils. *J. Biol. Chem.* 269: 3963-3967.
96. Hauser CJ, Fekete Z, Adams JM, Garced M, Livingston DH and Deitch EA. 2001. PAF-mediated Ca^{2+} influx in human neutrophils occurs via store-operated mechanisms. *J. Leukoc. Biol.* 69: 63-68.
97. Nube O, Serrander L, Foyouzi-Youssefi R, Monod A, Lew DP and Krause K-H. 1997. Store-operated Ca^{2+} influx and stimulation of exocytosis in HL-60 granulocytes. *J. Biol. Chem.* 272: 28360-28367.
98. Khalfi F, Gressier B, Dine T, Brunet C, Luyckx M, *et al.* 1998. Verapamil inhibits elastase release and superoxide anion production in human neutrophils. *Biol. Pharm. Bull.* 21: 109-112.
99. Krause K-H, Pittet D, Volpe P, Pozzan T, Meldolesi J and Lew DP. 1989. Calciosome, a sarcoplasmic reticulum-like organelle involved in intracellular Ca^{2+}-handling by non-muscle cells: Studies in human neutrophils and HL-60 cells. *Cell Calcium* 10: 351-361.
100. Pettit EJ and Fay FS. 1998. Cytosolic free calcium and the cytoskeleton in the control of leukocyte chemotaxis. *Physiol. Rev.* 78: 949-967.

101. Perraud AL, Fleig A, Dunn CA, Bagley LA, Launay P, *et al.* 2001. ADP-ribose gating of the calcium-permeable LTRPC2 channel revealed by Nudix motif homology. *Nature* 411: 595-599.
102. Sano Y, Inamura K, Miyake A, Mochizuki S, Yokoi H, *et al.* 2001. Immunocyte Ca^{2+} influx system mediated by LTRPC2. *Science* 293: 1327-1330.
103. Berg I, Potter BVL, Mayr GW and Guse AH. 2000. Nicotinic acid adenine dinucleotide phosphate ($NAADP^{+}$) is an essential regulator of T-lymphocyte Ca^{2+} signaling. *J. Cell Biol.* 150: 581-588.

Frances E. Lund, Trudeau Institute, PO Box 59,
Saranac Lake NY 12983, USA.
e-mail: flund@trudeauinstitute.org

Cyclic ADP-Ribose and NAADP. Structures, Metabolism and Functions. Lee HC (ed), pp. 241-267, Dordrecht: Kluwer.

Subcellular and Extracellular Trafficking of NAD^+ and Cyclic ADP-Ribose: A New Way for Regulating Intracellular Calcium Homeostasis

Antonio De Flora[1], Lucrezia Guida[1], Luisa Franco[2], Santina Bruzzone[1] And Elena Zocchi[1]

[1] Department of Experimental Medicine, Section of Biochemistry, and Center of Excellence for Biomedical Research, University of Genova;[2] G. Gaslini Institute, Genova, Italy

INTRODUCTION

Ectobiochemistry is an emerging branch of cell biology. It investigates at a molecular level all those processes that take place at the outer surface of cells. This localization can be functionally related on one hand to the interplay of solute-transporting and of ectoenzymatic activities occurring in an individual cell (autocrine model). On the other hand, ectocellular processes mediate many mechanisms of cell-to-cell communication (paracrine model), with some cells providing signal molecules to other cells and with biorecognition and molecular interactions between extracellular agonists and membrane receptors being transduced in the sensing cells to generate specific responses.

In both models, ectoenzymes seem to play an essential role which, however, is still poorly understood. In the past, they were mostly considered as markers of cell populations rather than receiving attention for specific functions. In some cases, these functions might simply be absent or irrelevant. This is in fact the case of ectoenzymes that, although potentially displaying discrete activities at the cell surface, are functionally switched off by virtual absence of their substrates, which are conversely present inside the same cell. Therefore, ectoenzymes may either be the remnant of molecular structures active at former maturation stages or, alternatively, they may reflect the need to exclude a biochemical reaction, which would be useless or even detrimental for the metabolism of that specific cell.

However, the lack-of-function hypothesis is not convincing in principle, in view of the high complexity and of selectivity of enzymatic catalysis in terms of substrate-active site recognition, dynamics, energy requirements and regulation. Accordingly, expression of an enzyme activity at a topologically unconventional site as the plasma membrane is more likely suggestive of as yet undefined roles and functions. These may be related to regulation of cell processes or to export of selected metabolites to neighbouring cells. Whatever the case, two major questions need to be addressed and answered: 1) Which are the sources of substrates of ectoenzymes and how can they be accessible to the corresponding catalytic sites? 2) Assuming catalytic operation of these ectoenzymes, where and how do the corresponding products display their functions?

In the last few years, CD38 has become an important paradigm of ectoenzymes and of their functional roles for many reasons. It proved to be an ectoenzyme after being identified as a leukocyte marker of differentiation and as a receptor/coreceptor [1]: it was demonstrated to catalyze the generation of two important calcium-mobilizing second messengers, cADPR and $NAADP^+$ [2-5]. In addition, CD38 was shown to be a multifunctional enzyme, being able to catalyze cyclization, hydrolysis and base-exchange reactions and featuring the unusual property of catalytically active transporter of its product cADPR [6]. Its functional interaction on the plasma membrane with hemichannels of connexin 43 (Cx43) proved to be the way by which NAD^+ can become available to its active site extracellularly, thus enabling display of its catalytic functions [7,8].

The consequent generation of cADPR at the outer cell surface proved to be followed both by its influx into the same $CD38^+$ cells across CD38 itself (autocrine mechanism) and by its occurrence in the extracellular space, where it can behave as a signal molecule for cADPR-responsive adjacent cells (paracrine mechanism): both loops, autocrine and paracrine, mediate calcium responses triggered by cADPR and modulation of calcium-regulated cell responses [9]. Therefore CD38 can potentially affect intracellular calcium homeostasis also at distant sites, including $CD38^-$ cells, provided they harbour cADPR-sensitive calcium stores.

Besides qualifying as a plasma membrane-bound system for regulation of intracellular calcium, CD38 proved to feature the same regulatory effects on cytosolic calcium ($[Ca^{2+}]_c$) levels when it is confined inside either exocytotic or endocytotic membrane vesicles. This property shows how an ectoenzyme (a nominally ectocellular enzyme protein) can express its functions also at intracellular sites during its export to the plasma membrane or when it undergoes internalization from the plasma membrane itself.

THE TOPOLOGICAL PARADOX OF THE CD38/cADPR SYSTEM

Phenomenological Data

As reviewed in detail in several chapters in this book, CD38 is the best characterized member of a family of mammalian ADP-ribosyl cyclases. Although the structure of native CD38 has not yet been elucidated, Lee and coworkers have used the crystallographic coordinates of the *Aplysia californica* monofunctional ADP-ribosyl cyclase (a homodimer) and the sequence homology data to human CD38 in order to model the structure of the catalytic domain of CD38 [10]. In addition, this structure was further dissected by site-directed mutagenesis of recombinant CD38 lacking the transmembrane and cytoplasmic domains. This allowed the identification of the structural determinants important in controlling the cyclase (on both substrates NAD^+ and NGD^+), the cADPR hydrolase and the base-exchange activities [11,12], thereby completing earlier studies [13].

Expression of all known enzyme activities of CD38 on its extracellular domain identifies this glycoprotein as a classical ectoenzyme. Native, i.e. transmembrane CD38, was shown to be a mixture of tetramers, dimers and monomers, although the precise proportion of the three species still escapes recognition because of the slightly denaturing conditions required to investigate CD38 structure by SDS-PAGE [14]. Homology with the *Aplysia* cyclase, however, may extend from the amino acid sequence to a preferred dimeric structure of transmembrane CD38. Assuming this to be the prevalent form of native CD38, as postulated in the first crystallographic study of the *Aplysia* cyclase [10], the two bean-shaped monomers with a central cleft are expected to associate in a head-to-head fashion, thereby yielding a doughnut-shaped molecule with a central cavity. Importantly, the size of such a cavity is comparable to that of cADPR.

Localization of the active site of CD38 at its ectocellular domain raised a number of questions, which characterize the topological paradox of the CD38/cADPR system [9, 15, 16]. This paradox concerns the ectocellular generation of a signal molecule - cADPR - whose only known function, i.e. calcium mobilization, is necessarily intracellular. The apparent inconsistency, from a topological point of view, is two-fold [9]: 1) since NAD^+ is intracellular, mostly intra-mitochondrial but in part also in the cytosol, this dinucleotide can hardly be expected to become available to the catalytic site of CD38 at the opposite side of the plasma membrane. 2) How can extracellularly generated cADPR reach its target calcium stores intracellularly?

The same inconsistencies seem to apply to that fraction of CD38 which does not reside in the plasma membrane but is bound to a number of subcellular membrane vesicles, still little characterized, which feature cargo functions [9]. While the relative proportions of plasma membrane-bound to intravesicular CD38 may constitutively differ among distinct cell types, exocytotic versus endocytotic CD38-containing membrane vesicles are likely to have a variable prevalence also during the lifespan of the same $CD38^+$ cell, depending on many factors. Thus, exocytosis, meant as the traffic of intravesicular CD38 to reach the plasma membrane, is expected to be quantitatively relevant under conditions leading to overexpression of CD38, as experimentally found following transfection of constitutively $CD38^-$ HeLa and 3T3 murine fibroblasts with human CD38 cDNA [17]. Endocytosis of plasma membrane-bound CD38 was recorded in two different experimental conditions: i) following ligation of surface CD38 with anti-CD38 monoclonal antibodies [18], and, ii) following exposure of $CD38^+$ cells (either constitutively or CD38-transfected) to some selected ligands that include NAD^+, GSH and N-acetylcysteine [19].

Whichever the destination of CD38-containing intracellular vesicles, they should apparently be incompetent for effective NAD^+ to cADPR conversion due to the same topological problems that apply to plasma membrane-bound CD38. Thus, since CD38 has certainly an opposite polarity in the plasma membrane and in the intracellular vesicles, respectively, the intravesicular active site of CD38 is not directly accessible to cytosolic NAD^+ [9]. Furthermore, even assuming catalytic functioning of vesicle-bound CD38, the cADPR so generated should be sequestered inside the vesicles and be therefore unable to target its responsive stores in the cells.

Since identification of this topological paradox, it became soon clear that $CD38^+$ cells can in fact overcome the two problems related to apparent compartmentation of NAD^+ and of cADPR. Thus, "de novo" expression of CD38 in constitutively $CD38^-$ cells was found to result in the following sequence of causally related events [17]: i) consumption of intracellular NAD^+ and NADH, ii) formation of intracellular cADPR (not detectable in the wild-type or in the corresponding CD38 antisense-transduced cells), iii) partial depletion of thapsigargin-sensitive calcium stores, iv) increase in basal $[Ca^{2+}]_c$, v) enhanced cell proliferation, due to, vi) shortening of the S phase of cell cycle.

On the other hand, the ligand-induced internalization of $CD38^+$ in Namalwa cells was shown to elicit intracellular overproduction of cADPR as compared to untreated cells [20]. Moreover, forced endocytosis of CD38, in $CD38^+$ (transduced) HeLa and 3T3 cells challenged with NAD^+, GSH and N-acetylcysteine, resulted in sustained and comparable time-dependent increases of $[Ca^{2+}]_c$. This increase was completely prevented by the cADPR analog and antagonist 8-N_3-cADPR [21], thus demonstrating a causal correlation between the intracellular cADPR production by internalized (i.e., intravesicularly oriented) CD38 and the $[Ca^{2+}]_c$ increase [20].

Thus, no matter whether dependent on exocytosis or endocytosis of CD38, an enhanced trafficking of vesicle-bound CD38 proved to correlate with (over)production of functionally active cADPR in the cytosol, as demonstrated by increased levels of $[Ca^{2+}]_c$.

Table 1 shows the levels of $[Ca^{2+}]_c$ in the native CD38⁻ 3T3 murine fibroblasts and, by comparison, in the same CD38⁺ transfected cells either as such (exocytosis) or following ligand-dependent CD38⁺ internalization (exocytosis + endocytosis). $[Ca^{2+}]_c$ was approximately doubled in the CD38-transfected 3T3 cells over the CD38⁻ cells and the association of exocytosis and endocytosis, as obtained by exposing of the CD38⁺ cells to either GSH or N-acetylcysteine (NAC), led to further increases of $[Ca^{2+}]_c$. These increases, however, never overcame values of 80-90 nM, despite a remarkable extent of CD38 internalization (reaching 60%-80%) and accordingly of a strongly enhanced trafficking of CD38-containing subcellular vesicles. Therefore, although a precise quantitative correlation between the intracellular density of endocytotic vesicles and the accompanying calcium responses was almost impossible to trace, nevertheless these preliminary data argued for some mechanisms of restriction of the corresponding imbalances of calcium homeostasis.

Table 1. **Levels of $[Ca^{2+}]_i$ in CD38⁻ versus CD38⁺ 3T3 murine fibroblasts.**

Cell type	CD38 exocytosis (expression)	CD38 endocytosis (ligand-induced)	$[Ca^{2+}]_c$ (nM)
CD38⁻ (native)	–	–	20±3
CD38⁺ (transduced)	+	–	38±6
CD38⁺ (transduced) +GSH/NAC	+	+	79±9

Recognition of the topological paradox of CD38/cADPR and of the fact that CD38⁺ cells can elude the apparent compartmentation of NAD⁺ and cADPR suggested an investigation of the underlying mechanisms. These were addressed by systematically seeking still unidentified transmembrane transporters involved in trafficking of NAD⁺ and of cADPR .

NAD⁺ Transport: Role of Connexin 43 Hemichannels

A putative NAD⁺ transporter making this dinucleotide available to the active site of CD38 should be equally effective in extruding cytosolic NAD⁺ from cells and in translocating NAD⁺ into the exocytotic and the endocytotic membrane vesicles as well. Accordingly, we addressed the occurrence of such a transporter in the plasma membrane of intact CD38⁻ HeLa and 3T3 cells [7]. The experiments designed to measure possible release of cytosolic NAD⁺ were carried out by deliberately creating an outward directed gradient of NAD⁺ concentrations by means of repeated washings of the cells. This condition resulted in the detectable release of cellular NAD^+ and NADH, which was measured both as progressive loss of cellular dinucleotide and as its stoichiometric appearance in the extracellular medium. Translocation of NAD^+ proved to occur according to an equilibrative mechanism, because influx into the same cells was observed at high extracellular concentrations of NAD^+. The properties of NAD^+ transport across the plasma membrane were then investigated in more detail. Transport was rather non-specific for pyridine dinucleotides, being recorded with NAD^+, NADH, $NADP^+$ and NADPH. It was reversibly inhibited by the NAD^+ analog 8-N_3-NAD^+. Saturation kinetics for NAD^+ influx were observed, with an apparent K_m value of approximately 15 mM and a high transport efficiency. No significant effect of temperature was found in influx and efflux experiments, this implicating a channel rather than a transporter oscillating between different conformations [7].

The NAD^+-translocating activity could be reconstituted in proteoliposomes, which allowed a demonstration of a clear dependence of translocation on time, on protein concentration and on pH, and also its inhibition by thiol reagents including N-ethylmaleimide and iodoacetamide. Moreover, NAD^+ transport in intact cells was markedly inhibited by both extracellular and intracellular Ca^{2+} [7]. All these properties were reminiscent of intercellular exchange of solutes across gap junction channels, suggesting a possible role of connexin hemichannels in mediating bidirectional NAD^+ transport across the plasma membrane of 3T3 and HeLa cells.

This hypothesis turned out to be correct on the basis of several lines of experimental evidence [7]: i) NAD^+ transport in the intact cells was markedly inhibited by known blockers of gap junctional channels; ii) "de novo" expression of connexin 43 (Cx43) in a Cx43-negative HeLa clone by means of treatment with 5-aza-2-deoxycytidine restored transmembrane NAD^+ translocation; iii) a selected Cx43 antisense (but not the corresponding sense) oligodeoxynucleotide completely prevented NAD^+ efflux from native, constitutively $Cx43^+$ 3T3 fibroblasts; iv) an anti-Cx43 monoclonal antibody completely inhibited influx of NAD^+ into 3T3 membrane protein-reconstituted proteoliposomes, while an anti-connexin 26 was totally ineffective; v) Cx43 immunopurified to homogeneity from NAD^+-transporting 3T3 cell membranes was reconstituted in unilamellar proteoliposomes which featured full NAD^+-transporting activity, while other proteoliposomes reconstituted with homogeneous connexin 32 were transport-negative. The Cx43-reconstituted proteoliposomes did not transport cADPR or ADP-ribose, this indicating a specific role of Cx43 for translocation of the CD38 substrate, but not of the two products of the cyclase and of the cADPR hydrolase reactions catalyzed by CD38 [7].

Role of CD38 as a Catalytically Active Transporter of cADPR

The model inferred by Lee and coworkers for the structure of CD38, based upon homology with the *Aplysia* ADP-ribosyl cyclase, shows a pocket bearing the active site close to the central cleft [10].

Assuming a homodimeric structure of transmembrane CD38 by homology with the *Aplysia* cyclase, as experimentally shown using SDS-PAGE experiments under mildly denaturing conditions [14], the active site-containing central cavity of CD38 has a size comparable to that of cADPR. This observation prompted us to start experiments aiming to investigate the "in situ" catalytic functioning of plasma membrane-bound CD38. Since human erythrocytes display the ectoenzymatic activities of CD38 [22], as also demonstrated by purification to homogeneity of CD38 itself from detergent-solubilized human erytrocyte membranes [23], we first used resealed red cell membranes as a model of CD38 functioning. Incubation of resealed ghosts with external NAD^+ or NGD^+ resulted in a steady internalization of cADPR or cGDPR, respectively, into these vesicles [6].

Conclusive evidence for the direct role of CD38 in active cADPR transport was obtained by using proteoliposomes reconstituted with homogeneous CD38 purified from human erythrocyte membranes [6]. Quality control assessment of these proteoliposomes showed that 99% of CD38 molecules were correctly oriented in these unilamellar vesicles which display the C-terminal region at the outer surface. Moreover, upon encapsulating NAD^+ into CD38-reconstituted proteoliposomes, its intravesicular concentration kept quite stable, this ruling out any outward directed transport of NAD^+ across CD38 and also any internal conversion of NAD^+ to cADPR. Under these conditions, exposure of the CD38-reconstituted proteoliposomes to external NAD^+ or NGD^+ resulted in the almost immediate appearance of cADPR or of cGDPR, respectively, in the intravesicular compartment. Careful quantitative analysis of CD38 substrate (NAD^+) and products (cADPR and ADP-ribose) in the extravesicular and in the intravesicular spaces clearly identified a concentrative mechanism for cADPR influx. During the first minutes of catalysis, this transport seems to spare, to some extent, cADPR by preventing its hydrolytic conversion to ADP-ribose.

Therefore, using both natural and artificial $CD38^+$ membranes where this glycoprotein exhibited monomeric, dimeric and tetrameric structures [6,14], it became clear that CD38 displays unique features of a catalytically active transporter, behaving as a generating system of cADPR and also as a vectorial transporter of cADPR itself. This unusual property, which adds to the multifunctional enzymatic activities of CD38, might be due on one hand to cleavage of the high-energy N-glycosidic bond between nicotinamide and ribose in the NAD^+ molecule and on the other to a channel-forming association of two or four CD38 monomers within the plasma membrane. A prediction of this model is that monomers should feature catalytic activity only, as experimentally checked on SDS-PAGE tracings [14], while the oligomeric forms of CD38 should be responsible for the two-step process of catalysis-transport leading to concentrative influx of cADPR into $CD38^+$ cells.

Other cADPR Transporters

Earlier studies had suggested that some cells are permeable to extracellular cADPR, added as such and not following its formation by CD38. This was observed with murine B lymphocytes [24] and with rat cerebellar granule cells [25], the corresponding cellular responses being enhanced proliferation and increased calcium responses to membrane depolarization, respectively. These responses, however, were not obviously interpretable as the consequence of cADPR influx because in principle cADPR could alternatively bind to surface receptors and then trigger signal-transducing processes resulting in calcium-mediated cell responses.

A first clue to identifying a genuine cADPR translocation process from the extracellular space was obtained in human hemopoietic progenitors exposed to externally added cADPR. In extracts from these cells, following extensive washings to remove all extracellular or surface-bound cADPR, the cyclic nucleotide was not detectable at short times (10 min), while it became detectable later at 24 h [26]. Another cell system apparently permissive for cADPR permeation was that of smooth myocytes from bovine trachea which responded to external cADPR with sustained and concentration-dependent increases of both basal $[Ca^{2+}]_c$ and Acetylcholine (ACh)-induced calcium responses [27].

Recently, evidence for cADPR influx was directly obtained using wild-type $CD38^-$ 3T3 murine fibroblasts [28]. Here, we observed a time-dependent association of cADPR to identically washed cells, while no association could be detected with the corresponding isolated cell membranes, this apparently ruling out ectocellular absorption of the cyclic nucleotide and supporting its influx into cells. Under the same conditions, a progressive increase of $[Ca^{2+}]_c$ was recorded, which was unaffected by oleamide, Cx43 sense – and antisense oligodeoxynucleotides and was abrogated by 8-Br-cADPR and 8-NH_2-cADPR [28]. Therefore, a cADPR-transporting system, unrelated to both CD38 and Cx43, is present in the plasma membrane of the 3T3 fibroblasts and, by inference, also of other cell types that respond to extracellular cADPR. Characterization of this transporter is underway in our laboratory and evidence is emerging for its heterogeneity since both equilibrative and concentrative mechanisms of cADPR transport have been observed (Guida L. *et al.*, in preparation).

In conclusion, while $CD38^+$ cells can actively pump cADPR into the cytosol soon after its ectocellular generation, either $CD38^-$ or $CD38^+$ cells are equipped with an additional family of transporters that, together with ryanodine receptors/channels, make them responsive to extracellular cADPR.

INTRACELLULAR TRAFFICKING OF NAD^+ AND cADPR: THE AUTOCRINE MODEL

Table 2 summarizes the various transporters for NAD^+ and cADPR, and their related mechanisms as well, which were identified and in part characterized in the plasma membrane.

Table 2. **Transporters of NAD^+ and of cADPR in the plasma membrane.**

Transported nucleotide	Transporter	Transport Mechanism
NAD^+	Cx43 hemichannels	Equilibrative
cADPR	CD38 dimers	Concentrative
cADPR	under study	Equilibrative
cADPR	Na^+/cADPR symport	Concentrative

These transporters should functionally interact in a cell expressing Cx43, CD38 and the additional cADPR transport systems [9]. The interplay is held to take place in the plasma membrane and also in subcellular membrane vesicles (irrespective of whether exocytotic or endocytotic). The latter assumption has not yet received unequivocal confirmation in terms of subcellular co-localization and awaits appropriate cell models where the three types of transporters are expressed in sufficient amounts to allow for confocal or EM characterization.

However, biochemical evidence for a strict functional crosstalk between CD38 and Cx43 is already available, which leads to the conclusion that cells expressing both transporters feature an intense intracellular trafficking of NAD^+ and of cADPR. In addition, evidence was recently obtained that this (di)nucleotide trafficking within a $CD38^+$ and $Cx43^+$ cell is an effective means for regulating intracellular calcium levels. This conclusion clearly emerged from experiments on 3T3 murine fibroblasts where both intracellular cADPR levels and $[Ca^{2+}]_c$ were measured in wild-type $CD38^-$ cells and, by comparison, in CD38-transfected cells either as such or Cx43-antisense treated. As mentioned, the antisense strategy had already demonstrated the involvement of Cx43 in the NAD^+ transport [7]. While the native, $CD38^-$ fibroblasts had no detectable cADPR and their $[Ca^{2+}]_c$ was as low as 20 nM, the corresponding $CD38^+$ cells had cADPR at 5.5 picomoles/mg protein and $[Ca^{2+}]_c$ at 40 nM. Pretreatment of the $CD38^+$ cells with a Cx43 antisense oligodeoxynucleotide, but not with the corresponding sense oligodeoxynucleotide, led to disappearance of intracellular cADPR, while $[Ca^{2+}]_c$ levels dropped to values (around 20 nM) comparable to those observed in the $CD38^-$ fibroblasts. Therefore, down-modulation of either CD38 or Cx43 at a time proved to remarkably decrease $[Ca^{2+}]_c$ as compared with cells expressing both proteins, and this effect was clearly related to disappearance of intracellular cADPR [8].

There are two lines of evidence supporting the view that, at least in the 3T3 fibroblasts, it is at the level of subcellular vesicles (either exocytotic or endocytotic), rather than in the plasma membrane, that the CD38-Cx43 crosstalk functionally operates so as to regulate their $[Ca^{2+}]_c$ levels via cADPR generation. First, exposure of cell cultures to extracellularly added NAD^+-glycohydrolase failed to determine significant decreases of $[Ca^{2+}]_c$ [28]. Secondly, in conditions of extensive ligand-dependent internalization of CD38, there is an approximate doubling of the $[Ca^{2+}]_c$ levels, this implicating the endocytotic vesicles as directly responsible for overproduction of intracellular cADPR [19]. Accordingly, these vesicles (and the exocytotic vesicles as well) seem to be able to regulate $[Ca^{2+}]_c$ as illustrated in Fig. 1. Cytosolic NAD^+ enters the vesicles across Cx43 hemichannels and is then converted by intravesicular CD38 to cADPR. This cyclic nucleotide can either be pumped outward by CD38 or be released from the vesicles through the family of cADPR transporters, in either case reaching the cytosol and becoming accessible to its target calcium stores, with consequent up-regulation of calcium-dependent cell processes.

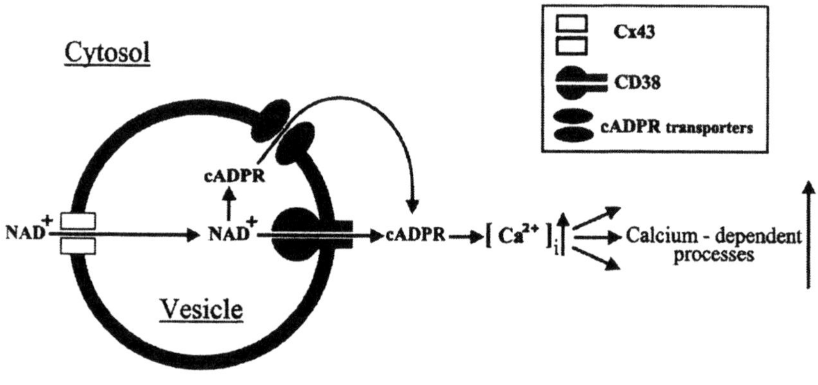

Figure 1. **Model of unrestricted intracellular trafficking of NAD^+ and cADPR between the cytosol and exocytotic or endocytotic membrane vesicles.**

The model shown in Fig. 1, though supported by the experimental data demonstrating a functional interaction between CD38 and Cx43 hemichannels, had two weak points. Since such crosstalk is reasonably occurring also at the plasma membrane level, then cytosolic NAD^+ should be released from Cx43-expressing cells in a way not obviously compatible with their survival. Secondly, as mentioned, under experimental conditions of forced, ligand-induced internalization of CD38, the increase of subcellular trafficking of NAD^+ and of cADPR (i.e., the extent of subcellular vesiculation) would be expected to result in increases of $[Ca^{2+}]_c$ much higher than those in fact recorded, never exceeding 80-90 nM (Table 1).

Both considerations suggested the investigation of whether and how these mechanisms of subcellular interaction between CD38 and Cx43 are under control, so as to avoid unrestricted losses of cellular NAD^+ on one hand and excessive accumulation of cytosolic calcium on the other. Calcium itself seemed to be a reasonable candidate to effect such down-regulation according to some feed-back mechanism decreasing the intracellular trafficking of NAD^+ and of cADPR. We had previously observed that NAD^+ transport across Cx43 hemichannels is inhibited in a concentration-dependent way by calcium and that this effect is mediated by a cytosolic Ca^{2+}-binding protein [7]. Based on this finding, we comparatively measured NAD^+-transporting activities in native, $CD38^-$ 3T3 fibroblasts (characterized by 20 nM $[Ca^{2+}]_c$ - see Table 1) and in the same $CD38^+$ cells (whose $[Ca^{2+}]_c$ is approximately doubled). These experiments showed that a sharp inverse correlation exists between NAD^+ translocation (both release from cells and influx into them) and $[Ca^{2+}]_c$ [8]. An identical conclusion was reached by measuring NAD^+ transport and $[Ca^{2+}]_c$ levels in $CD38^-$ 3T3 cells, where cADPR is undetectable, and by comparison in the same cells challenged with extracellular cADPR. Here, influx of external cADPR resulted in an increase of $[Ca^{2+}]_c$ from 20 to 50 nM and in the parallel decrease of NAD^+ transport across the plasma membrane.

Therefore, no matter whether dependent on endogenous cADPR generation or on supply of extracellular cADPR, high $[Ca^{2+}]_c$ 3T3 cells showed low NAD^+ transport and vice versa [8]. These variations of NAD^+-transporting activity were demostrated not to be related to differences in Cx43 expression, as shown by western blot experiments, by cytofluorimetric analyses and by semiquantitative PCR analyses. Accordingly, we sought calcium-dependent post-translational modifications of Cx43 in order to define the relationship between $[Ca^{2+}]_c$ and NAD^+ translocation.

Since Cx43 occurs in phosphorylated and nonphosphorylated forms [29] and it is known that Cx43 phosphorylation by protein kinase C (PKC) negatively affects its solute-exchange activity in gap junctions [30], we sought a possible calcium-stimulated phosphorylation of Cx43. Using an anti-Cx43 monoclonal antibody that specifically recognizes the nonphosphorylated Cx43 form, we found that the high $[Ca^{2+}]_c$ 3T3 cells (both the CD38-transfected and the cADPR-loaded $CD38^-$ ones), that exhibit a low NAD^+- transporting activity, have a higher content of PKC-phosphorylated Cx43 than the corresponding low $[Ca^{2+}]_c$, highly NAD^+-transporting $CD38^-$ 3T3 fibroblasts [8]. Treatments with okadaic acid abolished all responses of $CD38^+$ fibroblasts to the cADPR antagonist 8-NH_2-cADPR, i.e. decrease of $[Ca^{2+}]_c$, increase of NAD^+ transport and Cx43 dephosphorylation, this supporting a link between these events. Finally, experiments performed with a number of protein kinase inhibitors demonstrated that selective inhibition of PKC results in an increase of NAD^+ transport in cADPR-loaded $CD38^-$ 3T3 fibroblasts over that recorded in the same cells lacking PKC inhibitors [8].

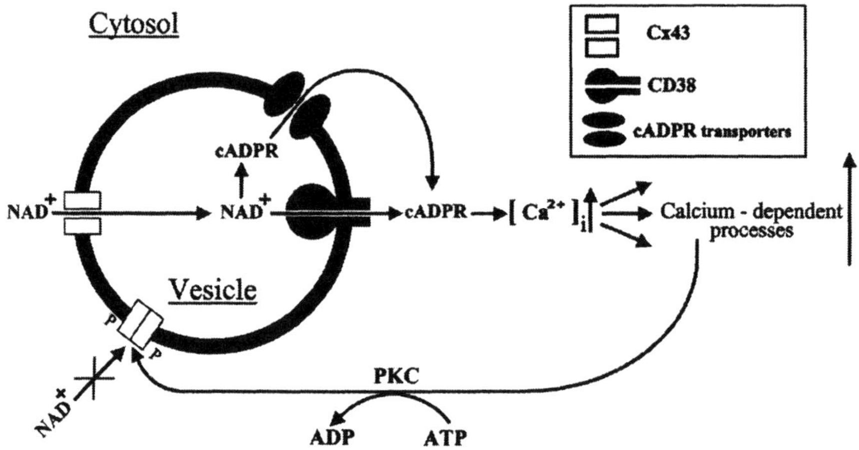

Figure 2. **Model of self-restricted intracellular trafficking of NAD^+ and cADPR between the cytosol and exocytotic or endocytotic membrane vesicles.** Reproduced from [8] with permission.

Thus, a complex interrelation emerged whereby the functional crosstalk occurring subcellularly between CD38 and Cx43 is in fact under a stringent control triggered by increased $[Ca^{2+}]_c$ and operated by PKC-catalysed phosphorylation of Cx43. As shown in Fig. 2, influx of NAD^+ into the intracellular membrane vesicles across Cx43 hemichannels is the target of this feed-back mechanism, being down-regulated by phosphorylation of Cx43 at PKC-selective sites, most likely serine 368 [31]. Such inhibition of NAD^- transport correlates with the known transition from an open to a closed state of Cx43, determined by its PKC-mediated phosphorylation and resulting in lower gap junctional coupling between adjacent cells [8]. Therefore, the model shown in Fig. 2 accomodates all experimental data available so far, while excluding the involvement of other Cx43-inhibiting mechanisms, e.g. via calcium/calmodulin [8]. Further studies are required, especially aiming to define a quantitative relationship, still rather elusive, between rates of NAD^+ translocation and ratios of nonphosphorylated to PKC-phosphorylated Cx43. Furthermore, the threshold and the dynamics of $[Ca^{2+}]_c$ triggering blockade of NAD^+ transport need to be elucidated. In any case, this seems to represent a system designed to finely tune the intracellular calcium homeostasis in $CD38^+$ and $Cx43^+$ cells equipped with cADPR-responsive ryanodine receptors/stores.

At the same time, the intracellular loop shown in Fig. 2 has the potential to avoid an excessive accumulation of $[Ca^{2+}]_c$ which, if unrestricted, could trigger cytotoxicity. Another beneficial consequence, strongly suggested by the fact that NAD^+-transporting activity was measured across the plasma membrane, seems to be the prevention of uncontrolled losses of cellular NAD^+.

These data point to the importance of cADPR in regulation of $[Ca^{2+}]_c$ levels in many types of cells. Comparison of Figs. 1 and 2 suggests that compartmentation of NAD^+ and cADPR would not be sufficient to control a potentially dangerous increase of cytosolic calcium and that the superimposed mechanism of Cx43 phosphorylation meets this essential requirement. Probably CD38 is not the only ADP-ribosyl cyclase in some mammalian cell types [16] and soluble cyclases have been reported to occur, for instance in T lymphocytes [32], in astrocytes [33] and in neuronal cells [34]. It is likely that even in these cases, where compartmentation should not play any regulatory role, other mechanisms may be involved, for instance cGMP-stimulated cyclase phosphorylation as shown in sea urchin eggs [35, 36] or PKA-mediated cyclase phosphorylation as observed in plants [37] and in sponges [38]. In any case, modulation of intracellular cADPR metabolism enables regulation of several calcium-stimulated cell processes under conditions fully compatible with cell viability.

EXTRACELLULAR TRAFFICKING OF NAD^+ AND cADPR: PARACRINE MECHANISMS

The transporters for NAD^+ and cADPR identified in our laboratory mediate an intracellular trafficking of both nucleotides which involves the complex system of subcellular membrane vesicles more than the plasma membrane where these transporters were directly observed and measured (see above). However, occurrence of the transporters in the plasma membrane appeared to suggest that they might play a role also in cell-to-cell communication processes. This role was first demonstrated in bovine tracheal smooth muscle, before identifying Cx43 as a NAD^+ transporter and before demonstrating influx of extracellular cADPR into cells [28].

Other cell systems, however, proved to feature an extracellular trafficking of NAD^+ and cADPR mediated by the corresponding plasma membrane-bound transporters and such a mechanism could be demonstrated to regulate intracellular calcium in target cells. These systems included co-cultures of "stromal" cells as generators of cADPR to be presented to "parenchymal" cells: i) $CD38^+/CD38^-$ 3T3 murine fibroblasts; ii) $CD38^+$ stromal cells/human hemopoietic progenitors; iii) $CD38^+$ rat hippocampal astrocytes/ neurons.

Interaction Between Mucosal Cells and Smooth Myocytes in Bovine Trachea

The studies on intact bovine trachea were suggested by the earlier demonstration that tracheal smooth myocytes respond to acetylcholine (ACh)-induced stimulation of muscarinic receptors by releasing calcium from ryanodine-sensitive stores expressed in the sarcoplasmic reticulum [39]. In addition, a role of cADPR in these intracellular calcium movements was demonstrated in permeabilized tracheal muscle myocytes. These data raised the possibility of cADPR being involved in the regulation of airway smooth muscle tone and prompted us to dissect the biochemical properties of the individual cell components of tracheal tissue, i.e. epithelial mucosal cells and smooth myocytes. CD38 was immunodetected in both epithelial mucosa cells and smooth myocytes, where its ectocellular cyclase and cADPR hydrolase activities were identified and measured. These activities were much higher in the epithelial cells than in smooth myocytes.

A second advancement was the demostration that extracellularly added cADPR elicits a sustained and concentration-dependent increase of $[Ca^{2+}]_c$ in intact myocytes, this effect being afforded also by nanomolar concentrations of the non-hydrolyzable analog 3-deaza-cADPR [40] and completely prevented by the cADPR antagonist 8-NH_2-cADPR. The same effects, i.e. $[Ca^{2+}]_c$ increases (cADPR and 3-deaza-cADPR) and blockade (8-NH_2-cADPR), were afforded on the ACh-induced calcium response recorded in intact myocytes [27]. Moreover, extracellular cADPR (or 3-deaza-cADPR at concentrations four logs lower) caused a significant increase of ACh-induced contraction measured in tracheal smooth muscle strips, while 8-NH_2-cADPR decreased it significantly. For example, extracellular cADPR, at 100 μM, reduced by one-half the ACh concentration that results in 50% of maximal force compared with the baseline curve [27].

Responsiveness of intact smooth myocytes to extracellular cADPR suggested an investigation of whether epithelial mucosal cells that overlay the myocytes "in vivo" can generate cADPR extracellularly. A first requirement was easily satisfied since tracheal epithelial cells were found to express on their plasma membrane the NAD^+-transporting system which was subsequently identified with Cx43 hemichannels. Furthermore, addition of intact mucosal strips to isolated smooth myocytes caused an increase of basal $[Ca^{2+}]$ in the latter cells from 111±4 to 147±12 nM ($p<0.05$). This increase was completely abrogated by addition of NAD^+-glycohydrolase to the medium, which underscores the central role that extracellular NAD^+, released by the mucosal cells, plays in the regulation of $[Ca^{2+}]_c$ levels in the adjacent smooth myocytes [27].

$CD38^{+}/CD38^{-}$ 3T3 Murine Fibroblasts

Following this study, we addressed a simpler model system for investigating the paracrine patterns of NAD^{+}/cADPR-mediated cell-to-cell communication. This was obtained by co-culturing in a transwell mode two types of 3T3 murine fibroblasts differing from each other exclusively for presence or absence of CD38. The CD38-transfected fibroblasts were thus used as feeders for the growth of native $CD38^{-}$cells to be investigated [28].

Table 3 summarizes the results obtained using both heterologous $CD38^{+}/CD38^{-}$ and homologous $CD38^{-}/CD38^{-}$ co-cultures (controls) and comparing them in terms of increased $[Ca^{2+}]_c$ levels and of concomitant effects on cell proliferation. Preliminary experiments showed in fact that the $[Ca^{2+}]_c$ of the $CD38^{-}$ target fibroblasts undergoes a sustained increase upon incubation of these cells on the $CD38^{+}$ feeders. Moreover, proliferation of the $CD38^{-}$ fibroblasts co-cultured on the $CD38^{+}$ cell feeders showed a significant increase ($p<0.05$) over that recorded with the same $CD38^{-}$ fibroblasts grown on homologous $CD38^{-}$ feeders [28]. This positive effect of the $CD38^{+}$ feeders on the growth of $CD38^{-}$ fibroblasts compares well with the enhanced cell proliferation previously observed in the CD38-transfected over the native $CD38^{-}$ fibroblasts [17].

Table 3. **Extracellular levels of NAD^{+} and cADPR, $[Ca^{2+}]_c$ shifts and variations of cell growth in various co-cultures of 3T3 murine fibroblasts.** Modified from [28].

Type of co-culture	Extracellular NAD^{+} (nM)	Extracellular cADPR (nM)	$[Ca^{2+}]_c$ increase	Cell proliferation increase
$CD38^{+}/CD38^{-}$	146 (±15)	6.0 (±8)	+++	+++
$CD38^{-}/CD38^{-}$	138 (±12)	ND	–	–
$CD38^{+}/CD38^{-}$ + NADase	120 (±13)	1.8 (±0.3)	+	+
$CD38^{-}/CD38^{-}$ + cyclase	136 (±16)	3.5 (±0.4)	++	++

Several lines of evidence conclusively support the mechanism of paracrine interaction illustrated in Fig. 3. Release of NAD^{+} from the $CD38^{+}$ feeders across Cx43 hemichannels was indicated by failure to record any sustained $[Ca^{2+}]_c$ increase in the co-cultured target $CD38^{-}$ cells following pre-incubation with Cx43 antisense, but not with sense, oligodeoxynucleotide. Another proof for the role of released NAD^{+} in this intracellular interaction was a significant attenuation of both responses - $[Ca^{2+}]_c$ increase and enhanced cell growth – in the target $CD38^{-}$ fibroblasts upon addition of

NAD$^+$-glycohydrolase to the co-culture medium. Importantly, supplementation of *Aplysia* monofunctional ADP-ribosyl cyclase to the medium of the homologous CD38$^-$/CD38$^-$ co-cultures resulted in detectable responses - [Ca^{2+}] increase and cell proliferation increase – as compared with the control co-culture settings [28].

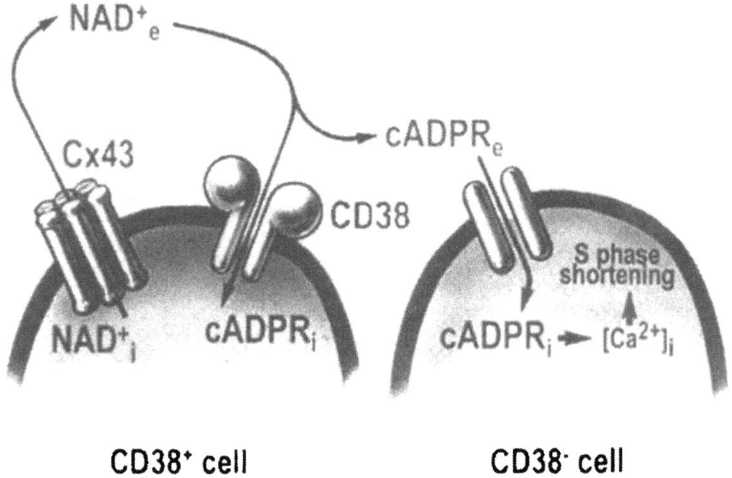

Figure 3. **NAD$^+$/cADPR-mediated paracrine interactions between CD38$^+$ and CD38$^-$ 3T3 fibroblasts.** Modified from [28].

That the extracellularly generated cADPR is involved in these cell responses was also demonstrated by several findings. Intracellular cADPR was undetectable in the CD38$^-$ target fibroblasts co-cultured with homologous CD38$^-$ feeders, but it was observed at 2.1±0.1 picomoles/mg in the CD38$^-$ cells grown on CD38$^+$ feeders, this indicating its influx from the extracellular medium across any of the cADPR transporters mentioned above. While no obvious quantitative correlation between concentrations of extracellular NAD$^+$ and cell responses was observed in any of the several experimental conditions investigated, such correlation was instead apparent with the levels of extracellular cADPR. Thus, these were not detectable in the control CD38$^-$/CD38$^-$ homologous co-cultures, but they became measurable at 3.5±0.4 nM cADPR upon supplementation of the *Aplysia* cyclase to the medium, which triggered both [Ca^{2+}]$_c$ increase and enhanced proliferation of the target CD38$^-$ fibroblasts. The highest extent of both responses was observed in the heterologous CD38$^+$/CD38$^-$ co-cultures where extracellular cADPR was 6.0±0.8 nM (Table 3). Finally, the role of cADPR as a mediator of the cell responses recorded in the CD38$^-$ fibroblasts grown on CD38$^+$ feeders was demonstrated by abrogation of both responses upon addition of 8-NH$_2$-cADPR or 8-Br-cADPR to the co-culture medium [28].

Thus, the process of paracrine interaction observed in this co-culture system was validated by experimentally dissecting it into its individual steps, each of which could be related to a specific nucleotide, NAD^+ or cADPR, respectively. The specific functions of Cx43, CD38 and the cADPR transporters in this paracrine loop based upon NAD^+/cADPR were established unambiguously.

Interactions Between $CD38^+$ Stromal Cells and Human Hemopoietic Progenitors

Bone marrow cells harbour both mammalian ADP-ribosyl cyclases known so far, CD38 (mature hemopoietic progenitors) and the GPI-anchored protein BST-1 (stromal cells) [41-43]. Therefore, the hemopoietic environment is certainly a good model system for investigating important functional effects of NAD^+ and cADPR. Current limitations of knowledge and of technologies required to achieve an efficient expansion of human hemopoietic progenitors for transplantation purposes represent a strong stimulus to these studies centered on NAD^+ and cADPR. These were started in our laboratory in view of the role that cADPR plays in the regulation of cell cycle progression in mammalian cells. Indeed, both in the CD38-transfected and in the native $CD38^-$ 3T3 fibroblasts grown on $CD38^+$ feeders, cADPR determines a significant enhancement of cell proliferation due to shortening of the S phase of cell cycle [17, 28].

A first result was achieved by exposing human committed hemopoietic progenitors (HP), the colony-forming cells (CFC), obtained from both cord blood and peripheral blood, to either 100 μM cADPR or 10 nM 3-deaza-cADPR for 24 h. This protocol of priming resulted in a significant increase of colony output and colony size over unprimed control HP [26]. Importantly, this effect proved to survive through multiple generations (up to four) over several weeks following exposure to cADPR (or 3-deaza-cADPR), indicating a genuine horizontal expansion of the HP. The increase of cADPR-induced output of CFC was observed with all colony types, i.e. CFU granulocyte macrophage, burst forming units erythroid and multipotential colony forming units [26].

The HP-expanding activity of extracellular cADPR was causally related to a slow and sustained increase of the $[Ca^{2+}]$ levels, most likely due to influx of cADPR as shown by its appearance within the cells at levels (0.55±0.16 nanomol/10^8 cells) higher than the hardly detectable levels measured in the same untreated $CD38^{low}$ HP and comparable with those of many $CD38^+$ cells. This correlation was demonstrated through the use of 8-NH_2-cADPR, which completely inhibited the calcium-releasing activity of cADPR in the HP [26]. The results obtained with extracellular cADPR and its antagonist 8-NH_2-cADPR lend support to an important role of the $[Ca^{2+}]_i$ levels in the control of growth of the HP which proved to be stimulated at higher than 20 nM $[Ca^{2+}]_c$, while cell death was conversely observed at lower than 20 nM $[Ca^{2+}]_c$ [26].

These promising results, strengthened by the same findings obtained with nanomolar concentrations of 3-deaza-cADPR (an analog of potential pharmacological significance [40]), led us to extend the studies on effects of extracellular cADPR to the most immature HP, i.e. the long-term culture initiating cells (LTC-IC). Interest toward LTC-IC comes from the fact that, unlike the more mature CFC, they include those HP capable of repopulating the irradiated host. Priming of the LTC-IC with 100 μM extracellular cADPR for 24 h enhanced their proliferation and self-renewal significantly over the untreated mononuclear cells (MNC) during the subsequent 5-weeks culture require to record colony growth [44].

This and the results obtained with the 3T3 murine fibroblasts (see above) suggested replacement of extracellularly added cADPR with a layer of murine stromal cell lines transfected with CD38 to produce extracellular cADPR. The latter system, operating in a transwell setting, mimics the physiology of the bone marrow microenvironment, where HP grow over stromal cells which express the ectocyclase activity of BST-1. A remarkable expansion (i.e. stimulation of self-renewal of clonogenic progenitors), was obtained upon coculturing for 24 h the LTC-IC (from cord blood MNC) with either of two murine stromal cell lines (3T3 and M210B4) transfected with human CD38 cDNA [44]. The experimental protocol of these experiments is summarized in Table 4. In addition, those LTC-IC either challenged for 24 h with extracellular cADPR or grown for 24 h over the confluent $CD38^+$ feeder layers produced colonies up to the fifth generation, while the unprimed progenitors never produced colonies beyond the second generation. The corresponding expansion factor of the LTC-derived colonies was 250 times higher than that of controls [44].

Therefore, the LTC-IC showed the same response, in terms of expansion, following a brief (24 h) exposure to either exogenous cADPR or to cADPR-generating $CD38^+$ feeder cells. The mechanism of these engineered stromal cells in releasing NAD^+ into the medium across Cx43 hemichannels and in converting NAD^+ to cADPR extracellularly with their "de novo" expressed CD38 cyclase activity was established in two ways. First, supplementation of NAD^+-glycohydrolase to the co-culture medium during the 24h of priming abrogated the stimulatory effect of the $CD38^+$ feeders on the LTC-C frequency (recorded after 5-week culture). In addition, a comparable increase in LTC-IC frequency was obtained upon incubating for 24 h the cord blood MNC over untrasfected $CD38^-$ murine feeders (lacking any ectocellular cyclase) supplemented with the *Aplysia* monofunctional cyclase and then grown for 5 weeks over $CD38^-$ feeders [44].

Table 4. **The experimental protocol for the expansion of LTC-IC by exogenously added or stroma-generated cADPR.** Modified from [44].

1. Priming: incubation of cord blood mononuclear cells (MNC) for 24 h with either 100 μM cADPR, or ADPR (control), or over confluent CD38-transfected (or $CD38^-$ as control) feeder layers
2. Transfer of the transwell on irradiated $CD38^-$ feeders
3. Long-term liquid culture on $CD38^-$ feeders (5 weeks)
4. Evaluation of colony growth in growth factor-supplemented semisolid medium
5. Evaluation of second (and subsequent generation) colonies in growth factor-supplemented semisolid medium

The positive effects of cADPR on self-renewal and proliferation of the immature human HP proved to be causally correlated with the sustained, 8-Br-cADPR (or 8-NH_2-cADPR)-inhibitable increase of $[Ca^{2+}]_c$ levels during the 24 h priming with the cyclic nucleotide [44].

When exposure of the LTC-IC to the cADPR-generating stroma was extended from 24 h to 5-week culture, unexpectedly, a dramatic decrease of their frequency was observed. The causal role of $CD38^+$ stromal cells in promoting such growth inhibition was demonstrated upon diluting the $CD38^+$ with the $CD38^-$ cells so as to decrease the levels of ectocyclase activity down to values observed in native (i.e., untransfected) human hemopoietic stroma, that expresses BST-1 only. The 5-week culture of the MNC under these conditions of low cyclase activity resulted in a LTC-IC output five times higher than that obtained in parallel control co-cultures over $CD38^-$ murine feeders [44].

These findings indicated that the inhibition of LTC-IC expansion observed after a continuous 5-week co-culture over the cADPR-generating $CD38^+$ feeders was due to a soluble factor released by the $CD38^+$ feeders but not by the native $CD38^-$ cell layers. Indeed, interferon-γ (IFN-γ), a potent hemopoiesis-inhibiting cytokine, was generated and released in the culture medium of the CD38-transfected, undiluted murine cell lines at concentrations five times higher than those measured in the media conditioned by $CD38^-$ feeders. Overproduction of IFN-γ was completely inhibited by 8-Br-cADPR, this implicating the increased $[Ca^{2+}]_c$ levels of the $CD38^+$ stromal cells as responsible for inhibition of HP cell growth [44].

Fig. 4 summarizes the complex interrelations that are mediated by NAD^+, cADPR and IFN-γ in the transwell co-culture of HP with $CD38^+$ stromal cells as investigated in our study. Unlike the simpler $CD38^+$/$CD38^-$ co-culture system developed with 3T3 murine fibroblasts (see Fig. 3), two paracrine loops are apparent. A first, stimulatory one, based upon CD38-catalyzed conversion of Cx43-released NAD^+ to extracellular cADPR and subsequent influx of the cyclic nucleotide into the HP, is responsible for calcium-mediated expansion of these cells.

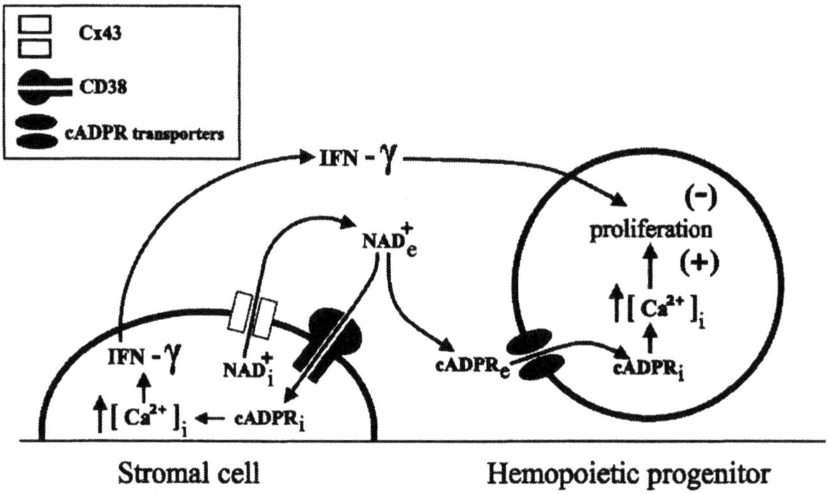

Figure 4. **Autocrine and paracrine interactions between a CD38-transfected stromal cell and a human hemopoietic progenitor as a model of bone marrow microenvironment.** From [44], with permission.

However, the cADPR-generating and channelling activity of CD38 expressed in the stromal cells triggers an autocrine loop responsible for the calcium-dependent overproduction and subsequent release of IFN-γ. Accordingly, an inhibitory paracrine loop takes place, which overwhelms the growth-promoting effect of the stimulatory, cADPR-mediated, paracrine loop.

It should be mentioned that native stromal cells, whose trophic role on the HP in the bone marrow environment is well known, express BST-1 instead of CD38 as an ectocyclase activity. Therefore, the co-culture system developed and studied in our laboratory might not be the ideal set-up for pursuing "in vitro" expansion of HP for transplantation purposes. BST-1, a GPI-anchored protein lacking a transmembrane channel-forming domain as conversely present in CD38 [45], may be the ADP-ribosyl cyclase best suited to provide the adjacent HP with extracellular cADPR, while avoiding the cADPR-induced autocrine overproduction of IFN-γ that inhibits HP expansion. This hypothesis is currently under investigation in our laboratory, with the aim to develop the best experimental system for achieving self-renewal and proliferation of the most immature HP. However, an added value of the co-culture system of human HP grown on CD38-transfected stromal cells may be a better knowledge of the still undefined mechanisms that underlie the pathogenesis of dysfunctional hemopoiesis in aplastic syndromes and poorly engrafting bone marrow transplants. A constant feature of these diseases is invasion of the bone marrow by activated, highly CD38-expressing lymphocytes. These might play an important role either generating extracellular cADPR which could enter stromal cells and trigger calcium-mediated IFN-γ overproduction, or directly releasing hemopoiesis-inhibiting IFN-γ through the autocrine loop shown in Fig. 4. Thus, elucidation of the autocrine-paracrine mechanisms related to the NAD^+/cADPR/IFN-γ system might lead to the design of new therapeutic strategies aimed to restore the physiological balance between stimulatory and inhibitory effects of stroma-generated signal molecules on HP proliferation "in vivo". Interestingly, there is increasing evidence that Cx43-mediated intercellular communication between bone marrow stroma and HP plays an important role in the physiology of hemopoiesis [46-48]. In addition, preliminary "in vivo" experiments indicate that "in vitro" priming of HP with cADPR and subsequent infusion into recipient NOD/SCID mice significantly improves short-and long-term engraftment [Podestà M. *et al.*, unpublished data].

Interactions between $CD38^+$ Rat Astrocytes and Hippocampal Neurons

Synapses are characterized by a number of functional interactions between glial cells and neurons. The most known among these interactions are mediated by an intercellular exchange of metabolites, the best recognized examples being those of glutamate/glutamine and of glucose/lactate, which have been recently shown to be reciprocally interrelated [49]. Specifically, in astrocytes, calcium oscillations and waves have been demonstrated to trigger glutamate release which impacts on adjacent neurons by inducing $[Ca^{2+}]_c$ responses [50, 51]. This relationship is bidirectional because $[Ca^{2+}]_c$ oscillations generated in neurons trigger an increased frequency of $[Ca^{2+}]_c$ oscillations is neighbouring astrocytes via the activation of metabotropic glutamate receptors expressed on the astrocyte membrane. Altogether, these data have led to the increasingly emerging concept of the importance of glial cells-neuron interactions in the central nervous system and of the involvement of Ca^{2+} in modulation of synaptic functions [52].

cADPR has been shown to elicit long-term synaptic depression in hippocampus, most likely acting at a presynaptic site [53]. Therefore, the possibility of NAD^+/cADPR-mediated paracrine interactions between astrocytes and neurons was experimentally addressed using the same experimental approaches as for the other heterologous cells systems.

A peculiar pattern of cell-to-cell interaction was found, which is illustrated in Fig. 5. Cultured hippocampal astrocytes express CD38 and Cx43 on their plasma membrane [33, 54]. In addition, they are able to respond to either extracellular NAD^+ or cADPR with 8-NH_2-inhibitable calcium responses [54]. This suggests a NAD^+ /cADPR-related autocrine loop resulting in a $[Ca^{2+}]$-mediated increase of glutamate release from astrocytes. This is similar to the autocrine interaction shown in Fig. 4 for the CD38-transfected 3T3 murine fibroblasts in which the eventual result is the overproduction (and release into the the extracellular medium) of IFN-γ and its inhibitory effects on expansion of co-cultured HP. In the astrocytes, conversely, the autocrinally stimulated, $[Ca^{2+}]_c$-mediated release of glutamate triggers a positive paracrine signalling to neurons nearby which respond with increased $[Ca^{2+}]_c$ levels. These data suggest the existence of a complex network of autocrine and paracrine loops between astrocytes and neurons: a first, NAD^+ /cADPR-mediated loop, having an apparently weak effect on adjacent neurons, and a second one, resulting from the autocrine processes in the astrocyte and related to consequent glutamate release, which represents an amplified signalling from astrocytes to neurons. Accordingly, cADPR seems to represent a built-in key molecule designed to up-regulate calcium signals between astrocytes and neurons, with potential outcomes on modulation of synaptic functions [54].

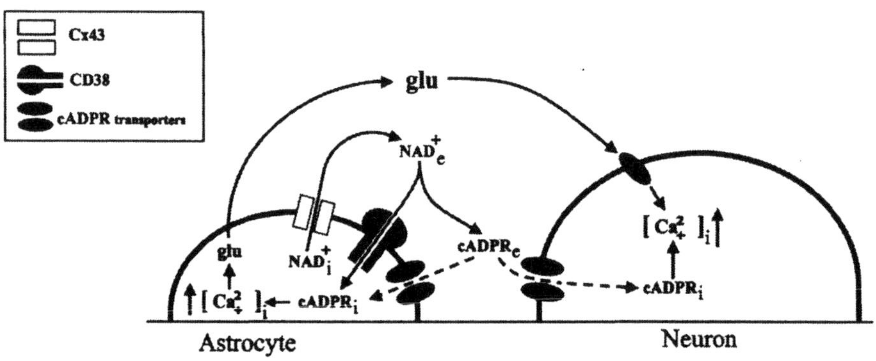

Figure 5. **Model of autocrine and paracrine interactions centered on NAD^+, cADPR and glutamate between an astrocyte and a neuron.** Modified from [54].

CONCLUSIONS

The autocrine and paracrine mechanisms that mediate regulation of intracellular calcium homeostasis by NAD^+ and cADPR in several cells, including cyclase-negative ones, demonstrate the potential "in vivo" relevance of this signalling system. Studies on intact cells are usually limited by loss of molecular details of the underlying mechanisms, but fortunately the NAD^+/cADPR system received, since its identification, remarkable attention in terms of biochemistry, molecular and cell biology and molecular pharmacology. Accordingly, considerable information has now been gathered on several important topics including: the mechanisms of calcium signalling triggered by cADPR and $NAADP^+$ [55], which seem to differ among specific cell types (e.g., T lymphocytes versus neurons); the organization of cADPR receptors [56-59]; the exceedingly complex correlations linking cADPR, $NAADP^+$ and inositol-triphosphate (IP_3) in modulating the dynamics of calcium movements in specific cells [55, 60,

61]; pathways and mechanisms of cADPR and $NAADP^+$ generation and related enzymology. At the same time the universal occurrence of this signalling system is witnessed by conservation of cADPR metabolism throughout evolution from porifera to mammals.

The topological paradox of CD38/cADPR [15] could be addressed and solved by demonstrating that some, in part already known proteins (yet unrecognized as NAD^+ or cADPR transporters) and in part so far undefined transporting systems can mediate an intense trafficking of NAD^+ and of cADPR across cell membranes. As discussed in this chapter, trafficking can occur intracellularly and extracellularly. A clear message coming from studies on intracellular trafficking is that exocytotic and endocytotic processes can remarkably affect the metabolism and the signalling functions of NAD^+ and of cADPR. A related outcome is that in mammalian cells compartmentation of NAD^+/cADPR metabolism needs to be refined and perfected by sophisticated regulatory mechanisms, e.g. PKC-mediated Cx43 phosphorylation, which enable proper control of calcium-dependent processes and hence cell survival [8]. Therefore, mechanisms of regulation of NAD^+/cADPR metabolism should be investigated in order to improve knowledge on calcium-modulated physiological processes in specific cell types and, furthermore, to identify molecular targets for new drugs aimed to control or to correct pathological processes linked to imbalances of cell calcium homeostasis. The discovery and characterization of ADP-ribosyl cyclases other than CD38 and BST-1, for instance monofunctional enzymes, or the identification of so far undefined mechanisms for regulating the balance between cADPR synthesis and hydrolysis featured by CD38 and BST-1 are relevant examples for both purposes.

The most surprising discovery that is related to identification of an extracellular trafficking of NAD^+ and cADPR is the conclusion that old (NAD^+) or adolescent (cADPR) molecules play concerted and new roles in some processes of cell-to-cell communication. Thus, in several cell systems, NAD^+ turns out to be a signal molecule functionally comparable to a hormone, or to a growth factor, while cADPR plays the role of a second messenger directly arising from NAD^+ itself and acting at an intracellular site despite its ectocellular generation. Intracellular calcium is a third messenger in this signalling system [27, 28, 44, 54] and it can be reasonably expected that distinctive modes of tuning $[Ca^{2+}]_c$ levels via different cADPR-generating enzymes, or via different mechanisms of regulation thereof, or via different kinetics and dynamics of cADPR production, can lead to distinctive responses in a single cell. DNA expression analyses and characterization of related proteins by means of high throughput technologies are obviously required to clarify these issues that appear to be critical for cell physiology.

ACKNOWLEDGEMENTS

This study was supported in part by the Italian Ministry of University and Scientific and Technological Research (MURST-PRIN 2000), by the Associazione Italiana per la Ricerca sul Cancro (AIRC), by the 5% Project on Biotechnology (MURST-CNR), by the CNR Target Project on Biotechnology, by the Cassa di Risparmio di Genova e Imperia and by the Interuniversity Consortium on Biotechnology (CIB).

REFERENCES

1. States DJ, Walseth TF and Lee HC. 1992. Similarities in amino acid sequences of *Aplysia* ADP-ribosyl cyclase and human lymphocyte antigen CD38. *Trends Biochem. Sci.* 17: 495.
2. Lee HC, Walseth TF, Bratt GT, Hayes RN and Clapper DL. 1989. Structural determination of a cyclic metabolite of NAD^+ with intracellular Ca^{2+}-mobilizing activity. *J.Biol. Chem.* 264: 1608-16015.
3. Lee HC and Aarhus R. 1995. A derivative of NADP mobilizes calcium stores insensitive to inositol trisphosphate and cyclic ADP-ribose. *J. Biol. Chem.* 270: 2152-2157.
4. Aarhus R, Graeff RM, Dickey DM, Walseth TF and Lee HC. 1995. ADP-ribosyl cyclase and CD38 catalyze the synthesis of a calcium-mobilizing metabolite from NADP. *J. Biol. Chem.* 270: 30327-30333.
5. Chini EN, Beers KW and Dousa TP. 1995. Nicotinate adenine dinucleotide phosphate (NAADP) triggers a specific calcium release system in sea urchin eggs. *J. Biol. Chem.* 270: 3216-3223.
6. Franco L, Guida L, Bruzzone S, Zocchi E, Usai C and De Flora A. 1998. The transmembrane glycoprotein CD38 is a catalytically active transporter responsible for generation and influx of the second messenger cyclic ADP-ribose across membranes. *FASEB J.* 12: 1507-1520.
7. Bruzzone S, Guida L, Zocchi E, Franco L and De Flora A. 2001. Connexin 43 hemichannels mediate Ca^{2+}-regulated transmembrane NAD^+ fluxes in intact cells. *FASEB J.* 15: 10-12.
8. Bruzzone S, Franco L, Guida L, Zocchi E, Contini P, Bisso A, Usai C and De Flora A. 2001. A self restricted CD38-connexin 43 cross-talk affects NAD^+ and cyclic ADP-ribose metabolism and regulates intracellular calcium in 3T3 fibroblasts. *J. Biol. Chem.* 276: 48300-48308.
9. De Flora A, Franco L, Guida L, Bruzzone S, Usai C and Zocchi E. 2000. Topology of CD38. *Chem. Immunol.* 75: 79-98.
10. Prasad GS, McRee DE, Stura EA, Levitt DG, Lee HC and Stout CD. 1996. Crystal structure of *Aplysia* ADP-ribosyl cyclase, a homologue of the bifunctional ectoenzyme CD38. *Nature Struct. Biol.* 3: 957-967.
11. Munshi C, Aarhus R, Graeff R, Walseth TF, Levitt D and Lee HC. 2000. Identification of the enzymatic active site of CD38 by site-directed mutagenesis. *J. Biol. Chem.* 275: 21566-21571.
12. Graeff R, Munshi C, Aarhus R, Johns M and Lee HC. 2001. A single residue at the active site of CD38 determines its NAD cyclizing and hydrolyzing activities. *J. Biol. Chem.* 276: 12169-12173.
13. Tohgo A, Takasawa S, Noguchi N, Koguma T, Nata K, Sugimoto T, Furuya Y, Yonekura, H and Okamoto H. 1994. Essential cysteine residues for cyclic ADP-ribose synthesis and hydrolysis by CD38. *J. Biol. Chem.* 269: 28555-28557.
14. Bruzzone S, Guida L, Franco L, Zocchi E, Corte G and De Flora A. 1998. Dimeric and tetrameric forms of catalytically active transmembrane CD38 in transfected HeLa cells. *FEBS Lett.* 433: 275-278.

15. De Flora A, Guida L, Franco L and Zocchi E. 1997. The CD38/Cyclic ADP-ribose system: A topological paradox. *Int. J. Biochem. Cell Biol.* 29: 1149-1166.
16. Lund FE, Cockayne DA, Randall TD, Solvason N, Schuber F and Howard MC. 1998. CD38: a new paradigm in lymphocyte activation and signal trasduction. *Immunol. Rev.* 161: 79-93.
17. Zocchi E, Daga A, Usai C, Franco L, Guida L, Bruzzone S, Costa A, Marchetti C and De Flora A. 1998. Expression of CD38 increases intracellular calcium concentration and reduces doubling time in HeLa and 3T3 cells. *J. Biol. Chem.* 273: 8017-8024.
18. Funaro A, Reinis M, Trubiani O, Santi S, Di Primio R and Malavasi F. 1998. CD38 functions are regulated through an internalization step. *J. Immunol.* 160: 2238-2247.
19. Zocchi E, Usai C, Guida L, Franco L, Bruzzone S, Passalacqua M and De Flora A. 1999. Ligand-induced internalization of CD38 results in intracellular Ca^{2+} mobilization. Role of NAD^+ transport across cell membranes. *FASEB J.* 13: 273-283.
20. Zocchi E, Franco L, Guida L, Piccini D, Tacchetti C and De Flora A. 1996. NAD^+-dependent internalization of the transmembrane glycoprotein CD38 in human Namalwa B cells. *FEBS Lett.* 396: 327-332.
21. Walseth TF and Lee HC. 1993. Synthesis and characterization of antagonists of cyclic-ADP-ribose-induced Ca^{2+} release. *Biochim. Biophys. Acta.* 1178: 235-242.
22. Lee HC, Zocchi E, Guida L, Franco L, Benatti U and De Flora A. 1993. Production and hydrolysis of cyclic ADP-ribose at the outer surface of human erythrocytes. *Biochem. Biophys. Res. Commun.* 191: 639-645.
23. Zocchi E, Franco L, Guida L, Benatti U, Bargellesi A, Malavasi F, Lee HC and De Flora A. 1993. A single protein immunologically identified as CD38 displays NAD^+ glycohydrolase, ADP-ribosyl cyclase and cyclic ADP-ribose hydrolase activities at the outer surface of human erythrocytes. *Biochem. Biophys. Res. Commun.* 196: 1459-1465.
24. Howard M, Grimaldi JC, Bazan JF, Lund FE, Santos AL, Parkhouse RM, Walseth TF and Lee HC. 1993. Formation and hydrolysis of cyclic ADP-ribose catalyzed by lymphocyte antigen CD38. *Science* 262: 1056-1059.
25. De Flora A, Guida L, Franco L, Zocchi E, Pestarino M, Usai C, Marchetti C, Fedele E, Fontana G and Raiteri M. 1996. Ectocellular *in vitro* and *in vivo* metabolism of cyclic ADP-ribose in cerebellum. *Biochem. J.* 320: 665-672.
26. Podestà M, Zocchi E, Pitto A, Usai C, Franco L, Bruzzone S, Guida L, Bacigalupo A, Scadden DT, Walseth TF, De Flora A and Daga A. 2000. Extracellular cyclic ADP-ribose increases intracelluair free calcium concentration and stimulates proliferation of human hemopoietic progenitors. *FASEB J.* 14: 680-690.
27. Franco L, Bruzzone S, Song P, Guida L, Zocchi E, Walseth TF, Crimi E, Usai C, De Flora A and Brusasco V. 2001. Extracellular Cyclic ADP-ribose potentiates ACh-induced contraction in bovine tracheal smooth muscle. *Am. J. Physiol. Lung Cell. Mol. Physiol.* 280: L98-L106.
28. Franco L, Zocchi E, Usai C, Guida L, Bruzzone S, Costa A and De Flora A. 2001. Paracrine roles of NAD^+ and Cyclic ADP-ribose in increasing intracellular calcium and enhancing cell proliferation of 3T3 fibroblasts. *J. Biol. Chem.* 276: 21642-21648.
29. Bruzzone R, White TW and Paul DL. 1996. Connections with connexins: the molecular basis of direct intercellular signaling. *Eur. J. Biochem.* 238: 1-27.
30. Lampe PD and Lau AF. 2000. Regulation of gap junctions by phosphorylation of connexins. *Arch. Biochem. Biophys.* 384: 205-215.
31. Lampe PD, TenBroek EM, Burt JM, Kurata WE, Johnson RG and Lau AF. 2000. Phosphorylation of connexin 43 on serine 368 by protein kinase C regulates gap junctional communication. *J. Cell Biol.* 149: 1503-1512.
32. Guse AH, da Silva CP, Berg I, Skapenko AL, Weber K, Heyer P, Hohenegger M, Ashamu GA, Schulze-Koops H, Potter BV and Mayr GW. 1999. Regulation of calcium signalling in T lymphocytes by the second messenger cyclic ADP-ribose. *Nature* 398: 70-73.

33. Hotta T. Asai K. Fujita K. Kato T and Higashida H. 2000. Membrane-bound form of ADP-ribosyl cyclase in rat cortical astrocytes in culture. *J. Neurochem.* 74: 669-675.
34. Higashida H. Hashii M. Yokoyama S. Hoshi N. Asai K and Kato T. 2001. Cyclic ADP-ribose as a potential second messenger for neuronal Ca^{2+} signaling. *J. Neurochem.* 76: 321-331.
35. Galione A. White A. Willmott N. Turner M. Potter BV and Watson SP. 1993. cGMP mobilizes intracellular Ca^{2+} in sea urchin eggs by stimulating cyclic ADP-ribose synthesis. *Nature* 365: 456-459.
36. Graeff RM. Franco L. De Flora A and Lee HC. 1998. Cyclic GMP dependent and independent effects on the synthesis of the calcium messengers cyclic ADP-ribose and NAADP. *J. Biol. Chem.* 273: 118-125.
37. Wu Y. Kuzma J. Marechal E. Graeff R. Lee HC. Foster R and Chua NH. 1997. Abscisic acid signaling through cyclic ADP-ribose in plants. *Science* 278: 2126-2130.
38. Zocchi E. Carpaneto A. Cerrano C. Bavestrello G. Giovine M. Bruzzone S. Guida L. Franco L and Usai C. 2001. The temperature-signaling cascade in sponges involves a heat-gated cation channel. abscisic acid. and cyclic ADP-ribose. *Proc. Natl. Acad. Sci. USA* 98: 14859-14864.
39. Prakash YS. Kannan MS. Walseth TF and Sieck GS. 1998. Role of cyclic ADP-ribose in the regulation of $[Ca^{2+}]_i$ in porcine tracheal smooth muscle. *Am. J. Physiol. Cell. Physiol.* 274: C1653-C1660.
40. Wong L. Aarhus R. Lee HC and Walseth TF. 1999. Cyclic 3-deaza-adenosine diphosphoribose: A potent and stable analog of cyclic ADP-ribose. *Biochim. Biophys. Acta.* 1472: 555-564.
41. Jackson DG. and Bell J. 1990. Isolation of a cDNA encoding the human CD38 (T10) molecule, a cell surface glycoprotein with an unusual discontinuous pattern of expression during lymphocyte differentiation. *J. Immunol.* 144: 2811-2815.
42. Mehta K and Cheema S. 1999. Retinoid-mediated signaling pathways in CD38 antigen expression in myeloid leukemia cells. *Leuk. Lymphoma* 32: 441-449.
43. Hirata Y, Kimura N, Sato K, Ohsugi Y, Takasawa S, Okamoto H, Ishikawa J, Kaisho T, Ishihara K and Hirano T. 1994. ADP-ribosyl cyclase activity of a novel bone marrow stromal cell surface molecule. BST-1. *FEBS Lett.* 356: 244-248.
44. Zocchi E. Podestà M. Pitto A. Usai C. Bruzzone S. Franco L. Guida L. Bacigalupo A and De Flora A. 2001. Paracrinally stimulated expansion of early human hemopoietic progenitors by stroma-generated cyclic ADP-ribose. *FASEB J.* 15: 1610-1612.
45. Ishihara K and Hirano T. 2000. BST-1/CD157 regulates the humoral immune responses in vivo. *Chem. Immunol.* 75: 235-255.
46. Rosendaal M, Green CR, Rahaman A and Morgan D. 1994. Up-regulation of the connexin 43-gap junction network in haemopoietic tissue before the growth of stem cells. *J. Cell Sci.* 107: 29-37.
47. Montecino-Rodriguez E, Leaders H and Dorshkind K. 2000. Expression of connexin 43 (Cx43) is critical for normal hematopoiesis. *Blood* 96: 917-924.
48. Cancelas JA. Koevoet WL, de Koning AE, Mayen AE, Rombouts EJ and Ploemacher RE. 2000. Connexin-43 gap junctions are involved in multiconnexin-expressing stromal support of hemopoietic progenitors and stem cells. *Blood* 96: 498-504.
49. Magistretti PJ. Pellerin L. Rothman DL. Shulman RG. 1999. Energy on demand. *Science* 283: 496-497.
50. Parpura V. Basarsky TA. Liu F. Jetnijia K. Jeftinijia S and Haydon PG. 1994. Glutamate-mediated astrocyte-neuron signalling. *Nature* 369: 744-747.
51. Bezzi P, Carmignoto G, Pasti L, Vesce S, Rossi D, Rizzini BL, Pozzan T and Volterra A. 1998. Prostaglandins stimulate calcium-dependent glutamate release in astrocytes. *Nature* 391: 281-285.
52. Bacci A, Verderio C, Pravettoni E and Matteoli M. 1999. The role of glial cells in synaptic function. *Philos. Trans. R. Soc. Lond. B. Biol. Sci.* 354: 403-409.

53. Reyes-Harde M, Empson R, Potter BV. Galione A and Stanton PK. 1999. Evidence of a role for cyclic ADP-ribose in long-term synaptic depression in hippocampus. *Proc. Natl. Acad. Sci. USA*. 96: 4061-4066.
54. Verderio C, Bruzzone S, Zocchi E, Fedele E, Schenk U, De Flora A and Matteoli M. 2001. Evidence of a role for cyclic ADP-ribose in calcium signalling and neurotransmitter release in cultured astrocytes. *J. Neurochem.* 78: 646-657.
55. Lee HC. 1997. Mechanism of calcium signaling by cyclic ADP-ribose and NAADP. *Physiol. Rev.* 77: 1134-1164.
56. Walseth TF. Aarhus R, Kerr JA and Lee HC. 1993. Identification of cyclic ADP-ribose-binding proteins by photoaffinity labeling. *J. Biol Chem.* 268: 26686-26691.
57. Lee HC, Aarhus R, Graeff R, Gurnack ME and Walseth TF. 1994. Cyclic ADP-ribose activation of the ryanodine receptor is mediated by calmodulin. *Nature* 370: 307-309.
58. Li PL, Tang WX, Valdivia HH, Zou AP and Campbell WB. 2001. cADP-ribose activates reconstituted ryanodine receptors from coronary arterial smooth muscle. *Am. J. Physiol, Heart Circ. Physiol.* 280: H208-H215.
59. Tang WX. Chen YF, Zou AP, Campbell WB and Li PL. 2002. Role of FKBP12.6 in cADPR-induced activation of reconstituted ryanodine receptors from arterial smooth muscle. *Am. Physiol. Heart Circ. Physiol.* 282: H1304-H1310.
60. Berridge MJ, Lipp P and Bootman MD. 2000. The versatility and universality of calcium signalling. *Nature Rev. Mol. Cell Biol.* 1: 11-21.
61. Bootman MD, Lipp P and Berridge MJ. 2001. The organisation and functions of local Ca^{2+} signals. *J. Cell Sci.* 114: 2213-2222.

Antonio De Flora, Department of Experimental Medicine,
Section of Biochemistry, and Center of Excellence
for Biomedical Research, University of Genova,
Genova, Italy.
e-mail: toninodf@unige.it

Cyclic ADP-Ribose and NAADP. Structures, Metabolism and Functions. Lee HC (ed), pp. 269-299, Dordrecht: Kluwer.

The CD38-Cyclic ADP-Ribose Signal System in Pancreatic ß-Cells

The Discovery and Biological Significance of a Novel Signal System in Mammalian Cells

Shin Takasawa and Hiroshi Okamoto

Department of Biochemistry, Tohoku University Graduate School of Medicine, Sendai 980-8575, Miyagi, Japan

HISTORICAL BACKGROUND

In the early 1980's, we showed that maintenance of the cellular NAD^+ level in β-cells of the islets of *Langerhans* is essential for the synthesis and secretion of insulin, and proposed a unifying model for the action of the diabetogenic agents alloxan and streptozotocin on pancreatic β-cells (Figure 1) [1-10]. Central to the model are breaks in the nuclear DNA of β-cells, resulting from either an accumulation of free radicals or from the alkylation of DNA. These breaks induce DNA repair involving the activation of poly(ADP-ribose) synthetase/polymerase (PARP), which uses cellular NAD^+ as a substrate. As a result, the intracellular levels of NAD^+ fall dramatically, which leads to energy depletion and the inhibition of cellular functions including insulin synthesis and secretion, and thus the β-cell ultimately dies.

If alloxan- and streptozotocin-induced β-cell damage is provoked by the mechanism shown in Figure 1, then, theoretically, it can be prevented by inhibiting the serial reactions that originate from the DNA strand breaks. One possibility would be by scavenging the radicals with such radical scavengers as superoxide dismutase and desferrioxamine. Another would be by inhibiting PARP with specific inhibitors such as nicotinamide and 3-aminobenzamide. In fact, the alloxan-induced and streptozotocin-induced NAD^+ depletions were prevented by PARP inhibitors such as nicotinamide and picolinamide [11-14]. The alloxan-induced and streptozotocin-induced decreases in insulin synthesis and secretion were also prevented by superoxide dismutase and catalase as well as by PARP inhibitors in a dose-dependent manner [13, 15].

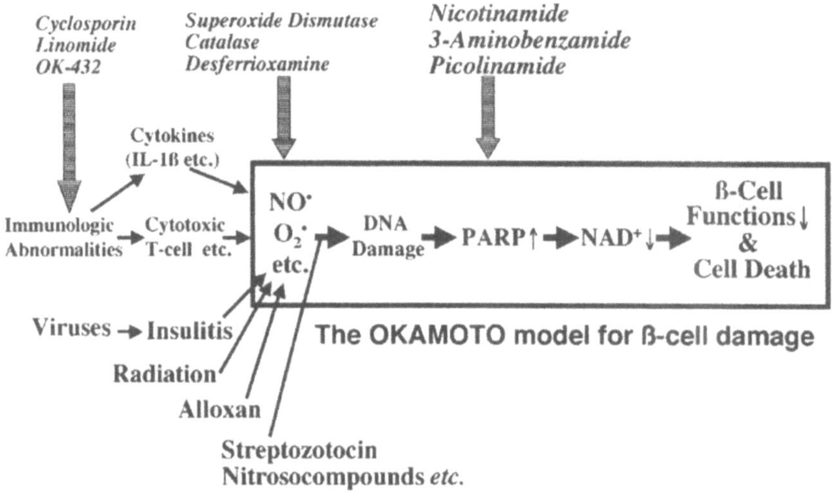

Figure 1. **A unifying model for ß-cell damage and its prevention in toxin- or virus-induced and immune diabetes (The OKAMOTO model)** (adapted from [3-10]). The ß-cell damage is theoretically preventable through inhibition of the serial reactions, as indicated by shaded arrows. One method is by inhibiting abnormal immune reactions with immunomodulators such as cyclosporin, linomide, and OK-432. Others include scavenging the radicals, which break DNA, by superoxide dismutase and other radical scavengers, and inhibiting the PARP by specific inhibitors such as nicotinamide, 3-aminobenzamide and picolinamide to prevent the decrease in the NAD^+ level. IL-1ß, interleukin-1ß. NO•, nitric oxide. O_2•, super oxide.

Interest in the model for the mechanism of action of alloxan and streptozotocin has been heightened by its possible extension to the effects of viruses and inflammation, especially immune-mediated events, on ß-cells [1-10]. Thus, since the early 1980s, we have proposed that, although type 1 (insulin-dependent) diabetes can be caused by many different agents such as immunologic abnormalities, inflammatory tissue damage, and ß-cytotoxic chemical substances, the final pathway for the toxic agents is the same (Figure 1). Therefore, type 1 (insulin-dependent) diabetes is theoretically preventable by suppressing immune reactions, scavenging free radicals, and inhibiting PARP.

Concerning nitric oxide (see Figure 1), we produced transgenic mice expressing nitric oxide synthase constitutively in pancreatic ß-cells and found that the ß-cell mass was markedly reduced and that the transgenic mice developed severe diabetes [16]. In 1999, using PARP deficient mice, three independent groups in Germany, Japan, and the U.S. provided irrefutable support for the model shown in Figure 1: PARP deficient mice were remarkably resistant to streptozotocin and did not show the ß-cell death [17-20]. More recently, many other tissues and cells such as in cerebral ischemia, myocardial ischemia, laryngeal injury, intestinal mucosal injury, organ injury by haemorrhagic shock, ischemic renal injury, and diabetic myocardial and endothelial injury have been reported to die by the same mechanism as in pancreatic ß-cell death [21-41].

	Cell Death	
	Necrosis (The OKAMOTO model)	Apoptosis
PARP	**Activated by DNA breaks**	**Cleaved by caspase(s)**
PARP inhibitors	**Effective**	**Ineffective**
Immunological Abnormality	**Trigger**	**Executor**
Occasion	**Pathological**	**Physiological**

Figure 2. **Comparison of the two types of cell death, necrosis and apoptosis.**

The cell death caused by the PARP activation described above is also involved in necrosis [3, 4, 9, 10, 42]. In apoptotic cell death, PARP is cleaved by caspases and inactivated. Therefore, PARP inhibitors may be effective in preventing necrosis but ineffective in preventing apoptosis (Figure 2). "Whether to die from necrosis or to die from apoptosis" may depend on the severity and duration of the cell damage, differences in death signals, and the species of cells. A recent report from Stauter *et al.* suggests that dendritic cells distinguish between two types of cell death, with necrosis providing a control that is critical for the initiation of immunity [43]. Therefore, immunological abnormalities, which are frequently observed in type 1 diabetes, may be triggered by the preceding necrotic cell death, followed by apoptotic death of ß-cells.

The question then arises as to why maintaining the cellular NAD^+ level is essential for β-cell functions such as insulin secretion. Cyclic ADP-ribose (cADPR) is a cyclic compound synthesized from NAD^+. This compound was first found in 1987 by Lee *et al.* when studying Ca^{2+} release in sea urchin eggs [44]. However, the physiological significance of cADPR in the mammalian system was not then fully understood. In 1992, we found that cADPR, but not inositol 1,4,5-trisphosphate (IP_3), releases Ca^{2+} from microsomes of the islets isolated from pancreases of normal rat (Wistar) and proposed a working hypothesis in which cADPR plays an essential role in

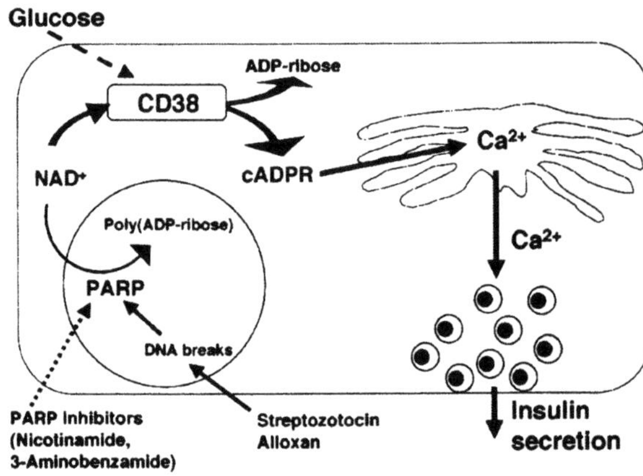

Figure 3. **A model showing cADPR mobilization of intracellular Ca^{2+} to induce islet ß-cells to secrete insulin** (adapted from [64]). cADPR is generated from NAD^+ in glucose-stimulated ß-cells by CD38. cADPR mobilizes Ca^{2+} from endoplasmic reticulum to secrete insulin. ß-cytotoxins such as streptozotocin and alloxan cause DNA strand breaks that activate nuclear PARP, leading to the NAD^+ depletion and the inhibition of the cADPR-mediated Ca^{2+} mobilization. PARP inhibitors such as nicotinamide and 3-aminobenzamide prevent the NAD^+ depletion and, therefore, the glucose-induced insulin secreting process continues to function. ß-cells that cannot mobilize Ca^{2+} by cADPR are unable to secrete insulin in response to glucose stimulation.

insulin secretion by glucose [45, 46], as shown in Figure 3, which can well explain both the reduction of insulin secretion and synthesis by various ß-cytotoxic agents such as streptozotocin and alloxan and the prevention of the reduction by PARP inhibitors nicotinamide and 3-aminobenzamide. Figure 3 also shows the formation of cADPR from NAD^+ by CD38 upon glucose stimulation in ß-cells.

THE CD38-CYCLIC ADP-RIBOSE SIGNAL SYSTEM IN INSULIN SECRETION

Glucose induces an increase in the intracellular Ca^{2+} concentration in pancreatic ß-cells of the islets of *Langerhans* to induce secretion of insulin. This increase in the Ca^{2+} concentration was first explained in 1984 by the hypothesis of Ashcroft *et al.* [47], in which Ca^{2+} is provided extracellularly. That is, millimolar concentrations of ATP, produced in the process of glucose metabolism, inhibit the potassium channel, inducing membrane depolarization and the opening of the voltage-dependent Ca^{2+} channels. In 1993 [46, 48], we proposed another model of insulin secretion by glucose via cADPR-mediated Ca^{2+} mobilization from an intracellular Ca^{2+} pool, the endoplasmic reticulum [5-10, 49-62], as shown in Figure 4. That is, ATP inhibits the cADPR hydrolase of CD38, causing the accumulation of cADPR, which acts as a second messenger for Ca^{2+} mobilization from the endoplasmic reticulum for insulin secretion.

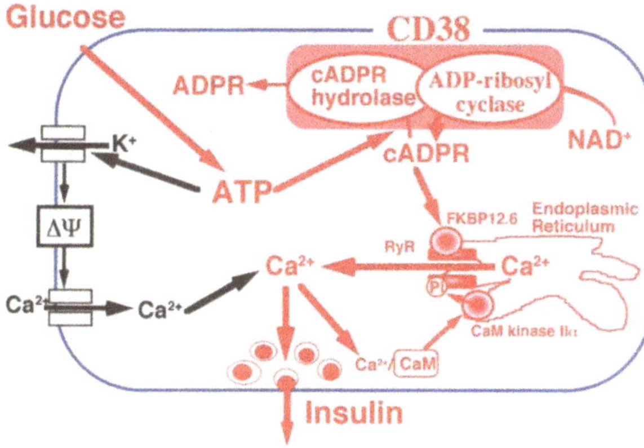

Figure 4. **Insulin secretion by glucose stimulation in β-cells** (adopted from [7]). The insulin secretion via the CD38-cADPR signal system is shown in red. cADPR binds to FKBP12.6 to release Ca^{2+}, dissociating FKBP12.6 from RyR [56]. CaM kinase II phosphorylates RyR to sensitize and activate the Ca^{2+} channel (Pi, phosphorylation of RyR by CaM kinase II) [54]. Ca^{2+}, released from intracellular stores and/or supplied from extracellular sources, further activates CaM kinase II and amplifies the process. In this way, Ca^{2+}-induced Ca^{2+} release (CICR) can be explained. The conventional insulin secretion mechanism by Ca^{2+} influx from extracellular sources [47] is shown in black. ADPR, ADP-ribose.

Cyclic ADP-Ribose Acts as a Second Messenger for Insulin Secretion by Glucose

The first important issue is whether the accumulation of cADPR is actually caused by glucose stimulation in pancreatic islets. We incubated normal rat (Wistar) and mouse (C57BL/6J) islets with low (2.8 mM) and high (20 mM) glucose, and assayed the cADPR content in the islets by radioimmunoassay using an anti-cADPR antibody. The cADPR content of islets incubated with high glucose was increased within 5 min, whereas the cADPR content of islets incubated with low glucose was not [59].

Next, we used rat islet microsomes as a cell-free system to study Ca^{2+} release and found that cADPR released Ca^{2+} from islet microsomes, as indicated by the observed prompt increase in fluo 3 fluorescence [46, 59] (Figure 5). IP_3 did not cause the release of Ca^{2+}, and after the addition of IP_3, the islet microsomes were still responsive to cADPR. On the other hand, IP_3 caused a release of Ca^{2+} from cerebellum microsomes, and cADPR also caused a release of Ca^{2+}. Heparin, an inhibitor of IP_3 binding to its receptor, blocked the IP_3-induced Ca^{2+} release from cerebellum microsomes, but did not block the cADPR-induced Ca^{2+} release. These results indicate that islet

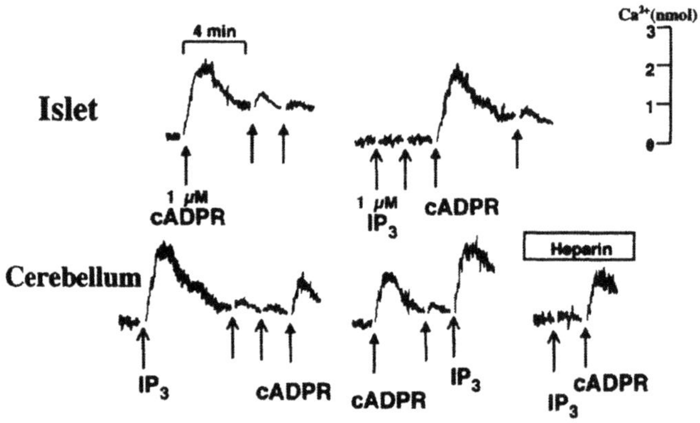

Figure 5. **Release of Ca^{2+} induced by cADPR and IP_3 from islet and cerebellum microsomes measured fluorometrically with fluo 3** (adapted from [46]).

microsomes respond to cADPR but not to IP_3. In contrast, cerebellum microsomes respond to both cADPR and IP_3, but cADPR induces the Ca^{2+} release via a different mechanism than that utilized by IP_3.

We then examined the effect of cADPR on insulin secretion using digitonin-permeabilized pancreatic islets. cADPR as well as Ca^{2+} induced insulin secretion, but IP_3 did not. The combined addition of cADPR and Ca^{2+} did not induce significantly more insulin secretion than the addition of cADPR or Ca^{2+} alone. The cADPR-induced insulin secretion was inhibited by the addition of EGTA. These results suggested that the cADPR-induced insulin secretion was mediated by Ca^{2+} mobilization from islet microsomes [46].

Thus, we proposed that glucose stimuli induce the cADPR formation from NAD^+. cADPR then mobilizes Ca^{2+} from the endoplasmic reticulum, serving as a second messenger for insulin secretion. There was a controversy about the second messenger role of cADPR in insulin secretion, and conflicting results were reported using diabetic ß-cells such as *ob/ob* mouse islets and RINm5F cells [7, 64-67]. No change in the islet cADPR content was reported in response to elevated glucose concentrations [68, 69]. Differences in the experimental condition may be responsible for the different results [7, 59]. Concerning the controversy, we will discuss in more detail in the following section of this Chapter ("The CD38-cADPR signal system in pathological conditions"). Nevertheless, recent reports have shown that Balb/c mouse islets showed distinct increases in intracellular cADPR, Ca^{2+}, and insulin secretion by glucose [70] and that MIN6 cells showed a dramatic Ca^{2+} mobilization in response to cADPR via the ryanodine receptor [71, 72] despite the fact that no response to IP_3 was observed [72]. These recent results, clearly support the CD38-cADPR signal system, described above, in insulin secretion in pancreatic ß-cells.

CD38 as a Major Enzyme for cADPR Metabolism in Mammalian Cells

The next question is why the glucose stimulus induces the formation of cADPR. CD38 is a 300-amino acid protein and was first recognized as a leukocyte antigen. We found that CD38 is expressed in a variety of tissues including pancreatic ß-cells [48, 49]. We and others have found that CD38 has both ADP-ribosyl cyclase, synthesizing cADPR from NAD^+, and cADPR hydrolase, producing ADP-ribose [48, 50, 73, 74]. We purified human CD38 protein and found that millimolar concentrations of ATP inhibit the cADPR hydrolase activity of CD38, competing with the substrate, cADPR [48, 57]. The competitive inhibition of the cADPR hydrolysis by ATP suggests that ATP and cADPR bind to the same site of CD38. We then labeled the purified CD38 with an ATP analogue, 5'-*p*-fluorosulfonylbenzoyladenosine, and identified the binding site for ATP and/or cADPR as the lysine-129 of CD38 [57]. From these results and other available evidence, we proposed that CD38 catalyzes the formation of cADPR from NAD^+ and also the hydrolysis of cADPR to ADP-ribose. As shown in Figure 6, lysine-129 of CD38 is the cADPR binding site, and ATP competes with cADPR for the binding site, resulting in the inhibition of the hydrolysis of cADPR and then, in the accumulation of cADPR [57, 63].

We assigned the human CD38 gene to band p15 of chromosome 4 as a single copy gene by fluorescence *in situ* hybridization [51]. We then isolated the human CD38 gene and determined its primary structure [58]. The CD38 gene contained 8 exons (Figure 7). Exon 1 encoded the 5'-untranslated region, translational start site, putative transmembrane domain and N-terminal region of CD38. Exon 2-8 encoded the remainder of CD38. Cysteines-119 and -201, which are essential for the hydrolysis of cADPR, were encoded in exon 2 and 5, respectively. Lysine-129, which is involved in the binding of cADPR, was encoded in exon 3. The 10 cysteine residues conserved among human, rat and mouse CD38s and *Aplysia* ADP-ribosyl cyclases were split into 6 exons [8, 58]. The stop codon and the 3'-untranslated region were located in exon 8.

By screening a human testis cDNA library, we isolated a novel cDNA in which nucleotides 364-499 of the CD38 cDNA were deleted, giving rise to a stop codon in the reading frame [58]. The cDNA thus encoded 122 amino acid residues and the predicted protein, designated as CD38 alt, had a molecular weight of 13.8 kilo-daltons. When expressed it in COS-7 cells, CD38 alt showed neither the cADPR-synthesizing nor -hydrolyzing activity. Both CD38 and CD38 alt mRNAs were detected in almost all human tissues and cell lines examined. Comparison of the nucleotide sequences of CD38 gene and CD38 alt cDNA revealed that the deleted sequence in the human CD38 alt cDNA corresponded exactly to exon 3 of the CD38 gene, indicating that CD38 alt mRNA is generated by alternative splicing.

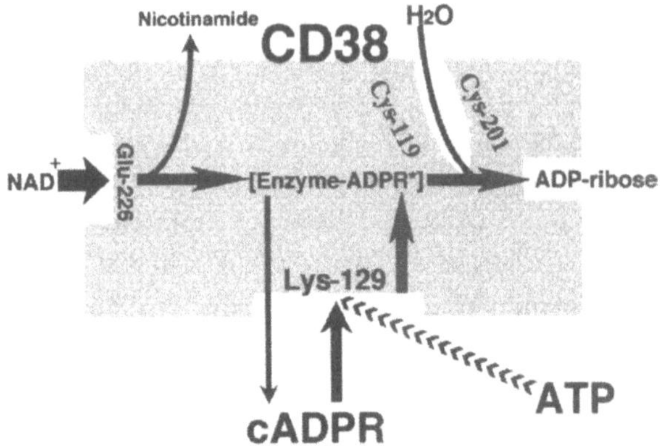

Figure 6. **Roles of amino acid residues of CD38 in enzymic activities** (adapted from [57]). Glu-226 is essential for ADP-ribosyl cyclase. Cys-119 and Cys-201 are essential for cADPR hydrolase. Lys-129 is the cADPR binding site and is essential for cADPR hydrolase. ATP competes with cADPR for the binding site (Lys-129), inhibiting the cADPR hydrolysis which causes the accumulation of cADPR. [Enzyme-ADPR*] is proposed as an enzyme-stabilized ADP-ribosyl oxocarbonium ion intermediate.

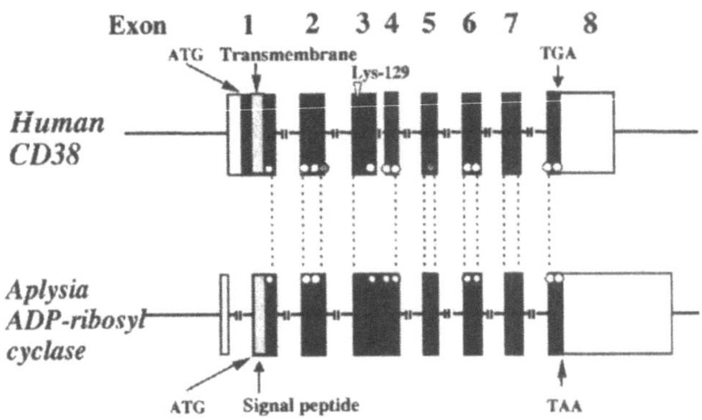

Figure 7. **Structural organization of the human CD38 [58] and *Aplysia kurodai* [52] ADP-ribosyl cyclase genes.** The exons are depicted as boxes; filled and open boxes represent protein-coding region and untranslated region, respectively. Hatched boxes represent transmembrane- or signal peptide-coding regions. The corresponding exon-intron junctions are depicted as broken lines.

Despite the lack of cADPR-metabolizing activity in CD38 alt, CD38 mRNA and CD38 alt mRNA are co-expressed in various tissues at different ratios between the two mRNAs. This suggests that the biosynthesis of CD38 can be regulated at the post-transcriptional process. We have already determined the primary structure of the *Aplysia kurodai* ADP-ribosyl cyclase gene [52]. The exon-intron organization of the *Aplysia* gene is very similar to that of human CD38 [58] (Figure 7).

Kaisho *et al.* [75] found that the amino acid sequence of bone marrow stromal antigen 1 (BST-1/CD157) had significant homology (30% identity) with those of CD38 and *Aplysia* ADP-ribosyl cyclase. They determined the gene structure of human CD157 [76]. Comparison of the gene organization reveals that CD157 gene has an exon-intron organization similar to those of CD38 and *Aplysia* cyclase genes except that human CD157 gene has an additional exon, exon 9, to code for a peptide that is removed upon attachment of the glycosyl phosphatidylinositol anchor. The similar exon-intron structure together with the fact that mouse CD157 (BP-3) is very close to the map site of the CD38 gene on mouse chromosome 5 [77], strongly suggests that the three genes (CD38, CD157 and *Aplysia* cyclase genes) have evolved from a common ancestor, from which CD38 and CD157 genes arose by gene duplication before the divergence of humans and rodents.

CD157 expression was observed in pancreatic ß-cells [78, 79]. However, CD157 knockout mice did not show any diabetic features [80]. Furthermore, CD157 exhibits neither cADPR-synthesizing nor -hydrolyzing activity in physiological conditions [81], suggesting that CD157 may play a role as a surface antigen rather than an enzyme in ß-cells.

Mechanism of Intracellular Ca^{2+} Mobilization by cADPR in ß-cells

An increase in the intracellular Ca^{2+} concentration, which then triggers insulin secretion, has conventionally been explained by Ca^{2+} influx from extracellular sources. On the other hand, in 1993, we described a novel mechanism of insulin secretion in which the Ca^{2+} release from the endoplasmic reticulum, an intracellular Ca^{2+} pool, induces insulin secretion. Rojas *et al.* [82] examined the Ca^{2+} influx from extracellular sources and the Ca^{2+} release from the intracellular pool in human β-cells, and showed that 42-75% of the increase in the intracellular Ca^{2+} concentration by glucose stimulation was due to the release of Ca^{2+} from the intracellular stores. Thus, the importance of Ca^{2+} release from the endoplasmic reticulum in insulin secretion by glucose has been recognized.

cADPR is thought to activate the ryanodine receptor to release Ca^{2+} from the intracellular stores, the endoplasmic reticulum [46, 72, 83]. We and others have shown that the type 2 ryanodine receptor is expressed in rat pancreatic islets [59, 84]. Our experiments indicated that cADPR did not bind directly to the ryanodine receptor but may act on the receptor through a mediator such as FK506-binding protein 12.6, FKBP12.6, to release Ca^{2+}[56].

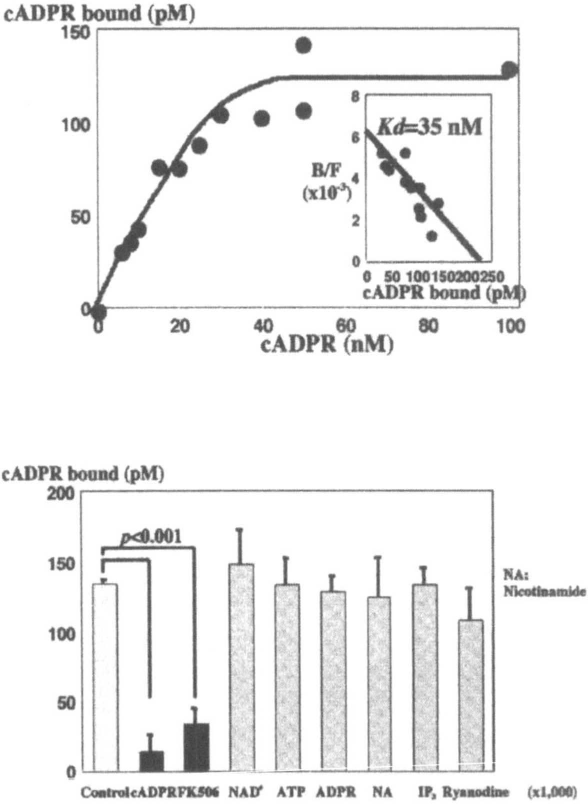

Figure 8. **cADPR binding activity of FKBP12.6** (adapted from [56]).

FK506 is one of the most widely used immunosuppressive agents. The cellular target for FK506 is thought to be FKBP12 and FKBP12.6. Rat FKBP12 is composed of 108 amino acids and is highly conserved among human, mouse, bovine, and rabbit. Rat FKBP12.6 is also a 108-amino acid protein as are human and bovine FKBP12.6. Rat islet microsomes contain FKBP12.6, but do not contain FKBP12. It is of great interest that cADPR was found to bind to FKBP12.6 at a *Kd* value of 35 nM. The binding of radiolabeled cADPR was inhibited by cold FK506 as well as cADPR and neither structurally nor functionally related analogues of cADPR inhibited the cADPR binding to FKBP12.6 [56] (Figure 8). These results indicate that FKBP12.6 acts as a cADPR-binding protein and strongly suggest that cADPR is the actual ligand for FKBP12.6 since FK506 does not normally exist in mammalian cells.

FKBP12.6 is present in rat islet microsomes. However, when rat islet microsomes were treated with cADPR, FKBP12.6 dissociated from the microsomes and moved to the supernatant, releasing Ca^{2+} from the intracellular stores [56]. From these results together with those of other experiments, it is strongly suggested that, when cADPR binds to FKBP12.6

in the ryanodine receptor and causes the dissociation of FKBP12.6 from the ryanodine receptor to form the FKBP12.6-cADPR complex, the channel activity of the ryanodine receptor is thereby increased to release Ca^{2+} from the endoplasmic reticulum.

As shown also in Figure 4, when FK506 is present, cADPR cannot act on the ryanodine receptor to release Ca^{2+} and the glucose-induced insulin secreting machinery ceases to function. In fact, when FK506 was used as an immuno-suppressant in kidney transplantation, hyperglycemia was observed in 20-35 per cent of the recipients [85, 86]. The diabetogenic side effect of FK506 may be explained by the mechanism shown in Figure 4.

Furthermore, in the presence of calmodulin (CaM), islet microsomes were sensitized to cADPR at much lower concentrations for Ca^{2+} release, and the Ca^{2+} release was greatly increased [54, 55]. Inhibitors for CaM-dependent protein kinase II (CaM kinase II) completely abolished the glucose-induced insulin secretion as well as the cADPR-mediated and CaM-activated Ca^{2+} mobilization. Western blot analysis revealed that rat microsomes contain CaM kinase IIα but do not contain CaM. When the active 30 kDa chymotryptic fragment of CaM kinase II was added to the microsomes, fully activated cADPR-mediated Ca^{2+} release was observed in the absence of

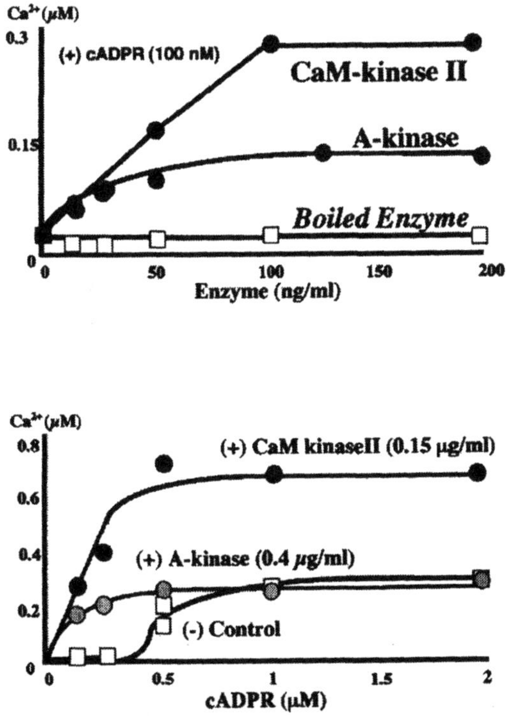

Figure 9. **Sensitization of cADPR-induced Ca^{2+} release from islet microsomes by CaM kinase II and A-kinase** (adapted from [54]).

CaM [53] (Figure 9). These results strongly suggest that CaM kinase II is required to phosphorylate and activate the ryanodine receptor for the cADPR-mediated Ca^{2+} release.

We also found that cyclic AMP-dependent protein kinase (A-kinase) activates the cADPR-mediated Ca^{2+} release from islet microsomes. In the absence of A-kinase, only a small amount of Ca^{2+} was released from the microsomes by low concentrations of cADPR. On the other hand, when the catalytic subunit of A-kinase was added to the islet microsomes, the Ca^{2+} release was sensitized at low concentrations of cADPR (Figure 9). As CaM kinase II was observed to be activated by glucose stimulation and several incretin peptide hormones such as VIP, PACAP, GLP-1, and GIP increase the intracellular concentrations of cAMP to activate A-kinase, it is quite possible the cADPR-mediated Ca^{2+} mobilization for insulin secretion is activated by CaM kinase II and A-kinase (Figure 10). Possibly, the activated kinases phosphorylate the ryanodine receptor to sensitize the Ca^{2+} channel for the cADPR signal.

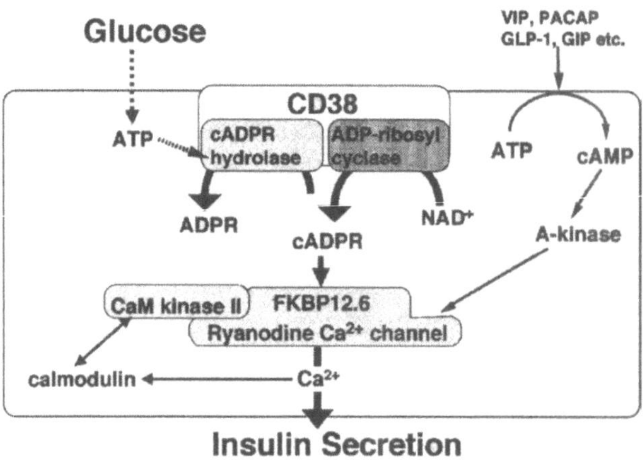

Figure 10. **Possible model for insulin secretion via the CD38-cADPR signal system and its enhancement by CaM kinase II and A-kinase.**

THE CD38-CYCLIC ADP-RIBOSE SIGNAL SYSTEM IN PATHOLOGICAL CONDITIONS

CD38 Knockout Mice

To verify our proposed mechanism of insulin secretion, the CD38/cADPR signal system, we created CD38 transgenic and knockout mice [53, 62]. We produced transgenic mice overexpressing human CD38 in pancreatic β-cells using a rat insulin promoter [53]. In immuno-histochemistry, pancreatic islets of the two transgenic lines were densely stained for human CD38. We isolated islets from the transgenic mouse pancreas. The insulin secretion in transgenic islets was significantly higher than that of the control islets at high concentrations of glucose. When islets were exposed to ketoisocaproate (KIC), which, like glucose, is metabolized to form ATP, the transgenic insulin secretion was significantly higher than the control. However, with tolbutamide and KCl, the transgenic insulin secretion was not altered as compared to the control. Tolbutamide blocks the ATP-sensitive K^+ channel and facilitates Ca^{2+} influx from extracellular sources through voltage-dependent Ca^{2+} channels, and KCl directly induces cell membrane depolarization resulting in Ca^{2+} influx [87]. In glucose-tolerance tests, the serum insulin level of CD38-transgenic mice was higher than that of control mice.

Furthermore, we created CD38 gene-disrupted mice [62]. $CD38^{-/-}$ mice

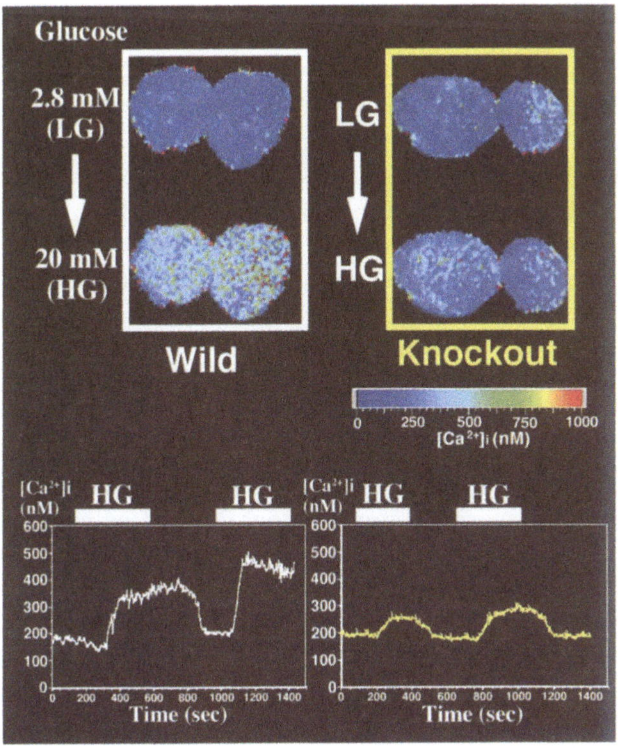

Figure 11. **Glucose-induced $[Ca^{2+}]_i$ changes in $CD38^{+/+}$ and $CD38^{-/-}$ islets** (adapted from [62]). Digital imaging of $[Ca^{2+}]_i$ in the islets and changes of $[Ca^{2+}]_i$ in response to glucose stimulation are shown. Islets were loaded with fura-2 by a 30-min incubation at 37 °C in Krebs-Ringer buffer containing 0.2% bovine serum albumin, 2.8 mM glucose, and 25 μM acetoxymethyl ester of fura-2. Islets were perfused with Krebs-Ringer buffer containing 2.8 mM (LG) or 20 mM (HG) at 37 °C at a rate of 2.5 ml/min. Fura-2 dual excitation (340 and 380 nm) and fluorescence detection (510 nm) were performed, and the 340/380 nm fluorescence ratio was converted to $[Ca^{2+}]_i$ using QuantiCell 900. The interval between successive images of recordings was 4 s.

developed normally but showed no increase in their glucose-induced production of cADPR in pancreatic islets. The glucose-induced $[Ca^{2+}]_i$ rise (Figure 11) and insulin secretion were both severely impaired in $CD38^{-/-}$ islets, whereas $CD38^{-/-}$ islets responded normally to the extracellular Ca^{2+} influx stimulants tolbutamide and KCl. $CD38^{-/-}$ mice showed impaired glucose tolerance, and the serum insulin level was lower than control. The blood glucose levels of $CD38^{-/-}$ mice at 15-60 min after insulin injection were essentially similar to those of $CD38^{+/+}$ mice, suggesting that the glucose intolerance in $CD38^{-/-}$ mice is the result of an attenuation of glucose-induced insulin secretion rather than an alteration in peripheral insulin resistance.

We further tested whether the observed phenotype could be rescued by pancreatic β-cell specific expression of CD38 cDNA. Transgenic mice carrying a human CD38 cDNA under the rat insulin promoter described above were crossed with $CD38^{-/-}$ mice. By inter-crossing their offsprings, $CD38^{-/-}$ mice carrying the human CD38 transgene were generated. The human CD38 transgene ameliorated the glucose intolerance and the decreased insulin secretion, suggesting that the observed phenotype is indeed caused by the absence of CD38 in pancreatic β-cells [62]. Thus, the results obtained with CD38-transgenic and CD38 gene-disrupted mice provide further evidence for the contribution of the CD38/cADPR signaling system to the release of Ca^{2+} from intracellular Ca^{2+} stores for insulin secretion.

CD38 Missense Mutation in Japanese Diabetic Patients

We analyzed the CD38 gene in 31 Japanese type 2 (non-insulin dependent) diabetes patients who had first and/or second-degree relative(s) with type 2 diabetes by single-stranded conformation polymorphism (SSCP) [60]. Two variant patterns were noted in exons 3 and 4. Sequence analysis showed that one of the variants was a silent mutation in the codon for Ile-168 of CD38 from AT<u>A</u> to AT<u>C</u>. The second change resulted in an amino acid substitution in which Arg-140 (<u>C</u>GG) was replaced by Trp (<u>T</u>GG). We screened these two variants in 90 non-diabetic controls with no family history of diabetes and normal glucose tolerance. The frequency of the Ile-168 (ATC) allele was 16% in type 2 diabetic patients as well as in controls. In contrast, the Trp-140 (TGG) allele was observed in 4 of 31 type 2 diabetic patients, but not in any of the controls. Blood samples from the family members of two patients with the Arg140Trp mutation were available and the genotype of the mutation at codon 140 of these subjects are shown in Figure 12. In the Arg140Trp mutant protein expressed in COS-7 cells, both ADP-ribosyl cyclase and cADPR hydrolase activities were reduced to 40-50% of the normal CD38 enzymic activity. The decreased function of the

CD38 mutant may have contributed to the impairment of glucose-stimulated insulin secretion in the type 2 patients.

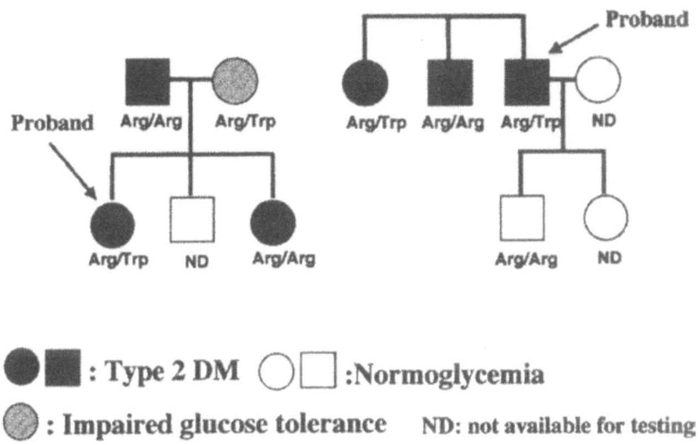

Figure 12. **Pedigree of the subjects with the polymorphism at codon 140**. Male, square; female circle; closed symbols, individuals with type 2 diabetes (DM); diagonal shading, individuals with impaired glucose tolerance; open symbols, individuals with normoglycemia; ND, individuals who were not available for testing; arrow, proband. Genotype is noted under each symbol.

Autoanibodies Against CD38 in Diabetic Patients

We previously reported that a monoclonal antibody against CD38, T16, inhibited the enzymic activity of CD38 [48]. The antibody inhibited the glucose-induced insulin secretion in a dose-dependent manner [61]. Therefore, if anti-CD38 antibodies are present, glucose-induced insulin secretion could be impaired and result in type 2 diabetes. We examined the existence of anti-CD38 antibodies in Japanese diabetic patients by Western blot [61]. Some type 2 diabetic sera reacted with CD38. When the diabetic serum was incubated with CD38 to absorb anti-CD38 antibodies, the signals were abolished, indicating that they contained anti-CD38 antibodies. We then screened sera for autoantibodies against CD38 in 377 Japanese type 2 patients and 75 non-diabetic controls with no family history of diabetes and normal fasting plasma glucose levels. In non-diabetic subjects, the relative antibody values were within the mean + 3 standard deviation (SD) except in one subject. On the other hand, the distribution of relative anti-CD38 values showed two clear peaks in type 2 patients: one was below the mean + 3 SD of non-diabetic controls and the other was over the mean + 3 SD (Figure 13). Therefore, it is reasonable to consider the value of mean + 3 SD (8.02) of non-diabetic controls as a cut off value, and 52 of 377 (13.8%) patients with type 2 diabetes had anti-CD38 antibodies.

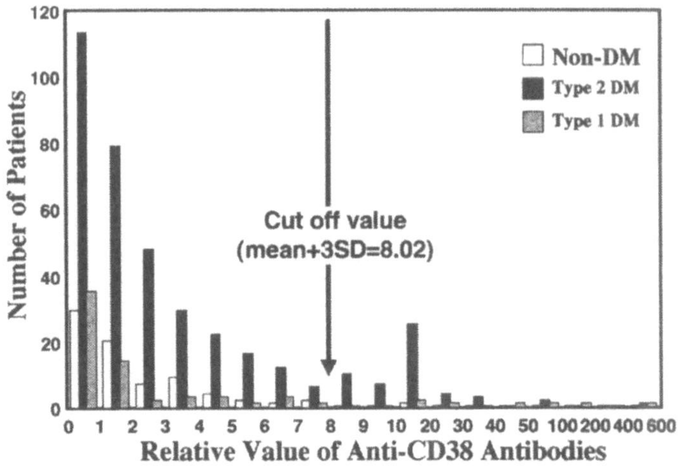

Figure 13. **Relative value of anti-CD38 autoantibodies in type 1 and type 2 diabetic and non-diabetic sera** (adapted from [61]). Arrow indicates mean + 3 SD of non-diabetic control. The levels of CD38 antibodies expressed as an index value relative to a standard serum. Values above mean + 3 SD of the control samples were regarded as antibody positive.

Although the general characteristics did not show significant differences between the anti-CD38 antibody positive and negative type 2 patients, there was an inverse correlation between the age of diabetes onset and the anti-CD38 antibody value [61]. Of 52 anti-CD38 positive diabetic sera, 7 diabetic sera exhibited a positive reaction to rat CD38. We examined the effects of the diabetic sera on glucose-induced insulin secretion from isolated rat islets. All of the 7 diabetic sera with a positive reaction to rat CD38 significantly and dose-dependently inhibited the glucose-induced insulin secretion, whereas the control sera did not exhibit such an inhibitory effect on the glucose-induced insulin secretion [61]. The addition of CD38 abolished the inhibitory effect of the diabetic sera. Furthermore, the cADPR levels in rat islets incubated with the diabetic sera were significantly reduced and the reduction was attenuated by the addition of CD38 [61], suggesting that the inhibition of insulin secretion is due to changes in the cADPR levels regulated by CD38. By the addition of diabetic sera, the ADP-ribosyl cyclase activity of CD38 was significantly inhibited and the cyclic ADP-ribose hydrolase activity of CD38 was rather activated. These results explain well why the diabetic sera inhibited the glucose-induced insulin secretion and strongly suggest that the presence of anti-CD38 autoantibodies in diabetic patients could be one of the major causes of impaired glucose-induced insulin secretion in type 2 diabetes.

Recently, autoantibodies against CD38 were also found in Caucasian diabetic patients and some of their sera showed increased intracellular Ca^{2+} concentrations and glucose-independent insulin secretion from human islets

[88-90]. In some cases, the effects of the antibodies are opposite: Some antibodies show agonistic effects and others antagonistic effects (Figure 14). In CD38 autoantibodies in diabetic patients, differences of the effects between Japanese and Caucasian antibodies may be explained by the nature of the antibodies as described above as well as the difference in species (rat and human) of the islets in the analyses of insulin secretion.

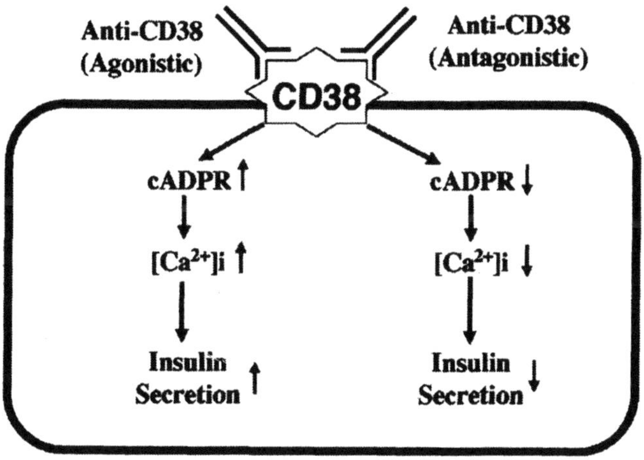

Figure 14. **Effects of autoantibodies against CD38.** Anti-CD38 antibodies found in Japanese type 2 diabetes patients act antagonistic, and antibodies found in Caucasian patients act agonistic.

cADPR and IP_3 as Alternate Second Messengers for Intracellular Ca^{2+} Mobilization in Normal and Diabetic ß-cells

The CD38/cADPR signal system for insulin secretion is different from the conventional hypothesis [47] in which Ca^{2+} influx from extracellular sources plays a role in insulin secretion by glucose. Furthermore, the CD38/cADPR signal system is also different from the hypothesis proposed by Berridge and Irvine [91], in which IP_3 induces Ca^{2+} release from the intracellular pool, the endoplasmic reticulum. In this context, the CD38/cADPR signal system was the focus of intense debate [7, 64-67]. Controversial results were observed using diabetic ß-cells such as *ob/ob* mouse islets and RINm5F cells. We found that the Ca^{2+} release responses of these diabetic ß-cell microsomes were quite different from those of normal islet microsomes [59]. Microsomes from normal C57BL mouse islets released Ca^{2+} in response to cADPR but scarcely in response to IP_3. This response to cADPR was completely attenuated by the prior addition of 8-amino-cADPR, an antagonist of cADPR [92] (Figure 15, inset). In contrast to normal islet microsomes, *ob/ob* mouse islet microsomes released only a small amount of Ca^{2+} by cADPR but released much Ca^{2+} by IP_3. RINm5F cell microsomes responded well to IP_3 to release Ca^{2+} but did not respond to cADPR (Figure 15). RINm5F cells are rat insulinoma-derived immortal cells and show almost no glucose-induced insulin secreting ability.

Concerning intracellular Ca^{2+} release channels, the mRNA expression of type 2 rynodine receptor (RyR-2), which is postulated to be a Ca^{2+} release channel for cADPR [56, 83], was clearly detected in normal islets but not in *ob/ob* islets. In contrast, IP_3 receptor (IP_3R-1, IP_3R-2, IP_3R-4 and IP_3R-5) mRNAs were not detected in normal islets but were clearly detected in *ob/ob* islets, and although IP_3R-3 mRNA was slightly detected in normal islets, the mRNA was significantly increased in *ob/ob* islets (Figure 15), well fitting with the observation that IP_3-induced Ca^{2+} mobilization preferentially worked in *ob/ob* islet microsomes.

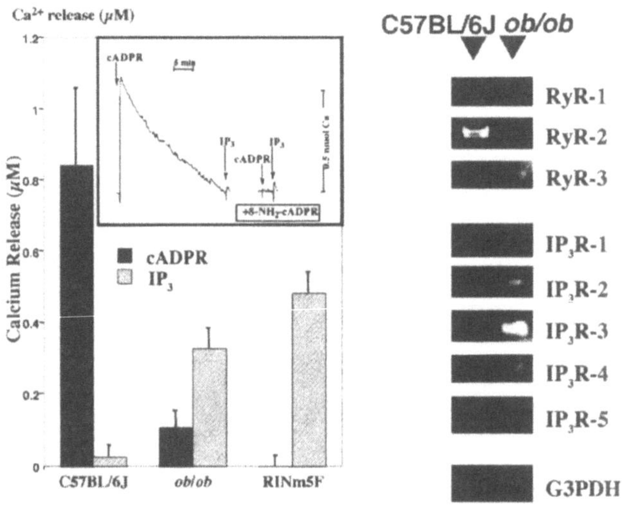

Figure 15. **Changes of intracellular Ca^{2+} mobilization of C57BL/6J, ob/ob islets and RINm5F ß-cells and expression of the Ca^{2+} mobilizing channels in C57BL/6J and ob/ob islets** (adapted from [59]). Ca^{2+} release from microsomes (10 μg protein) was induced by the addition of 100 nM cADPR or 1 μM IP_3 in the presence of 7 μg/ml calmodulin. Values are mean± S.E. of triplicate experiments. The inset shows a typical result of Ca^{2+} release from C57BL/6J mouse islet microsomes.

Furthermore, the CD38 mRNA level was decreased significantly in *ob/ob* islets [59], and in RINm5F cells, CD38 mRNA was not expressed [49] (Figure 16). The decrease of CD38 mRNA in *ob/ob* islets may explain the low response in the cADPR content of ß-cells by glucose stimulation because CD38 has both the ADP-ribosyl cyclase and the cADPR hydrolase activities, and the increase of cADPR by glucose stimulation is achieved by inhibition of the cADPR hydrolase activity of CD38. Decreased CD38 mRNA was also reported in islets of Goto-Kakizaki (GK) diabetic rats [93], which show impaired glucose-induced insulin secretion. These results indicate that the CD38/cADPR signal system for insulin secretion is used under normal physiological conditions, and is replaced by the IP_3 system in diabetic ß-cells such as *ob/ob* mouse islets and RINm5F cells. In fact, it was more recently observed that normal Balb/c mouse islets showed distinct increases in intracellular cADPR, Ca^{2+}, and insulin secretion by glucose [70], and that MIN6 cells showed a dramatic Ca^{2+} mobilization in response to cADPR [71, 72] via the ryanodine receptor despite the fact that no response to IP_3 was observed [72].

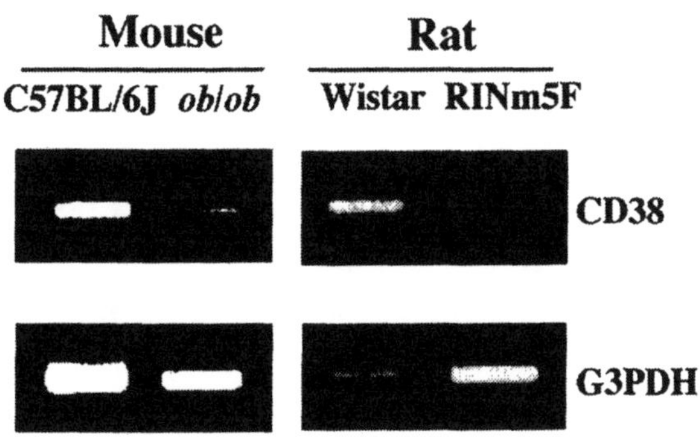

Figure 16. **Expression of CD38 mRNA in normal (C57BL/6J and Wistar islet) and diabetic cells (ob/ob islet and RINm5F).**

Although IP_3 is thought to be a second messenger for Ca^{2+} mobilization from intracellular stores, we first observed in 1993 that cADPR induces Ca^{2+} release from pancreatic islet microsomes but IP_3 does not [46]. In cerebellum microsomes, both cADPR and IP_3 induced Ca^{2+} release. Recently, various physiological phenomena from animal to plant cells become understandable in terms of the CD38/cADPR signal system as described in this chapter [94-118].

It is of great interest that, in pancreatic acinar cells of CD38-knockout mice, the acetylcholine-induced Ca^{2+} oscillation was greatly reduced or completely disappeared under a physiological concentration of acetylcholine [117] (Figure 17). Furthermore, acetylcholine induced the cADPR formation in normal acinar cells, but not in CD38 knockout acinar cells. The IP_3 formation was very small in the presence of a physiological concentration of acetylcholine and there was no difference between normal and CD38 knockout cells [117]. Probably, acetylcholine induces the cADPR formation via the G-protein coupled CD38 system [119] (Figure 18). Therefore, cells can utilize at least two second messengers, IP_3 and cADPR, for Ca^{2+} mobilization, depending on the species of cells as well as differences in cellular conditions, physiological or pathological, performing a variety of cellular functions (Figure 19). In pancreatic ß-cells, glucose is metabolized and induces the CD38-cADPR system to secrete insulin. In many other cells, hormones and neurotransmitters may regulate the CD38-cADPR signal system in a receptor-coupled manner, such as in a G-protein-coupled manner, to express various physiological responses (Figure 18).

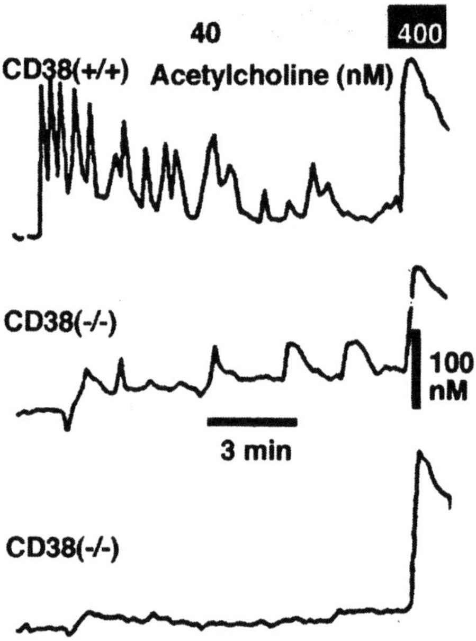

Figure 17. **Acetylcholine-evoked Ca^{2+} responses from single $CD38^{+/+}$ and $CD38^{-/-}$ pancreatic acinar cells** (adapted from [117]).

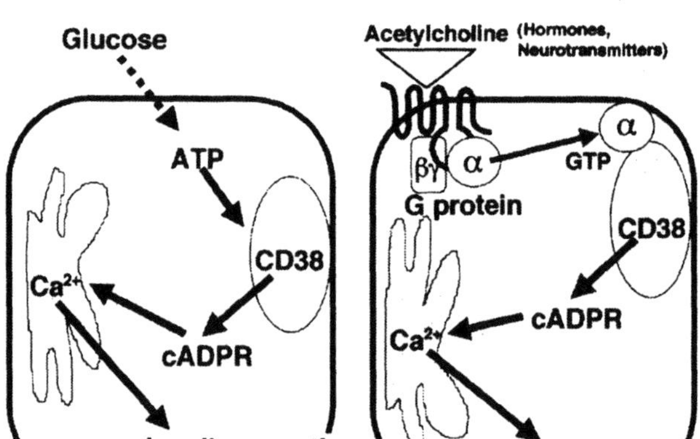

Figure 18. **The CD38-cADPR signal system by glucose (Metabolic type, left) and by acetylcholine (Receptor-G protein coupled type, right).** Glucose is transported into ß-cells and metabolized to generate ATP. ATP regulates cADPR hydrolase of CD38 to increase cADPR. Acetylcholine and other hormones/neurotransmitters regulate the enzymic activity of CD38 to increase cADPR via the receptor-G protein coupled system.

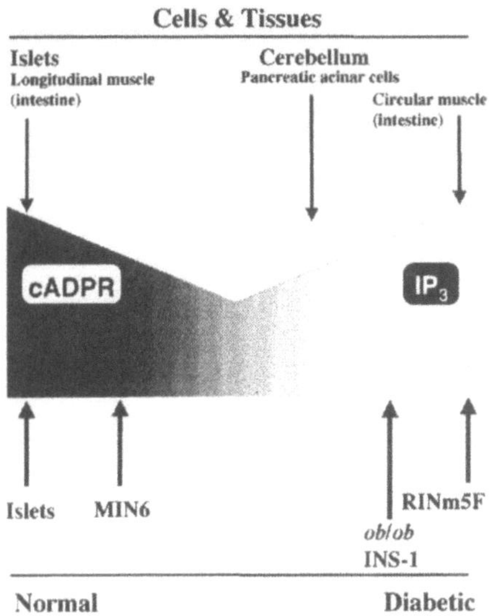

Figure 19. **Alternative usage of two Ca^{2+} mobilizing second messengers depending on differences in the cell types and on physiological/pathological conditions.**

NAD$^+$ METABOLISM IN BETA-CELLS AND CONCLUDING REMARKS

In the first part of this chapter, we described that PARP activation causes NAD$^+$ depletion to form poly(ADP-ribose), resulting in necrotic ß-cell death. More recently, accumulating evidence has confirmed that the mechanism proposed for ß-cell death is involved in the process of the cell death of many tissues and cells. Recently, we found that PARP acts as a transcription factor for a ß-cell regeneration factor gene, *Reg* (**Re**generating **g**ene) [120-125] and that the active transcriptional complex for *Reg* gene is not formed when PARP is activated and auto-poly(ADP-ribosyl)ated [126] (Figure 20). Accumulating evidence indicates that Reg proteins [123] also act as growth factors for Schwann cells, neuronal cells, and gastrointestinal cells [127-131]. As mainly described in this chapter, the cADPR formation from NAD$^+$ is essential for insulin secretion by glucose in ß-cells as well as for various physiological phenomena from animal to plant cells. Thus, NAD$^+$ metabolism by PARP activation results in at least three important events in pancreatic ß-cells (cell death, regeneration, and insulin secretion) and possibly in many other cells (Figure 21). Therefore, PARP inhibitors prevent the necrotic cell death, keep PARP active as a transcription factor for cell regeneration, and maintain the formation of a second messenger, cADPR, to achieve the cell function.

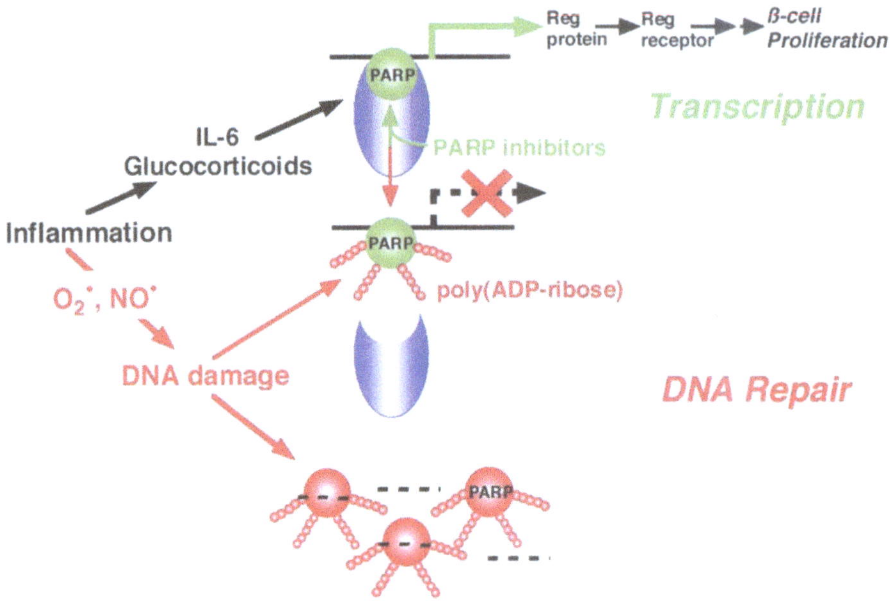

Figure 20. **Representation of the unified role of PARP in the Reg gene transcription and DNA repair** (adapted from [126]). β-cells are affected by many agents such as immunological abnormalities, virus infections, irradiation, and chemical substances (see Figure 1), leading to local inflammation in and/or around pancreatic islets. Inflammatory mediators such as IL-6 and glucocorticoids are produced in the inflammation process. IL-6/glucocorticoid stimulation induces the formation of an active transcriptional complex for Reg, a β -cell regenerating factor gene, in which PARP is involved, and when the PARP is not poly(ADP-ribosyl)ated in the presence of PARP inhibitors, the transcriptional complex is stabilized and the Reg gene transcription is maintained. Reg protein then produced in β -cells acts as an autocrine/paracrine growth factor on β -cells via the Reg receptor. DNA replication in β -cells thus occurs and the β -cell regeneration is achieved. DNA damaging substances such as superoxide (O_2•) and nitric oxide (NO•) are frequently produced in inflammatory processes. When the DNA is damaged, PARP senses DNA nicks and auto-poly(ADP-ribosyl)ates itself for the DNA repair. Once PARP is auto-poly(ADP-ribosyl)ated, the formation of the Reg gene transcriptional complex is inhibited, interfering with the interaction between PARP and other nuclear proteins necessary for the active complex, and therefore the transcription of Reg gene stops. When DNA is massively damaged, PARP is rapidly activated to repair the DNA and the complex for Reg gene transcription is not formed at all.

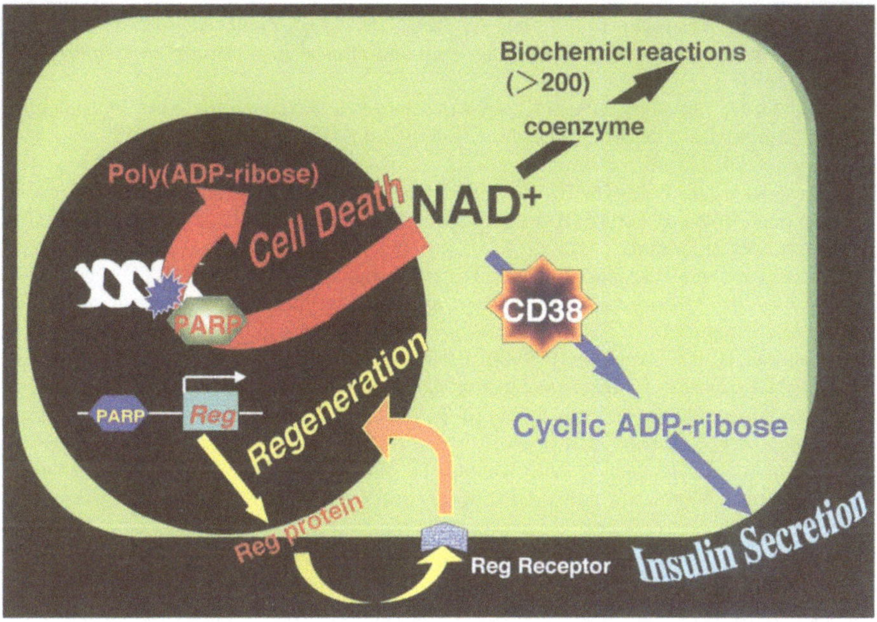

Figure 21. **Importance of NAD$^+$ metabolism in β-cell death, regeneration, and functioning.** The mechanism depicted in this model may explain much more cellular phenomena in many other cells.

ACKNOWLEDGEMENTS

The authors thank Brent Bell for valuable assistance in preparing the manuscript for publication. This work was supported in part by grants-in-aid from the Ministry of Education, Culture, Sports, Science, Sports and Technology, Japan.

REFERENCES

1. Okamoto H. 1981. Regulation of proinsulin synthesis in pancreatic islets and a new aspect to insulin-dependent diabetes. *Mol. Cell. Biochem.* 37: 43-61.
2. Okamoto H. 1985. Molecular basis of experimental diabetes: degeneration, oncogenesis, and regeneration of pancreatic B-cells of islets of Langerhans. *BioEssays* 2: 15-21.
3. Okamoto H, Yamamoto H, Takasawa S, Inoue C, Terazono K, Shiga K and Kitagawa M. 1988. Molecular mechanism of degeneration, oncogenesis and regeneration of pancreatic B-cells of islets of Langerhans. In *Lessons from Animal Diabetes II*, ed. E Shafrir, AE Renold, pp. 149-157. London: John Libbey & Company Ltd.
4. Okamoto H. 1990. The molecular basis of experimental diabetes. In *Molecular biology of the islets of Langerhans*, ed. H Okamoto, pp.209-231. Cambridge: Cambridge University Press.
5. Okamoto H, Takasawa S and Tohgo A. 1995. New aspects of the physiological significance of NAD, poly ADP-ribose and cyclic ADP-ribose. *Biochimie* 77: 356-363.
6. Okamoto H. 1996. Okamoto model for B-cell damage: Recent advances. In *Lessons from animal diabetes VI*, ed. E Shafrir, pp. 97-111. Boston: Birkhäuser.

7. Okamoto H, Takasawa S and Nata K. 1997. The CD38-cyclic ADP-ribose signalling system in insulin secretion: Molecular basis and clinical implications. *Diabetologia* 40: 1485-1491.
8. Okamoto H, Takasawa S, Nata K, Kato I, Tohgo A and Noguchi N. 2000. Physiological and pathological significance of the CD38-cyclic ADP-ribose signaling system. *Chem. Immunol.* 75: 121-45.
9. Takasawa S and Okamoto H. 2002. Pancreatic ß-Cell death, regeneration and insulin secretion: Roles of poly(ADP-ribose) polymerase and cyclic ADP-ribose. *Int. J. Exp. Diabet. Res.* (in press).
10. Okamoto H and Takasawa S. 2002. Recent advances in the OKAMOTO model: The CD38-cyclic ADP-ribose signal system and the Reg-Reg receptor system in ß-cells. *Diabetes* (in press).
11. Yamamoto H and Okamoto H. 1980. Protection by picolinamide, a novel inhibitor of poly(ADP-ribose) synthetase, against both streptozotocin-induced depression of proinsulin synthesis and reduction of NAD content in pancreatic islets. *Biochem. Biophys. Res. Commun.* 95: 474-481.
12. Yamamoto H, Uchigata Y and Okamoto H. 1981. Streptozotocin and alloxan induce DNA strand breaks and poly(ADP-ribose) synthetase in pancreatic islets. *Nature* 294: 284-286.
13. Uchigata Y, Yamamoto H. Kawamura A and Okamoto H. 1982. Protection by superoxide dismutase, catalase, and poly(ADP-ribose) synthetase inhibitors against alloxan- and streptozotocin-induced islet DNA strand breaks and against the inhibition of proinsulin synthesis. *J. Biol. Chem.* 257: 6084-6088.
14. Uchigata Y, Yamamoto H, Nagai H and Okamoto H. 1983. Effect of poly(ADP-ribose) synthetase inhibitor administration to rats before and after injection of alloxan and streptozotocin on islet proinsulin synthesis. *Diabetes* 32: 316-318.
15. Shima K. Hirota M, Sato M, Numoto S and Oshima I. 1987. Effect of poly(ADP-ribose) synthetase inhibitor administration to streptozotocin-induced diabetic rats on insulin and glucagon contents in their pancreas. *Diabetes Res. Clin. Practice* 3: 135-142.
16. Takamura T, Kato I, Kimura N, Nakazawa T, Yonekura H, Takasawa S and Okamoto H. 1998. Transgenic mice overexpressing type 2 nitric oxide synthase in pancreatic ß cells develop insulin-dependent diabetes without insulitis. *J. Biol. Chem.* 273: 2493-2496.
17. Charron MJ and Bonner-Weir S. 1999. Implicating PARP and NAD^+ depletion in type I diabetes. *Nature Med.* 5: 269-270
18. Burkart V, Wang ZQ, Radons J, Heller B, Herceg Z, Stingl L, Wagner EF and Kolb H. 1999. Mice lacking the poly(ADP-ribose) polymerase gene are resistant to pancreatic beta-cell destruction and diabetes development induced by streptozocin. *Nature Med.* 5: 314-319.
19. Masutani M, Suzuki H, Kamada N, Watanabe M, Ueda O, Nozaki T, Jishage K, Watanabe T, Sugimoto T, Nakagama H, Ochiya T and Sugimura T. 1999. Poly(ADP-ribose) polymerase gene disruption conferred mice resistant to streptozotocin-induced diabetes. *Proc. Natl. Acad. Sci. USA* 96: 2301-2304.
20. Pieper AA. Brat DJ. Krug DK, Watkins CC, Gupta A, Blackshaw S, Verma A, Wang ZQ and Snyder SH. 1999. Poly(ADP-ribose) polymerase-deficient mice are protected from streptozotocin-induced diabetes. *Proc. Natl. Acad. Sci. USA* 96: 3059-3064.
21. Eliasson MJ, Sampei K, Mandir AS, Hurn PD, Traystman RJ, Bao J, Pieper A, Wang ZQ, Dawson TM, Snyder SH and Dawson VL. 1997. Poly(ADP-ribose) polymerase gene disruption renders mice resistant to cerebral ischemia. *Nature Med.* 3: 1089-1095.
22. Szabó C, Cuzzocrea S, Zingarelli B, O'Connor M and Salzman AL. 1997. Endothelial dysfunction in a rat model of endotoxic shock: Importance of the activation of poly(ADP-ribose) synthetase by peroxynitrite. *J. Clin. Invest.* 100: 723-735.

23. Zingarelli B, Salzman AL and Szabo C. 1998. Genetic disruption of poly (ADP-ribose) synthetase inhibits the expression of P-selectin and intercellular adhesion molecule-1 in myocardial ischemia/reperfusion injury. *Circ. Res.* 83: 85-94.
24. Szabó C, Virág L, Cuzzocrea S, Scott GS, Hake P, O'Connor MP, Zingarelli B, Salzman A and Kun E. 1998. Protection against peroxynitrite-induced fibroblast injury and arthritis development by inhibition of poly(ADP-ribose) synthetase. *Proc. Natl. Acad. Sci. USA* 95: 3867-3872.
25. Mandir AS, Przedborski S, Jackson-Lewis V, Wang ZQ, Simbulan-Rosenthal CM, Smulson ME, Hoffman BE, Guastella DB, Dawson VL and Dawson TM. 1999. Poly(ADP-ribose) polymerase activation mediates 1-methyl-4-phenyl-1,2,3,6-tetrahydropyridine (MPTP)-induced parkinsonism. *Proc. Natl. Acad. Sci. USA* 96: 5774-5779.
26. Stern Y, Salzman A, Cotton RT and Zingarelli B. 1999. Protective effect of 3-aminobenzamide, an inhibitor of poly(ADP-ribose) synthetase, against laryngeal injury in rats. *Am. J. Resir. Crit. Care Med.* 160: 1743-1749.
27. Bowes J, McDonald MC, Piper J and Thiemermann C. 1999. Inhibitors of poly(ADP-ribose) synthetase protect rat cardiomyocytes against oxidant stress. *Cardiovasc. Res.* 41: 126-134.
28. Filipovic DM, Meng X and Reeves WB. 1999. Inhibition of PARP prevents oxidant-induced necrosis but not apoptosis in LLC-PK1 cells. *Am. J. Physiol.* 277: F428-F436.
29. Oliver FJ, Ménissier-de Murcia J, Nacci C, Decker P, Andriantsitohaina R, Muller S, de la Rubia G, Stoclet JC and de Murcia G. 1999. Resistance to endotoxic shock as a consequence of defective NF-κB activation in poly(ADP-ribose) polymerase-1 deficient mice. *EMBO J.* 18: 4446-4454.
30. Zingarelli B, Szabó C and Salzman AL. 1999. Blockade of poly(ADP-ribose) synthetase inhibits neutrophil recruitment, oxidant generation, and mucosal injury in murine colitis. *Gastroenterology* 116: 335-345.
31. Tsao BP, Cantor RM, Grossman JM, Shen N, Teophilov NT, Wallace DJ, Arnett FC, Hartung K, Goldstein R, Kalunian KC, Hahn BH and Rotter JI. 1999. PARP alleles within the linked chromosomal region are associated with systemic lupus erythematosus. *J. Clin. Invest.* 103:1135-1140.
32. Plaschke K, Kopitz J, Weigand MA, Martin E and Bardenheuer HJ. 2000. The neuroprotective effect of cerebral poly(ADP-ribose) polymerase inhibition in a rat model of global ischemia. *Neurosi. Lett.* 284: 109-112.
33. Ducrocq S, Benjelloun N, Plotkine M, Ben-Ari Y and Charriaut-Marlangue C. 2000. Poly(ADP-ribose) synthase inhibition reduces ischemic injury and inflammation in neonatal rat brain. *J. Neurochem.* 74: 2504-2511.
34. Mandir AS, Poitras MF, Berliner AR, Herring WJ, Guastella DB, Feldman A, Poirier GG, Wang ZQ, Dawson TM and Dawson VL. 2000. NMDA but not non-NMDA excitotoxicity is mediated by Poly(ADP-ribose) polymerase. *J. Neurosci.* 20: 8005-8011.
35. Pieper AA, Walles T, Wei G, Clements EE, Verma A, Snyder SH and Zweier JL. 2000. Myocardial postischemic injury is reduced by polyADPribose polymerase-1 gene disruption. *Mol. Med.* 6: 271-282.
36. Liaudet L, Soriano FG, Szabó E, Virag L, Mabley JG, Salzman AL and Szabó C. 2000. Protection against hemorrhagic shock in mice genetically deficient in poly(ADP-ribose)polymerase. *Proc. Natl. Acad. Sci. USA* 97: 10203-10208.
37. McDonald MC, Mota-Filipe H, Wright JA, Abdelrahman M, Threadgill MD, Thompson AS and Thiemermann C. 2000. Effects of 5-aminoisoquinolinone, a water-soluble, potent inhibitor of the activity of poly(ADP-ribose) polymerase on the organ injury and dysfunction caused by haemorrhagic shock. *Br. J. Pharmacol.* 130: 843-850.
38. Jijon HB, Churchill T, Malfair D, Wessler A, Jewell LD, Parsons HG and Madsen KL. 2000. Inhibition of poly(ADP-ribose) polymerase attenuates inflammation in a model of chronic colitis. *Am. J. Physiol.* 279: G641-G651.

39. Martin DR, Lewington AJ, Hammerman MR and Padanilam BJ. 2000. Inhibition of poly(ADP-ribose) polymerase attenuates ischemic renal injury in rats. *Am. J. Physiol.* 279: R1834-R1840.
40. Soriano FG, Virág L, Jagtap P, Szabó E, Mabley JG, Liaudet L, Marton A, Hoyt DG, Murthy KGK, Salzman AL, Southan GJ and Szabó C. 2001. Diabetic endothelial dysfunction: the role of poly(ADP-ribose) polymerase activation. *Nature Med.* 7: 108-113.
41. Pacher P, Liaudet L, Soriano FG, Mabley JG, Szabó É and Szabó C. 2002. The role of poly(ADP-ribose) polymerase activation in the development of myocardial and endothelial dysfunction in diabetes. *Diabetes* 51: 514-521.
42. Germain M, Scovassi I and Poirier GG. 2000. Role of poly(ADP-ribose) polymerase in apoptosis. In *Cell Death—The Role of Poly(ADP-ribose) polymerase*, ed. C. Szabó, pp. 209-225. Boca Raton: CRC Press.
43. Sauter B, Albert ML, Francisco L, Larsson M, Somersan S and Bhardwaj N. 2000. Consequences of cell death: Exposure to necrotic tumor cells, but not primary tissue cells or apoptotic cells, induces the maturation of immunostimulatory dendritic cells. *J. Exp. Med.* 191: 423-433.
44. Clapper DL, Walseth TF, Dargie PJ and Lee HC. 1987. Pyridine nucleotide metabolites stimulate calcium release from sea urchin egg microsomes desensitized to inositol trisphosphate. *J. Biol. Chem.* 262: 9561-9568.
45. Okamoto H, Takasawa S, Nata K and Yonekura H. 1992. Cyclic ADP-ribose, a novel second messenger for intracellular Ca^{++} mobilization in pancreatic islets. In *First IUBMB Conference: Biochemistry of Diseases*, ed. I Yamashina, pp. 218. Nagoya: Bridge Ltd.
46. Takasawa S, Nata K, Yonekura H and Okamoto H. 1993. Cyclic ADP-ribose in insulin secretion from pancreatic β cells. *Science* 259: 370-373.
47. Ashcroft FM, Harrison DE and Ashcroft SJH. 1984. Glucose induced closure of single potassium channels in isolated rat pancreatic ß-cells. *Nature* 312: 446-448.
48. Takasawa S, Tohgo A, Noguchi N, Koguma T, Nata K, Sugimoto T, Yonekura H and Okamoto H. 1993. Synthesis and hydrolysis of cyclic ADP-ribose by human leukocyte antigen CD38 and inhibition of the hydrolysis by ATP. *J. Biol. Chem.* 268: 26052-26054.
49. Koguma T, Takasawa S, Tohgo A, Karasawa T, Furuya Y, Yonekura H and Okamoto H. 1994. Cloning and characterization of cDNA encoding rat ADP-ribosyl cyclase/cyclic ADP-ribose hydrolase (homologue to human CD38) from islets of Langerhans. *Biochim. Biophys. Acta* 1223: 160-162.
50. Tohgo A, Takasawa S, Noguchi N, Koguma T, Nata K, Sugimoto T, Furuya Y, Yonekura H and Okamoto H. 1994. Essential cysteine residues for cyclic ADP-ribose synthesis and hydrolysis by CD38. *J. Biol. Chem.* 269: 28555-28557.
51. Nakagawara K, Mori M, Takasawa S, Nata K, Takamura T, Berlova A, Tohgo A, Karasawa T, Yonekura H, Takeuchi T and Okamoto H. 1995. Assignment of CD38, the gene encoding human leukocyte antigen CD38 (ADP-ribosyl cyclase/cyclic ADP-ribose hydrolase), to chromosome 4p15. *Cytogenet. Cell Genet.* 69: 38-39.
52. Nata K, Sugimoto T, Tohgo A, Takamura T, Noguchi N, Matsuoka A, Numakunai T, Shikama K, Yonekura H, Takasawa S and Okamoto H. 1995. The structure of the *Aplysia kurodai* gene encoding ADP-ribosyl cyclase, a second-messenger enzyme. *Gene* 158: 213-218.
53. Kato I, Takasawa S, Akabane A, Tanaka O, Abe H, Takamura T, Suzuki Y, Nata K, Yonekura H, Yoshimoto T and Okamoto H. 1995. Regulatory role of CD38 (ADP-ribosyl cyclase/cyclic ADP-ribose hydrolase) in insulin secretion by glucose in pancreatic ß cells: Enhanced insulin secretion in CD38 transgenic mice. *J. Biol. Chem.* 270: 30045-30050.
54. Takasawa S, Ishida A, Nata K, Nakagawa K, Noguchi N, Tohgo A, Kato I, Yonekura H, Fujisawa H and Okamoto H. 1995. Requirement of calmodulin-dependent protein kinase

II in cyclic ADP-ribose-mediated intracellular Ca^{2+} mobilization. *J. Biol. Chem.* 270: 30257-30259.

55. Kato I, Suzuki Y, Akabane A, Yonekura H, Tanaka O, Kondo H, Takasawa S, Yoshimoto T and Okamoto H. 1996. Enhancement of glucose-induced insulin secretion in transgenic mice overexpressing human VIP gene in pancreatic ß cells. *Ann. NY Acad. Sci.* 805: 232-243.
56. Noguchi N, Takasawa S, Nata K, Tohgo A, Kato I, Ikehata F, Yonekura H and Okamoto H. 1997. Cyclic ADP-ribose binds to FK506-binding protein 12.6 to release Ca^{2+} from islet microsomes. *J. Biol. Chem.* 272: 3133-3136.
57. Tohgo A, Munakata H, Takasawa S, Nata K, Akiyama T, Hayashi, N and Okamoto H. 1997. Lysine 129 of CD38 (ADP-ribosyl cyclase/cyclic ADP-ribose hydrolase) participates in the binding of ATP to inhibit the cyclic ADP-ribose hydrolase. *J. Biol. Chem.* 272: 3879-3882.
58. Nata K, Takamura T, Karasawa T, Kumagai T, Hashioka W, Tohgo A, Yonekura H, Takasawa S, Nakamura S and Okamoto H. 1997. Human gene encoding CD38 (ADP-ribosyl cyclase/cyclic ADP-ribose hydrolase): organization, nucleotide sequence and alternative splicing. *Gene* 186: 285-292.
59. Takasawa S, Akiyama T, Nata K, Kuroki M, Tohgo A, Noguchi N, Kobayashi K, Kato I, Katada T and Okamoto H. 1998. Cyclic ADP-ribose and inositol 1,4,5-trisphosphate as alternate second messengers for intracellular Ca^{2+} mobilization in normal and diabetic ß-cells. *J. Biol. Chem.* 273: 2497-2500.
60. Yagui K, Shimada F, Miura M, Hashimoto N, Suzuki Y, Tokuyama Y, Nata K, Tohgo A, Ikehata F, Takasawa S, Okamoto H, Makino H, Saito Y and Kanatsuka A. 1998. A missense mutation in the CD38 gene, a novel factor for insulin secretion: Association with Type II diabetes mellitus in Japanese subjects and evidence of abnormal function when expressed in vitro. *Diabetologia* 41: 1024-1028.
61. Ikehata F, Satoh J, Nata K, Tohgo A, Nakazawa T, Kato I, Kobayashi S, Akiyama T, Takasawa S, Toyota T and Okamoto H. 1998. Autoantibodies against CD38 (ADP-ribosyl cyclase/cyclic ADP-ribose hydrolase) which impair glucose-induced insulin secretion in non-insulin dependent diabetes patients. *J. Clin. Invest.* 102: 395-401.
62. Kato I, Yamamoto Y, Fujimura M, Noguchi N, Takasawa S and Okamoto H. 1999. CD38 disruption impairs glucose-induced increases in cyclic ADP-ribose, $[Ca^{2+}]_i$ and insulin secretion. *J. Biol. Chem.* 274: 1869-1872.
63. Okamoto H and Takasawa S. 2002. CD38. In *Encyclopedia of Molecular Medicine*, ed. TE Creighton, pp. 601-604. New York: John Wiley & Sons, Inc.
64. Islam MS, Larsson O, Berggren PO, Takasawa S, Nata K, Yonekura H and Okamoto H, Galione A. 1993. Cyclic ADP-ribose in ß cells. *Science* 262: 584-586.
65. Rutter GA, Theler J-M and Wollheim CB. 1994. Ca^{2+} stores in insulin-secreting cells: lack of effect of cADP ribose. *Cell Calcium* 16: 71-80.
66. Webb D-L, Islam MS, Efanov AM, Brown G, Kohler M, Larsson O and Berggren P-O. 1996. Insulin exocytosis and glucose-mediated increase in cytoplasmic free Ca^{2+} concentration in the pancreatic ß-cell are independent of cyclic ADP-ribose. *J. Biol. Chem.* 271: 19074-19079.
67. Islam MS and Berggren PO. 1997. Cyclic ADP-ribose and the pancreatic beta cell: where do we stand? *Diabetologia* 40: 1480-1484.
68. Malaisse WJ, Kanda Y, Inageda K, Scruel O, Sener A and Katada T. 1997. Cyclic ADP-ribose measurements in rat pancreatic islets. *Biochem. Biophys. Res. Commun.* 231: 546-548.
69. Scruel O, Wada T, Kontani K, Sener A, Katada T and Malaisse WJ. 1998. Effects of D-glucose and starvation upon the cyclic ADP-ribose content of rat pancreatic islets. *Biochem. Mol. Biol. Int.* 45: 783-790.

70. An NH, Han MK, Um C, Park BH, Park BJ, Kim HK and Kim UH. 2001. Significance of ecto-cyclase activity of CD38 in insulin secretion of mouse pancreatic islet cells. *Biochem. Biophys. Res. Commun.* 282: 781-786.
71. Varadi A and Rutter GA. 2002. Dynamic imaging of endoplasmic reticulum Ca^{2+} concentration in insulin-secreting MIN6 cells using recombinant targeted cameleons: Roles of sarco(endo)plasmic reticulum Ca^{2+}-ATPase (SERCA)-2 and ryanodine receptors. *Diabetes* 51: S190-S201.
72. Mitchell KJ, Pinton P, Varadi A, Tacchetti C, Ainscow EK, Pozzan T, Rizzuto R and Rutter GA. 2001. Dense core secretory vesicles revealed as a dynamic Ca^{2+} store in neuroendocrine cells with a vesicle-associated membrane protein aequorin chimaera. *J. Cell Biol.* 155: 41-51.
73. Howard M, Grimaldi JC, Bazan JF, Bazan JF, Lund FE, Santos-Argumedo L, Parkhouse RM, Walseth TF and Lee HC. 1993. Formation and hydrolysis of cyclic ADP-ribose catalyzed by lymphocyte antigen CD38. *Science* 262: 1056-1059.
74. Zocchi E, Franco L, Guida L, Benatti U, Bargellesi A, Malavasi F, Lee HC and De Flora A. 1993. A single protein immunologically identified as CD38 displays NAD^+ glycohydrolase, ADP-ribosyl cyclase and cyclic ADP-ribose hydrolase activities at the outer surface of human erythrocytes. *Biochem. Biophys. Res. Commun.* 196: 1459-1465.
75. Kaisho T, Ishikawa J, Oritani K, Inazawa J, Tomizawa H, Muraoka O, Ochi T and Hirano T. 1994. BST-1, a surface molecule of bone marrow stromal cell lines that facilitates pre-B-cell growth. *Proc. Natl. Acad. Sci. USA* 91: 5325-5329.
76. Muraoka O, Tanaka H, Itoh M, Ishihara K and Hirano T. 1996. Genomic structure of human BST-1. *Immunology Lett.* 54: 1-4.
77. Dong C, Willerford D, Alt FW and Cooper MD. 1996. Genomic organization and chromosomal localization of the mouse Bp3 gene, a member of the CD38/ADP-ribosyl cyclase family. *Immunogenetics* 45: 35-43.
78. Furuya Y, Takasawa S, Yonekura H, Tanaka T, Takahara J and Okamoto, H. 1995. Cloning of a cDNA encoding rat bone marrow stromal cell antigen 1 (BST-1) from the islets of Langerhans. *Gene* 165: 329-330.
79. Kajimoto Y, Miyagawa J, Ishihara K, Okuyama Y, Fujitani Y, Itoh M, Yoshida H, Kaisho T, Matsuoka T, Watada H, Hanafusa T, Yamasaki Y, Kamada T, Matsuzawa Y and Hirano T. 1996. Pancreatic islet cells express BST-1, a CD38-like surface molecule having ADP-Ribosyl cyclase activity. *Biochem. Biophys. Res. Commun.* 219: 941-946.
80. Itoh M, Ishihara K, Hiroi T, Lee BO, Maeda H, Iijima H, Yanagita M, Kiyono H and Hirano T. 1998. Deletion of bone marrow stromal cell antigen-1 (CD157) gene impaired systemic thymus independent-2 antigen-induced IgG3 and mucosal TD antigen-elicited IgA responses. *J. Immunol.* 161: 3974-3983.
81. Hirata Y, Kimura N, Sato K, Ohsugi Y, Takasawa S, Okamoto H, Ishikawa J, Kaisho T, Ishihara K and Hirano T. 1994. ADP ribosyl cyclase activity of a novel bone marrow stromal cell surface molecule, BST-1. *FEBS Lett.* 356: 244-248.
82. Rojas E, Carroll PB, Ricordi C, Boschero AC, Stojilkovic SS and Atwater I. 1994. Control of cytosolic free calcium in cultured human pancreatic β-cells occurs by external calcium-dependent and independent mechanisms. *Endocrinology* 134: 1771-1781.
83. Mészáros LG, Bak J and Chu A. 1993. Cyclic ADP-ribose as an endogenous regulator of the non-skeletal type ryanodine receptor Ca^{2+} channel. *Nature* 364: 76-79.
84. Holz GG, Leech GA, Heller RS, Castonguay M and Habener JF. 1999. cAMP-dependent mobilization of intracellular Ca^{2+} stores by activation of ryanodine receptors in pancreatic beta-cells: A Ca^{2+} signaling system stimulated by insulinotropic hormone glucagon-like peptide-1-(7-37). *J. Biol. Chem.* 274: 14147-14156.
85. Fukao K, Ochiai T, Takahashi K, Endo T, Yokoyama I, Uchida K, Oshima S, Ishibashi M, Takahara S, Iwasaki Y, Ota K, Takagi H and Sonoda T. 1994. A late phase II study of FK506 (Tacrolimus) on kidney transplantation. *Jps. J. Transplantation* 29: 614-631.

86. Pirsch JD, Miller J, Deierhoi MH, Vincenti F and Filo RS. 1997. A comparison of tacrolimus (FK506) and cyclosporine for immunosuppression after cadaveric renal transplantation: FK506 Kidney Transplant Study Group. *Transplantation* 63: 977-983.
87. Ashcroft FM and Ashcroft SJH. 1992. Mechanism of insulin secretion. In *Insulin: Molecular biology to pathology*, ed. FM Ashcroft, SJH Ashcroft, pp. 97-150. Oxford: IRL Press.
88. Pupilli C, Giannini S, Marchetti P, Lupi R, Antonelli A, Malavasi F, Takasawa S, Okamoto H and Ferrannini E. 1999. Autoanitibodies to CD38 (ADP-ribosyl cyclase/cyclic ADP-ribose hydrolase) in Caucasian patients with diabetes: Effects on insulin release from human islets. *Diabetes* 48: 2309-2315.
89. Mallone R, Ortolan E, Baj G, Funaro A, Giunti S, Lillaz E, Saccucci F, Cassader M, Cavallo-Perin P and Malavasi F. 2001. Autoantibody response to CD38 in Caucasian patients with type 1 and type 2 diabetes: immunological and genetic characterization. *Diabetes* 50: 752-762.
90. Antonelli A, Baj G, Marchetti P, Fallahi P, Surico N, Pupilli C, Malavasi F and Ferrannini E. 2001. Human anti-CD38 autoantibodies raise intracellular calcium and stimulate insulin release in human pancreatic islets. *Diabetes* 50: 985-991
91. Berridge MJ and Irvine RF. 1984. Inositol trisphosphate, a novel second messenger in cellular signal transduction. *Nature* 312, 315-321.
92. Walseth TF and Lee HC. 1993. Synthesis and characterization of antagonists of cyclic-ADP-ribose-induced Ca^{2+} release. *Biochim. Biophys. Acta* 1178: 235-242.
93. Matsuoka T, Kajimoto Y, Watada H, Umayahara Y, Kubota M, Kawamori R, Yamasaki Y and Kamada T. 1995. Expression of CD38 gene, but not of mitochondrial glycerol-3-phosphate dehydrogenase gene, is impaired in pancreatic islets of GK rats. *Biochem. Biophys. Res. Commun.* 214: 239-246.
94. Galione A. 1993. Cyclic ADP-ribose: A new way to control calcium. *Science* 259: 325-326.
95. Sasaki T, Shimura S, Takasawa S, Nagaki M, Satoh M, Okamoto H and Shirato K. 1993. Cyclic ADP-ribose, a candidate for a novel Ca^{2+}-mobilizing second messenger, induced Ca^{2+}-dependent current responses in airway submucosal gland cells. *Am. Rev. Resp. Dis.* 147: A936.
96. Hua S-Y, Tokimasa T, Takasawa S, Furuya Y, Nohmi M, Okamoto H and Kuba K. 1994. Cyclic ADP-ribose modulates Ca^{2+} release channels for activation by physiological Ca^{2+} entry in bullfrog sympathetic neurons. *Neuron* 12: 1073-1079.
97. Thorn P, Gerashimenko O and Petersen OH. 1994. Cyclic ADP-ribose regulation of ryanodine receptors involved in agonist evoked cytosolic Ca^{2+} oscillations in pancreatic acinar cells. *EMBO J.* 13: 2038-2043.
98. Lee HC, Aarhus R, Graeff R, Gurnack ME and Walseth TF. 1994. Cyclic ADP ribose activation of the ryanodine receptor is mediated by calmodulin. *Nature* 370: 307-309.
99. Tanaka Y and Tashjian AHJr. 1995. Calmodulin is a selective mediator of Ca^{2+}-induced Ca^{2+} release via the ryanodine receptor-like Ca^{2+} channel triggered by cyclic ADP-ribose. *Proc. Natl. Acad. Sci. USA* 92: 3244-3248.
100. Higashida H, Robbins J, Egorova A, Noda M, Taketo M, Ishizaka N, Takasawa S, Okamoto H and Brown DA. 1995. Nicotinamide-adenine dinucleotide regulates muscarinic receptor-coupled K^{+}(M) channels in rodent NG108-15 cells. *J. Physiol.* 482: 317-323.
101. Allen GJ, Muir SR and Sanders D. 1995. Release of Ca^{2+} from individual plant vacuoles by both $InsP_3$ and cyclic ADP-ribose. *Science* 268: 735-737.
102. Kuemmerle JF and Makhlouf GM. 1995. Agonist-stimulated cyclic ADP ribose: Endogenous modulator of Ca^{2+}-induced Ca^{2+} release in intestinal longitudinal muscle. *J. Biol. Chem.* 270: 25488-25494.

103. Gromada J, Jørgensen TD and Dissing S. 1995. Cyclic ADP-ribose and inositol 1,4,5-trisphosphate mobilizes Ca^{2+} from distinct intracellular pools in permeabilized lacrimal acinar cells. *FEBS Lett.* 360: 303-306.
104. Rakovic S, Galione A, Ashamu GA, Potter BVL and Terrar DA. 1996. A specific cyclic ADP-ribose antagonist inhibits cardiac excitation-contraction coupling. *Curr. Biol.* 6: 989-996.
105. Ebihara S, Sasaki T, Hida W, Kikuchi Y, Oshiro T, Shimura S, Takasawa S, Okamoto H, Nishiyama A, Akaike N and Shirato K. 1997. Role of cyclic ADP-ribose in ATP-activated potassium currents in alveolar macrophages. *J. Biol. Chem.* 272:16023-16029.
106. Yamaki H, Morita K, Kitayama S, Imai Y, Itadani K, Akagawa Y and Doi T. 1998. Cyclic ADP-ribose induces Ca^{2+} release from caffeine-insensitive Ca^{2+} pools in canine salivary gland cells. *J. Dental Res.* 77: 1807-1816.
107. Prakash YS, Kannan MS, Walseth TF and Sieck GC. 1998. Role of cyclic ADP-ribose in the regulation of $[Ca^{2+}]_i$ in porcine tracheal smooth muscle. *Am. J. Physiol.* 43: C1653-C1660.
108. Mothet JP, Fossier P, Meunier FM, Stinnakre J, Tauc L and Baux G. 1998. Cyclic ADP-ribose and calcium-induced calcium release regulate neurotransmitter release at a cholinergic synapse of *Aplysia*. *J. Physiol.* 507: 405-414.
109. Li PL, Zou AP and Campbell WB. 1998. Regulation of K-Ca-channel activity by cyclic ADP-ribose and ADP-ribose in coronary arterial smooth muscle. *Am. J. Physiol.* 44: H1002-H1010.
110. Sun L, Adebanjo OA, Moonga BS, Corisdeo S, Anandatheerthavarada HK, Biswas G, Arakawa T, Hakeda Y, Koval A, Sodam B, Bevis PJ, Moser AJ, Lai FA, Epstein S, Troen BR, Kumegawa M and Zaidi M. 1999. CD38/ADP-ribosyl cyclase: A new role in the regulation of osteoclastic bone resorption. *J. Cell Biol.* 146: 1161-1172.
111. Reyes-Harde M, Potter BV, Galione A and Stanton PK. 1999. Induction of hippocampal LTD requires nitric-oxide-stimulated PKG activity and Ca^{2+} release from cyclic ADP-ribose-sensitive stores. *J. Neurophysiol.* 82: 1569-1576.
112. Rakovic S, Cui Y, Iino S, Galione A, Ashamu GA, Potter BV and Terrar DA. 1999. An antagonist of cADP-ribose inhibits arrhythmogenic oscillations of intracellular Ca^{2+} in heart cells. *J. Biol. Chem.* 274: 17820-17827.
113. Inngjerdingen M, Al-Aoukaty A, Damaj B and Maghazachi AA. 1999. Differential utilization of cyclic ADP-ribose pathway by chemokines to induce the mobilization of intracellular calcium in NK cells. *Biochem. Biophys. Res. Commun.* 262: 467-472.
114. Guse AH, da Silva CP, Berg I, Skapenko AL, Weber K, Heyer P, Hohenegger M, Ashamu GA, Schulze-Koops H, Potter BV and Mayr GW. 1999. Regulation of calcium signalling in T lymphocytes by the second messenger cyclic ADP-ribose. *Nature* 398: 70-73.
115. Khoo KM, Han MK, Park JB, Chae SW, Kim UH, Lee HC, Bay BH and Chang CF. 2000. Localization of the cyclic ADP-ribose dependent calcium signaling pathway in hepatocyte nucleus. *J. Biol. Chem.* 275: 24807-24817.
116. Han MK, Cho YS, Kim YS, Yim CY and Kim UH. 2000. Interaction of two classes of ADP-ribose transfer reactions in immune signalling. *J. Biol. Chem.* 275: 20799-20805.
117. Fukushi Y, Kato I, Takasawa S, Sasaki T, Ong BH, Ohsaga A, Sato K, Shirato K, Okamoto H and Maruyama Y. 2001. Identification of cyclic ADP-ribose-dependent mechanisms in pancreatic muscarinic Ca^{2+} signaling using CD38 knockout mice. *J. Biol. Chem.* 276: 649-655.
118. Partida-Sánchez S, Coockayne DA, Monard S, Jacobson EL, Oppenheimer N, Garvy B, Kusser K, Goodrich S, Howard M, Harmsen A, Randall TD and Lund FE. 2001. Cyclic ADP-ribose production by CD38 regulates intracellular calcium release, extracellular calcium influx and chemotaxis in neutrophils and is required for bacterial clearance in vivo. *Nature Med.* 7: 1209-1216.

119. Higashida H, Yokoyama S, Hashii M, Taketo M, Higashida M, Takayasu T, Ohshima T, Takasawa S, Okamoto H and Noda M. 1997. Muscarinic receptor-mediated dual regulation of ADP-ribosyl cyclase in NG108-15 neuronal cell membranes analyzed by thin layer chromatography. *J. Biol. Chem.* 272: 31272-31277.
120. Terazono K, Yamamoto H, Takasawa S, Shiga K, Yonemura Y, Tochino Y and Okamoto H. 1988. A novel gene activated in regenerating islets. *J. Biol. Chem.* 263: 2111-2114.
121. Watanabe T, Yonekura H, Terazono K, Yamamoto H and Okamoto H. 1990. Complete nucleotide sequence of human *reg* gene and its expression in normal and tumoral tissues: The *reg* protein, pancreatic stone protein, and pancreatic thread protein are one and the same product of the gene. *J. Biol. Chem.* 265: 7432-7439.
122. Watanabe T, Yonemura Y, Yonekura H, Suzuki Y, Miyashita H, Sugiyama K, Moriizumi S, Unno M, Tanaka O, Kondo H, Bone AJ, Takasawa S and Okamoto H. 1994. Pancreatic beta-cell replication and amelioration of surgical diabetes by Reg protein. *Proc. Natl. Acad. Sci. USA* 91: 3589-3592.
123. Okamoto H. 1999. The *Reg* gene family and Reg proteins: With special attention to the regeneration of pancreatic ß-cells. *J. Hepatobiliary Pancreat. Surg.* 6: 254-262.
124. Kobayashi S, Akiyama T, Nata K, Abe M, Tajima M, Shervani NJ, Unno M, Matsuno S, Sasaki H, Takasawa S and Okamoto H. 2000. Identification of a receptor for *Reg* (*Regenerating Gene*) protein, a pancreatic ß-cell regeneration factor. *J. Biol. Chem.* 275: 10723-10726.
125. Unno M, Nata K, Noguchi N, Narushima Y, Akiyama T, Ikeda T, Nakagawa K, Takasawa S and Okamoto H. 2002. Production and characterization of *Reg* knockout mice: Reduced proliferation of pancreatic ß-cells in *Reg* knockout mice. *Diabetes* (in press).
126. Akiyama T, Takasawa S, Nata K, Kobayashi S, Abe M, Shervani NJ, Ikeda T, Nakagawa K, Unno M, Matsuno S and Okamoto H. 2001. Activation of *Reg* gene, a gene for insulin-producing ß-cell regeneration: Poly(ADP-ribose) polymerase binds *Reg* promoter and regulates the transcription by autopoly(ADP-ribosyl)ation. *Proc. Natl. Acad. Sci. USA* 98: 48-53.
127. Livesey JF, O'Brien AJ, Li M, Smith GA, Murphy JL and Hunt PS. 1997. A Schwann cell mitogen accompanying regeneration of motor neurons. *Nature* 390: 614-618.
128. Nishimune H, Vasseur S, Wiese S, Birling MC, Holtmann B, Sendtner M, Iovanna JL and Henderson CE. 2000. Reg-2 is a motoneuron neurotrophic factor and a signalling intermediate in the CNTF survival pathway. *Nature Cell Biol.* 2: 906-914.
129. Asahara M, Mushiake S, Shimada S, Fukui H, Kinoshita Y, Kawakami C, Watanabe T, Tanaka T, Ichikawa A, Uchiyama Y, Narushima Y, Takasawa S, Okamoto H, Tohyama M and Chiba T. 1996. *Reg* gene expression is increased in rat gastric enterochromaffin-like cells following water immersion stress. *Gastroenterology* 111: 45-55.
130. Fukui H, Kinoshita Y, Maekawa T, Okada A, Waki S, Hassan S, Okamoto H and Chiba T. 1998. Regenerating gene protein may mediate gastric mucosal proliferation induced by hypergastrinemia in rats. *Gastroenterology* 115: 1483-1493.
131. Kazumori H, Ishirara S, Hoshino E, Kawashima K, Moriyama N, Suetsugu H, Sato H, Adachi K, Fukuda R, Watanabe M, Takasawa S, Okamoto H, Fukui H, Chiba T and Kinoshita Y. 2000. Neutrophil chemoattractant-2ß regulates the expression of the Reg gene in injured gastric mucosa in rats. *Gastroenterology* 119: 1610-1622.

Hiroshi Okamoto, Department of Biochemistry,
Tohoku University Graduate School of Medicine,
Sendai 980-8575,Miyagi, Japan.
e-mail: okamotoh@mail.cc.tohoku.ac.jp

Cyclic ADP-Ribose and NAADP. Structures, Metabolism and Functions. Lee HC (ed), pp. 301-318, Dordrecht: Kluwer.

The Role of cADPR and NAADP in T Cell Calcium Signaling and Activation

Andreas H. Guse
University Hospital Hamburg-Eppendorf, Institute for Medical Biochemistry and Molecular Biology, Division of Cellular Signal Transduction, Martinistr. 52, D-20246 Hamburg, Germany.

INTRODUCTION

Ca^{2+} signaling in T-lymphocytes involves Ca^{2+} release from intracellular stores as well as Ca^{2+}entry from the extracellular space [reviewed in 1-3]. Ten years ago many researchers believed that Ca^{2+} release in T cells is exclusively operated by D-*myo*-inositol 1,4,5-trisphosphate ($InsP_3$; reviewed in ref. 4) while the capacitative mechanism of Ca^{2+} entry proposed by Putney [5] was thought to play a central role in the influx process.

Two observations from my own work during that period of time led to the idea that more players might be involved in the process of T cell Ca^{2+} signaling. First, analyzing the $InsP_3$ content of Jurkat lymphoblastoid T cells, a T cell line widely used in the field, revealed a rapid increase of $InsP_3$ within the first minutes upon T cell receptor/CD3 (TCR/CD3) stimulation, followed by a decrease to almost baseline levels within the next 15 min or so [6,7]. Similar results have also been obtained by other groups [8,9]. Since T cells display an extremely long phase of Ca^{2+} entry (at least a couple of hours, but some authors have measured inflow of $^{45}Ca^{2+}$ for more than 48 h [10]), it became clear that either additional mechanisms of Ca^{2+} entry, not requiring previous pool depletion for activation, must exist, or that additional Ca^{2+} release mechanisms are present in T cells somehow covering the late phase of Ca^{2+} signaling. The second observation was Ca^{2+} release from a non-$InsP_3$ sensitive Ca^{2+} pool by caffeine, a pharmacological activator of the ryanodine receptor (RyR; ref. 11).

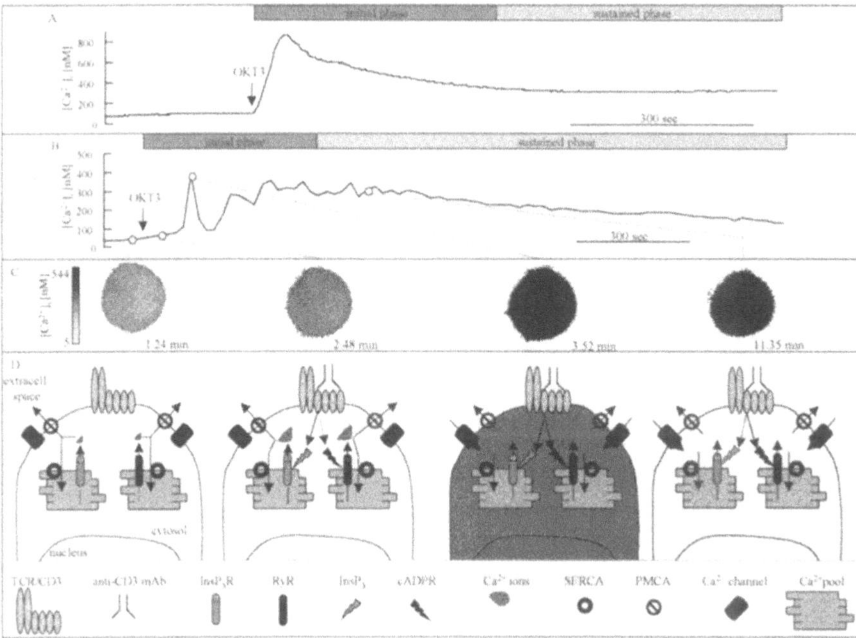

Figure 1. **Model of T cell Ca^{2+} signaling.** $[Ca^{2+}]_i$ of Fura2-loaded Jurkat T cells was measured either in suspension in a fluorescence photometer (**A**; see ref. 50 for experimental details) or by single cell Ca^{2+} imaging at 100-fold magnification (**B, C**). OKT3 (10 µg/ml (**A**) or 1 µg/ml (**B,C**)) was added as indicated to stimulate signaling via the TCR/CD3 complex. In (**C**) selected ratiometric greyscale images demonstrate the distribution of $[Ca^{2+}]_i$ in a single cell during the different phases of T cell Ca^{2+} signaling. (**D**) illustrates our current working model for the different phases of T cell Ca^{2+} signaling.

These observations initialized a series of experiments in which we (and others) investigated the role of the messengers cADPR and NAADP in T cell Ca^{2+} signaling and activation [reviewed in 12-16]. Taking together all these data obtained in the last 10 years, the following working model for Ca^{2+} signaling was developed (Fig. 1).

Ca^{2+} signaling of Fura2-loaded Jurkat T cells, either measured in suspension (Fig. 1A) or by single cell imaging (Fig. 1B, C), displays a biphasic pattern starting with a rapid increase followed by a somewhat slower decrease to a plateau phase. The latter stays more or less constant for a long period of time (Fig. 1A). In single cell imaging, individual cells show a lot of diversity regarding their spatiotemporal Ca^{2+} pattern; however, the cell displayed in Fig. 1B appears a good representative to illustrate the signaling events on the single cell level. During the resting phase (Fig. 1C and D, left image) the Ca^{2+} pump activity of sarcoplasmic/endoplasmic reticulum Ca^{2+}-ATPase (SERCA) and plasma membrane Ca^{2+}-ATPase (PMCA) overrides most of the spontaneous release and influx events resulting in very low intracellular Ca^{2+} concentrations (at least when measured by single cell imaging). Although the concentration of Ca^{2+} releasing messengers cADPR and $InsP_3$ is low (see paragraph "RECEPTOR-MEDIATED FORMATION OF cADPR" later in this chapter), in a number of cells spontaneous Ca^{2+} signals can be observed; these signals are often locally restricted, but sometimes also culminate into a global Ca^{2+} signal. Are these local Ca^{2+} signals important for the whole machinery of Ca^{2+} signaling in T cells, e.g. by providing sufficient trigger Ca^{2+} to support rapid signal onset once a cell surface receptor is activated? And secondly, are the known Ca^{2+} mobilizing messengers, $InsP_3$, cADPR and NAADP, or additional so far not recognized mechanisms involved in these spontaneous release events? Actually, we do not know the answers, but future research may bring novel insights.

Global Ca^{2+} signals are preceded by a pacemaker phase (Fig. 1C and D, second left image). During this phase local Ca^{2+} release events can be monitored by confocal Ca^{2+} imaging in Jurkat T cells (S. Heidbrink & A.H. Guse, unpublished results). There is evidence that in this pacemaker phase of TCR/CD3 complex-mediated Ca^{2+} signaling both $InsP_3$ and cADPR are the Ca^{2+} releasing molecules, but the concentration of $InsP_3$ reaches its maximum in this phase [6-9] while the concentration of cADPR is already elevated, but still relatively low [17].

Once a critical threshold $[Ca^{2+}]$ is passed, the local events culminate into a global response apparently involving all parts of the cell (Fig. 1B,C and D, second right image). In the initial phase of stimulation at least one global Ca^{2+} spike occurs in the absence of extracellular Ca^{2+}; however, subsequent Ca^{2+} spikes, if any, are much smaller in amplitude and duration (not shown). This suggests that Ca^{2+} pool depletion occurs rapidly in T-lymphocytes, even in the large Jurkat T-lymphoblasts. In the presence of extracellular Ca^{2+}, Ca^{2+} entry comes into play at this point. The fact that both $InsP_3$ and cADPR-sensitive Ca^{2+} pools are (partially) depleted, indicates that capacitative Ca^{2+} entry is activated (Fig. 1 D, second right image). However, additional receptor-modulated and/or ligand-activated Ca^{2+} channels cannot be ruled out at present.

In the longlasting phase of Ca^{2+} signaling (Fig. 1C and D, right image), there is constant inflow of Ca^{2+} from the extracellular space. In this phase, cADPR is at its highest concentration and accomplishes, in concert with the still somewhat elevated levels of $InsP_3$, a continuous pool depletion. Thus, the constant Ca^{2+} entry may be simply explained by capacitative Ca^{2+} entry. Again, it is difficults to judge at present whether additional influx pathways not related to store depletion are involved.

Presently, NAADP is not really integrated into our working model of T cell Ca^{2+} signaling. The main reason is the lack of knowledge of its intracellular concentration and its kinectics upon TCR/CD3 complex (or other cell surface receptors`) ligation. However, we have shown previously that auto-inactivation of the NAADP system abolished the Ca^{2+} mobilizing effects of both cADPR and $InsP_3$ [18]. Furthermore, TCR/CD3 complex-mediated Ca^{2+} signaling was also completely blocked by prior auto-inactivation of the NAADP system [18] indicating an important role for NAADP in one (or more) of the phases of T cell Ca^{2+} signaling described above (see also paragraph "NAADP AND T CELL Ca^{2+} SIGNALING" later in this chapter).

What are the experimental data underlying our working model of T cell Ca^{2+} signaling?

RECEPTOR-MEDIATED FORMATION OF cADPR

Analysis of intracellular cADPR using a newly developed 2-step HPLC system (Fig. 2, ref. 19) revealed a stimulatory effect of the TCR/CD3 complex [17]. In contrast to the rapid, but transient elevation of $InsP_3$, cADPR increased slowly within 10 min, but then remained constant for at least the next 50 min [17]. The basal cADPR levels measured are in accordance with data from other T cell lines analyzed by a radioimmunoassay specific for cADPR [20]. Importantly, the increase in cADPR was only observed with TCR/CD3 ligation, whereas artificial means to elevate $[Ca^{2+}]_i$, e.g. the SERCA inhibitor thapsigargin or the Ca^{2+} ionophore ionomycin, were insufficient to raise cADPR levels [17]. In accordance with a receptor-regulated increase in cADPR, ADP-ribosyl cyclase (ADPRC) activities were detected in all subcellular fractions of T cells (2,000xg-, 10,000xg-, 100,000xg-pellet, 100,000xg-supernatant [21]), including a very weak activity in the soluble fraction (100,000xg supernatant) of Jurkat T cells [17, 21].

Despite many efforts to purify or to clone the ADPRC from the soluble fraction, the molecular identity of this enzyme in T cells is still not resolved. However, it is clear that Jurkat T cells express CD38 [22], a multifunctional enzyme displaying both ADPRC, NAD-glycohydrolase (NADase) and cADPR-hydrolase activity (see chapters by Lee *et al.*, and by Ferrero and Malavasi in this book, and references therein). However, the cannonical CD38 is an ectoenzyme and the nucleotide cADPR is not membrane permeant. In addition, the extracellular concentration of the substrate NAD usually is low, but may, however, rise under certain circumstances, e.g. lysis of neighbouring cells. Last, but not least, CD38 mainly produces the linear adenosine diphosphoribose (ADPR). These data argue against a role for ecto-CD38 in the formation of intracellular cADPR in T cells. To solve the "topological paradox" De Flora and colleagues have put forward the idea that the extracellular catalytic site of CD38 is connected to the cytosol by specific transport systems (see the chapter by De Flora *et al.*, in this book, and references therein). They showed that connexin43 hemichannels transport NAD from the cytosol to the extracellular space [23] and that

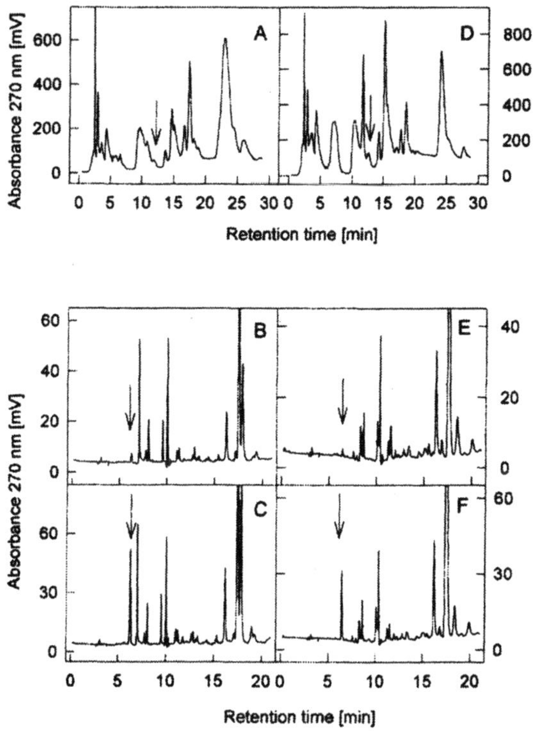

Figure 2. **HPLC analysis of cADPR in T cell lines.** A perchloric acid extract from Jurkat cells (8x10^7 cells; **A-C**) or HPB.ALL cells (1x10^8 cells, **D-F**) was analysed by (**A,D**) strong anion-exchange HPLC on a PRPX 100 column using a gradient from 1 mM Tris, pH 8.0, to 150 mM trifluoroacetic acid at a flow rate of 1.5 ml/min. Material co-eluting with standard cADPR was collected, lyophilized and subsequently analysed by (B,E) reversed-phase ion-pair HPLC on a Hypersil column using the conditions described in [19]. Co-chromatography of cellular material with standard cADPR (0.5 nmol) on reversed-phase HPLC gave chromatogramms of the type shown in (**C,F**). The retention time of standard cADPR is indicated by the arrow. The tracings are representative of at least eight independent experiments. Reprinted by permission from J. Chromatography, Series B 707, 43-50, copyright 1998, Elsevier Science, Amsterdam.

cADPR is transported back into the cell via CD38 or an additional transport system [24]. Alternatively, it was also proposed that this whole cycle may procede between cytosol and endocytotic vesicles upon internalization of cell surface ecto-CD38 [25]. Experiments in intact quiescent Jurkat T cells showed that incubation with NAD resulted in ecto-CD38 catalyzed formation of cADPR, but only on the cell surface; transport of cADPR into the cells was neither detectable by HPLC nor by using Ca^{2+} signaling as a reporter assay [22]. A similar result was obtained with the surrogate substrate nicotinamide guanine dinucleotide (NGD) [22]. Furthermore, the substrate NAD alone did not induce internalization of CD38 in Jurkat T cells

[22]. These data indicate that Jurkat T cells do not use the pathway described by De Flora and colleagues, at least not when the cells are not activated [22].

Recent studies in different cell types demonstrated expression of CD38 in intracellular membranes, e.g. in the nucleus [26, 27]; in addition, an ADPRC/NADase with similar properties as CD38 was detected in mitochondria of bovine liver [28]. Thus, it may be possible that intracellular variants of CD38 are involved in the formation of cADPR in T cells, too.

Ca^{2+} MOBILIZING EFFECTS OF cADPR

Ca^{2+} Release by cADPR–In 1995 two independent studies showed an effect of cADPR on Ca^{2-} release in T cells. Bourguignon and colleagues demonstrated in mouse T-lymphoma cells [^{3}H]ryanodine binding to and cADPR-mediated Ca^{2-} release from a specific membrane vesicle fraction purified by sucrose gradient centrifugation [29]. In addition, expression of RyR was shown for the first time in T cells by western blotting [29]. In the same year we published cADPR-mediated Ca^{2+} release in permeabilized human Jurkat T-lymphocytes [30], a finding confirmed by Takemura and colleagues one year later [31]. The effect of cADPR was dose-dependent and could be specifically inhibited by 8-NH_2-cADPR and ruthenium red [30]. Ca^{2-} release of cADPR is regulated in T cells by inorganic phosphate and Mg^{2-} ions; the latter is, as shown for ruthenium red inhibition, in accordance with the concept of RyRs being involved in cADPR-mediated Ca^{2+} release. The naturally occuring 2′-phospho-cADPR [32] released Ca^{2+} with a similar pharmacology as shown for cADPR indicating that it may fulfill a similar role in T cell Ca^{2-} signaling [33].

Ca^{2+} Entry by cADPR–Though cell homogenates or permeabilized cell preparations are a standard model to investigate the effects of non-membrane permeant Ca^{2+} mobilizing compounds, it is an artificial system that may give rise to systematical errors, e.g. loss or partial loss of important soluble (co-) factors during the preparation process. Thus, in 1995/96 we established the microinjection technique and used it to inject Ca^{2+} mobilizing agonists (and antagonists) while $[Ca^{2+}]_i$ of the cell was monitored continuously by digital Ca^{2-} imaging [34]. Using this experimental approach the Ca^{2+} mobilizing effects of cADPR in T cells were confirmed. Microinjection of cADPR concentration-dependently stimulated longlasting Ca^{2+} spikes (Fig. 3; ref. 34]. The effect was specific since co-injection of a 10-fold excess of the cADPR antagonist 8-NH_2-cADPR completely blocked Ca^{2+} signaling [34]. Similar experiments carried out in the absence of extracellular Ca^{2+}, or in the presence of Ca^{2-} channel inhibitors, e.g. Zn^{2+}or SK-F 96365, largely abolished the Ca^{2+} signals indicating that cADPR-mediated Ca^{2+} release subsequently activated Ca^{2+} entry via store-operated Ca^{2+} channels [34]. Thus, in addition to Ca^{2+} release, cADPR can also promote Ca^{2+} entry in T cells; the underlying mechanism is likely to be store-operated Ca^{2+} entry secondary to Ca^{2+} pool depletion via RyR, but other mechanisms, e.g. a direct effect of cADPR on plasma membrane Ca^{2+} channels cannot be ruled out at present.

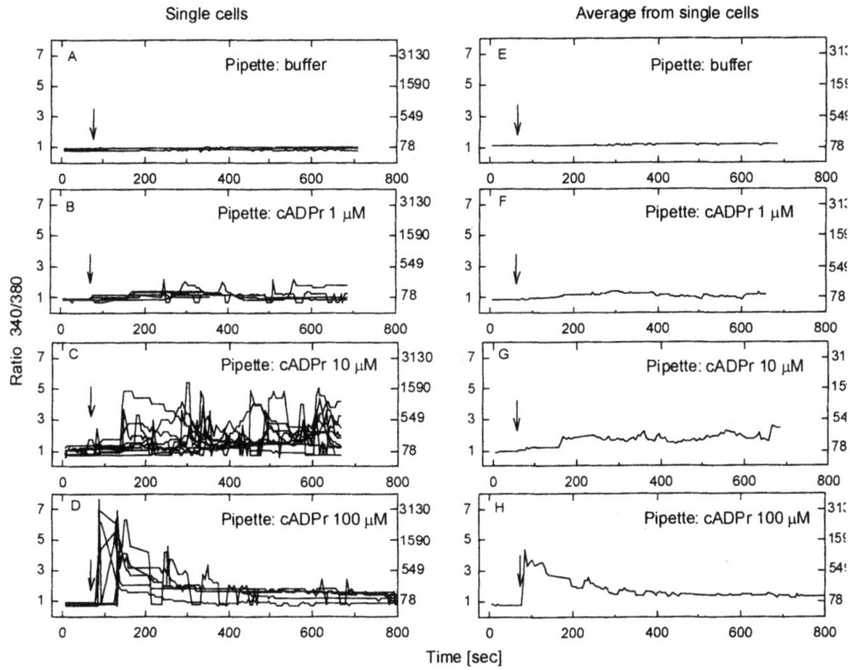

Figure 3. **Ca^{2+}-signals in single intact T-lymphocytes microinjected with cADPR.** Jurkat T-lymphocytes were loaded with Fura2/AM and $[Ca^{2+}]_i$ was measured as detailed in [34]. During re-analysis of the image data, regions of interest were set to cover the whole cell; data obtained from these regions of interest are plotted against time. Cells were injected as described in Experimental procedures of [34] in the presence of 1 mM extracellular Ca^{2+} with: (**A**) intracellular buffer (number of cells displayed, n=3), or increasing concentrations of cADPR as indicated (**B-D**, n=4-10). Averages from these measurements are shown in (**E-H**). Arrows indicate the time point of microinjection. Reprinted by permission from Journal of Biological Chemistry 272, 8546-8550, copyright 1997, American Society for Biochemistry and Molecular Biology.

RyR: The Receptor for cADPR?–First evidence for expression of RyR in T cells came from cloning and sequence analysis of Jurkat T cell cDNA: Hakamata and colleagues specifically detected fragments of the type 3 RyR [35]. Protein expression of RyR was shown by [^{3}H]ryanodine binding in mouse T-lymphoma BW5147 cells [29], and in human Jurkat T cells [36, 17]. In addition, western blotting using an anti-RyR_{common} monoclonal antibody demonstrated specific staining of a protein band of approx. 550 kD in both mouse T-lymphoma BW5147 cells [29], and in human Jurkat T cells [17]. RT-PCR analysis from total RNA of Jurkat T cells revealed the expression of the type 3 RyR [17]; however, the expression of additional types of RyR (1 and/or 2) can not be ruled out at present.

Functional evidence for expression of RyR in T cells was obtained in Ca^{2+} signaling and proliferation studies. Caffeine mediated Ca^{2+} signaling in Jurkat T cells in suspension [11, 36] and at the single cell level [36]. Ricard and colleagues also demonstrated Ca^{2+} signaling activated by ryanodine in both Jurkat T cell suspensions and single cells [36]. Ca^{2+} signaling involving RyR was also observed in response to microinjection of the RyR agonist suramin in Jurkat T cells [37]. In addition, Hakamata *et al.* reported a growth promoting effect of ryanodine in T cell proliferation experiments [35].

Pharmacological evidence for RyR being involved in cADPR-mediated Ca^{2+} release was obtained in many cellular systems (see chapters by Lee, Li *et al.*, and Walseth and Lee in this book and references therein). In accordance with these results, in Jurkat T cells inhibition of cADPR-mediated Ca^{2+} release was observed by ruthenium red [30] and by high Mg^{2+} concentrations [38].

Recent work using stable anti-RyR antisense transfected Jurkat T cell subclones indicates involvement of RyR in Ca^{2+} signaling in T cells (N. Schwarzmann & A.H. Guse, unpublished results), thereby confirming pharmacological data obtained with a membrane-permeant cADPR antagonist (ref. 17; see also the "Pharmacological studies" section below).

Pharmacological Studies–Analogues of cADPR were analyzed for their agonistic or antagonistic properties in the T cell system. Synthesis of cADPR analogues was done either by chemical synthesis or chemical modification of the corresponding linear NAD analogue and subsequent cyclization by *Aplysia californica* ADPRC or, in case of cyclic adenosine diphospho-carbocyclic-ribose (cADPcR) by total chemical synthesis; the analogues were prepared by Prof. Potter (Bath, UK), except 2′′-NH_2-cADPR which was made by Prof. Schuber (Illkirch, France). To clarify structural modifications of the two ribose moities, the ribose originally linked to adenine in the precursor NAD is called the "southern" ribose and the C atoms are labeled " ′ ", whereas the ribose linked to nicotinamide in NAD is called "northern" ribose and the C atoms are labeled " ′′ " (Fig. 4).

Three compounds tested so far had agonistic properties similar to or better than cADPR in T cells. 2′-Phospho-cADPR showed a very similar pharmacological profile as compared to cADPR [33], whereas both 2′′-NH_2-cADPR and cyclic aristeromycin diphosphoribose (cArisDPR) were somewhat more efficient in releasing Ca^{2+} in permeabilized T cells [39].

Five compounds were neither agonists nor antagonists. Four of them were derivatized in the "southern" ribose, 3′-phospho- and 3′-deoxy-cADPR, 2′-deoxy-cADPR, and 2′,3′-cyclic-phospho-cADPR (Fig. 4; ref. 33, and A.H. Guse & B.V.L. Potter, unpublished results). The carbocyclic derivative of

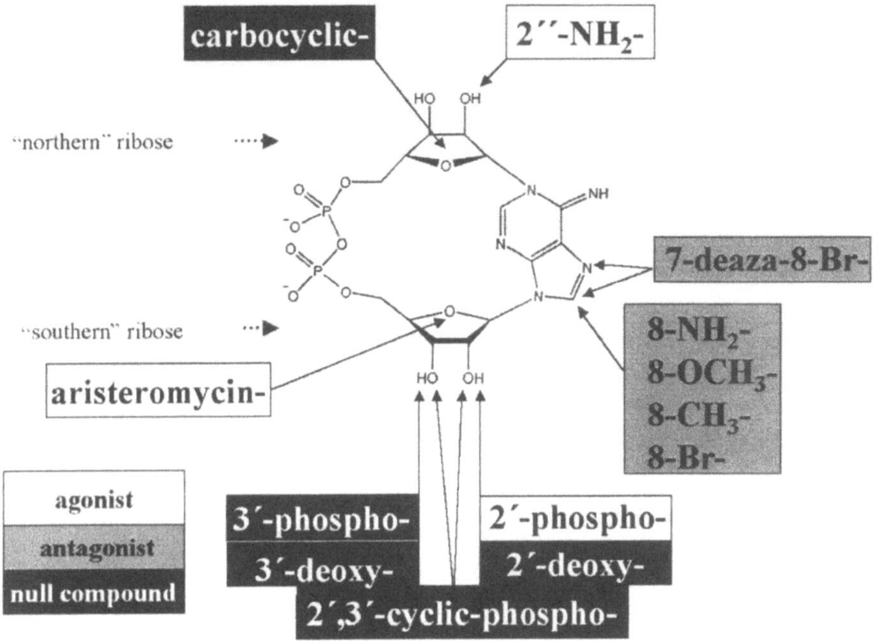

Figure 4. **Pharmacology of cADPR in T cells.**

the "northern" ribose was a very weak agonist showing considerable Ca^{2+} release only at 50 to 100 μM (Fig. 4; ref. 39).

It is of considerable interest that some, but not all of the compounds have very different effects in higher vs. lower eukaryotes. While cArisDPR was active in both higher and lower eukaryotes, 2′-phospho-cADPR was inactive in sea urchin eggs [40], but highly active in permeabilized Jurkat T cells [33] and in different brain microsome preparations [41, 42]. On the other hand, 2′-deoxy-cADPR and cADPcR were agonists in sea urchin eggs [43, 44], but not in T cells (A.H. Guse & B.V.L. Potter, unpublished results, and ref. 39).

8-Substituted analogues are well known for their antagonistic properties (see chapter by Walseth & Lee in this book and references therein). 8-NH_2-cADPR and 8-Br-cADPR, first evaluated in the sea urchin egg system [45], were antagonists for cADPR-mediated Ca^{2+} signaling, too [30, 34]. In addition, 8-OCH_3-cADPR and 8-CH_3-cADPR were successfully used to antagonize Ca^{2+} signaling effects of cADPR in both permeabilized and microinjected intact T cells [17, 33, 37].

In addition to the pharmacological evidence for RyR and cADPR as parts of the Ca^{2+} signaling machinery mentioned in previous paragraphs (see section "RyR: the receptor for cADPR?" above), an important study was carried out using the membrane-permeant cADPR antagonist 7-deaza-8-Br-cADPR (Fig. 4; ref. 46). Jurkat T cells were preincubated with increasing

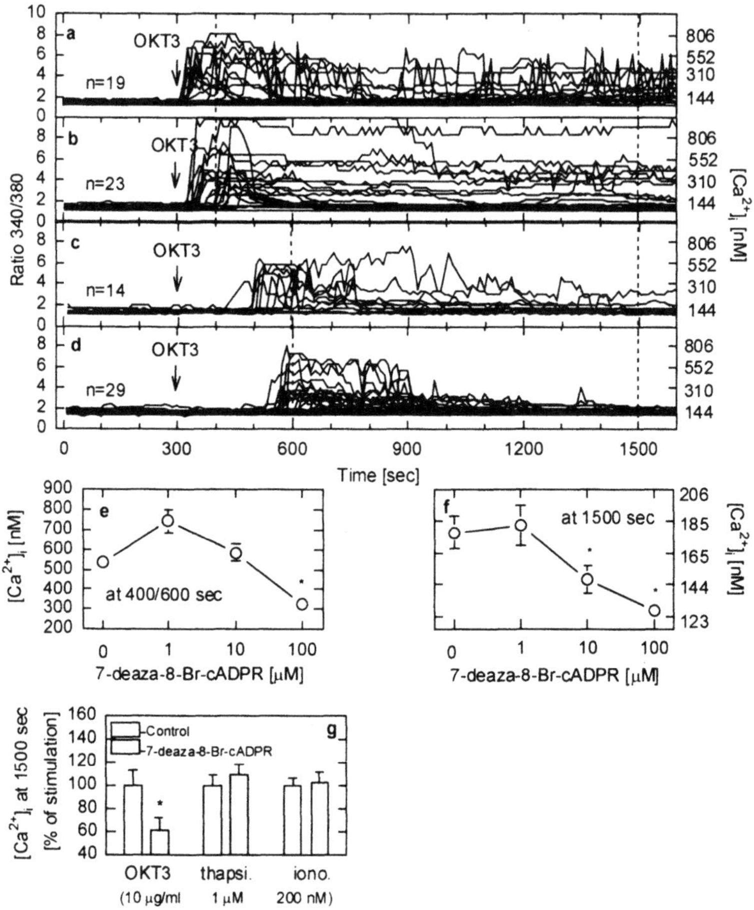

Figure 5. **Inhibition of longlasting Ca^{2+}-signalling by 7-deaza-8-Br-cADPR.** $[Ca^{2+}]_i$ was measured in fura-2-loaded Jurkat T cells using single cell Ca^{2+}-imaging [17, 34]. Cells were either untreated (**a**) or treated with increasing concentrations of 7-deaza-8-Br-cADPR for 20 min (**b** 1 μM, **c** 10 μM, **d** 100 μM) prior to stimulation with anti-CD3 OKT3 (10 μg/ml). **e** and **f** show $[Ca^{2+}]_i$ at 400/600 or 1500 sec after OKT3 addition as combined data obtained in 5 independent experiments (mean±SEM, n =108 to 203 cells for each concentration of 7-deaza-8-Br-cADPR; * p > 0.95). **g** Effect of 7-deaza-8-Br-cADPR on cells stimulated by either OKT3, thapsigargin or ionomycin (mean±SD, 3-4 independent experiments, 91 to 104 individual cells investigated: * p > 0.95). Reprinted by permission from Nature 398, 70-73, copyright 1999, Macmillan Publishers Ltd.

concentrations of 7-deaza-8-Br-cADPR and $[Ca^{2+}]_i$ was determined on the single cell level by Ca^{2+} imaging studies [17]. Both the initial phase of Ca^{2+} signaling as well as longlasting Ca^{2+} signaling were antagonized concentration-dependently and specifically by 7-deaza-8-Br-cADPR (Fig. 5). Most importantly, in human peripheral blood T cells, Ca^{2+} dependent and more downstream occurring activation events, e.g. CD25 and HLA-DR expression and proliferation, were effectively antagonized by 7-deaza-8-Br-cADPR [17]. In addition, 7-deaza-8-Br-cADPR also partially blocked Ca^{2+} signaling mediated via ligation of α6ß1-integrin [47] indicating that the cADPR/Ca^{2+} signaling system is not restricted to TCR/CD3 ligation, but is used by at least two T cell surface receptors. Taken together these data suggest that activation of the cADPR/Ca^{2+} signaling pathway is essential for T cell activation. Moreover, antagonists of the cADPR/Ca^{2+} signaling system display immunosuppressive properties opening the principal possibility for use in the therapy of autoimmune diseases.

As discussed earlier in this chapter there is kinetic evidence for $InsP_3$ being involved mainly in the initial phase of T cell Ca^{2+} signaling while cADPR is involved mainly in the sustained phase. To test this hypothesis pharmacologically, inhibitors of $InsP_3$- and/or cADPR-mediated Ca^{2+} signaling were microinjected either before ligation of the TCR/CD3 complex, or a few minutes after ligation [17]. Importantly, D-*myo*-inositol

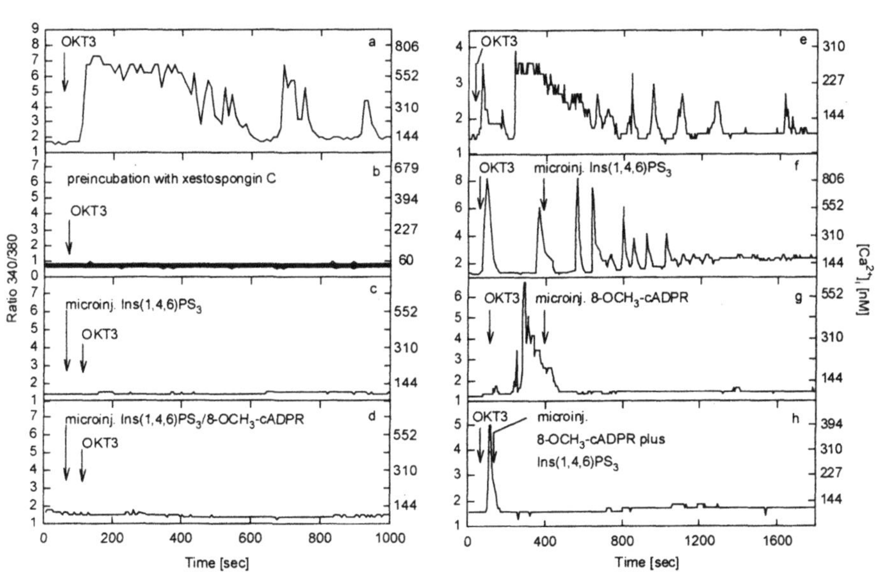

Figure 6. **Effect of antagonists against Ins(1,4,5)P_3 and cADPR on Ca^{2+}-signaling in single Jurkat T cells.** $[Ca^{2+}]_i$ was measured as above. T cells were stimulated by OKT3 and either left untreated (**a**, **e**; n >100), or preincubated with the $InsP_3$ receptor antagonist xestospongin C (2 μM, 20 min) before OKT3 addition (n = 16; **b**), microinjected [17, 34] with 80 μM Ins(1,4,6)PS_3 (n = 3; **c**) or with 80 μM Ins(1,4,6)PS_3 plus 100 μM 8-methoxy cADPR before OKT3 addition (n = 3; **d**), or microinjected after the first OKT3-dependent Ca^{2+}-spikes were observed with 80 μM Ins(1,4,6)PS_3 (**f**; n=10), 100 μM 8-methoxy cADPR (**g**; n=11), or a combination of both (**h**; n=12). In **b**) 16 cells monitored in 1 experiment are displayed as an overlay. Reprinted by permission from Nature 398, 70-73, copyright 1999, Macmillan Publishers Ltd.

1,4,6-phosphorothioate, a specific antagonist of $InsP_3$, inhibited Ca^{2+} signaling only when microinjected before TCR/CD3 complex ligation, whereas 8-OCH_3-cADPR, a specific antagonist of cADPR also was effective when microinjected at later timepoints (Fig. 6).

Taken together the pharmacological and kinetic data support the model displayed in Fig. 1 where individual global Ca^{2+} spikes are initiated by subcellular pacemaker signals. The latter involve, depending on the phase of Ca^{2+} signaling, either more $InsP_3$ and less cADPR (initial phase), or more cADPR and less $InsP_3$ (sustained phase). However, the different appearance of the individual Ca^{2+} signals between initial and sustained phase indicate additional differences in fine tuning of the two systems, or the involvement of additional mechanisms not yet fully recognized. One element of fine tuning is likely to be phosphorylation of $InsP_3$ receptors and RyR on tyrosine residues [48, 49]. In addition, the NAADP/Ca^{2+} signaling system may be such a mechanism which must be integrated in the model for T cell mediated Ca^{2+} signaling in the future (see also paragraph "NAADP AND T CELL Ca^{2+} SIGNALING" below).

NAADP AND T CELL Ca^{2+} SIGNALING

Microinjection of very low amounts of NAADP, e.g. 10 nM NAADP in intracellular solution, evoked longlasting Ca^{2+} spikes of low amplitude in a number of Jurkat T cells [18]. The concentration-response relationship of NAADP in T cells has a characteristic bell-shaped curve with a maximal response occurring between 50 and 100 nM (Fig. 7). At these concentrations usually one (or very few) high global Ca^{2+} spikes were observed, which were then followed by trains of much smaller Ca^{2+} spikes [18]. Importantly, injection of higher concentrations of NAADP ($\geq$ 10 μM) almost completely abolished T cell Ca^{2+} signaling [18]. Such behaviour of the NAADP/Ca^{2+} system has been described for other cells previously (see chapters on NAADP in this book and references therein) and is referred to as "auto-inactivation". In further experiments in T cells the NAADP/Ca^{2+} system was "auto-inactivated" by injection of 10 μM NAADP, but activating concentrations of cADPR or $InsP_3$ were co-injected. Surprisingly, most of the Ca^{2+} signaling effects of cADPR and $InsP_3$ were abolished when the NAADP/Ca^{2+} system was inactivated indicating that an intact NAADP/Ca^{2+} system is a prerequisite for cADPR and $InsP_3$ to work properly [18]. Knowing that both cADPR and $InsP_3$ are essential second messengers during T cell activation, the NAADP/Ca^{2+} system was "auto-inactivated" by microinjection of 10 μM NAADP and shortly thereafter Jurkat T cells were stimulated by TCR/CD3 ligation; under these conditions TCR/CD3 ligation did not activate any Ca^{2+} signaling indicating that an intact NAADP/Ca^{2+} system is essential for T cell Ca^{2+} signaling and activation (Fig. 8; ref. 18).

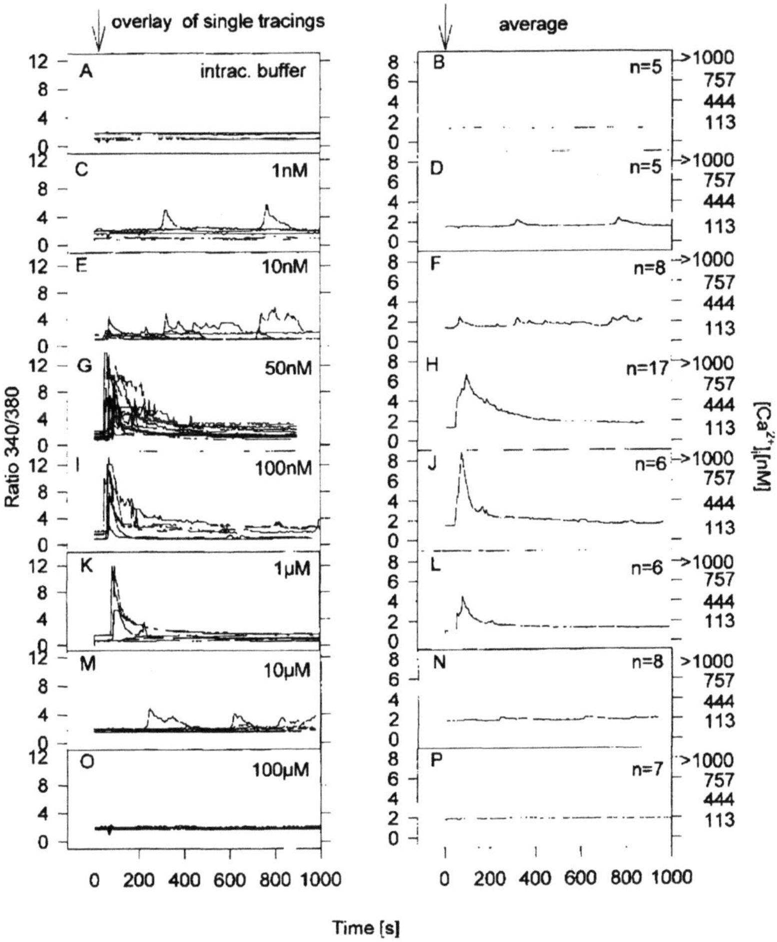

Figure 7. **Concentration-response curves of Ca^{2+} signals in single intact T-lymphocytes microinjected with $NAADP^{+}$.** Jurkat T-lymphocytes were loaded with Fura2/AM and Ca^{2+} was measured as detailed under 'Experimental Procedures' in [18]. Cells were injected as described in 'Experimental Procedures' in [18] in the presence of 1mM extracellular Ca^{2+}. Data are presented as overlays of single tracings of individual cells (left panel). The right panel shows the corresponding averages from these measurements (number of cells displayed, n = 5 to 17). As a control vehicle buffer without $NAADP^{+}$ was injected (**A,B**). The time point of microinjection is indicated by arrows. Reproduced from The Journal of Cell Biology, 2000, 150, 581-588, by copyright permission of The Rockefeller University Press.

CONCLUSION

To regulate intracellular Ca^{2+} signaling during cell activation, T-lymphocytes make use of multiple mechanisms including Ca^{2+} release by $InsP_3$, cADPR and NAADP. In addition, Ca^{2+} entry via the capacitative

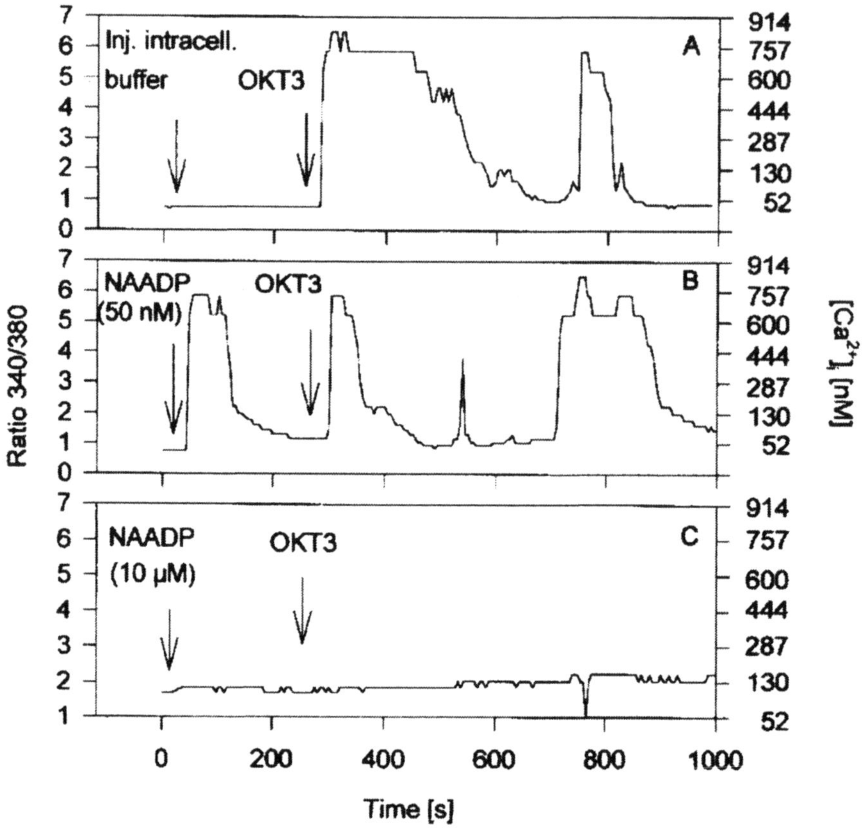

Figure 8. **Effect of NAADP+ on OKT3-induced Ca^{2+} signaling in single Jurkat T-lymphocytes.** Jurkat T-lymphocytes were loaded with Fura2/AM and ratiometric Ca^{2+} imaging and parallel microinjection in the presence of 1mM extracellular Ca^{2+} was carried out as detailed under 'Experimental Procedures' in [18]. The cells were injected with different concentrations of $NAADP^+$ and then OKT3 (10 μg/ml) was added. Injection of intracellular buffer (**A**), $NAADP^+$ (50 nM (**B**) and 10 μM (**C**)), and addition of OKT3 is indicated by arrows. Data are presented as a typical tracing from one individual cell, for each condition at least 3 experiments were carried out. Reproduced from The Journal of Cell Biology, 2000, 150, 581-588, by copyright permission of The Rockefeller University Press.

pathway and perhaps additional mechanisms refill intracellular Ca^{2+} pools and thereby serve as the source for the sustained phase of Ca^{2+} signaling. Several aspects of the cADPR/Ca^{2+} signaling system have been analyzed successfully in T cells, among others the endogenous concentration and kinetics of cADPR upon T cell activation, and the involvement of RyR in cADPR mediated Ca^{2+} signaling. However, still many aspects are unclear and await future investigations, e.g. the identity of the cADPR forming enzyme certainly is one of the key questions which should be solved in the future. In addition to the cADPR/Ca^{2+} signaling system the NAADP/Ca^{2+}

signaling system was found to be essential for T cell activation, too. As for other cellular systems many aspects of the NAADP/Ca^{2+} signaling system are still mysterious, e.g. mode of formation, endogenous concentrations, and the nature of the putative NAADP receptor. All these questions need to be addressed to fully integrate the NAADP/Ca^{2+} signaling system into our current view of T cell Ca^{2+} signaling. However, the principal appearance of both the cADPR/Ca^{2+} signaling system and the NAADP/Ca^{2+} signaling system in addition to the $InsP_3$/Ca^{2+} signaling system during T cell activation is meanwhile well documented and underlines the broad evolutionary distribution of these systems ranging from primitive eukaryotes up to human cell systems.

ACKNOWLEDGEMENTS

Thank are expressed to Drs Inge Berg and Cristina da Silva, several hard-working PhD and diploma students, and my technician Karin Weber. Without their engagement for the cADPR and NAADP projects the results summarized in this chapter would not have been obtained. In addtion, collaborative work with Drs B. Potter (Bath, UK), M. Hohenegger (Vienna, Austria), P. Deterre (Paris, France), F. Schuber (Strasbourg, France), T. Walseth (Minneapolis, USA) and many others largely supported the projects on T cell Ca^{2+} signaling. I would also like to thank Svenja Heidbrink for help with preparation of Figure 1 and Georg W. Mayr (Hamburg) for continuous support. Work in my laboratory was and is supported by the Deutsche Foschungsgemeinschaft (grants GU 360/2-4 and 2-5, and GU 360/7-1, jointly with G.W. Mayr, Hamburg), the Werner-Otto-Foundation (Hamburg), the Wellcome Trust (Biomedical Research Collaboration Grant no. 051326, jointly with B.V.L. Potter, Bath, UK), the Deutsche Akademische Austauschdienst (PROCOPE program, grant no D/9822786, jointly with P. Deterre, Paris, F; VIGONI program, grant no 314-vigoni-dr, jointly with A. De Flora, Genova, I), and the Alexander-von-Humboldt-Foundation (Bonn).

REFERENCES

1. Guse AH. 1998. Ca^{2+}-signaling in T-lymphocytes. *Crit. Rev. Immunol.* 18: 419-448.
2. Lewis RS. 2001, Calcium signaling mechanisms in T lymphocytes. *Annu. Rev. Immunol.* 19: 497-521.
3. Grafton G and Thwaite L. 2001. Calcium channels in lymphocytes. *Immunology* 104: 119-126.
4. Premack BA and Gardner P.1992. Signal transduction by T-cell receptors: mobilization of Ca^{2+} and regulation of Ca^{2+}-dependent effector molecules. *Am. J. Physiol.* 263: C1119-C1140.
5. Putney JW Jr. 1986. A model for receptor-regulated calcium entry. *Cell Calcium* 7: 1-12.
6. Guse AH and Emmrich F. 1991. T-cell receptor-mediated metabolism of inositol polyphosphates in Jurkat T-lymphocytes. *J. Biol. Chem.* 266: 24498-24502.
7. Guse AH, Goldwich A., Weber K and Mayr GW. 1995. Non-radioactive, isomer-specific inositol phosphate mass determinations: Micro Metal-Dye detection-HPLC strongly improves speed and sensitivity of analyses from cells and micro-enzyme assays. *J. Chromatogr. B* 672: 189-198.
8. Ng J, Gustavsson J, Jondal M and Andersson T. 1990. Regulation of calcium influx across the plasma membrane of the human T-leukemic cell line, JURKAT: dependence on a rise in cytosolic free calcium can be dissociated from formation of inositol phosphates. *Biochim. Biophys. Acta.* 1053: 97-105.

9. Brattsand G, Cantrell DA, Ward S. Ivars F and Gullberg M. 1990. Signal transduction through the T cell receptor-CD3 complex. Evidence for heterogeneity in receptor coupling. *J. Immunol.* 144: 3651-3658.
10. Wolff C, Akerman KE and Andersson LC, 1985. Kinetics of long-term (72hr) calcium content during mitogen activation of cultured human T-lymphocytes. *J. Cell. Physiol.* 123: 46-50.
11. Guse AH, Roth E and Emmrich F. 1993. Intracellular Ca^{2+}-pools in Jurkat T-lymphocytes. *Biochem. J.* 291: 447-451.
12. Guse AH. 1999. Cyclic ADP-ribose: a novel Ca^{2+}-mobilizing second messenger. *Cell. Signalling* 11: 309-316.
13. Guse AH. 2000. Cyclic ADP-ribose. *J. Mol. Med.* 78: 26-35.
14. Guse AH. 2000.The calcium mobilising second messenger cyclic ADP-ribose. In *Calcium: the molecular basis of calcium action in biology and medicine*, eds R. Pochet, R. Donato, J. Haiech. C. Heizmann. V. Gerke. pp.109-128, Dordrecht, Kluwer Academic Publishers.
15. Guse AH and DaSilva CP. 2000. Intracellular Ca^{2+} release mechanisms: multiple pathways having multiple functions within the same cell type? *Biochim. Biophys. Acta* (Molecular Cell Research Special Issue 2000) 1498: 122-133.
16. Guse AH. 2002. Cyclic ADP-ribose (cADPR) and nicotinic acid adenine dinucleotide phosphate (NAADP): novel regulators of Ca^{2+} signaling and cell functions. *Curr. Mol. Med.* (in press).
17. Guse AH. da Silva CP, Berg I. Skapenko AL, Weber K, Heyer P, Hohenegger M, Ashamu GA, Schulze-Koops H. Potter BVL and Mayr GW. 1999. Regulation of Ca^{2+}-signaling in T-lymphocytes by the second messenger cyclic ADP-ribose. *Nature* 398: 70-73.
18. Berg I. Potter BVL. Mayr GW and Guse AH. 2000. Nicotinic acid adenine dinucleotide phosphate ($NAADP^{+}$) is an essential regulator of T-lymphocyte Ca^{2+}-signaling. *J. Cell Biol.* 150: 581-588.
19. da Silva CP. Potter BVL. Mayr GW and Guse AH. 1998. Quantification of intracellular levels of cyclic ADP-ribose by high-performance liquid chromatography. *J. Chromatogr. B* 707: 43-50.
20. Takahasi K. Kukimoto I. Tokita K-i, Inageda K, Inoue S-i, Kontani K, Hoshino S-i, Nishina H, Kanaho Y and Takada T. 1995. Accumulation of cyclic ADP-ribose measured by a specific radioimmunoassay in differentiated human leukemic HL-60 cells with all-trans-retinoic acid. *FEBS Lett.* 371: 204-208.
21. Schweitzer K, Mayr GW, and Guse AH. 2001. Assay for ADP-ribosyl cyclase by reverse-phase HPLC. *Anal. Biochem.* 15: 218-226.
22. da Silva CP, Heyer P, Schweitzer K, Malavasi F, Mayr GW, and Guse AH. 1998. Ectocellular CD38-catalyzed synthesis and intracellular Ca^{2+}-signalling activity of cyclic ADP-ribose in T-lymphocytes are not functionally related. *FEBS Lett.* 439: 291-296.
23. Bruzzone S. Guida L. Zocchi E. Franco L and De Flora A. 2001. Connexin 43 hemi channels mediate Ca^{2+}-regulated transmembrane NAD^{+} fluxes in intact cells. *FASEB J.* 15:10-12.
24. Franco L. Guida L. Bruzzone S, Zocchi E, Usai C and De Flora A. 1998. The transmembrane glycoprotein CD38 is a catalytically active transporter responsible for generation and influx of the second messenger cyclic ADP-ribose across membranes. *FASEB J.* 12: 1507-1520.
25. Zocchi E, Usai C, Guida L, Franco L, Bruzzone S, Passalacqua M and De Flora A. 1999. Ligand-induced internalization of CD38 results in intracellular Ca^{2+} mobilization: role of NAD^{+} transport across cell membranes. *FASEB J.* 13: 273-283.
26. Adebanjo OA, Anandatheerthavarada HK, Koval AP, Moonga BS, Biswas G, Sun L, Sodam BR, Bevis PJ, Huang CL, Epstein S, Lai FA, Avadhani NG and Zaidi M. 1999. A

new function for CD38/ADP-ribosyl cyclase in nuclear Ca^{2+} homeostasis. *Nature Cell Biol.* 1: 409-414.

27. Khoo KM, Han MK. Park JB, Chae SW, Kim UH, Lee HC, Bay BH and Chang CF. 2000. Localization of the cyclic ADP-ribose-dependent calcium signaling pathway in hepatocyte nucleus. *J. Biol. Chem.* 275: 24807-24817.
28. Ziegler M, Jorcke D and Schweiger M. 1997. Identification of bovine liver mitochondrial NAD^+ glycohydrolase as ADP-ribosyl cyclase. *Biochem. J.* 326: 401-405.
29. Bourguignon LY, Chu A, Jin H and Brandt NR. 1995. Ryanodine receptor-ankyrin interaction regulates internal Ca^{2+} release in mouse T-lymphoma cells. *J. Biol. Chem.* 270: 17917-17922.
30. Guse AH, da Silva CP, Emmrich F, Ashamu GA, Potter BVL and Mayr GW. 1995. Characterization of cyclic adenosine diphosphate-ribose-induced Ca^{2+}-release in T-lymphocyte cell lines. *J. Immunol.* 155: 3353-3359.
31. Takemura H, Imoto K, Sakano S. Kaneko M and Ohshika H. 1996. Lysophosphatidic acid-sensitive intracellular Ca^{2+} store does not regulate Ca^{2+} entry at plasma membrane in Jurkat human T-cells. *Biochem. J.* 319: 393-397.
32. Vu CQ, Coyle DL and Jacobson MK. 1997. Natural occurrence of 2'-phospho-cyclic ADP ribose in mammalian tissues. *Biochem. Biophys. Res. Commun.* 236: 723-726.
33. Guse AH, da Silva CP, Weber K, Armah C, Ashamu GA, Schulze C, Potter BVL, Mayr GW and Hilz H. 1997. 1-(5-Phopho-ß-D-ribosyl)2′-phosphoadenosine 5′-phosphate cyclic anhydride-induced Ca^{2+}-release in human T cell lines. *Eur. J. Biochem.* 245: 411-417.
34. Guse AH, Berg I, da Silva CP, Potter BVL and Mayr GW. 1997. Ca^{2+}-entry induced by cyclic ADP-ribose in intact T-lymphocytes. *J. Biol. Chem.* 272: 8546-8550.
35. Hakamata Y, Nishimura S, Nakai J, Nakashima Y, Kita T and Imoto K. 1994. Involvement of the brain type of ryanodine receptor in T-cell proliferation. *FEBS Lett.* 352: 206-210.
36. Ricard I, Martel J, Dupuis L, Dupuis G and Payet MD. 1997. A caffeine/ryanodine-sensitive Ca^{2+} pool is involved in triggering spontaneous variations of Ca^{2+} in Jurkat T lymphocytes by a Ca^{2+}-induced Ca^{2+} release (CICR) mechanism. *Cell. Signal.* 9: 197-206.
37. Hohenegger M, Berg I, Weigl L, Mayr GW, Potter BVL and Guse AH. 1999. Pharmacological activation of the ryanodine receptor in Jurkat T-lymphocytes. *Brit. J. Pharmacol.* 128: 1235-1240.
38. Guse AH, da Silva CP, Weber K, Ashamu GA, Potter BVL and Mayr GW. 1996. Regulation of cyclic ADP-ribose-induced Ca^{2+}-release by Mg^{2+} and inorganic phosphate. *J. Biol. Chem.* 271: 23946-23954.
39. Guse AH, Cakir-Kiefer C, Fukuoka M, Shuto S, Weber K, Bailey VC, Matsuda A, Mayr GW, Oppenheimer N, Schuber F and Potter BVL. 2002. Novel hydrolysis-resistant analogues of cyclic ADP-ribose: modification of the "northern" ribose and calcium release activity. *Biochemistry* (in press).
40. Aarhus R, Graeff RM, Dickey DM, Walseth TF and Lee HC. 1995. ADP-ribosyl cyclase and CD38 catalyze the synthesis of a calcium-mobilizing metabolite from NADP. *J. Biol. Chem.* 270: 30327-30333.
41. Zhang FJ, Gu QM, Jing P and Sih CJ. 1995. Enzymatic cyclization of nicotinamide adenine dinucleotide phosphate (NADP). *Bioorg. Med. Chem. Lett.* 5: 2267-2272.
42. Vu CQ, Lu PJ, Chen CS and Jacobson MK. 1996. 2'-Phospho-cyclic ADP-ribose, a calcium-mobilizing agent derived from NADP. *J. Biol. Chem.* 271: 4747-4754.
43. Ashamu GA, Sethi JK, Galione A and Potter BV. 1997. Roles for adenosine ribose hydroxyl groups in cyclic adenosine 5'-diphosphate ribose-mediated Ca^{2+} release. *Biochemistry* 36: 9509-9517.
44. Shuto S, Fukuoka M, Manikowsky A, Ueno Y, Nakano T, Kuroda R, Kuroda H and Matsuda A. 2001. Total synthesis of cyclic ADP-carbocyclic-ribose, a stable mimic of

Ca^{2+}-mobilizing second messenger cyclic ADP-ribose. *J. Am. Chem. Soc.* 123: 8750-8759.
45. Walseth TF and Lee HC. 1993. Synthesis and characterization of antagonists of cyclic-ADP-ribose-induced Ca^{2+} release. *Biochim. Biophys. Acta* 1178: 235-242.
46. Sethi JK, Empson RM, Bailey VC, Potter BV and Galione A. 1997. 7-Deaza-8-bromo-cyclic ADP-ribose, the first membrane-permeant, hydrolysis-resistant cyclic ADP-ribose antagonist. *J. Biol. Chem.* 272: 16358-16363.
47. Schöttelndreier H, Potter BVL, Mayr GW and Guse AH. 2001. Mechanisms involved in α6ß1-integrin-mediated Ca^{2+} signalling. *Cell. Signalling* 13: 895-899.
48. Jayaraman T, Ondrias K, Ondriasova E and Marks AR. 1996. Regulation of the inositol 1,4,5-trisphosphate receptor by tyrosine phosphorylation. *Science* 272: 1492-1494.
49. Guse AH, Tsygankov AY, Weber K and Mayr GW. 2001. Transient tyrosine phosphorylation of human ryanodine receptor upon T cell stimulation. *J. Biol. Chem.* 276: 34722-34727.
50. Guse AH, de Wit C, Klokow T, Schweitzer K, and Mayr GW. 1997. Unique properties of the capacitative Ca^{2+}-entry antagonist LU 52396: Its inhibitory activity depends on the activation state of the cells. *Cell Calcium* 22: 91-97.

Andreas H. Guse, University Hospital Hamburg-Eppendorf,
Institute for Medical Biochemistry and Molecular Biology,
Division of Cellular Signal Transduction, Martinistr. 52,
D-20246 Hamburg, Germany.
e-mail: guse@uke.uni-hamburg.de

Cyclic ADP-Ribose and NAADP. Structures, Metabolism and Functions. Lee HC (ed), pp. 319-341, Dordrecht: Kluwer.

Calcium Signaling by cADPR in Cardiac Myocytes

Stevan Rakovic and Derek A Terrar
Department of Pharmacology, Oxford University, Mansfield Road, Oxford OX1 3QT, UK.

INTRODUCTION

The heart is a muscular organ, composed of electrically connected excitable cells that act in a coordinated, syncytial fashion to pump blood to the other tissues of the body. Since Ringer's observation in 1883 that Ca^{2+} is essential for contraction of the heart [1], major advancements have been made in elucidating the mechanisms by which electrical excitation of the myocardium (the cardiac action potential, initiated by the pacemaker region) is linked to mechanical contraction of the heart, and it is now well established that Ca^{2+} ions play the fundamental role in this coupling of excitation to contraction.

A simplified scheme illustrating excitation-contraction coupling is presented in Figure 1. Ca^{2+} entry across the sarcolemma (external cell membrane) during the cardiac action potential, mainly through L-type voltage-gated Ca^{2+} channels, is believed to trigger the release of a more substantial quantity of Ca^{2+} from the sarcoplasmic reticulum (SR), an intracellular membrane-delimited store of Ca^{2+} ions, via a mechanism known as Ca^{2+}-induced Ca^{2+} release (CICR) [2]. The resulting brief elevation of intracellular Ca^{2+}, from a resting level of approximately 100 nM to a value in excess of 1 μM, is the signal that initiates the process of mechanical contraction, through an interaction of Ca^{2+} with the myofilament protein troponin C.

The majority of the Ca^{2+} for triggering contraction comes from the SR, an extensive intracellular membrane network that has the capacity to sequester Ca^{2+} (through the action of a Ca^{2+}-ATPase), store it (buffered by intra-SR proteins such as calsequestrin and calreticulin), and release it via ryanodine receptor Ca^{2+} release channels (RyRs) [2-5]. These Ca^{2+} release channels are large homotetrameric proteins (four identical 565 kDa subunits) that span the

SR membrane [6], and are particularly concentrated in the terminal cisternae, expanded regions of the SR specialized for Ca^{2+} release. Three RyR isoforms are believed to be expressed in mammals, encoded by different genes, the predominant cardiac subtype being RyR2 [7]. This isoform mediates the CICR that underlies excitation-contraction coupling in the heart, Ca^{2+} itself acting as the ligand that binds to the RyR and enhances its probability of opening [8].

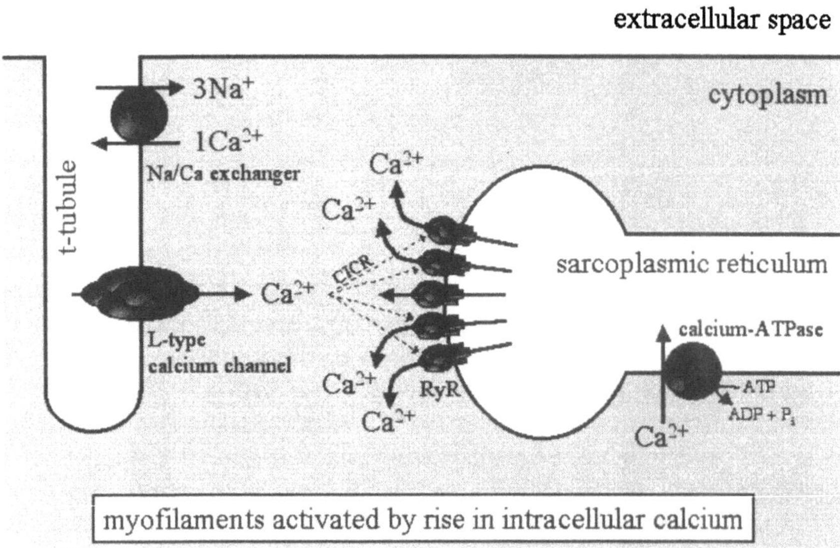

Figure 1. Excitation-contraction coupling in the heart. Ca^{2+} entry through L-type Ca^{2+} channels (which are voltage-gated, and open during the plateau phase of the action potential) raises the local Ca^{2+} concentration in the vicinity of nearby RyRs (located in the terminal cisternae of the SR), causing them to open, and hence resulting in Ca^{2+}-induced release of Ca^{2+} from the SR. Ca^{2+} from both sources diffuses to the myofilaments, initiating contraction by binding to the protein troponin C. Subsequently, this Ca^{2+} is transported back into the SR by a Ca^{2+}-ATPase, and extruded from the cell by Na^+/Ca^{2+} exchange (with a minor additional contribution made by a sarcolemmal Ca^{2+}-ATPase), so that relaxation occurs.

The large cytoplasmic region of the RyR provides a potential opportunity for RyR regulation by intracellular proteins and molecules, and indeed it is known that many such substances can interact with and influence the cardiac RyR – these include magnesium, adenine nucleotides, calmodulin, FK506 binding protein 12.6 (FKBP12.6) and sorcin; the RyR may also be phosphorylated by several protein kinases, including cAMP-dependent protein kinase and calmodulin-dependent protein kinase [5, 9]. The discovery of cADPR as a molecule that induces Ca^{2+} release from endo/sarcoplasmic reticulum stores in many cells types via a ryanodine-sensitive mechanism (as discussed in other chapters in this book) has raised the possibility that cADPR may also play a role in the heart as a regulator of Ca^{2+} release from the SR. This chapter will consider the evidence that cADPR is an endogenous modulator of CICR in the heart, describe what is known about the regulation of cADPR levels in cardiac myocytes, and discuss the debate surrounding the mechanism of action of cADPR in heart cells.

ACTIONS OF cADPR AND ITS ANTAGONISTS IN INTACT MYOCYTES

Effects of the cADPR Antagonist, 8-amino-cADPR

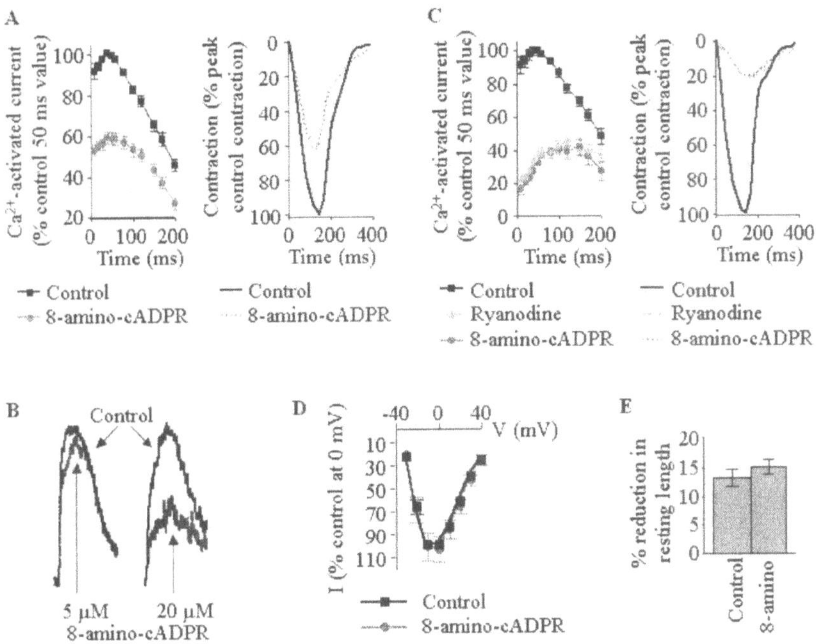

Figure 2. **Effects of 8-amino-cADPR in guinea-pig ventricular myocytes stimulated to fire action potentials. A:** 20 µM 8-amino-cADPR reduces the magnitudes of Ca^{2+} transients (constructed from the magnitudes of Ca^{2+}-activated currents) and cell contractions (measured from the video image of cells using an edge-detection system). **B:** 8-amino-cADPR suppresses Ca^{2+} transients measured using fura-2 as a Ca^{2+} indicator. **C:** In the presence of ryanodine to interfere with the function of the SR, 20 µM 8-amino-cADPR is without significant effect on Ca^{2+} transients and contractions. **D:** 20 µM 8-amino-cADPR is without significant effect on the current-voltage relationship for L-type Ca^{2+} current, measured under voltage-clamp conditions. **E:** The contractile response to rapid application of 20 mM caffeine is not reduced by 20 µM 8-amino-cADPR, indicating that the inhibitory effects of 8-amino-cADPR on Ca^{2+} transients and contractions are not due to a reduction in SR Ca^{2+} load.

The first evidence that cADPR may modulate excitation-contraction coupling in intact cardiomyocytes was provided by Rakovic *et al.* [10], who reported inhibitory effects of 8-amino-cADPR, a competitive antagonist of cADPR, on the function of guinea-pig ventricular myocytes. In cells stimulated to fire action potentials, application to the cell cytoplasm of 8-amino-cADPR resulted in a reduction in the magnitudes of Ca^{2+} transients (measured using Ca^{2+}-activated currents) and contractions without reducing Ca^{2+} entry through L-type Ca^{2+} current (Figure 2). These effects of 8-amino-cADPR were not observed in the presence of either ryanodine (to interfere with SR Ca^{2+} release) or thapsigargin (to inhibit Ca^{2+} sequestration by the SR), consistent with an action of the cADPR antagonist to suppress Ca^{2+} release from the SR. Furthermore, the extent to which the SR was loaded with Ca^{2+} (assessed from the contractile response to rapid application of 20 mM caffeine) was not reduced by 8-amino-cADPR, indicating that the reductions in the magnitudes of the Ca^{2+} transients and contractions were not a consequence of a decrease in the SR Ca^{2+} load. In view of evidence in the sea-urchin egg that 8-amino-cADPR is a competitive antagonist of cADPR binding and action [11], the most likely explanation for these findings is that 8-amino-cADPR was acting to inhibit the effects of endogenous levels of cADPR, which were sensitizing the CICR mechanism to Ca^{2+}.

Effects of cADPR in Guinea-Pig Ventricular Myocytes

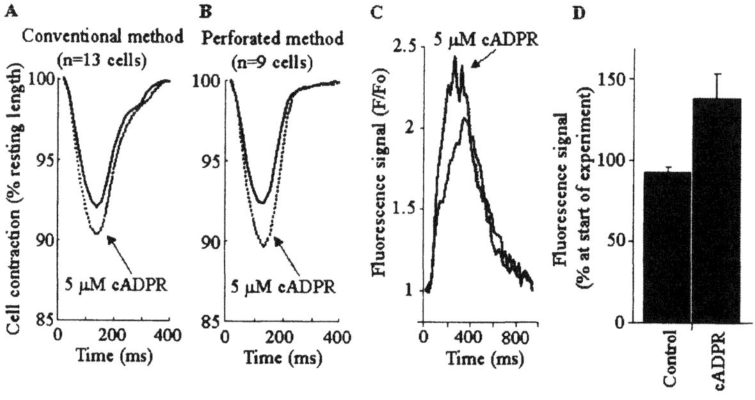

Figure 3. **cADPR (5 µM) enhances Ca^{2+} transients and contractions in guinea-pig ventricular myocytes stimulated to fire action potentials. A:** cADPR, applied to the cytoplasm via patch electrodes (whole-cell configuration), increased cell contraction by 25 ± 7 % (control measurements made 30 s after membrane rupture, when little cADPR should have diffused into the cell). **B:** In these experiments, control measurements were made in the amphotericin-perforated patch configuration, after which physical rupture of the membrane was carried out to allow diffusion of cADPR into the cytoplasm. The mean increase was 30 ± 5 %. **C:** cADPR enhanced Ca^{2+} transients, measured using fura-2 as the Ca^{2+} indicator. **D:** Bar graph showing the mean effect of cADPR on the Ca^{2+} transient in a series of experiments.

Iino *et al.* (1997) were the first to report effects of exogenously applied cADPR in intact heart cells under quasi-physiological conditions [12]. In guinea-pig ventricular myocytes stimulated to fire action potentials at 1 Hz, application of 5 μM cADPR to the cytoplasm via a patch-pipette (in the whole-cell patch configuration) enhanced the magnitudes of both Ca^{2+} transients (measured using fura-2 as the Ca^{2+} indicator, with fluorescence detected with an image-intensified CCD camera) and contractions (Figure 3). The augmenting effect of cADPR on cell contraction was found to occur in a concentration-dependent manner, with an EC_{50} of 2 – 3 μM, and was not observed in the presence of inhibitors of SR function (ryanodine and thapsigargin). Furthermore, these actions of cADPR could be prevented by two competitive antagonists of cADPR, 8-amino-cADPR and 8-bromo-cADPR, and this antagonism could be overcome by increasing the concentration of cADPR. These data were thus consistent with an enhancement of SR Ca^{2+} release by cADPR, perhaps achieved by a sensitization of the CICR mechanism to Ca^{2+}, as has been reported for other tissues.

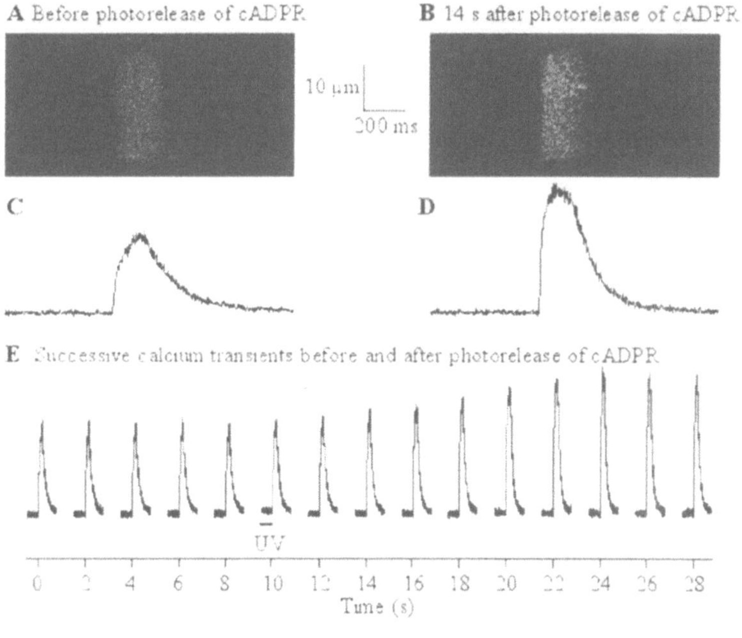

Figure 4. **UV-photolysis of caged cADPR (1 mM in the patch pipette) enhances Ca^{2+} transients in guinea-pig ventricular myocytes imaged with confocal microscopy (fluo-3). A:** Linescan image showing a Ca^{2+} transient accompanying an action potential. **B:** Photorelease of cADPR increases the Ca^{2+} transient magnitude (brighter fluo-3 fluorescence). **C, D:** More quantitative representations of the Ca^{2+} transients shown above in panels A and B, obtained by averaging the magnitude of fluorescence across the cell. **E:** The effect of photoreleased cADPR on Ca^{2+} transients develops slowly, taking 14 s for development of full effect in this cell.

A similar potentiation of Ca^{2+} transients by UV-photolysis of caged cADPR has been reported by Cui and coworkers [13] in guinea-pig ventricular myocytes, imaged using scanning laser confocal microscopy (with fluo-3 as the Ca^{2+} indicator). In cells stimulated to fire action potentials, uncaging of NPE-caged cADPR resulted in a 29 % increase in the magnitude of the Ca^{2+} transient, and this could be prevented by inclusion of the cADPR antagonist, 8-amino-cADPR. Interestingly, this effect of photoreleased cADPR was not immediate in onset, requiring approximately 15 seconds to reach the peak effect (Figure 4); this may reflect an indirect action of cADPR on the ryanodine receptor, perhaps involving accessory proteins or even an enzymatic step (see later).

Effects of cADPR in Rat Ventricular Myocytes

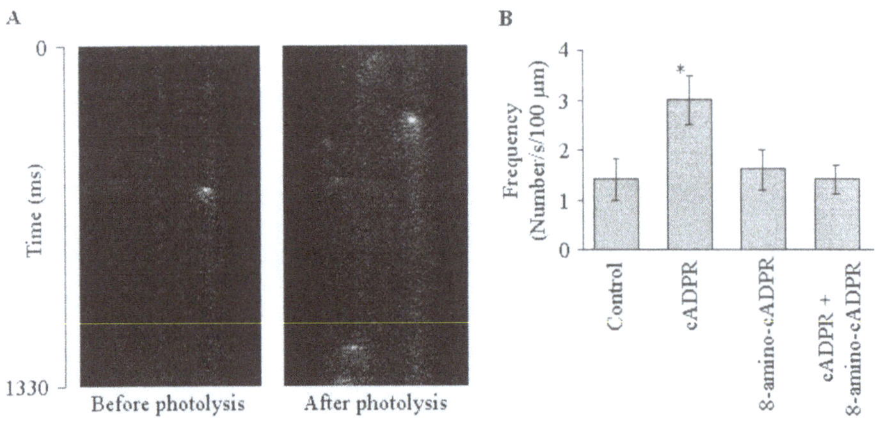

Figure 5. **UV-photolysis of caged cADPR (1 mM in the patch pipette) increases Ca^{2+} spark frequency in rat ventricular myocytes, imaged using confocal microscopy (fluo-3). A:** Line-scan images of a rat cell before (left) and after (right) photorelease of cADPR. **B:** Mean changes in spark frequency induced by cADPR in the presence and absence of 8-amino-cADPR (a competitive antagonist). In the absence of 8-amino-cADPR, photorelease of cADPR enhanced spark frequency approximately two-fold. In contrast, no significant effects of photoreleased cADPR on spark frequency were observed in the presence of 8-amino-cADPR.

Augmenting actions of cADPR have also been reported in rat heart cells by Cui *et al.* [13]. In rat ventricular myocytes stimulated to fire action potentials, and imaged using confocal microscopy, UV-photolysis of caged cADPR was associated with a 16 % enhancement in the Ca^{2+} transient magnitude; as in guinea-pig myocytes, this effect required approximately 15 seconds to fully develop. In rat myocytes at rest, UV photolysis of caged cADPR was found to increase the frequency of occurrence of Ca^{2+} sparks (spatially localized Ca^{2+} release events from clusters of ryanodine receptors), as shown in Figure 5; the individual spark characteristics (time to peak, rate of decay and width at half-maximum amplitude) were not significantly altered. These effects of cADPR on spark frequency were not observed in the presence of the cADPR antagonist, 8-amino-cADPR. Since Ca^{2+} transients are believed to result from the summation of many Ca^{2+} sparks, the action of cADPR to increase the probability of spark occurrence may underlie the augmentation of the whole-cell Ca^{2+} transient by cADPR, since more sparks would occur in response to the same trigger Ca^{2+}, amounting to an increase in the gain of CICR.

Prakash and colleagues [14] have also observed effects of cADPR in rat ventricular myocytes imaged using confocal microscopy. Microinjection of cADPR (0.1 – 3 μM) caused concentration-dependent increases in intracellular Ca^{2+} levels in resting myocytes, increased the magnitudes of Ca^{2+} transients in electrically paced cells (0.5 Hz), and increased the amplitude and frequency of Ca^{2+} waves in cells showing spontaneous activity (presumably due to Ca^{2+} overload). Interestingly, at higher concentrations (1 and 3 μM), cADPR also increased the duration of Ca^{2+} transients in paced cells (14 and 26 % respectively), an observation not found in the other studies described above. The cADPR antagonist 8-amino-cADPR reduced the amplitude and frequency of spontaneous Ca^{2+} waves, and decreased the magnitude of the Ca^{2+} transient by 78 % in electrically paced cells (a response greater than that observed in guinea-pig cells, as might be expected from the greater contribution of SR Ca^{2+} release to the Ca^{2+} transient in the rat); however, a substantial response to caffeine could still be evoked, illustrating a significant degree of SR Ca^{2+} loading. Interestingly, the authors also reported that a small response to cADPR persisted in the presence of ryanodine (to inhibit SR function), despite a lack of response to caffeine. This was interpreted in terms of cADPR having ryanodine-insensitive, as well as ryanodine-sensitive mechanisms of action.

The Importance of Temperature for the Actions of cADPR in the Heart

In contrast to these studies, Guo and colleagues [15] have reported no effect of cADPR or 8-amino-cADPR on Ca^{2+} transients in rat ventricular cells under voltage-clamp conditions. UV flash photolysis of caged cADPR (100 μM in the patch pipette), both in the presence and absence of calmodulin, was without effect on intracellular Ca^{2+} levels, whereas local application of caffeine (2.5 mM) or flash photolysis of caged Ca^{2+} (1 mM) was effective in provoking increases in intracellular Ca^{2+} in these cells; furthermore, photolysis of caged cADPR was able to evoke Ca^{2+} transients in sea urchin eggs. Flash photolysis of caged cADPR was also unable to affect Ca^{2+} transients accompanying step depolarisations to 0 mV (in order to activate Ca^{2+} entry through L-type Ca^{2+} current, and hence trigger CICR from the SR). In addition, inclusion of 100 μM cADPR or 10 μM 8-amino-cADPR in the patch pipette did not alter the relationship between Ca^{2+} transient magnitude and peak Ca^{2+} current, consistent with a lack of effect of these agents on CICR.

However, these experiments were carried out at room temperature, and subsequently it was reported by Iino and colleagues in guinea-pig ventricular myocytes that the effects of cADPR and its antagonists were critically dependent on temperature [12]. At 36° C, 5 μM cADPR (applied via a patch pipette) was found to significantly enhance the magnitude of cell contraction, and both 8-amino-cADPR (5 μM) and 8-bromo-cADPR (50 μM) were found to reduce contraction; however, at room temperature, no significant effects of cADPR, 8-amino-cADPR or 8-bromo-cADPR were observed. The precise mechanisms underlying such temperature dependence of the actions of cADPR and its antagonists remain to be elucidated, but it is possible that this reflects a temperature sensitivity of the binding of cADPR to the postulated binding proteins, or of the interaction between the binding protein-ligand complex and the RyR.

cADPR UNDER CONDITIONS OF SR CALCIUM OVERLOAD AND ARRHYTHMOGENESIS

In addition to the effects of cADPR on the stimulated contraction of healthy myocytes, Rakovic *et al.* have provided evidence that sensitization of the CICR mechanism by cADPR may also be important during abnormal conditions (arrhythmias) associated with SR Ca^{2+} overload and spontaneous oscillations of cell Ca^{2+} and membrane potential [16]. In this study, 8-amino-cADPR was found to suppress spontaneous activity in guinea-pig ventricular myocytes exposed to high concentrations of the ß-adrenoceptor agonist isoproterenol or the Na^+/K^+-ATPase inhibitor ouabain (to provoke SR Ca^{2+} overload and arrhythmogenic activity). The spontaneous events suppressed by 8-amino-cADPR included intracellular Ca^{2+} waves, spontaneous action potentials, afterdepolarisations and transient inward currents (Figure 6). Another antagonist of cADPR, 8-bromo-cADPR, was also effective in suppressing isoproterenol-induced oscillatory activity. Furthermore, in the presence of ouabain under conditions in which there was no arrhythmogenesis, exogenous cADPR was found to be capable of triggering spontaneous contractile and electrical activity. It therefore seems possible that cADPR may exert an important influence on arrhythmogenic activity in the heart, particularly under conditions where loading of the SR with Ca^{2+} is high, and hence compounds which reduce the actions of endogenous cADPR may prove useful in the treatment of certain cardiac arrhythmias.

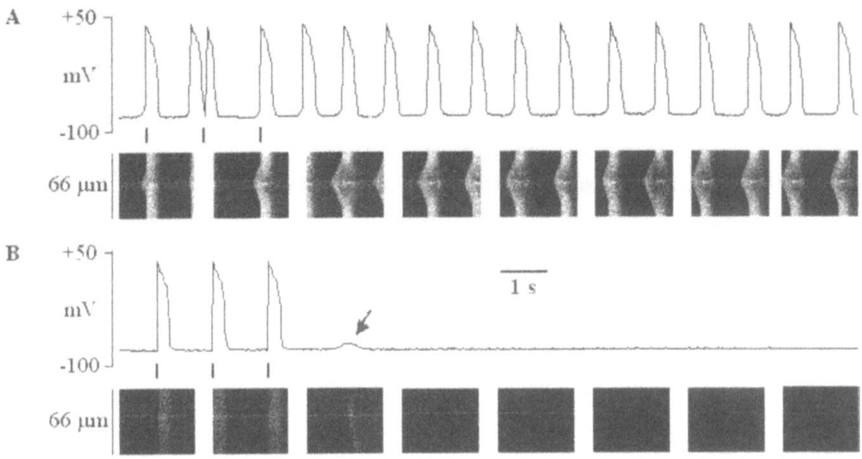

Figure 6. **Effects of 8-amino-cADPR on isoproterenol-provoked Ca^{2+} waves in guinea-pig ventricular myocytes, imaged using confocal microscopy (fluo-3). A:** Membrane potential records (upper panel) and line scan images (lower panel) obtained from a single cell superfused for 3 minutes with isoproterenol (50 nM). Current stimuli were applied at the vertical lines. In this cell, isoproterenol provoked the generation of spontaneous action potentials, each of which was associated with elevations of cytosolic Ca^{2+}. Preceding the upstroke of each spontaneous action potential was a Ca^{2+} wave, initiated at a locus near the centre of the scanned line, and subsequently spreading in opposite directions along the scanned line towards both edges of the cell. The wave was terminated by a global increase in Ca^{2+} that was associated with the upstroke of the spontaneous action potential. **B:** Data from the same cell, recorded 5 minutes after infusion of 8-amino-cADPR (20 µM) in the continued presence of isoproterenol. Spontaneous action potential generation had been abolished, although some spontaneous activity persisted in the form of a single afterdepolarisation (arrow, upper panel) accompanied by a Ca^{2+} wave.

REQUIREMENT OF CALMODULIN AND CALMODULIN-DEPENDENT PROTEIN KINASE II FOR EFFECTS OF cADPR IN THE HEART

In the sea urchin egg, it has been demonstrated that calmodulin is essential for normal cADPR responsiveness [17-19]. Furthermore, there is also evidence that phosphorylation of the RyR by calmodulin-dependent protein kinase II (CAM-KII) may be required to confer cADPR sensitivity in pancreatic islets [20]. We have recently obtained evidence that the enhancing effects of cADPR on Ca^{2+} transients and contractions of guinea-pig ventricular myocytes are sensitive to inhibitors of calmodulin and CAM-KII [21]. The calmodulin antagonists calmidazolium and W7 were both found to reduce the magnitude of cell contraction; in the presence of either of these calmodulin antagonists, cADPR (applied via a patch pipette) was found to have no significant further effect on cell contraction (Figure 7). Similarly, in cells imaged with confocal microscopy, UV-photolysis of caged cADPR was without effect on the Ca^{2+} transient in the presence of either calmidazolium, or the CAM-KII inhibitor KN93 (Figure 7). In agreement with these observations, Okabe and colleagues have recently reported that cADPR-induced Ca^{2+} release from cardiac SR vesicles only occurred when the release mechanism was sensitized by calmodulin [22-23].

These studies thus provide evidence for a role of calmodulin and CAM-KII in the actions of cADPR, and are hence consistent with observations in sea urchin eggs and other tissues. However, the precise mechanism of action of cADPR, and the nature of the involvement of calmodulin-dependent processes, remains to be established.

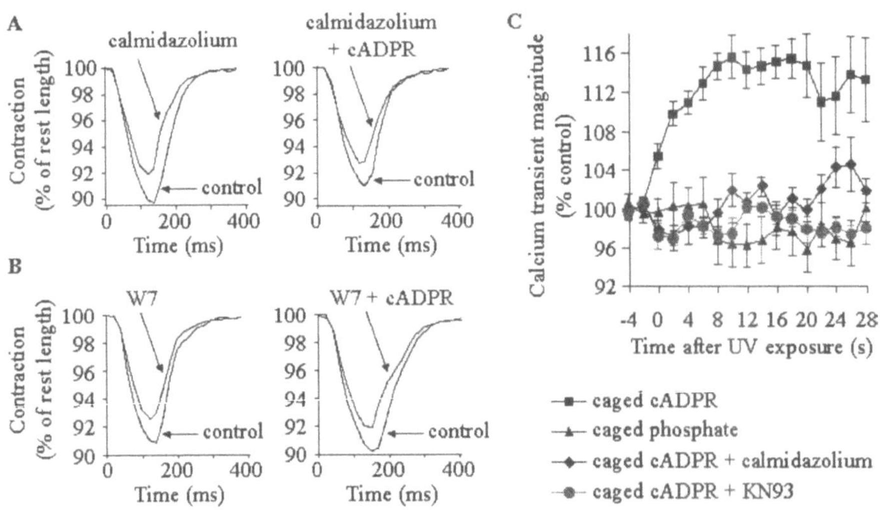

Figure 7. **Requirement of calmodulin and CAM-KII for the actions of cADPR in the heart. A:** The calmodulin antagonist calmidazolium reduces the magnitude of cell contraction in guinea-pig ventricular myocytes stimulated to fire action potentials (left). The reduction in contraction observed with co-application of cADPR and calmidazolium is not significantly different from that seen with calmidazolium alone, i.e. in the presence of calmidazolium, cADPR is without further effect. **B:** Similar results are obtained with another calmodulin antagonist, W7, that is structurally different to calmidazolium: the reduction in contraction caused by co-application of cADPR with W7 is not significantly different from that of W7 alone. **C:** In guinea-pig ventricular myocytes stimulated to fire action potentials, and imaged using confocal microscopy (fluo-4), UV-photolysis of caged cADPR (1 mM) results in an enhancement of the Ca^{2+} transient, which requires approximately 10 s to fully develop. In contrast, photorelease of cADPR is without significant effect in the presence of calmidazolium (a calmodulin inhibitor) or KN93 (an inhibitor of CAM-KII). UV-photolysis of caged phosphate was used as a control (no significant effects on Ca^{2+} transients).

ROLE OF FK506 BINDING PROTEIN 12.6 (FKBP12.6) IN THE ACTIONS OF cADPR

FKBP12.6 is a low molecular mass protein that is known to interact with cardiac RyRs, and has been proposed to regulate Ca^{2+} release from the SR through this mechanism [24-26]. FKBP12.6 is thought to stabilize the RyR in a closed configuration, such that its removal from the RyR (for example, induced by agents such as FK506 or rapamycin) would tend to enhance RyR opening and hence SR Ca^{2+} release. In pancreatic islet tissue, Noguchi and colleagues have reported that cADPR promotes the dissociation of FKBP12.6 from the RyR2, and have suggested that this underlies the mechanism by which cADPR causes Ca^{2+} release from pancreatic islet SR [27].

Recently, evidence has been presented that a similar mechanism may occur in the heart. Xin and colleagues have reported [28] that cADPR exerts an enhancing effect on Ca^{2+} sparks in cardiac myocytes in wild type mice, but not in those of FKBP12.6 knockout mice; furthermore, the spark characteristics in wild type cells dialysed with cADPR were similar to those of the FKBP12.6 knockout mice. The authors suggested that cADPR may modify CICR in cardiac myocytes by releasing FKBP12.6 from the RyR2, as has been proposed for the pancreas. Interestingly, the exchange reaction between soluble FKBP12 and the FKBP-RyR complex has been shown to be temperature sensitive [29], and this may explain the temperature dependence of cADPR and its antagonists described by Iino and colleagues [12]. Whether cADPR interacts directly with FKBP12.6, or does so via intermediate proteins/factors, remains to be resolved.

It is interesting to note that McCall *et al.* [25] have reported effects of FK506 in rat ventricular myocytes that resemble those of cADPR we have observed in our preparations [10, 12-13]. In their studies, FK506 was found to enhance cell contraction even in myocytes in which the SR Ca^{2+} load had not increased (indicative of an augmentation of SR Ca^{2+} release), and to increase Ca^{2+} transient magnitude without significantly influencing its decay (suggesting a lack of effect on SR Ca^{2+} uptake), with no apparent change in L-type Ca^{2+} current; furthermore, Ca^{2+} spark frequency in resting myocytes was greatly enhanced, with no significant change in their characteristics. These observations would thus be consistent with the hypothesis that cADPR may exert its actions through a mechanism involving FKBP12.6.

EFFECTS OF cADPR IN PURIFIED CARDIAC MEMBRANE PREPARATIONS

The first work to show a possible effect of cADPR on cardiac muscle was that of Meszaros, Bak and Chu in 1993 [30], which demonstrated that cADPR could potentiate the efflux of $^{45}Ca^{2+}$ from passively loaded cardiac (but not skeletal muscle) SR vesicles with a half-maximal concentration of 150 nM. Furthermore, in studies of single cardiac RyRs incorporated into planar lipid bilayers, 1 μM cADPR was observed to greatly increase the RyR open probability at Ca^{2+} concentrations on the cytoplasmic face of the channel of 10 and 100 nM, but not at Ca^{2+} concentrations of 1 μM and above. This was taken as evidence that cADPR could act directly to sensitise the cardiac isoform of the RyR to Ca^{2+} concentrations in the diastolic range.

However, Sitsapesan, McGarry and Williams subsequently reported that, in the presence of millimolar concentrations of ATP, cADPR was unable to open RyRs isolated from sheep heart and studied in planar lipid bilayers [31-32]. Their studies demonstrated a potentiating effect of cADPR in the absence of ATP that was similar to that of Meszaros and colleagues; however, in the presence of low concentrations of ATP (100 μM), there was a slight decrease in the channel open probability in response to cADPR. As similar effects to cADPR were observed with ß-NAD and ADPR, the authors concluded that the action of cADPR may involve competition with ATP for the nucleotide binding site of the RyR; since the cellular concentration of ATP is significantly higher than that of cADPR, Sitsapesan and colleagues argued that, under physiological conditions, cADPR could not act as a modulator of the RyR [31-32]. In support of this was the observation that, in the presence of 10 mM ATP and 0.5 mM magnesium, cADPR was found to have no effect on the cardiac RyR under the conditions of their experiments.

Interestingly, a recent report from Bak *et al.* [33], in a less purifed cardiac SR membrane preparation, has observed an action of cADPR on Ca^{2+} release in the presence of millimolar concentrations of ATP. cADPR (and also NAADP, another putative modulator of intracellular Ca^{2+} release) was found to enhance $^{45}Ca^{2+}$ efflux from passively loaded rabbit cardiac microsomes, with an EC_{50} in the region of 200 nM. This effect of cADPR could be inhibited by 8-amino-cADPR (a competitive cADPR antagonist) and ruthenium red (an RyR blocker), consistent with an action of cADPR, either direct or indirect, at the RyR. However, this stimulation of Ca^{2+} release by cADPR was observed both in the presence and absence of 1 mM ATP. This is unlikely to be due to an action of cADPR at the nucleotide binding site, since the ATP concentration in this study was 500-fold higher. It thus may be that in addition to the action of cADPR at the nucleotide binding site seen in highly purified membrane preparations, there is also another effect of cADPR which becomes evident in less purified systems, and which may be of physiological relevance.

In contrast to the aforementioned findings, Fruen and colleagues have observed no influence of cADPR on RyR function [34]. cADPR (0.1 – 5 μM), ADPR (0.1 – 5 μM) and ß-NAD (0.1 – 30 mM) were found not to influence ^{3}H-ryanodine binding to porcine cardiac SR vesicles, and 1 μM cADPR was without effect on single RyRs studied in lipid bilayers.

Similarly, Copello and coworkers [35] have reported no effects of cADPR on cardiac RyRs incorporated into lipid bilayers, studied both at room temperature and at 37°C, and either in the presence or absence of calmodulin. In this study, cADPR was also found not to affect FKBP12.6 association to cardiac SR membranes, and no ^{32}P-cADPR binding was detected in SR microsomes. The authors concluded that the observed actions of cADPR in intact myocytes may be due to indirect modulation of RyRs, or RyR-independent release systems.

The differences in these data are difficult to reconcile, but a possible explanation is that cADPR does not act directly on the RyR but via one or more intermediate proteins, and that purification of RyRs for the purposes of bilayer studies results in the loss of one or more factors that are required to confer cADPR sensitivity. Several pieces of evidence are consistent with this explanation. Firstly, Lee and co-workers have demonstrated that purified sea urchin egg microsomes are no longer responsive to cADPR, calmodulin (absent in most of the above experiments) being essential for cADPR action [17-18]. Secondly, Ca^{2+} mobilization by cADPR in the sea urchin egg occurs in the presence of ATP; Graeff *et al.* have found no effects of ATP on cADPR-induced Ca^{2+} release [36], which would not support an action of cADPR via the nucleotide binding site, as proposed by Sitsapesan and coworkers [31-32]. Thirdly, photoaffinity labeling studies of Walseth and colleagues have revealed two cADPR-specific binding proteins in cardiac muscle (of masses 45 kDa and 55 kDa) that do not correspond in size to RyRs [37]; furthermore, the binding of cADPR to these proteins is not affected by ATP. It has also been reported recently, in both pancreatic islets and the heart, that FKBP12.6 may play a role in mediating the effects of cADPR [27-28]. It therefore remains possible that the influence of cADPR on the RyR may be altered when the channel is removed from its native environment; for example, a suitable RyR phosphorylation state, such as phosphorylation by CAM-KII [20], or the presence of one or more accessory proteins/molecules, such as calmodulin or FKBP12.6, may be required to confer cADPR sensitivity.

REGULATION OF SR CALCIUM SEQUESTRATION BY cADPR

cADPR may Enhance Calcium Uptake by the SR

In 1999, Lukyanenko and Gyorke reported that 5 μM cADPR enhanced the frequency of Ca^{2+} spark occurrence, without affecting spark characteristics (amplitude, time to peak, rate of decay), in saponin-permeabilised rat ventricular myocytes imaged using confocal microscopy [38]. This action was seen at room temperature, developed over a 2 – 6 minute period, and was reversible. Subsequently, the same group reported that this effect of cADPR was due not to an action at the RyR, but was the consequence of enhanced Ca^{2+} uptake by the SR [39]. The gradual enhancement of Ca^{2+} spark activity by cADPR was accompanied by an increase in the SR Ca^{2+} content, assessed from the Ca^{2+} transient observed in response to a rapid application of a high concentration of caffeine. In contrast to cADPR, lower concentrations of caffeine (a known agonist at the RyR) caused a rapid enhancement of spark frequency, followed by a decrease in SR content. Partial inhibition of the SR Ca^{2+}-ATPase by thapsigargin prevented cADPR from exerting an effect on spark activity. Furthermore, cADPR enhanced Ca^{2+} uptake by SR microsomes, but was without effect on RyRs incorporated into lipid bilayers. The authors concluded that cADPR exerts its actions by enhancing Ca^{2+} uptake by the SR, causing luminal-dependent activation of RyRs, rather than by a direct action on the RyR.

Such a mechanism of action of cADPR would offer an explanation for the lack of effect of cADPR in several studies using purified cardiac membrane preparations, particularly those examining the direct influence of cADPR on RyRs in planar lipid bilayers [34-35]. Furthermore, an action of cADPR to enhance CICR through an augmentation of Ca^{2+} uptake would potentially account for the delay required for cADPR to reach its peak effect, both in the studies of Lukyanenko and colleagues in permeabilised cells [38-39], and those from our laboratory investigating UV-photolysis of caged cADPR in intact myocytes [13, 21]. It would also be consistent with the argument (which appears particularly convincing in the case of the rat, which does not have a prolonged positive plateau of the action potential) that a substance that only increases Ca^{2+} release from the SR cannot have a sustained augmenting effect on Ca^{2+} transients and associated contractions, due to a compensatory reduction in the loading of the SR [40].

Evidence Against Enhanced SR Calcium Uptake being the only Mechanism of Action of cADPR

There are several pieces of evidence which point to the conclusion that an enhanced uptake of Ca^{2+} into the SR cannot be the only mechanism by which cADPR enhances CICR in the heart. If the only action of cADPR was to increase Ca^{2+} sequestration, it would be expected that this would be associated with an acceleration of the decay phase of the Ca^{2+} transient; however, several studies [10, 12, 13, 21] have failed to observe any changes in the decay of the Ca^{2+} transient for both cADPR and its antagonists, despite alterations in the rising phase of the transient (consistent with an effect on Ca^{2+} release). Furthermore, Prakash *et al.* [14] observed an increase in the duration of the Ca^{2+} transient in their experiments with rat cells, an effect opposite to that which would be predicted.

We have recently carried out fluorimetry-based experiments, similar to those of Lukyanenko and coworkers [39], to investigate the effects of cADPR on thapsigargin/cyclopiazonic acid-dependent Ca^{2+} uptake by guinea-pig cardiac microsomes (in the presence of ruthenium red to inhibit any complicating effects of Ca^{2+} release). In 6 experiments carried out, 10 μM cADPR was without effect on Ca^{2+} uptake by these microsomes [41], in contrast to the observations of Lukyanenko and colleagues in rat microsomes [39]. However, it cannot be excluded that factors were missing from our preparation that are required to confer cADPR sensitivity to the Ca^{2+} uptake mechanism, or that the differences arise from species variation.

Experiments examining the Ca^{2+} content of the SR in intact myocytes have also yielded results inconsistent with a sole mechanism of action of cADPR at the level of Ca^{2+} sequestration. Rakovic *et al.* [10] have reported that the suppressing effects of 8-amino-cADPR on cell contraction are not associated with a reduction in SR Ca^{2+} load. Furthermore, in preliminary experiments [42], we have also compared the SR Ca^{2+} content in guinea-pig ventricular myocytes before and after application of cADPR to the cytoplasm to assess whether the increase in cell contraction evoked by cADPR is associated with a change in SR loading: in 4 cells exposed to cADPR, the magnitude of cell contraction increased by 28 ± 5 % despite little or no change (a 2 ± 4 % decrease) in the SR Ca^{2+} load (assessed from the integral of the Na/Ca exchange current, measured under voltage-clamp conditions, following rapid application of 20 mM caffeine).

Taken together, it perhaps seems unlikely that the action of cADPR is limited to an enhancement of Ca^{2+} uptake by the SR, particularly in view of evidence in other tissues that cADPR appears to be acting primarily through a ryanodine-sensitive release mechanism (as described in other chapters in this book), rather than through alterations in Ca^{2+} uptake. It may be that cADPR has dual actions in the heart, enhancing SR Ca^{2+} release through an effect, perhaps indirect, on the RyR, and also augmenting SR Ca^{2+} sequestration in order to prevent a depletion of SR content that may occur as a result of enhanced Ca^{2+} release. Indeed, it cannot be ruled out that other mechanisms may also participate. McCall *et al.* have reported that FK506 causes a sustained enhancement of Ca^{2+} transients and contractions (even in cells in which SR loading was not increased) through a mechanism predominantly involving an augmentation of Ca^{2+} release; in addition, they observed an inhibition of Na^{+}/Ca^{2+} exchange by FK506, which would contribute to the maintenance of SR loading in the face of enhanced SR Ca^{2+} release [25]. If cADPR does indeed act through an FKBP12.6-dependent mechanism [27-28], this too may contribute to the actions of cADPR.

OTHER SUGGESTED ACTIONS OF cADPR

Lahouratate *et al.* [43] have observed an enhancement of Ca^{2+} release from canine cardiac SR by cADPR, but via a mechanism that appeared independent of RyR activation. cADPR (but not ß-NAD or ADPR) stimulated Ca^{2+} release in a concentration-dependent manner, but this Ca^{2+} release could not be suppressed by RyR inhibitors. In addition, Ca^{2+} release evoked by cADPR and a known RyR activator (9-Me-7-Br-eudistomin-D) were additive, and no cross-desensitization between these two molecules was seen, in contrast to observations of cross-desensitization between cADPR and RyR agonists in sea urchin egg microsomes [44]. A similar phenomenon has been observed by Morrissette *et al.* in skeletal muscle SR preparations [45]: cADPR, but not ß-NAD or ADPR, was found to release Ca^{2+} from rabbit skeletal muscle SR, and this release was not sensitive to the RyR blockers ruthenium red and buthotus 1-1 scorpion toxin; furthermore, cADPR did not increase the open probability of RyRs studied in planar lipid bilayers. The nature of these Ca^{2+} release mechanisms that appear to be independent of RyR activation remain to be established, but such reports have raised the possibility that cADPR may act through Ca^{2+} release pathways other than those involving the RyR.

SYNTHESIS AND HYDROLYSIS OF cADPR

It is now clear that metabolic pathways exist in many mammalian cell types, including heart cells, which are capable of synthesizing cADPR from ß-NAD, and degrading it to ADPR. Rusinko and Lee were the first to provide evidence that synthesis of cADPR from ß-NAD occurred in cardiac cells: incubation of ß-NAD with extracts prepared from a variety of rabbit tissues, including heart, resulted in the synthesis of cADPR [46]. Walseth and colleagues subsequently measured endogenous concentrations of cADPR in a variety of rat tissues, and reported levels in rat heart of 1.04 pmol/mg protein, which (assuming that the tissues were 20 % protein by mass) would correspond to a cytoplasmic concentration of approximately 200 nM [47]. Subsequent studies have confirmed and extended these observations.

Meszaros and colleagues have studied the kinetics of cADPR formation in rat heart homogenates, and have reported that the synthesis of cADPR from ß-NAD begins 2-3 minutes after addition of ß-NAD, in contrast to the formation of ADPR which appears to commence immediately on addition of the substrate [48]. The authors also observed a mono-ADP-ribosylation of cardiac proteins during the first few minutes after addition of ß-NAD, and have suggested that such ADP-ribosylation may cause activation of ADP-ribosyl cyclase; future studies may reveal whether or not such a mechanism of regulation exists.

Regulation of cADPR Synthesis by Protein Kinase C

For cADPR to be considered as a possible intracellular messenger in the heart, mechanisms must exist by which the enzymes that synthesise and degrade cADPR are under the control of extracellular stimuli. The first study to provide such evidence was that of Meszaros, Wrenn & Varadi in 1997 [49], which reported the existence of an SR-associated ADP-ribosyl cyclase that was inhibited by protein kinase C (PKC)-dependent phosphorylation. Two types of ADP-ribosyl cyclase activity were identified in both dog and rat cardiac tissue, one associated with the sarcolemma and another with the SR; the sarcolemmal ADP-ribosyl cyclase resembled CD38 in that it was sensitive to thiols and lectins, and recognized by an anti-CD38 antibody, whereas this was not the case for the SR-associated enzyme. Interestingly, the SR ADP-ribosyl cyclase was inhibited by phorbol esters (activators of PKC), an effect that was prevented by two PKC inhibitors (chelerythrine or staurosporine); the presence of endogenous PKC activity in their cardiac preparations was also confirmed. This therefore raises the possibility that PKC-dependent phosphorylation of cardiac ADP-ribosyl cyclase may be an endogenous mechanism by which cADPR synthesis may be regulated.

Regulation of cADPR Synthesis by ß-Adrenoceptor Agonists and Protein Kinase A (PKA)

Catecholamines (norepinephrine released from sympathetic nerves, and circulating epinephrine and norepinephrine from the adrenal medullae) play an important role in the physiological regulation of the heart, increasing the rate (positive chronotropy) and force (positive inotropy) of cardiac contraction through their action on ß-adrenergic receptors. It is well established that two important mechanisms contributing to this positive inotropic action are an enhancement of Ca^{2+} entry through L-type Ca^{2+} channels (secondary to PKA-dependent phosphorylation), and an augmentation of Ca^{2+} sequestration by the SR, due in part to PKA-dependent phosphorylation of phospholamban, which removes its inhibitory effect on the SR Ca^{2+}-ATPase [9]. In view of the evidence that cADPR may increase the gain of CICR, the possibility has been raised that changes in cADPR levels may also contribute to the mechanisms by which ß-adrenoceptor agonists increase the force of contraction of the heart, and there is now a growing body of evidence that this indeed may be the case.

Higashida and colleagues have reported recently that the ß-adrenoceptor agonist isoproterenol enhances the rate of cADPR synthesis in rat ventricular myocytes [50]. ADP-ribosyl cyclase activity in a crude membrane fraction of rat ventricular myocytes was monitored either from the synthesis of [^{3}H]cADPR formation from [^{3}H]NAD, or from the synthesis of the fluorescent cADPR analogue, cGDPR, from ß-NGD. Isoproterenol was found to enhance ADP-ribosyl cyclase activity in a concentration-dependent manner, and this effect was inhibited by the ß-adrenoceptor antagonist, propranolol. The stimulatory effect of isoproterenol was mimicked by low concentrations of GTP or by an analogue, GTP-γ-S, but this was not synergistic or additive with the actions of isoproterenol, consistent with a shared mechanism of action of these agents. Furthermore, activation of ADP-ribosyl cyclase by isoproterenol or GTP was blocked by cholera toxin, suggesting the involvement of the G_s G-protein in the coupling of the ß-adrenoceptor to ADP-ribosyl cyclase.

We have also observed ADP-ribosyl cyclase activity in SR-enriched microsomal preparations of guinea-pig ventricular myocytes, using cuvette-based fluorimetry to monitor the synthesis of cGDPR, a fluorescent cADPR analogue, from ß-NGD [51]. A representative experiment is illustrated in Figure 8, in which 200 μM ß-NGD was added to a suspension of cardiac microsomes: addition of ß-NGD resulted in an initially linear increase in fluorescence which subsequently tailed off to reach a plateau after approximately 20 minutes. As has been observed in sea urchin egg microsomes [52], we have found that the rate of cGDPR synthesis may be inhibited by nicotinamide [53]. Using this preparation, we have found that the catalytic subunit of PKA enhanced the rate of cGDPR synthesis in a concentration dependent manner (Figure 8), with the highest concentration of PKA used approximately doubling the rate of cGDPR synthesis [51]. This enhancement of ADP-ribosyl cyclase activity by PKA was associated with increases in both the K_m and V_{max} of the enzyme. Furthermore, the effects of PKA were potentiated by the phosphatase inhibitor okadeic acid, suggesting the presence of phosphatases in our preparation that limit the effects of exogenous PKA. These data are thus consistent with the existence of an SR-associated ADP-ribosyl cyclase in guinea-pig heart that may be stimulated by PKA-mediated phosphorylation.

Similar studies in permeabilised ventricular myocytes have provided additional evidence supporting the proposition that ADP-ribosyl cyclase is regulated through PKA-dependent phosphorylation mechanisms. In both guinea-pig and rat ventricular myocytes permeabilised with either ß-escin or saponin (to allow access of ß-NGD and PKA to the intracellular compartment), we have found that the catalytic subunit of PKA causes an enhancement in the rate of cGDPR synthesis [54].

The most direct evidence that cardiac cADPR levels are regulated physiologically by catecholamines acting through ß-adrenoceptors has been provided recently by the studies of Thomas and colleagues [55]. Using [^{32}P]cADPR binding to sea urchin egg homogenate as a radioreceptor assay, endogenous levels of cADPR were measured as being 12 pmol/mg protein. Isoproterenol, applied to a Llangendorff-perfused heart, was observed to increase cADPR levels in a concentration-dependent manner, with the half-maximal increase occurring at approximately 5 nM isoproterenol (Figure 8). Furthermore, this effect of isoproterenol was abolished by co-perfusion with propranolol, a ß-adrenoceptor antagonist, and reduced by co-perfusion with H-89, an inhibitor of PKA.

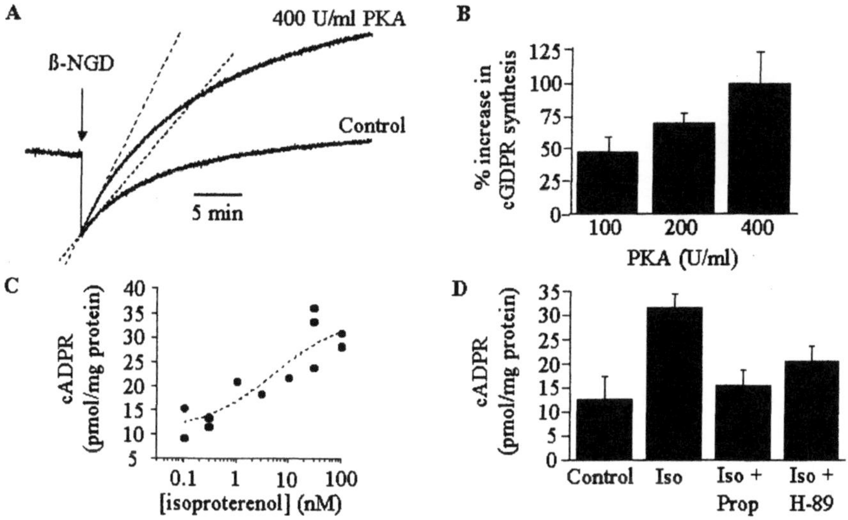

Figure 8. **A:** Enhancement of ADP-ribosyl cyclase activity (assessed fluorimetrically from the synthesis of cGDPR from ß-NGD) by PKA in guinea-pig cardiac ventricular microsomes. Addition of 200 μM ß-NGD results in the gradual synthesis of the autofluorescent compound cGDPR, as shown by the gradual increase in fluorescence intensity that eventually reaches a plateau (control trace). In the presence of the catalytic subunit of PKA, the rate of cGDPR synthesis is accelerated. **B:** Concentration-dependent enhancement of the rate of cGDPR synthesis in guinea-pig ventricular microsomes by the catalytic subunit of PKA. **C:** Concentration-dependent enhancement of cADPR levels in guinea-pig heart by the ß-adrenoceptor agonist isoproterenol. Following perfusion for 3 minutes with varying concentrations of isoproterenol, acutely isolated hearts were snap frozen in liquid nitrogen. Following acid extraction, absolute cADPR levels were assessed using the sea urchin egg based radioreceptor assay. Each point represents the mean cADPR value from 1 heart. **D:** The enhancement of whole-heart cADPR levels by isoproterenol (30 nM, 3 min) is abolished by co-perfusion with the competitive ß-adrenoceptor antagonist, propranolol (100 nM) and reduced by co-perfusion with the PKA inhibitor, H-89 (1 μM).

It thus appears that modulation of cADPR levels may be an additional novel mechanism by which the sympathetic nervous system regulates the contractile state of the heart. Although the precise mechanism by which ß-agonists elevate cADPR levels has not been determined, it is likely that this involves the cAMP/PKA signaling pathway, perhaps through phosphorylation of ADP-ribosyl cyclase itself; further work will be needed to determine if this is the case.

CONCLUSIONS

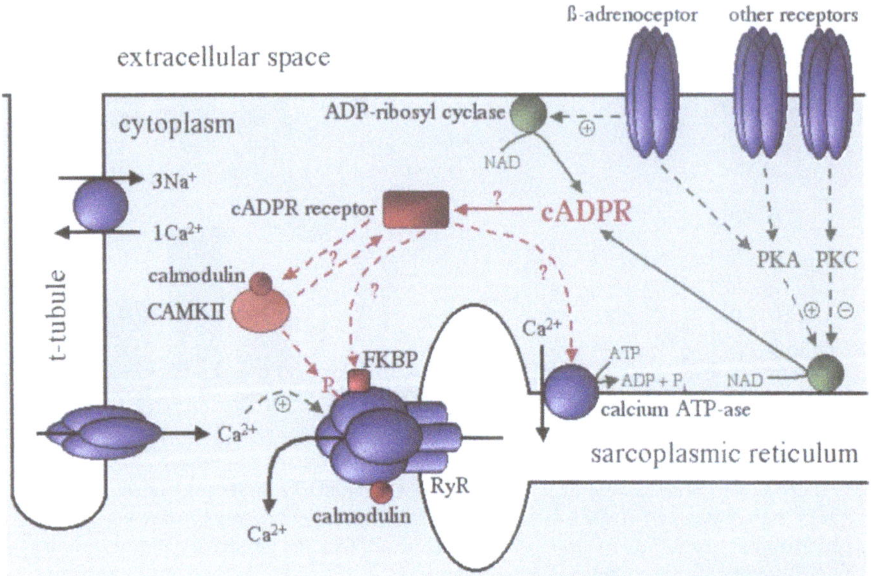

Figure 9. **Possible role for cADPR in the modulation of cardiac EC-coupling.** cADPR, synthesized from ß-NAD by both sarcolemmal and SR-associated ADP-ribosyl cyclase, may enhance CICR through an action on the RyR, the SR Ca^{2+}-ATPase, or both. This effect of cADPR may be direct, but is perhaps more likely to involve intermediate proteins, such as the putative cADPR-binding protein. The actions of cADPR appear to require both calmodulin and CAM-KII; it is not known whether cADPR acts through a calmodulin-CAM-KII pathway to exert its effects, or whether calmodulin association to, or CAM-KII phosphorylation of, target proteins (e.g. the RyR or the cADPR binding protein) is required to have a permissive role. The dissociation of FKBP12.6 from the RyR may also play a role in mediating the effects of cADPR on Ca^{2+} release from the SR. ADP-ribosyl cyclase may be regulated by endogenous hormones and transmitters. ß-adrenoceptor agonists (and perhaps agonists at other receptors) enhance cADPR synthesis, possibly through PKA-dependent mechanisms; other receptors may be linked to suppression of cADPR levels, via a PKC-mediated action.

Since its discovery in the sea-urchin egg, cADPR has been shown to influence Ca^{2+} release in many different cell types, including mammalian cells such as neurons, smooth muscle cells and pancreatic cells. There is now a substantial body of evidence demonstrating that cADPR is also an endogenous regulator of CICR in the heart, and that its cellular levels may be regulated by hormones (Figure 9). In particular, catecholamines appear to enhance cADPR levels through their action on ß-adrenergic receptors, and this may contribute to their positive inotropic effects. Furthermore, although increased cellular cADPR levels would enhance cell contraction, excessive levels would tend to predispose the myocyte to arrhythmias. Many questions still remain concerning the mechanisms by which cADPR exerts its actions, although future work should identify the putative cADPR receptor and increase our understanding of the interaction of proteins such as calmodulin, CAM-KII and FKBP12.6 in the actions of cADPR in the heart

REFERENCES

1. Ringer S. 1883. A further contribution regarding the influence of the different constituents of the blood on the contraction of the heart. *J. Physiol.* 4: 29-42.
2. Bers DM. 2001. *Excitation-contraction coupling and cardiac contractile force* (2nd ed). Kluwer Academic, Dordrecht, Norwell, MA.
3. Tada M, Yamada M, Kadoma M, Inui M and Ohmori F. 1982. Ca^{2+} transport by cardiac sarcoplasmic reticulum and phosphorylation of phospholamban. *Mol. Cell. Biochem.* 46: 74-95.
4. Scott BT, Simmerman HKB, Collins JH, Nadal-Ginard B and Jones LR. 1988. Complete amino acid sequence of canine cardiac calsequestrin deduced by cDNA cloning. *J. Biol. Chem.* 263: 8958-8964.
5. Zucchi R and Ronca-Testoni S. 1997. The sarcoplasmic reticulum Ca^{2+} channel/ryanodine receptor: modulation by endogenous effectors, drugs and disease states. *Pharmacol. Rev.* 49: 1-51.
6. Otsu K, Willard HF, Khanna VK, Zorzato F, Green NM and MacLennan DH. 1990. Molecular cloning of cDNA encoding the Ca^{2+} release channel (ryanodine receptor) of rabbit cardiac muscle sarcoplasmic reticulum. *J. Biol. Chem.* 265: 13472-13483.
7. Sorrentino V and Volpe P. 1993. Ryanodine receptors: how many, where and why? *Trends Pharmacol. Sci.* 14: 98-103.
8. Rousseau E and Meissner G. 1989. Single cardiac sarcoplasmic reticulum Ca^{2+}-release channel: activation by caffeine. *Am. J. Physiol.* 256: H328-H333.
9. Rapundalo ST. 1998. Cardiac protein phosphorylation: functional and pathophysiological correlates. *Cardiovasc. Res.* 38: 559-588.
10. Rakovic S, Galione A, Ashamu GA, Potter BVL and Terrar DA. 1996. A specific cyclic ADP-ribose antagonist inhibits cardiac excitation-contraction coupling. *Curr. Biol.* 6: 989-996.
11. Walseth TF and Lee HC. 1993. Synthesis and characterization of antagonists of cyclic-ADP-ribose-induced Ca^{2+} release. *Biochim. Biophys. Acta* 1178: 235-242.
12. Iino S, Cui Y, Galione A and Terrar DA. 1997. Actions of cADP-ribose and its antagonists on contraction in guinea pig isolated ventricular myocytes. Influence of temperature. *Circ. Res.* 81: 879-884.
13. Cui Y, Galione A and Terrar DA. 1999. Effects of photoreleased cADP-ribose on calcium transients and calcium sparks in myocytes isolated from guinea-pig and rat ventricle. *Biochem. J.* 342: 269-273.
14. Prakash YS, Kannan MS, Walseth TF and Sieck GC. 2000. cADP ribose and $[Ca^{2+}]_i$ regulation in rat cardiac myocytes. *Am. J. Physiol. Heart Circ. Physiol.* 279: H1482-H1489.
15. Guo X, Laflamme MA and Becker PL. 1996. Cyclic ADP-ribose does not regulate sarcoplasmic reticulum Ca^{2+} release in intact cardiac myocytes. *Circ. Res.* 79: 147-151.
16. Rakovic S, Cui Y, Iino S, Galione A, Ashamu GA, Potter BVL and Terrar DA. 1999. An antagonist of cADP-ribose inhibits arrhythmogenic oscillations of intracellular Ca^{2+} in heart cells. *J. Biol. Chem.* 274: 17820-17827.
17. Lee HC, Aarhus R, Graeff RM, Gurnack ME and Walseth TF. 1994. Cyclic ADP-ribose activation of the ryanodine receptor is mediated by calmodulin. *Nature* 370: 307-309.
18. Lee HC, Aarhus R and Graeff RM. 1995. Sensitization of calcium-induced calcium release by cyclic ADP-ribose and calmodulin. *J. Biol. Chem.* 270: 9060-9066.

19. Tanaka Y and Tashjian A. 1995. Calmodulin is a selective mediator of Ca^{2+}-induced Ca^{2+} release via the ryanodine receptor-like Ca^{2+} channel triggered by cyclic ADP-ribose. *Proc. Natl. Acad. Sci. USA* 92: 3244-3248.
20. Takasawa S, Ishida A, Nata K, Nakagawa K, Noguchi N, Tohgo A, Kato I, Yonekura H, Fujisawa H and Okamoto H. 1995. Requirement of calmodulin-dependent protein kinase II in cyclic ADP-ribose-mediated intracellular Ca^{2+} mobilization. *J. Biol. Chem.* 270: 30257-30259.
21. Rakovic S, Cui Y, Galione A and Terrar DA. 2002. Requirement of calmodulin for actions of cADPR in guinea-pig ventricular myocytes. *Biophys. J.* 82: 57a.
22. Okabe E, Tsujimoto Y and Kobayashi Y. 2000. Calmodulin and cyclic ADP-ribose interaction in Ca^{2+} signaling related to cardiac sarcoplasmic reticulum: superoxide anion radical-triggered Ca^{2+} release. *Antioxid. Redox. Signal* 2: 47-54.
23. Kumasaka S, Shoji H and Okabe E. 1999. Novel mechanisms involved in superoxide anion radical-triggered Ca^{2+} release from cardiac sarcoplasmic reticulum linked to cyclic ADP-ribose stimulation. *Antioxid. Redox. Signal* 1: 55-69.
24. Timerman AP, Jayaraman T, Wiederrecht G, Onoue H, Marks AR and Fleischer S. 1994. The ryanodine receptor from canine heart sarcoplasmic reticulum is associated with a novel FK-506 binding protein. *Biochem. Biophys. Res. Commun.* 198: 701-706.
25. McCall E, Li L, Satoh H, Shannon TR, Blatter LA and Bers DM. 1996. Effects of FK-506 on contraction and Ca^{2+} transients in rat cardiac myocytes. *Circ. Res.* 79: 1110-1121.
26. Xiao RP, Valdivia HH, Bogdanov K, Valdivia C, Lakatta EG and Cheng H. 1997. The immunophilin FK506-binding protein modulates Ca^{2+} release channel closure in rat heart. *J. Physiol.* 500: 343-354.
27. Noguchi N, Takasawa S, Nata K, Tohgo A, Kato I, Ikehata F, Yonekura H and Okamoto H. 1997. Cyclic ADP-ribose binds to FK506-binding protein 12.6 to release Ca^{2+} from islet microsomes. *J. Biol. Chem.* 272: 3133-3136.
28. Xin H-B, Ji GJ, Deng K-Y, Fleischer S and Kotlikoff MI. 2002. FK506 binding protein 12.6 plays a central role in cyclic ADP-ribose triggered Ca^{2+}-induced Ca^{2+} release via type 2 ryanodine receptor in mouse cardiac myocytes. *Biophys. J.* 82: 59a.
29. Timerman AP, Wiederrecht G, Marcy A and Fleischer S. 1995. Characterization of an exchange reaction between soluble FKBP-12 and the FKBP ryanodine receptor complex. *J. Biol. Chem.* 270: 2451-2459.
30. Meszaros LG, Bak J and Chu A. 1993. Cyclic ADP-ribose as an endogenous regulator of the non-skeletal type ryanodine receptor Ca^{2+} channel. *Nature* 364: 76-79.
31. Sitsapesan R, McGarry SJ and Williams AJ. 1994. Cyclic ADP-ribose competes with ATP for the adenine nucleotide binding site on the cardiac ryanodine receptor Ca^{2+}-release channel. *Circ. Res.* 75: 596-600.
32. Sitsapesan R, McGarry SJ and Williams AJ. 1995. Cyclic ADP-ribose, the ryanodine receptor and Ca^{2+} release. *Trends Pharmacol. Sci.* 16: 386-391.
33. Bak J, Billington RA, Timar G, Dutton AC and Genazzani AA. 2001. NAADP receptors are present and functional in the heart. *Curr. Biol.* 11: 987-990.
34. Fruen BR, Mickelson JR, Shomer NH, Velez P and Louis CF. 1994. Cyclic ADP-ribose does not affect cardiac or skeletal muscle ryanodine receptors. *FEBS Lett.* 352: 123-126.
35. Copello JA, Qi Y, Jeyakumar LH, Ogunbunmi E and Fleischer S. 2001. Lack of effect of cADP-ribose and NAADP on the activity of skeletal muscle and heart ryanodine receptors. *Cell Calcium* 30: 269-284.
36. Graeff RM, Podein RJ, Aarhus R and Lee HC. 1995. Magnesium ions but not ATP inhibit cyclic ADP-ribose-induced calcium release. *Biochem. Biophys. Res. Commun.* 206: 786-791.
37. Walseth TF, Aarhus R, Kerr JA and Lee HC. 1993. Identification of cyclic ADP-ribose-binding proteins by photoaffinity labeling. *J. Biol. Chem.* 268: 26686-26691.
38. Lukyanenko V and Gyorke S. 1999. Ca^{2+} sparks and Ca^{2+} waves in saponin-permeabilized rat ventricular myocytes. *J. Physiol.* 521: 575-585.

39. Lukyanenko V, Gyorke I, Wiesner TF and Gyorke S. 2001. Potentiation of Ca^{2+} release by cADP-ribose in the heart is mediated by enhanced SR Ca^{2+} uptake into the sarcoplasmic reticulum. *Circ. Res.* 89: 614-622.
40. Trafford AW, Diaz ME and Eisner DA. 1998. Stimulation of Ca^{2+}-induced Ca^{2+} release only transiently increases the systolic Ca^{2+} transient: measurements of Ca^{2+} fluxes and sarcoplasmic reticulum Ca^{2+}. *Cardiovasc. Res.* 37: 710-717.
41. Rakovic S, Galione A and Terrar DA. Unpublished observations.
42. MacGregor AT, Rakovic S, Galione A and Terrar DA. Unpublished observations.
43. Lahouratate P, Guibert J and Faivre JF. 1997. cADP-ribose releases Ca^{2+} from cardiac sarcoplasmic reticulum independently of ryanodine receptor. *Am. J. Physiol.* 273: H1082-H1089.
44. Galione A, Lee HC and Busa WB. 1991. Ca^{2+}-induced Ca^{2+} release in sea urchin egg homogenates: modulation by cyclic ADP-ribose. *Science* 253: 1143-1146.
45. Morrissette J, Heisermann G, Cleary J, Ruoho A and Coronado R. 1993. Cyclic ADP-ribose induced Ca^{2+} release in rabbit skeletal muscle sarcoplasmic reticulum. *FEBS Lett.* 330: 270-274.
46. Rusinko N and Lee HC. 1989. Widespread occurrence in animal tissue of an enzyme catalyzing the conversion of NAD^+ into a cyclic metabolite with intracellular Ca^{2+} mobilizing activity. *J. Biol. Chem.* 264: 11725-11731.
47. Walseth TF, Aarhus R, Zeleznikar RJ Jr and Lee HC. 1991. Determination of endogenous levels of cyclic ADP-ribose in rat tissues. *Biochim. Biophys. Acta* 1094: 113-120.
48. Meszaros V, Socci R and Meszaros LG. 1995. The kinetics of cyclic ADP-ribose formation in heart muscle. *Biochem. Biophys. Res. Commun.* 210: 452-456.
49. Meszaros LG, Wrenn RW and Varadi G. 1997. Sarcoplasmic reticulum-associated and protein kinase C-regulated ADP-ribosyl cyclase in cardiac muscle. *Biochem. Biophys. Res. Commun.* 234: 252-256.
50. Higashida H, Egorova A, Higashida C, Zhong ZG, Yokoyama S, Noda M and Zhang JS. 1999. Sympathetic potentiation of cyclic ADP-ribose formation in rat cardiac myocytes. *J. Biol. Chem.* 274: 33348-33354.
51. Heath BM, Galione A and Terrar DA. 1999. Regulation of cyclic ADP-ribose (cADPR) production by protein kinase A in cardiac ventricular microsomes prepared from guinea-pig heart. *Br. J. Pharmacol.* 126: 204P.
52. Sethi JK, Empson RM and Galione A. 1996. Nicotinamide inhibits cyclic ADP-ribose-mediated calcium signalling in sea urchin eggs. *Biochem. J.* 319: 613-617.
53. Rakovic S, Galione A, Terrar DA. Unpublished observations.
54. Lancashire H, Nankivell P, Rakovic S, Galione A and Terrar DA. Unpublished observations.
55. Thomas JM, Churchill GC, Terrar DA and Galione A. Unpublished observations.

Derek A Terrar, Department of Pharmacology,
Oxford University, Mansfield Road,
Oxford OX1 3QT, UK.
e-mail: derek.terrar@pharmacology.oxford.ac.uk

Cyclic ADP-Ribose and NAADP. Structures, Metabolism and Functions. Lee HC (ed), pp. 343-363, Dordrecht: Kluwer.

Cyclic ADP-Ribose and Vasomotor Response

Pin-Lan Li, Ai-Ping Zou and William B. Campbell
Medical College of Wisconsin, Milwaukee, WI53226, USA.

INTRODUCTION

Vascular smooth muscle (VSM) usually exists in a partially contracted state, from which it can contract further or relax in response to different physiological or pathological stimulations. This contracted state of blood vessels and their contracting and relaxing response to stimuli are often referred to as "vascular tone" and "vasomotor response," respectively. Numerous studies indicate that cytosolic free calcium concentrations ($[Ca^{2+}]$) in VSM cells play an essential role in mediating or modulating both vascular tone and vasomotor response in a variety of blood vessels. It is well recognized that the rise in intracellular $[Ca^{2+}]$ initiates and maintains contraction in VSM, which importantly determines the peripheral vascular resistance and blood pressure [1-5].

Intracellular $[Ca^{2+}]$ and Ca^{2+}-mediated excitation-contraction coupling in VSM cells are primarily controlled by the influx of extracellular Ca^{2+} and the mobilization of Ca^{2+} from intracellular stores. It is generally accepted that the influx of extracellular Ca^{2+} may directly produce vasoconstriction or stimulate Ca^{2+} release from the sarcoplasmic reticulum (SR) through Ca^{2+}-induced Ca^{2+} release (CICR) and thereby lead to vasoconstriction. In addition, the Ca^{2+} mobilization importantly mediates the vasomotor response to many vasoactive substances such as hormones, autocrine or paracrine, and others stimuli. This intracellular Ca^{2+} release and consequent vasomotor response play a critical role in the homeostasis of blood circulation. In this regard, inositol 1,4,5-tris-phosphate (IP_3) has been recognized as an intracellular second messenger mediating the vasomotor effects of various agonists such as norepinephrine (NE), angiotensin II (ANGII), vasopressin

(ADH), endothelin (ET), thromboxane A_2 and others. In this process, the agonists activate phospholipase C, which converts phosphatidylinositol-4,5,-biphosphate (PIP_2) into IP_3 and diacylglycerol (DAG). IP_3 induces Ca^{2+} release through IP_3 receptors (IP_3R) on the SR and thereby leads to the rise of intracellular [Ca^{2+}], producing vasoconstriction [3-7].

Although IP_3 as a second messenger mediates the effects of a number of endogenous or exogenous vasoactive substances and importantly participates in the control of vascular tone and mediates vasomotor responses, there are many other agonists or stimuli that induce intracellular Ca^{2+} release independently of the IP_3 pathway. For example, caffeine, Ca^{2+}, 5-HT, acetylcholine (Ach) and prostaglandin $F_{2\alpha}$ produce the Ca^{2+} mobilization from the SR through activation of ryanodine receptors (RyR) on the SR. Recently, the RyR have been well characterized in the VSM cells and RyR-mediated Ca^{2+} mobilization has been considered as another important mechanism mediating vasomotor response in VSM. Cyclic ADP-ribose (cADPR) is a novel Ca^{2+} mobilizing second messenger that has been reported to participate in the control of vascular tone and vasomotor response through activation of RyR. cADPR is involved in the generation or modulation of intracellular Ca^{2+} sparks, Ca^{2+} waves or oscillations, CICR and spontaneous transient outward currents (STOCs) [8-10]. This cADPR-RyR-mediated Ca^{2+} signaling is emerging as a fundamental mechanism regulating cardiovascular function and is considered as a new therapeutic target for the treatment of cardiovascular diseases. The major focus of this chapter is to discuss the physiological roles of cADPR-mediated Ca^{2+} signaling in the control of vascular tone and vasomotor response and to summarize studies on the underlying mechanism mediating the actions of cADPR in VSM cells.

PRODUCTION OF cADPR IN VSM

cADPR was first detected in sea urchin eggs [11] and then in a variety of mammalian tissues or cells such as heart, liver, spleen, brain, red blood cells, lymphocytes, pituitary cells and cultured renal epithelial cells [12-16]. Basal concentrations of cADPR in cardiac muscle, liver and brain are estimated as 100-200 nM [17]. Recent studies have shown that homogenates prepared from dissected small bovine coronary arteries and cultured arterial smooth muscle cells produced cADPR and its metabolite ADPR when incubated with NAD [8, 18-20]. Tissue cADPR concentration in the coronary smooth muscle is about 150 nM [20]. The cADPR extracted from the reaction mixture of NAD with cultured smooth muscle cells or arterial homogenates is capable of stimulating Ca^{2+} release *in vitro* using single cell Ca^{2+} fluorospectrometry.

Enzymatic Pathway of cADPR Production and Metabolism

By HPLC analysis, the production of cADPR and ADPR is greater in the microsomal fraction compared to the cytosolic fraction from bovine coronary arteries, suggesting that the enzymes responsible for both production and metabolism of cADPR are primarily present on the cell membrane. This is consistent with the findings that ADP-ribosylcyclase, NAD glycohydrolase and cADPR hydrolase are membrane-bound enzymes in a wide range of mammalian tissues [17, 21-23].

CD38, a lymphocyte differentiation antigen, has multiple activities including NAD glycohydrolase, ADP-ribosyl cyclase and cADPR hydrolase activities [21, 23]. This protein is postulated to function as a molecular switch to regulate the cellular levels of cADPR in many cells. The cellular cADPR levels could be elevated or decreased by balancing the synthetic and hydrolytic activities of this enzyme, respectively [17]. In response to stimuli, this multi-functional enzyme can aggregate and accumulate in the cytoplasm and thereby more efficiently produce or metabolize cADPR [24, 25]. By Western blot analysis and RT-PCR, CD38 was detected in arterial smooth muscle. Two immunoreactive bands with molecular sizes of 42 and 90 kDa were recognized by a monoclonal antibody against CD38 in arterial homogenates and microsomes. Removal of CD38 by immunoprecipitation significantly decreased the production and catabolism of cADPR in these arterial homogenates [25, 26]. It is clear that CD38 present in coronary arterial smooth muscle is at least one of the enzymes responsible for the production and metabolism of cADPR in these cells.

In VSM cells, the production of ADPR is usually accompanied by a decrease in cADPR concentrations when NAD is used as a substrate. This suggests that cADPR can be dynamically converted into ADPR and that cADPR hydrolase is also of importance in determining the concentrations of cADPR [20]. Taken together, it is now clear that an enzymatic pathway responsible for the formation and metabolism of cADPR is present in VSM cells. cADPR is formed from NAD by ADP-ribosylcyclase, and the production of ADPR may be associated with both NAD glycohydrolase and cADPR hydrolase in VSM cells. [19, 20, 26, 27].

A recent study illustrates the distribution of the enzyme activity along vascular tree in the kidney [18]. ADP-ribosyl cyclase activity can be detected in the arteries or arterioles, even in the smallest renal arteriole with diameters less than 10 μm such as afferent arterioles and vasa recta. In these studies, interestingly, preglomerular arteries or arterioles exhibited a greater activity of ADP-ribosyl cyclase compared to postglomerular vasa recta. This suggests that the VSM is a major tissue type to produce cADPR in these renal vessels. This view is supported by a direct measurement of cADPR production in smooth muscle cells freshly dissociated from interlobular and afferent arterioles [18].

Regulation of cADPR Levels in VSM

The production and hydrolysis of cADPR are regulated by a number of intracellular factors such as ATP, ADP, pH, cGMP, polyamines, Ca^{2+} and redox status [17, 27-33]. There are three important regulatory mechanisms, which are probably involved in the action of this signaling pathway in the regulation of vasomotor response.

cGMP-Independence of ADP-Ribosyl Cyclase Activity–Increases in cGMP levels enhance the production of cADPR in nonvascular tissues. This cGMP-induced enhancement of cADPR synthesis may contribute to the effects of nitric oxide (NO) to increase $[Ca^{2+}]_i$ in these non-muscle tissues [17, 30, 34]. However, this cADPR-induced increase in $[Ca^{2+}]_i$ could not mediate the vasodilator effect of NO, since NO has been reported to decrease $[Ca^{2+}]_i$ [9, 10]. There is evidence that both 8-bromo-cGMP and the guanylate cyclase inhibitor, ODQ are without effect on ADP-ribosyl cyclase activity in VSM. In addition, recent studies have indicated that the NO donor, sodium nitroprusside (SNP) decreases ADP-ribosyl cyclase activity in coronary arterial smooth muscle cells. This inhibitory effect of NO on ADP-ribosyl cyclase occurs by a cGMP-independent mechanism [10]. Therefore, in VSM cells cGMP does not mediate or modulate the production of cADPR.

Redox Regulation of ADP-Ribosyl Cyclase Activity–ADP-ribosyl cyclase activity is regulated by redox status in many tissues or cells. cADPR production is increased by intracellular oxidants. Based on recent findings, it has been proposed that ADP-ribosyl cyclase serves as a link between intracellular oxidants and $[Ca^{2+}]_i$ [35-37]. Oxidants activate ADP-ribosyl cyclase and result in the conversion of NAD into cADPR, thereby increasing $[Ca^{2+}]_i$ [28, 30]. There is substantial evidence that $[Ca^{2+}]_i$ is increased by oxidative stress in the cells [38, 39]. Increase in cADPR may be importantly involved in this oxidant-induced Ca^{2+} mobilization [35, 36, 40]. This oxidant-induced activation of ADP-ribosyl cyclase may participate in the effect of NO on the production of cADPR in VSM cells, since a rapid interaction of NO with superoxide may decrease intracellular oxidants, resulting in reduction of cADPR production.

The redox regulation of cADPR production is dependent on a dimerization of ADP-ribosyl cyclase or CD38. The dimerization of ADP-ribosyl cyclase enhances its catalytic activity [21, 28, 41-44]. Specific cysteine residues in the CD38 or ADP-ribosyl cyclase determine the ability of this enzyme to function as ADP-ribosyl cyclase or cADPR hydrolase [45]. The oxidation of cysteine molecules may lead to the formation of one or several disulfide bonds, which induces the dimerization of the enzyme protein [28]. This dimerized ADP-ribosyl cyclase is present in the tissue even under resting condition. The activity of ADP-ribosyl cyclase is strongly inhibited by disulfide bond reducing reagents [28, 30, 44, 46]. Oxidants increase the formation of ADP-ribosyl cyclase dimers, thereby increasing ADP-ribosyl cyclase activity [28].

Role of GTP-binding Proteins (G-proteins)–G-proteins play an important role in linking cell surface receptors to intracellular second messengers. G proteins may also participate in the signaling between surface receptors and the production of cADPR [47]. In this context, G-proteins link Ach receptors to ADP-ribosyl cyclase and Ach stimulates the production of cADPR through G protein activation. Recently, we found that GTP-γS increased, but GDP-βS decreased the ADP-ribosyl cyclase activity of coronary arterial homogenates (unpublished data). It appears that G protein activation is involved in the production of cADPR in VSM cells. However, the conclusion cannot be drawn until further studies confirm a direct structural or functional connection between G proteins and any enzyme determining intracellular cADPR levels.

Ca^{2+} MOBILIZING ACTION OF cADPR IN VSM CELLS

cADPR Directly Stimulates Ca^{2+} Release

As discussed above, the role of cADPR in the intracellular Ca^{2+} mobilization has been extensively studied in a variety of tissues and cells [13, 16, 48, 49]. In VSM cells, this nucleotide also stimulates Ca^{2+} release from intracellular Ca^{2+} stores [8, 10, 18, 50]. In this regard, Kannan *et al.* [50] reported that cADPR induces the SR Ca^{2+} release in β-escin-permeabilized smooth muscle cells freshly isolated from porcine coronary arteries. In studies using α-toxin permeabilized cells, we have found that cADPR produces Ca^{2+} release in both cultured and freshly dissociated cow coronary and rat renal VSM cells [8, 10, 18]. This cADPR-induced Ca^{2+} release from the SR can be completely blocked by cADPR antagonist, 8-bromo-cADPR, but not by IP_3R blockers [10]. It is concluded that cADPR mobilizes intracellular Ca^{2+} through a mechanism independent of IP_3 in VSM cells.

Role of Endogenous cADPR in KCl-Induced Ca^{2+} Release

KCl is often used to depolarize the cell membrane, thereby producing vascular vasoconstriction. This KCl-induced membrane depolarization leads to the activation of Ca^{2+} channels on the cell membrane and influx of extracellular Ca^{2+}. There is a general agreement that the influx of extracellular Ca^{2+} can result in Ca^{2+} mobilization from the SR and consequent global increase in intracellular $[Ca^{2+}]$, producing vasoconstriction. Recent studies have demonstrated that this Ca^{2+} influx-induced Ca^{2+} release is associated with the production and action of cADPR

[8, 18]. Inhibition of ADP-ribosyl cyclase by nicotinamide, blockade of cADPR action by cell-permeable antagonist 8-bromo-cADPR and RyR blockade by ryanodine all suppress KCl-induced Ca^{2+} mobilization in VSM cells [10, 51]. These results suggest that endogenous cADPR participates in the production of Ca^{2+} influx-induced Ca^{2+} release response and thereby plays an important in the control of intracellular $[Ca^{2+}]$ in VSM cells.

Ca^{2+} Sparks and STOCs

Ca^{2+} sparks are local and spontaneous transient Ca^{2+}release through the RyR on the SR. Ca^{2+} sparks occur in relation to Ca^{2+}-dependent K (K_{Ca}) channel activation and muscle relaxation in VSM. A spontaneous Ca^{2+} spark in a superficial area close to the cell membrane activates 10-100 K_{Ca} channels (namely, spontaneous transient outward currents (STOCs)) and induces membrane hyperpolarization, which reduces Ca^{2+} channel activity, resulting in decrease in vascular tone [52-55]. Ca^{2+} sparks occur by spontaneous opening of RyR/Ca^{2+} release channels in the smooth muscle cells. Therefore, the sensitivity of RyR may largely determine the occurrance of Ca^{2+} sparks. Since cADPR is also reported to sensitize the RyR on the SR, it is not surprising that cADPR participates in the Ca^{2+} sparks. Recent studies have reported that exogenous cADPR increases Ca^{2+} sparks in cardiac cells [43]. However, in VSM cells, ryanodine and cADPR transiently enhances Ca^{2+} sparks and then produces a sustained inhibition of Ca^{2+} sparks. This sustained inhibition of Ca^{2+} sparks may be due to the depletion of Ca^{2+} stores [54, 56, 57]. In addition to its effect to increase global $[Ca^{2+}]$, cADPR-induced Ca^{2+} depletion of the superficial SR, inhibition of Ca^{2+} sparks and consequent membrane depolarization may be importantly involved in the control of vascular tone and vasomotor response [58, 59].

CICR and Ca^{2+} Waves

Although there is evidence that a Ca^{2+} buffering mechanism is present in VSM cells, a large Ca^{2+} influx and rise of Ca^{2+} around the deep SR lead to CICR. As summarized by Berridge, the entry of Ca^{2+} through voltage-operated channels is further amplified by CICR in VSM, thereby giving rise to a global Ca^{2+} signal throughout the cytoplasm and nucleus and leading to vasoconstriction [1]. CICR is not only an important mechanism for increasing global $[Ca^{2+}]_i$, but also participates in the formation of Ca^{2+} waves or oscillations in these cells [60, 61]. Repeating cycles of Ca^{2+} uptake and release in the SR constitute intracellular Ca^{2+} waves or oscillations. By Ca^{2+} waves, Ca^{2+} is guided from the plasma membrane to myofilaments, instead of direct diffusion from the plasma membrane through the cytoplasm [61]. Therefore, the Ca^{2+} oscillations may importantly contribute to the tension development in VSM in response to different agonists such as phenylephrine [60, 61].

There is considerable evidence that cADPR participates in CICR in nonvascular tissues [14, 62-66]. With respect to the mechanism of CICR, a small increase in intracellular [Ca^{2+}] by influx results in cADPR production and subsequent activation of cADPR-mediated Ca^{2+} mobilization or in the augmentation of the sensitivity of Ca^{2+} pool to cADPR, leading to Ca^{2+} release and large global increase in $[Ca^{2+}]_i$ throughout the cytoplasm and nucleus. This CICR is a fundamental mechanism in the control of basal vascular tone and in mediating vasomotor response [1, 67-70]. In the cerebral circulation, RyR-mediated CICR largely contributes to the vasoconstrictor response of cerebral resistance arteries to stimuli [67]. In the pulmonary vascular bed, CICR is crucial for the development of basal vascular tone and production of hypoxic vasoconstriction [68, 71]. More recently, we have reported that cADPR participates in KCl, $CaCl_2$, Bay K 8644 (Ca^{2+} channel activator) and caffeine-induced Ca^{2+} release response in coronary and renal VSM cells; strongly suggesting that cADPR contributes to CICR [8, 18, 51, 69]. In these smooth muscle cells, we also found that high extracellular Ca^{2+} (5 mM $CaCl_2$) and agonist Ach produced 1-1.5 Hz oscillations, which are blocked by the CICR inhibitor, tetracaine and the cADPR antagonist 8-bromo-cADPR (Fig. 1). Kannan *et al.* have also reported that cADPR increases Ach-induced Ca^{2+} oscillations and cADPR blocker, 8-amino-cADPR diminished Ach-induced $[Ca^{2+}]_i$ oscillations in porcine tracheal smooth muscle cells [72]. Taken together, these results provide direct evidence that cADPR is necessary for CICR and intracellular Ca^{2+} oscillation and RyR are the mechanistic link between cADPR and CICR or Ca^{2+} oscillations.

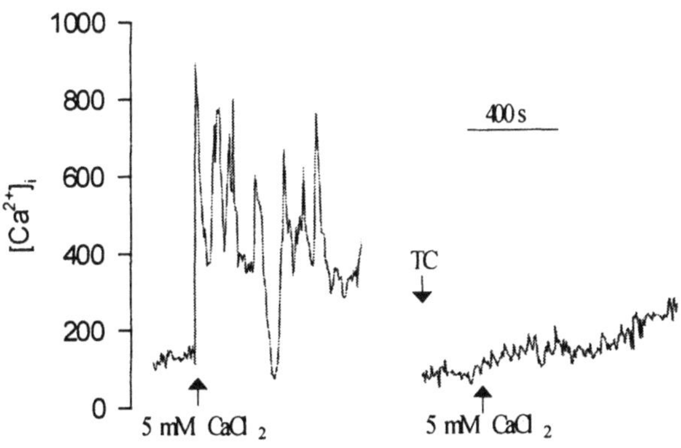

Figure 1. **$CaCl_2$-induced Ca^{2+} oscillations in freshly dissociated coronary arterial smooth muscle cells.** Fura-2 fluorescence was monitored by fluorespectrometry. TC indicates tetracaine.

ROLE OF cADPR IN VASOCONSTRICTION

cADPR and Basic Vascular Tone

As discussed at the beginning of this chapter, VSM usually operates in a contracted state, which is referred to as vascular tone. Intracellular [Ca^{2+}] importantly contributes to the production of this "resting" vascular tone. It is well known that under resting conditions, [Ca^{2+}]$_i$ in VSM is dependent upon the Ca^{2+} influx, spontaneous and brief releasing bursts of Ca^{2+} from the SR into the cytoplasm and induce CICR. cADPR participates in the control of the resting Ca^{2+} levels in these smooth muscle cells through RyR and CICR [8, 18, 51, 58, 72]. Therefore, cADPR may play an important role in the development of basic vascular tone. There is conceivable evidence supporting this view. In the isolated, perfused and pressurized small coronary arteries, basic vascular tone or spontaneous tension can be developed during a 1.5-hour equilibration period. Under this condition, the SR Ca^{2+}-ATPase inhibitor, thapsigargin (TSG) decreases the arterial diameter, and CICR blocker, tetracaine and cADPR antagonist, 8-bromo-cADPR slightly dilates these arteries, suggesting that [Ca^{2+}] associated with cADPR-RyR signaling pathway and CICR is one of the determinants of the basic vascular tone [69].

cADPR in Vasoconstriction Associated with Ca^{2+} Influx

It is well known that membrane depolarization by KCl and activation of L-type Ca^{2+} channels produces the influx of extracellular Ca^{2+} and thereby increases intracellular [Ca^{2+}], leading to vasoconsritction. In this process, the entry of Ca^{2+} is further amplified by CICR, thereby producing an increase in global [Ca^{2+}] throughout the cytoplasm and nucleus and consequent vasoconstriction [1]. In spite of the essential effect of Ca^{2+} influx, the amplification of Ca^{2+} signal or Ca^{2+} propagation within VSM cells also plays a critical role in the vasoconstriction in response to membrane depolarization and activation of membrane Ca^{2+} channels. The SR Ca^{2+}-ATPase inhibitor, TSG significantly blocks KCl-induced vasoconstriction in coronary arteries, suggesting that Ca^{2+} refilling and release in the SR mediated by this enzyme are critical for KCl-induced Ca^{2+} release and vasoconstriction. Similarly, the CICR blocker tetracaine and the cADPR antagonist 8-bromo-cADPR also significantly attenuate KCl-induced decrease in arterial diameters. However, a cell permeable IP_3 blocker, xestospongin C at a concentration that attenuated U46619-induced contraction, has no effect on KCl-induced vasoconstriction in the same small resistant coronary arteries. Using a Ca^{2+} channel activator, Bay K8644, and $CaCl_2$ to stimulate Ca^{2+} influx, the vasoconstrictor effects are also attenuated by TSG, tetracaine or 8-bromo-cADPR. This role of CICR in vasoconstrictor response to Ca^{2+} influx has been documented in other vascular beds such as aorta, cerebral, renal and pulmonary arteries [67-71]. It is concluded that cADPR-mediated Ca^{2+} signaling is of importance in amplification of intracellular Ca^{2+} signal and vasoconstriction through CICR.

cADPR in Vasoconstrictor Response to Agonists

cADPR mediates the effects of Ach receptors in adrenal chromaffin cells, estrogen receptors in uterus, 5-HT 2B receptor in arterial endothelial cells and retinoic acid in renal tubular cells and aortic smooth muscle [70, 72-76]. In VSM, the contribution of the cADPR-mediated signaling pathway to Ach-induced contraction has been intensively studied [73]. There are three lines of evidence supporting the role of cADPR in mediating or modulating the vasoconstrictor effect of Ach or M_1 agonist [77].

First, vasoconstriction induced by Ach or M_1 agonist, oxotremorine (OXO), is attenuated by the inhibition of cADPR formation or blockade of cADPR action, suggesting that cADPR is probably linked to M_1 mAChRs and mediates the vasoconstrictor action of this subtype of mAChRs in VSM cells [73, 77-79]. The action of cADPR is associated with RyR activation and Ca^{2+} release from Ca^{2+}-ATPase-dependent stores, since ryanodine also blocks OXO-induced vasoconstriction in these arteries. Although some studies have demonstrated that activation of recombinant M_1 mAChRs reconstituted with cerebral or hepatic Gq/11 protein stimulates phospholipase C-β to produce IP_3 [76], it seems that cADPR-mediated Ca^{2+} signaling is a major pathway mediating activation of M_1 mAchR in arterial smooth muscle.

Second, OXO-induced vasoconstriction is directly associated with cADPR-mediated Ca^{2+} release from the SR in VSM cells isolated from small coronary arteries. OXO produces a rapid Ca^{2+} release in single VMS cells bathed with Ca^{2+}-free solution. Inhibition of cADPR production by nicotinamide and blockade of cADPR action by 8-bromo-cADPR substantially blocks OXO-induced Ca^{2+} release. Moreover, ryanodine also blocks OXO-induced Ca^{2+} release in these VSM cells [77]. These results provide direct evidence that endogenous cADPR and RyR contribute to OXO-induced Ca^{2+} mobilization in VSM cells. In other studies, cADPR has also been reported to serve as a second messenger to mediate ACh-induced Ca^{2+} release from the SR of tracheal smooth muscle cells [50, 73], further supporting the view that cADPR mediates the action of M_1 mAchR activation.

Third, OXO markedly enhances the activity of ADP-ribosyl cyclase and increases the production of cADPR in cultured coronary VSM cells, which is blockable by the M_1 mAChR blocker, pirenzepine (PIR) and by the ADP-ribosyl cyclase inhibitor, nicotinamide [77]. In other studies, ACh stimulates the production of cADPR in adrenal chromaffin cells and NG108-15 neuronal cells. It seems that ADP-ribosylcylcase is directly coupled to M_1 mAChRs through G-proteins. When these receptors are activated, the production of cADPR increases, whereby Ca^{2+} is released from the SR through RyR [47, 75, 79].

ROLE OF cADPR IN VASODILATION

Nicotinamide-Induced Vasodilation

Nicotinamide, an amide derivative of vitamin B_3, is a potent vasodilator [80]. Numerous studies have demonstrated that nicotinamide increases the perfusion and oxygenation of tumors, thereby increasing the sensitivity of the tumors to radio- or chemical therapies [51]. However, the mechanism by which nicotinamide produces vasodilation and increases tissue perfusion is poorly understood. Recently, nicotinamide has been reported to inhibit the activity of purified ADP-ribosylcyclase, decrease cADPR production and block the Ca^{2+}-mobilizing action of β-NAD^+, cGMP and NO in nonvascular cells [81]. In VSM, this effect of nicotinamide on ADP-ribosyl cyclase activity may produce vasodilation. Therefore, cADPR-mediated Ca^{2+} mobilization may be a target for the vasodilator effect of nicotinamide. Consistent with this hypothesis, we have shown that this ADP-ribosyl cyclase inhibitor, nicotinamide produces a concentration-dependent relaxation in U46619-precontracted coronary arteries. This relaxation can be markedly blocked by pretreatment of coronary arteries with either 8-bromo-cADPR or ryanodine. It has been proposed that cADPR-mediated Ca^{2+} signaling may be a therapeutic target for the action of vasodilator compounds like nicotinamide [51].

NO-Induced Vasodilation

NO, an endothelium-derived relaxing factor plays an important role in the control of vascular tone [8]. Although NO-induced cGMP production is considered as an important mechanism mediating its vasodilator effects [8, 10], many studies have shown that NO-induced vasodilation in some vascular beds is not completely blocked by guanylyl cyclase inhibitors. This has led to the view that cGMP-independent mechanisms may also participate in the vasodilator effect of NO [82]. In this regard, activation of K_{Ca} channels, inhibition of vasoconstrictor, 20-HETE, production and direct effect on Ca^{2+} influx and Ca^{2+} release have been reported to be possible mechanisms mediating the action of NO on VSM [83, 84]. However, it is still poorly understood how NO alters Ca^{2+} activation and subsequent tension development in VSM.

There is evidence that NO affects Ca^{2+} release from ryanodine-sensitive stores through the cGMP/PKG pathway in nonvascular tissues. NO activates guanylyl cyclase and increases cGMP levels, which promotes cADPR production and consequently increases Ca^{2+} release from the SR [17, 30, 34, 85]. Obviously, this NO-induced rise in $[Ca^{2+}]_i$ could not contribute to its vasodilator effect, since NO decreases $[Ca^{2+}]_i$ in VSM cells. NO may increase cADPR hydrolase activity and in this way reduce the concentrations of cADPR, leading to a decrease in $[Ca^{2+}]_i$ [10]. Therefore, endogenous cADPR-induced Ca^{2+} release in the resting state probably contributes to the control of basal $[Ca^{2+}]_i$ in VSM cells, and NO decreases intracellular cADPR levels and thereby lowers $[Ca^{2+}]_i$. This cADPR-dependent changes in $[Ca^{2+}]_i$ may be one of the mechanisms mediating NO-induced vasodilation [8, 10].

Indeed, NO donors such as SNP and deta NONOate produce a concentration-dependent vasodilation in coronary arteries, but this vasodilator response to NO donors is not completely blocked by a guanylyl cyclase inhibitor, ODQ, unless a cADPR antagonist is added in combination. Thus, a cADPR-mediated mechanism is involved in NO-induced vasodilation [8]. By determining Ca^{2+} release response in single VSM cells, NO has no direct effect on either cADPR- or IP_3-induced Ca^{2+} release response in permeabilized cells, but inhibits KCl- and caffeine-induced Ca^{2+} release in intact cells [8, 10, 67]. It appears that NO decreases $[Ca^{2+}]_i$ by inhibiting endogenous cADPR-mediated Ca^{2+} release, but not by a direct effect on the Ca^{2+} release machinery on the SR of VSM cells. This inhibitory effect of NO on cADPR-induced Ca^{2+} release has also been confirmed in porcine tracheal smooth muscle cells [9, 50].

The mechanism by which NO decreases $[Ca^{2+}]_i$ in VSM cells and porcine tracheal smooth muscle cells, but increases $[Ca^{2+}]_i$ in some other cells remains unknown. It has been assumed that NO may have different effects on the enzyme activities responsible for the production or degradation of cADPR in different cells [10]. NO decreases the production of cADPR in a concentration-dependent manner by inhibition of ADP-ribosyl cyclase, but it has no effect on cADPR hydrolase [10]. Based on these results, it is concluded that the inhibitory effect of NO on the production of endogenous cADPR plays an important role in mediating NO-induced vasodilation.

EETs-Induced Vasodilation

Epoxyeicosatrienoic acids (EETs), endothelium-derived hyperpolarizing factors (EDHF), also produce vasodilation in a variety of vascular beds and thus play an important role in the regulation of vascular tone [86]. Recently, the mechanism mediating the vasodilator response to this EDHF has been intensively studied. EETs induce vasodilation through the activation of K_{Ca} channels in the VSM cells. ADPR, a metabolite of cADPR, directly stimulates K_{Ca} channels in coronary VSM cells [20]. Interestingly, EETs can activate both NAD glycohydrolase and cADPR hydrolase, increasing ADPR levels in VSM cells. Increased intracellular ADPR may serve as an intracellular signaling molecule mediating the effect of EETs on K_{Ca} channels [26].

Patch clamp studies have shown that selective inhibition of NAD glycohydrolase by 3GA or inhibition of cADPR hydrolase by novobiocin completely abolished the activation of the K_{Ca} channels induced by 11,12-EET in coronary VSM cells. It appears that EETs-induced activation of K_{Ca} channels is associated with the production of ADPR. Consistent with these findings in patch clamp studies, 11,12-EET-induced vasorelaxation can be attenuated by 3GA or novobiocin in epicardial coronary arteries and small resistance coronary arteries [26]. These results indicate that the NAD or cADPR metabolite, ADPR, may serve as a signaling molecule mediating the vasodilator effects of EDHF. This role of ADPR may represent a new signaling pathway regulating the vasomotor response.

MECHANISMS MEDIATING cADPR ACTION

As discussed above, cADPR mobilizes intracellular Ca^{2+} by a mechanism completely independent of IP_3 [10, 17, 87]. Despite controversial findings [50, 88, 89], most studies have indicated that cADPR can activate RyR, producing Ca^{2+} release from the SR or ER in different tissues or cells [90]. Although there are different isoforms of RyR, an important property of these receptors is that they can be opened by Ca^{2+} and can mediate CICR [2, 30, 63, 64, 91]. In VSM cells, CICR and RyR activation have been recognized as an important mechanism regulating vascular tone or vasomotor response [1, 58, 59, 67, 91-93]. Now, two mechanistic models have been proposed to elucidate the role of endogenous cADPR in mediating vascular activity. Considering cADPR as a mediator, it is proposed that the agonists or stimuli activate ADP-ribosyl cyclase to produce cADPR, leading to activation of Ca^{2+} release from the SR through the RyR. This activation of ADP-ribosyl cyclase may also occur when intracellular Ca^{2+} levels increase even slightly, thereby resulting in CICR. Another model considers cADPR as a modulator of CICR or RyR activity. In this way, cytosolic cADPR sensitizes the RyR, enhancing CICR activated by agonists or Ca^{2+} influx. The relative contribution of these two mechanisms to the vascular reactivity may vary depending upon the concentrations of intracellular cADPR, $[Ca^{2+}]$, and calmodulin and the functional status of RyR in different cells.

RyR and Action of cADPR in VSM

RyR are widespread in many cell types and mediate Ca^{2+} release from the SR or ER in a variety of tissues or cells [2, 63, 94, 95]. A RyR complex is a homotetramer with a molecular weight of approximately 550 kDa for each monomer. Four monomers comprise a high conductance Ca^{2+} channel pore on the membrane of the SR, which is responsible for Ca^{2+} release. Three

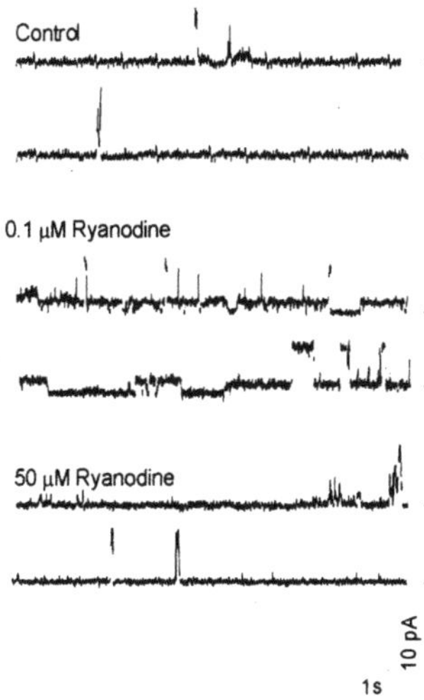

Figure 2. **Representative Recordings of reconstituted RyR/Ca^{2+} release channels from coronary VSM in the absence and presence of ryanodine.**

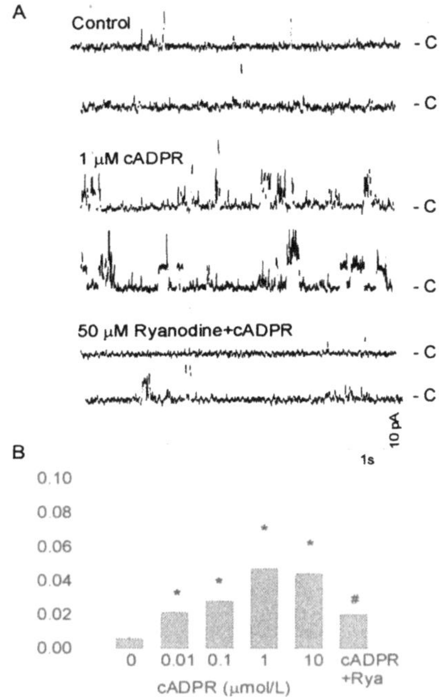

Figure 3. **The effects of cADPR on the activity of coronary arterial RyR/Ca^{2+} release channels. A.** Representative recordings of reconstituted RyR/Ca^{2+} release channels under condition and after addition of cADPR in *cis*-solution. **B**. Summarized data showing cADPR-induced activation of these RyR/Ca^{2+} release channels. Rya indicates ryanodine.

subtypes of RyR with specific tissue distributions are now recognized and named as RyR1, RyR2, and RyR3. Studies using ligand binding and molecular approaches demonstrate that RyR3 is a predominant form on the SR of VSM cells [96]. Cytosolic Ca^{2+} is a primary ligand activating these RyR, and the binding of Ca^{2+} to RyR activates Ca^{2+} release. Therefore, RyRs are basically the Ca^{2+}-gated Ca^{2+} channels, which may mediate CICR and play an important role in the regulation of vascular tone [88, 91, 93, 96-98].

Using electrophysiological methods, the RyR/Ca^{2+} release channels can be reconstituted into a planar lipid bilayer [58, 59, 99, 100] and visualized with electrical stimulation or clamp. In coronary arterial smooth muscle, a calcium channel with 245 pS conductance is present on the SR membrane (Fig. 2). The conductance of this RyR/Ca^{2+} release channels in VSM is relatively smaller compared with 500-1000 pS channels found in either skeletal or cardiac muscle [96, 97, 99, 100]. The RyR ligand, ryanodine at low concentrations significantly reduces the current amplitude to a sub-conductance state and increased the open probability (NP_O), but a high concentration of ryanodine (50 μM) inactivates these channels [58, 96].

In this reconstituted RyR preparation, cADPR increases the NP_O of RyR/Ca^{2+} release channels in a concentration-dependent manner. In the presence of ryanodine (50 μM), cADPR-induced activation of these channels is completely abolished (Fig. 3). These results provide direct evidence that cADPR activates RyR and therefore may serve as an endogenous activator or modulator of the RyR in these VSM cells.

However, there is evidence indicating that cADPR releases Ca^{2+} independently of RyR in neurons and cells from the myocardium and other smooth muscle [50, 88, 89]. The reason for these discrepancies remains unknown. It is possible that there is a tissue specific effect of cADPR on the RyR that may be associated with the intermediate proteins or accessory proteins that regulate RyR activity. For example, calmodulin is essential for cADPR-induced Ca^{2+} release in sea urchin eggs [90, 98]. Recent studies have demonstrated that FK 506 binding protein (FKBP) may have a binding site for cADPR, and that cADPR activates the RyR through binding to this accessory protein [59, 101, 102].

FKBP 12.6 Protein and Action of cADPR in VSM

There is increasing evidence indicating that FKBP plays an important role in the activation of the RyR and Ca^{2+} release from the SR [67, 96]. FKBP 12.6 is a ubiquitous 12-kDa cytosolic protein. As an accessory protein, each FKBP 12 can bind to one RyR monomer [102, 103, 104], and its activity can be inhibited by the immunosuppressant drugs, FK506 and rapamycin [104, 105, 106]. Ca^{2+} release from the SR is inhibited when FKBP 12 is bound to the RyR and dissociation of FKBP 12.6 from the RyR releases Ca^{2+} [101].

Recent studies in our laboratory and by others have shown that cADPR can bind to FKBP 12.6 and thereby activate the RyR on the SR [59, 101]. This 12.6 kDa protein is expressed in coronary arterial smooth muscle and cultured coronary arterial smooth muscle cells, as measured by Western blot analysis. Blockade, dissociation or removal of FKBP 12.6 protein from the RyR substantially abolishes cADPR-induced activation of RyR/Ca^{2+} release channels on lipid bilayer membrane [59]. Clearly, FKBP 12.6 plays a critical role in mediating cADPR-induced activation of RyR/Ca^{2+} release channels in the SR of VSM cells. FKBP 12.6 may be dissociated with the RyR complex by binding cADPR, thereby resulting in Ca^{2+} release from the SR in VSM cells.

CONCLUSION

As summarized in Fig. 4, a membrane-bound enzyme system is present in VSM cells, which can synthesize and metabolize cADPR. cADPR serves a second messenger to stimulate Ca^{2+} release from the SR via RyR and participates in the regulation of CICR and consequent Ca^{2+} waves, producing global Ca^{2+} increase. cADPR may activate the RyR by binding to FKBP12.6

and resulting in dissociation of this accessory protein from RyR. This cADPR-mediated Ca^{2+} signaling plays a critical role in the Ca^{2+} release and vasoconstrictor response to membrane depolarization, Ca^{2+} influx and different agonists such as Ach and OXO. Moreover, a decrease in this cADPR-mediated Ca^{2+} mobilization importantly contributes to vasodilation induced by NO, EETs and nicotinamide.

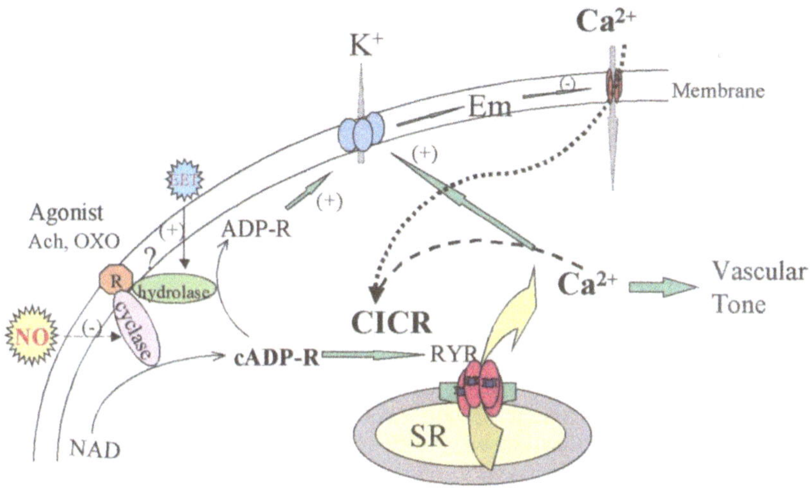

Figure 4. **A diagram showing the production and actions of cADPR in VSM cells.** (+) indicates activation or enhancement; (-) indicates inhibition or attenuation.

ACKNOWLEDGMENTS

These studies were supported by grants from the National Heart, Lung and Blood Institute (HL-57244, HL51055 and DK54924) and AHA (#96007310). Dr. Pin-Lan Li is a recipient of Established Investigator Award from American Heart Association (#9940167N).

REFERENCES

1. Berridge MJ. 1997. Elementary and global aspects of calcium signalling. *J. Physiol.* 499: 291-306.
2. Berridge MJ. 1994. The biology and medicine of calcium signaling. *Mol. Cellul. Endocrinol.* 98: 119-124.
3. Galione A and Sethi J. 1996. Cyclic ADP-ribose and calcium signaling. In *Biochemistry of smooth muscle contraction*. pp. 295-305. ed. M Barany. Academic Press.
4. Himpens B, Missiaen L and Casteels R. 1995. Ca^{2+} homeostasis in vascular smooth muscle. *J. Vasc. Res.* 32: 207-219.
5. Nelson MT, Patlak JB, Worley JF and Standen NB. 1990. Calcium channels, potassium channels, and voltage dependence of arterial smooth muscle tone. *Am. J. Physiol.* 259: C3-C18.
6. Fasolato C, Innocenti B and Pozzan T. 1994. Receptor-activated Ca^{2+} influx: how many mechanisms for how many channels? *Trends Pharmacol. Sci.* 15: 77-83.
7. Petersen OH. 1996. New aspects of cytosolic calcium signaling. *News Physiol. Sci.* 11:

13-17.
8. Li N, Zou AP, Ge ZD, Campbell WB and Li P-L. 2001. Effect of nitric oxide on calcium-induced calcium release in coronary arterial smooth muscle. *Gen. Pharmacol. Vasc. System* 35: 37-45.
9. Kannan MS, Prakash YS, Johnson DE and Sieck GC. 1997. Nitric oxide inhibits calcium release from sarcoplasmic reticulum of porcine tracheal smooth muscle cells. *Am. J. Physiol.* 272: L1-L7.
10. Yu J-Z, Zhang DX, Zou AP, Campbell WB and Li P-L. 2000. Nitric oxide inhibits intracellular calcium mobilization through the cyclic ADP-ribose signaling pathway in bovine coronary arterial smooth muscle cells. *Am. J. Physiol.* 279: 873-881.
11. Lee HC, Walseth TF, Bratt GT, Hayes RN and Clapper DL. 1989. Structural determination of a cyclic metabolite of NAD^+ with intracellular Ca^{2+} mobilizing activity. *J. Biol. Chem.* 264: 1608-1615.
12. Beers K, Chini EN, Lee HC and Dousa TP. 1995. Metabolism of cyclic ADP-ribose in opossum kidney renal epithelial cells. *Am. J. Physiol.* 268: C741-C746.
13. Koshiyama HH., Lee HC and Tashjian AH. 1991. Novel mechanism of intracellular calcium release in pituitary cell. *J. Biol. Chem.* 266:.16985-16988.
14. Galione A, Lee HC and Busa WB. 1991. Ca^{2+}-induced Ca^{2+} release in sea urchin egg homogenates and its modulation by cyclic ADP-ribose. *Science* 253: 1143-1146.
15. Lee HC and Aarhus R. 1993. Wide distribution of an enzyme that catalyzes the hydrolysis of cyclic ADP-ribose. *Biochim. Biophys. Acta* 1164: 68-74.
16. Takesawa S, Nata K, Yonekura H and Okamoto H. 1993. Cyclic ADP-ribose in insulin secretion from pancreatic β cells. *Science* 259: 370-373.
17. Lee HC. 1994. A signaling pathway involving cyclic ADP-ribose, cGMP, and nitric oxide. *News Physiol. Sci.* 9: 134-137.
18. Li N, Teggatz EG, Li P-L, Allaire R and Zou AP. 2000. Formation and actions of cyclic ADP-ribose in renal microvessels. *Microvasc. Res.* 60: 149-159.
19. Li P-L, Zou AP and Campbell WB. 1997. Metabolism and actions of ADP-ribose in coronary arterial smooth muscle. *Adv. Exp. Med. Biol.* 419: 437-441.
20. Li P-L, Zou AP and Campbell WB. 1998. Regulation of the K_{Ca} channel activity by cyclic ADP-riboses and ADP-ribose in bovine coronary arterial smooth muscle. *Am. J. Physiol.* 275: H1002-1010.
21. Franco L, Zocchi E, Calder L, Guida L, Benatti U and De Flora A. 1994. Self-aggregation of the transmembrane glycoprotein CD38 purified from human erythrocytes. *Biochem. Biophys. Res. Commun.* 202: 1710-1715.
22. Malawasi F, Funaro A, Roggero S, Horenstein A, Calosso L and Mehta K. 1994. Human CD38: a glycoprotein in search of a function. *Immunol. Today* 15: 95-97.
23. Zocchi E, Franco L, Guida L, Benatti U, Bragellesi MF, Lee HC and De Flora A. 1993. A single protein immunologically identified as CD38 displays NAD glycohydrolase, ADP-ribosyl cyclase and cyclic ADP-ribose hydrolase activities at the outer surface of human erythrocytes. *Biochem. Biophys. Res. Commun.* 196: 1459-1465.
24. Zocchi E, Franco L, Guida L, Calder L and De Flora A. 1995. Self-aggregation of purified and membrane-bound erythrocyte CD38 induces extensive decrease of its ADP-ribosyl cyclase activity. *FEBS Lett.* 359: 35-40.
25. Zocchi E, Usai C, Guida L, Franco L, Bruzzone S, Passalacqua M and De Flora A. 1999. Ligand-induced internalization of CD38 results in intracellular Ca^{2+} mobilization role of NAD^+ transport across cell membranes. *FASEB J.* 13: 273-283.
26. Li P-L, Zhang DX, Zhi-Dong Ge and Campbell WB. 2002. Role of ADP-ribose in mediating 11,12-EET-induced activation of K_{Ca} channels in coronary arterial smooth muscle cells. *Am. J. Physiol.-Heart Circ. Physiol.* 282: (in press).
27. Zhang DX, Zou AP and P-L Li. 2001. ADP-ribose dilates coronary arteries through apyrase-mediated metabolism. *J. Vasc. Res.* 38: 64-72,
28. Berruet L, Muller-Steffner H and Schuber F. 1998. Occurrence of bovine spleen

CD38/NAD$^+$ glycohydrolase disulfide-linked dimer. *Biochem. Mol. Biol. Int.* 46: 847-855.

29. Chini EN, Beers KW, Chini CS and Dousa TP. 1995. Specific modulation of cyclic ADP-ribose-induced Ca^{2+} release by polyamines. *Am. J. Physiol.* 269: C1042-C1047.
30. Galione A, White H, Willmott N, Turner M, Potter BVL and Watson SP. 1993. cGMP mobilizes intracellular Ca^{2+} in sea urchin eggs by stimulating cyclic ADP-ribose synthesis. *Nature* 365: 456-459.
31. Takasawa S, Tohgo A, Noguchi N, Koguma T, Nata K, Sugimoto T, Yonekura H and Okamoto H. 1993. Synthesis and hydrolysis of cyclic ADP-ribose by human leukocyte antigen CD38 and inhibition of the hydrolysis by ATP. *J. Biol. Chem.* 268: 26052-26054.
32. Tohgo A, Munakata H, Takasawa S, Nata K, Akiyama T, Hayashi N and Okamoto H. 1997. Lysine 129 of CD38 (ADP-ribosyl cyclase/cyclic ADP-ribose hydrolase) participates in the binding of ATP to inhibit the cyclic ADP-ribose hydrolase. *J. Biol. Chem.* 272: 3879-3882.
33. Vu CQ, Coyle DL, Tai H-H, Jacobson EL and Jacobson MK. 1997. Intramolecular ADP-ribose transfer reaction and calcium signalling. *Adv. Exp. Med. Biol.* 419: 381-388.
34. Whalley T, McDougall A, Crossley L, Swann K and Whitaker M. 1992. Internal calcium release and activation of sea urchin eggs by cGMP are independent of the phosphoinositide signaling pathway. *Mol. Biol. Cell 3: 373*-383.
35. Chakraborti T, Ghosh SK, Michael JR, Batabyal SK and Chakraborti S. 1998. Targets of oxidative stress in cardiovascular system. *Mol. Cell. Biochem.* 187: 1-10.
36. Clementi E. 1998. Role of nitric oxide and its intracellular signalling pathways in the control of Ca^{2+} homeostasis. *Biochem. Pharmacol.* 55: 713-718.
37. Van Zwieten PA and Doods HN. 1995. Muscarinic receptors and drugs in cardiovascular medicine. *Cardiovas. Drugs Ther.* 9: 159-167.
38. Darley-Usmar VM, McAndrew J, Patel R, Moellering D, Lincoln TM, Jo H, Cornwell T, Digerness S and White CR. 1997. Nitric oxide, free radicals and cell signalling in cardiovascular disease. *Biochem. Soc. Transact.* 25: 925-929.
39. Suzuki YJ, Forman HJ and Sevanian A. 1997. Oxidants as stimulators of signal transduction. *Free Radical Biol. Med.* 22: 269-285.
40. Kawakami M and Okabe E. 1998. Superoxide anion radical-triggered Ca^{2+} release from cardiac sarcoplasmic reticulum through ryanodine receptor Ca^{2+} channel. *Mol. Pharmacol.* 53: 497-503.
41. Chidambaram N, Wong ET and Chang CF. 1998. Differential oligomerization of membrane-bound CD38/ADP-ribosyl cyclase in porcine heart microsomes. *Biochem. Mol. Biol. Intl.* 4: 1225-1233.
42. Guida L, Franco L, Zocchi E and De Flora A. 1995. Structural role of disulfide Bridges in the cyclic ADP-ribose related bifunctional ectoenzyme CD38. *FEBS Lett.* 369: 481-484.
43. Munshi C, Braumann C, Levitt D, Bloomfield VA and Lee HC. 1998. The homo-dimeric form of ADP-ribosyl cyclase in solution. *Biochim. Biophys. Acta* 1388: 428-436.
44. Shubinsky G and Schlesinger M. 1997. The CD38 lymphocyte differentiation marker: new insight into its ectoenzymatic activity and its role as a signal transducer. *Immunity* 7: 315-324,
45. Tohgo A, Takasawa S, Noguchi N, Koguma T, Nata K, Sugimoto T, Furuya Y, Yonekura H and Okamoto H. 1994. Essential cysteine residues for cyclic ADP-ribose synthesis and hydrolysis by CD38. *J. Biol. Chem.* 269: 28555-28557.
46. Prasad GS, McRee DE., Stura EA, Levitt DG, Lee HC and Stout CD. 1996. Crystal structure of *Aplysia* ADP ribosyl cyclase, a homologue of the bifunctional ecto-enzyme CD38. *Nature Struct. Biol.* 3: 957-964.
47. Higashida H. 1997. ADP-ribosyl cyclase coupled with receptors via G-proteins. *FEBS Lett.* 418: 355-356.

48. Lee HC, Aarhus R and Walseth TF. 1993. Calcium mobilization by dual receptors during fertilization of sea urchin eggs. *Science* 261: 352-355.
49. Lee HC, Walseth TF, Aarhus R and Levitt D. 1993. Cyclic ADP-ribose as a second messenger for mobilizing intracellular calcium stores. *J. Reprod. Devel.* 39: 64-69.
50. Kannan MS, Fenton AM, Prakash YS and Sieck GC. 1996. Cyclic ADP-ribose stimulates sarcoplasmic reticulum calcium release in porcine coronary artery smooth muscle. *Am. J. Physiol.* 270: H801-H805.
51. Geiger J, Zou AP, Campbell WB and Li P-L. 2000. Inhibition of cyclic ADP-ribose formation produces vasodilation in bovine coronary arteries. *Hypertension* 35: 397-402.
52. Imaizumi Y, Ohi Y, Yamamura H, Ohya S, Muraki K and Watanabe M. 1999. Ca^{2+} spark as a regulator of ion channel activity. *Jpn. J. Pharmacol.* 80: 1-8.
53. Jaggar JH, Wellman GC, Heppner TJ, Porter VA, Perez GJ, Gollasch M, Kleppisch T, Rubart M, Stevenson AS, Lederer WJ, Knot HJ, Bonev AD and Nelson MT. 1998. Ca^{2+} channels, ryanodine receptors and Ca^{2+}-activated K^+ channels: a functional unit for regulating arterial tone. *Acta Physiol. Scand.* 164: 577-587.
54. Jaggar JH, Porter VA, Lederer WJ and Nelson MT. 2000. Calcium sparks in smooth muscle. *Am. J. Physiol. (Cell Physiol.)* 278: C235-C256.
55. Nelson MT, Cheng H, Rubart M, Santana LF, Bonev AD, Knot HJ and Lederer WJ. 1995. Relaxation of arterial smooth muscle by calcium sparks. *Science* 270: 633-637.
56. Cui Y, Galione A and Terrar DA. 1999. Effects of photoreleased cADP-ribose on calcium transients and calcium sparks in myocytes isolated from guinea-pig and rat ventricle. *Biochem. J.* 342: 269-273.
57. Bychkov R, Gollasch M, Ried C, Luft FC and Haller H. 1997. Regulation of spontaneous transient outward potassium currents in human coronary arteries. *Circ.* 95: 503-510.
58. Li P-L, Tang WX, Valdivia HH, Zou AP and Campbell WB. 2001. Cyclic ADP-ribose activates reconstituted ryanodine receptors from coronary arterial smooth muscle. *Am. J. Physiol.* 280: H208-H215.
59. Tang WX, Cheng YF, Zou AP, Campbell WB and Li P-L. 2002. Role of FKBP 12.6 in cADPR-induced activation of reconstituted ryanodine receptors from coronary arterial smooth muscle. *Am. J. Physiol.* (in press).
60. Bolton TB and Gordienko DV. 1998. Confocal imaging of calcium release events in single smooth muscle cells. *Acta Physiol. Scand.* 164: 567-575.
61. Ruehlmann DO, Lee CH, Poburko D and van Breemen C, 2000. Asynchronous Ca^{2+} waves in intact venous smooth muscle. *Circ. Res.* 86: E72-E79.
62. Empson RM and Galione A. 1997. Cyclic ADP-ribose enhances coupling between voltage-gated Ca^{2+} entry and intracellular Ca^{2+} release. *J. Biol. Chem.* 272: 20967-20970.
63. Galione A. 1992. Ca^{2+}-induced Ca^{2+} release and its modulation by cyclic ADP-ribose. *Trends Pharmacol. Sci.* 13: 304-306.
64. Galione A, Lee HC and Busa WB. 1991. Ca^{2+}-induced Ca^{2+} release in sea urchin egg homogenates: modulation by cyclic ADP-ribose. *Science* 253: 1143-1146.
65. Lee HC. 1993. Potentiation of calcium- and caffeine-induced calcium release by cyclic ADP-ribose. *J. Biol. Chem.* 268: 293-299.
66. Rakovic S, Galione A, Ashamu GA, Ptter BVL and Terrar DA. 1996. A specific cyclic ADP-ribose antagonist inhibits cardiac excitation-contraction coupling. *Curr. Biol.* 6: 989-996.
67. Kamishima T and McCarron JG. 1997. Regulation of the cytosolic Ca^{2+} concentration by Ca^{2+} stores in single smooth muscle cells from rat cerebral arteries. *J. Physiol.* 501: 497-508.
68. Vandier C., Delpech M., Rebocho M and Bonnet P. Hypoxia enhances agonist-induced pulmonary arterial contraction by increasing calcium sequestration. *Am. J. Physiol.* 1997; 273, H1075-1081.
69. Zhang DX, MD Harrison and P-L Li. 2001. Role of calcium-induced calcium release in

the control of intracellular calcium concentration and contraction of coronary arterial smooth muscle. *FASEB J.* 15: A1115.

70. Beers KW, Chini EN and Dousa TP. 1995. All-trans-retinoic acid stimulates synthesis of cyclic ADP-ribose in renal LLC-PK1 cells. *J. Clin. Invest.* 95:2395-2390.
71. Jabr RI, Toland H, Gelband CH, Wang XX and Hume JR. 1997. Prominent role of intracellular Ca^{2+} release in hypoxic vasoconstriction of canine pulmonary artery. *Br. J. Pharmacol.* 122: 21-30.
72. Chini EN, de Toledo FG, Thompson MA and Dousa TP. 1997. The effect of estrogen upon cyclic ADP-ribose metabolism: beta-estradiol stimulates ADP ribosyl cyclase in rat uterus. *Proc. Natl. Acad. Sci. USA* 94: 5872-5876.
73. Kannan MS, Prakash YS, Brenner T, Mickelson JR and Sieck GC. 1997. Role of ryanodine receptor channels in Ca^{2+} oscillations of porcine tracheal smooth muscle. *Am. J. Physiol.* 272: L659-L664.
74. De Toledo FG, Cheng J and Dousa TP. 1997. Retinoic acid and triiodothyronine stimulate ADP-ribosyl cyclase activity in rat vascular smooth muscle cells. *Biochem. Biophy. Res. Commun.* 238: 847-850.
75. Morita K, Kitayama S and Dohi T. 1997. Stimulation of cyclic ADP-ribose synthesis by acetylcholine and its role in catecholamine release in bovine adrenal chromaffin cells. *J. Biol. Chem.* 272: 21002-21009.
76. Tripathy A, Resch W, Xu L, Valdivia HH and Meissner G. 1998. Imperatoxin A induced subconductance states in Ca^{2+} release channels (ryanodine receptors) of cardiac and skeletal muscle. *J. Gen. Physiol.* 111: 679-690.
77. Ge ZD, Zou AP, Campbell WB and Li P-L. 2001. Cyclic ADP-ribose mediates the action of muscarinic m_1 receptors in coronary arterial smooth muscle. *Hypertension* 38: 160 (Abstract).
78. Bernstein G. 1992. Reconstitution of agonist-stimulated phosphatidyl-inositol 4,5-biphosphate hydrolysis using purified m1 muscarinic receptor, Gq/11, and phospholipase C-beta 1. *J. Biol. Chem.* 267: 8081-8088.
79. Higashida H, Yokoyama S, Hashii M, Taketo M, Higashida M, Takayasu T, Ohshima T, Takasawa S, Okamoto H and Noda M. 1997. Muscarinic receptor-mediated dual regulation of ADP-ribosyl cyclase in NG108-15 neuronal cell membranes. *J. Biol. Chem.* 272: 31272-31277.
80. Burns DM, Ruddock MW, Walker MD, Allen JM, Kennovin GD and Hirst DG. 1999. Nicotinamide-inhibited vasoconstriction: Lack of dependence on agonist signaling pathways. *Eur. J. Pharmacol.* 374: 213-220.
81. Graeff RM, Franco L, De Flora A and Lee HC. 1998. Cyclic GMP-dependent and -independent effects on the synthesis of the calcium messengers cyclic ADP-ribose and nicotinic acid adenine dinucleotide phosphate. *J. Biol. Chem.* 273: 118-125.
82. Carrier GO, Fuchs LC, Winecoff AP, Giulumian AD and White RE. 1997. Nitrovasodilators relax messenteric microvessels by cGMP-induced stimulation of Ca-activated K channels. *Am. J. Physiol.* 273: H76-H84.
83. Bolotina VM, Najibi S, Palacino JJ, Pagano PJ and Cohen RA. 1994. Nitric oxide directly activates potassium channels in vascular smooth muscle. *Nature.* 368: 850-853.
84. Sun CW, Alonso-Galicia M, Taheri MR, Falck JR, Harder DR and Roman RJ. 1998. Nitric oxide-20-hydroxy-eicosatetraenoic acid interaction in the regulation of K^+ channel activity and vascular tone in renal arterioles. *Circ. Res.* 83: 1069-1079.
85. Willmott N, Sethi JK, Walseth TF, Lee HC, White AM and Galione A. 1996. Nitric oxide-induced mobilization of intracellular calcium via the cyclic ADP-ribose signaling pathway. *J. Biol. Chem.* 271: 3699-3705.
86. Campbell WB, Gebremedhin D, Pratt PF and Harder DR. 1996. Identification of epoxyeicosatrienoic acid as endothelium-derived hyperpolarizing factors. *Circ. Res.* 78: 415-423.

87. Galione A, Cui Y, Empson R, Iino S. Wilson H and Terrar D. 1998. Cyclic ADP-ribose and the regulation of calcium-induced calcium release in eggs and cardiac myocytes. *Cell Biochem. Biophy.* 28: 19-30.
88. Lahouratate P, Guibert J and Faivre JF. 1997. cADP-ribose releases Ca^{2+} from cardiac sarcoplasmic reticulum independently of ryanodine receptor. *Am. J. Physiol.* 273: H1082-H1089.
89. Sitsapesan R, McGarry SJ and William AJ. 1994. Cyclic ADP-ribose compete with ATP for the adenine nucleotide binding site on the cardiac ryanodine receptor Ca^{2+}-release channel. *Circ. Res.* 75: 596-600.
90. Lee HC, Aarhus R, Gurnack ME and Walseth TF. 1994. Cyclic ADP-ribose activation of the ryanodine receptor is mediated by calmodulin. *Nature* 370: 307-309.
91. Ito K, Ikemoto T and Takakura S. 1991. Involvement of Ca^{2+} influx-induced Ca^{2+} release in contractions of intact vascular smooth muscles. *Am. J. Physiol.* 261: H1464-H1470.
92. Asano M, Kuwako M, Nomura Y, Suzuki Y, Shibuya M, Sugita K and Ito K. 1996. Possible mechanism of the potent vasoconstrictor responses to ryanodine in dog cerebral arteries. *Eur. J. Pharmacol.* 311: 53-60.
93. Knot HJ, Standen NB and Nelson MT. 1998. Ryanodine receptors regulate arterial diameter and wall $[Ca^{2+}]$ in cerebral arteries of rat via Ca^{2+}-dependent K^+ channels. *J. Physiol.* 508: 211-221.
94. Lesh RE, Nixon GF, Fleischer S, Airey JA, Somlyo AP and Somlyo AV. 1998. Localization of ryanodine receptors in smooth muscle. *Circ. Res.* 82: 175-185.
95. Meszaros LG, Bak J and Chiu A. 1993. Cyclic ADP-ribose as an endogenous regulator of the non-skeletal type ryanodine receptor Ca^{2+} channel. *Nature* 364: 76-79,
96. Franzini-Armstrong C and Protasi F. 1997. Ryanodine receptors of striated muscles: A complex channel capable of multiple interactions. *Pharmacol. Rev.* 77: 699-729.
97. Herrmann-Frank A, Darling E and Meissner G. 1991. Functional characterization of the Ca^{2+}-gated Ca^{2+} release channel of vascular smooth muscle sarcoplasmic reticulum. *Pfluger Arch.* 418: 353-359.
98. Lee HC, Aarhus R and Gruff RM. 1995. Sensitization of calcium-induced calcium release by cyclic ADP-ribose and calmodulin. *J. Biol. Chem.* 270: 9060-9066.
99. Lokuta AJ, Meyers MB, Sander PR, Fishman GI and Valdivia HH. 1997. Modulation of cardiac ryanodine receptors by sorcin. *J. Biol. Chem.* 272: 25333-25338
100. Valdivia, HH. 1998. Modulation of intracellular Ca^{2+} level in the heart by sorcin and FKBP 12, two accessory proteins of ryanodine receptors. *Trends Pharmacol. Sci.* 19: 479-482.
101. Noguchi N, Takasawa S, Nata K, Tohgo A, Kato I, Ikehata F, Yonekura H and Okamoto H. 1997. Cyclic ADP-ribose binding to FK 506-binding protein 12.6 to release Ca^{2+} from islet microsomes. *J. Biol. Chem.* 272: 3133-3136.
102. Ahern GP, Junankar PR and Dulhunty AF. 1997. Subconductance states in single-channel activity of skeletal muscle ryanodine receptors after removal of FKBP 12. *Biophys. J.* 72: 146-62.
103. Timerman AP, Ogunbumni E, Freund E, Wiederrecht G, Marks AR and Fleischer S. 1993. The calcium release channel of sarcoplasmic reticulum is modulated by FK506-binding protein. Dissociation and reconstitution of FKBP-12 to the calcium release channel of skeletal muscle sarcoplasmic reticulum. *J. Biol. Chem.* 268: 22992-22999.
104. Timerman AP, Wiederrecht G, Marcy A and Fleischer S. 1995. Characterization of an exchange reaction between soluble FKBP-12 and the FKBP. Ryanodine receptor complex. Modulation by FKBP mutants deficient in peptidyl-prolyl isomerase activity. *J. Biol. Chem.* 270: 2451-2459.

105. Ahern GP, Junankar PR and Dulhunty AF. 1997. Ryanodine receptors from rabbit skeletal muscle are reversibly activated by rapamycin. *Neurosci Lett.* 225: 81-84.
106. Harding MW, Galat A, Uehling DE and Schreiber SL. 1989. A receptor for the immunosuppressant FK506 is a *cis-trans* peptidyl-prolyl isomerase. *Nature*. 341: 758-760.

Pin-Lan Li, Medical College of Wisconsin,
Milwaukee, WI 53226, USA.
e-mail: pli@mcw.edu

Cyclic ADP-Ribose and NAADP. Structures, Metabolism and
Functions. Lee HC (ed), pp. 365-379, Dordrecht: Kluwer.

Coordination of Calcium Signaling by cADPR and NAADP in Pancreatic Acinar Cells

Jose Manuel Cancela
Affiliation Laboratoire de Neurobiologie Cellulaire et Moléculaire. CNRS. 1,av de la terrasse. 91 198 Gif-sur-Yvette cédex. France.

Ca^{2+} OSCILLATIONS EVOKED BY CCK AND ACh

In mouse pancreatic acinar cells, the hormone cholecystokinin (CCK) and the neurotransmitter acetylcholine (ACh) are the most important secretagogues and both of them elicit specific Ca^{2+} signatures (Fig. 1) [1-4]. At low physiological concentration, CCK evokes a mixture of short-lasting local Ca^{2+} spikes in the secretory pole of the cell and long-lasting global Ca^{2+} spikes, whereas ACh at low concentration elicits local Ca^{2+} spikes in the secretory pole of the cell. At higher concentrations, CCK and ACh evoke global Ca^{2+} transients. Despite intensive research, the mechanisms underlying the complex Ca^{2+} oscillations remain unclear. It has been shown that heparin, an IP_3-receptor antagonist, blocked the Ca^{2+} response elicited by ACh and CCK [5-6]. These pharmacological data suggested that the IP_3 receptors are involved in the secretagogue-evoked Ca^{2+} spikes and a two-pool model has been proposed for both ACh and CCK. In this model IP_3 induces primary Ca^{2+} release, which then releases Ca^{2+} from an IP_3-insensitive pool by a Ca^{2+}-induced Ca^{2+} release (CICR) process [1,5]. However, this two-pool model relies on IP_3 generation by CCK and ACh, which is not supported by biochemical data on the entire pancreatic acinar cell population [7]. In contrast to ACh, a physiological concentration of CCK does not generate detectable IP_3. All together, these data from Matozaki *et al.*, 1990 [7] led to the search for additional mechanisms responsible for the CCK Ca^{2+} signature. Alternative proposals have been made to explain the discrepancy between the pharmacological and biochemical data. For example, IP_3 receptors can be sensitized by phosphorylation and therefore a small undetectable variation of IP_3 levels may well be generated and sufficient to elicit a Ca^{2+} signal (8-10).

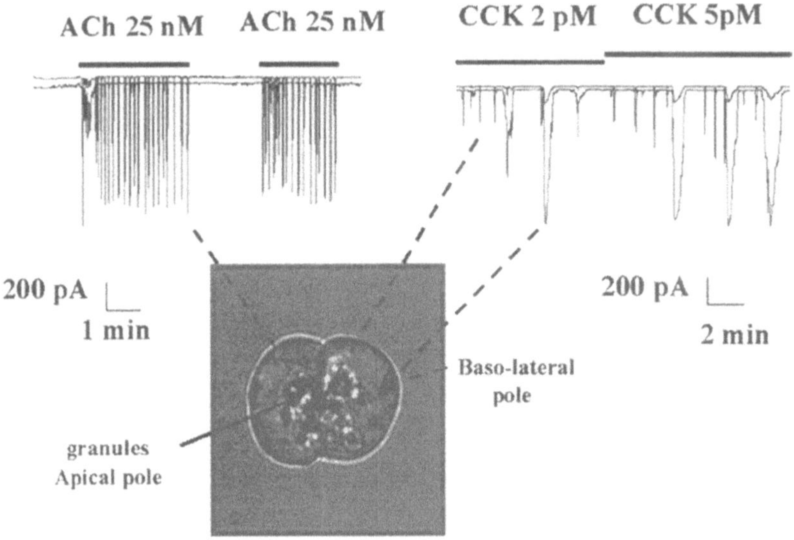

Figure 1. **CCK and ACh evoked specific Ca^{2+} signatures.** The patch-clamp recording in whole-cell configuration was used to record the Ca^{2+} dependent currents. At -30 mV we obtained a measure of the Ca^{2+}-dependent currents, which are an index of the cytosolic Ca^{2+} changes [22,23]. A low physiological concentration of ACh (25 nM) evokes short-lasting Ca^{2+} spikes in the apical pole of the cell. Physiological concentrations of CCK evoke a mixture of short-lasting Ca^{2+} spikes localized in the apical pole and global long-lasting Ca^{2+} spikes localised in whole cell. Modified and reprinted, with permission, from the Annual Review of Physiology, Volume 63, ©2001, by Annual Reviews, www.AnnualReviews.org [24].

In addition to the IP_3-sensitive store, ACh and CCK mobilize Ca^{2+} from a CICR sensitive pool, which has been identified as a ryanodine-sensitive store since both ACh and CCK responses were blocked by ryanodine [3, 5, 11, 12]. The mechanisms by which ryanodine receptors are recruited in pancreatic acinar cells are not fully understood. In many cell types, Ca^{2+} and cyclic ADP-ribose (cADPR), act as the endogenous stimulator of the ryanodine receptor [13-15].

cADPR IS AN ENDOGENOUS RYANODINE RECEPTOR STIMULATOR

Lee made the important discovery that cADPR, a metabolite of NAD^+, releases Ca^{2+} from sea urchin microsomes [15]. It was shown later that cADPR was mobilizing Ca^{2+} from a ryanodine sensitive pool suggesting that cADPR is an endogenous modulator of the ryanodine receptors [13]. The ryanodine receptors type II and III are sensitive to cADPR but not the type I [13-16]. It has been proposed that in sea urchin eggs, cADPR-mediated Ca^{2+} release is absolutely dependent on an accessory protein identified as calmodulin [15]. This result led to the proposal that although the ryanodine receptor is the final effector, cADPR may act through an accessory protein [15].

The enzyme responsible for cADPR synthesis is the ADP-ribosyl cyclase [15, 17]. The *Aplysia* cyclase has been purified, cloned and the active site characterized [17, 18]. Three proteins have been identified as ADP-ribosyl cyclases, the *Aplysia* cyclase, the leukocyte surface antigen CD38 and BST-1 found on bone marrow stromal cells [17-21]. The ADP-ribosyl cyclase has been found widespread in mammalian tissues [15, 17] and is localized in the plasma membrane, in the cytosol, nucleus and in mitochondria [21, 25-27]. However, how a cell surface receptor stimulates the ADP-ribosyl cyclase is poorly understood. Recent work reported that cell surface membrane receptors control the ADP-ribosyl cyclase activity and consequently the levels of cADPR in different tissues [15, 17, 20, 21]. For example, in rat pancreatic islet, glucose stimulates insulin secretion by activation of cADPR production [28]. In longitudinal intestinal smooth muscle cells, CCK has been found to control the ADP-ribosyl cyclase activity [29].

cADPR and Pancreatic Acinar Cells

Thorn *et al.* 1994 [11] showed that cADPR elicits short-lasting Ca^{2+} spikes in the secretory pole of the pancreatic acinar cells. Ryanodine (10μM) and, more surprisingly, heparin blocked the cADPR-evoked spikes. These results suggested that the Ca^{2+} spikes elicited by cADPR were primarily due to ryanodine receptor activation, followed by the recruitment of the IP_3-receptors by a CICR mechanism. Finally, whether cADPR is generally involved in agonist-evoked Ca^{2+} release or linked to a specific agonist has also been investigated [2, 3, 30, 31].

We first investigated the role of cADPR in local Ca^{2+} spiking evoked by low agonist concentrations. In pancreatic acinar cells, physiological concentrations of CCK from 1 to 10 pM, evoke local Ca^{2+} spikes in the secretory pole of the cell, which may trigger, from time to time, longer-lasting global Ca^{2+} transients. ACh, at low concentrations, evokes repetitive local Ca^{2+} spikes in the secretory pole without triggering a global Ca^{2+} transient [1]. To investigate whether cADPR is involved in the agonist-evoked Ca^{2+} spikes or not, 8-amino-cADPR, a specific cADPR antagonist was used [30, 32]. In pancreatic acinar cells, 8-amino-cADPR blocked specifically cADPR-evoked Ca^{2+} spikes but had no or little effects on IP_3-evoked Ca^{2+} spiking. Experiments performed with infusion of 8-amino-cADPR were rewarding, since the responses to physiological concentrations of CCK (2-5pM) were specifically blocked [3, 30]. Importantly, the response to a low concentration of ACh (25 nM) was not blocked by 8-amino-cADPR, although ryanodine can block the ACh-response [3]. This important result indicated that cADPR is a relevant Ca^{2+}-releasing messenger of CCK in pancreatic acinar cells. In contrast, to generate local Ca^{2+} spiking, low concentrations of ACh recruit the ryanodine receptors likely by a CICR process [3].

At this particular point, a surprising discovery highlighted the redundancy of Ca^{2+} signalling. We have observed that intracellular infusion of glucose at a concentration as low as 300 µM inhibited the cADPR-evoked Ca^{2+} spikes. In contrast glucose potentiated IP_3-evoked Ca^{2+} spikes [31]. Surprisingly, infusion of glucose, which inhibits the cADPR pathway, had little or no effects on CCK-evoked Ca^{2+} spikes [31]. The simplest hypothesis is that glucose compensated its inhibitory effect on cADPR by stimulating the IP_3-sensitive store. In this view, it is interesting to note that it has been reported that in the β-cell, glucose increases the Ca^{2+} uptake into the IP_3-sensitive Ca^{2+} stores of the endoplasmic reticulum and thereby sensitizes those stores [33, 34]. Finally, these finding, also highlighted that glucose may have an amplifier role for the IP_3-receptor mediated activities.

There is no information available on glucose resting levels in pancreatic acinar cells. Work on other cells types suggests resting glucose concentrations between 0 and 300 µM. It is also interesting to note that, similar to other hormones and neurotransmitters, physiological concentrations of CCK stimulate glucose uptake [35]. Other glycolytic metabolites such as fructose 1,6 bisphosphate have recently been reported to potentiate cADPR-evoked Ca^{2+} release and inhibit IP_3-evoked Ca^{2+} release [36]. Further study on the control of Ca^{2+} homeostasis by glycolytic metabolites may well be very rewarding in the future.

Finally, the glucose experiment led to investigations of a role for additional messengers. One interesting candidate is nicotinic acid adenine dinucleotide phosphate (NAADP) since the ADP-ribosyl cyclase, which forms cADPR can also form NAADP, a Ca^{2+} releasing messenger in sea urchin eggs [15, 37, 38].

NAADP A NEW Ca^{2+} RELEASING MESSENGER

In the presence of $NADP^+$ and nicotinic acid, the ADP-ribosyl cyclase, by a base-exchange reaction, forms NAADP. NAADP has been first reported to evoke Ca^{2+} release in sea urchin egg microsomes [38]. NAADP is one of the most potent Ca^{2+} releasing messengers found so far [17, 39, 40]. The NAADP receptor has not yet been purified, but evidence suggests its existence. NAADP releases Ca^{2+} from a thapsigargin-insensitive Ca^{2+} store. In addition, this NAADP sensitive Ca^{2+} store can be physically separated from those mobilized by IP_3 and cADPR [17, 37, 41]. An important property of NAADP signalling is self-desensitization induced by sub-threshold NAADP concentrations. Finally, in contrast to ryanodine and IP_3-receptors, it has been found that the NAADP receptor does not behave as a Ca^{2+}-induced Ca^{2+} release channel [42].

NAADP and Pancreatic Acinar Cells

In mouse pancreatic acinar cells, it has been reported that NAADP releases Ca^{2+} and is involved in the CCK-response [40]. In that study, the patch-clamp technique in whole cell configuration was used, which permitted intracellular infusion of NAADP and recording of the resultant Ca^{2+} sensitive currents. Surprisingly, the NAADP-effect appears to be biphasic; at low concentrations, NAADP evokes Ca^{2+} spikes, while at high concentrations it does not generate Ca^{2+} spikes. NAADP, at 50 nM, evokes repetitive short-lasting Ca^{2+} spikes mainly localized in the apical part of the cell [43]. However, a more complex mixture of short-lasting local Ca^{2+} spikes and long-lasting global Ca^{2+} spikes also occurred occasionally [40]. When the NAADP concentrations were in the micromolar range, no Ca^{2+} release was detected, suggesting that NAADP receptors were desensitized. This self-desensitisation property of NAADP is specific to NAADP-evoked Ca^{2+} spikes, since IP_3- and cADPR-evoked Ca^{2+} spikes were not altered by desensitizing concentrations of NAADP [40].

Work on the Ca^{2+} stores recruited by NAADP revealed a crucial cooperation between different intracellular receptors. NAADP-evoked Ca^{2+} spikes are inhibited by 8-amino-cADPR and also by heparin, suggesting that Ca^{2+} released by activation of the NAADP receptors is the primary response, which then sensitizes the neighbouring IP_3 and cADPR receptors to give rise to cytosolic Ca^{2+} spikes. Importantly, NAADP has been proposed to be a second messenger of the CCK-mediated Ca^{2+} spikes and therefore is physiologically relevant in pancreatic acinar cells. The proposal is based on the inhibitory effect of desensitizing concentrations of NAADP, which specifically blocked or partially abolished the response to CCK, whereas the ACh-response and the bombesin-response were not affected [3, 40, 44]. Thus, it appears that the Ca^{2+} response to physiological concentrations of CCK and NAADP share the same pharmacology, since both are inhibited by desensitizing concentrations of NAADP, 8-amino-cADPR and heparin. Finally, in support of a role of cADPR and NAADP in pancreatic acinar cells, it is important to note that an ADP-ribosyl cyclase activity has been found in pancreatic acinar cells, indicating that the machinery necessary to produce these messengers is present [40, 45].

SPATIAL ORGANIZATION OF THE Ca^{2+} SIGNALS EVOKED BY CCK AND ACh

From recent comparative studies on specific Ca^{2+} signatures evoked by ACh or CCK, a new concept has emerged [2, 9, 12, 40, 43, 46]. Signals induced by different extracellular agonists are encoded by several Ca^{2+} releasing messengers, which may be recruited at different agonist concentrations.

Ca^{2+} signals evoked by CCK are of a complex nature. Low physiological concentrations of CCK elicit mostly short-lasting Ca^{2+} spikes localized in the apical pole of the cell. Occasionally, long-lasting Ca^{2+} spikes initiated at the apical pole and spreading toward the basolateral part of the cell are also seen. At higher CCK concentrations, the occurrence of long-lasting Ca^{2+} transients is more frequent. In contrast, ACh, at low concentrations, mainly evokes repetitive and short lasting Ca^{2+} spikes in the apical part of the cell, whereas higher ACh concentrations evoke global Ca^{2+} waves initiated at the apical pole [47]. These local Ca^{2+} spikes are sufficient to elicit exocytosis, as assessed by increases in membrane capacitance [48] and fluid secretion. The latter was assessed by activation of the Ca^{2+}-dependent Cl^- channels in the apical membranes [23, 49]. However, higher agonist concentrations elicit global Ca^{2+} transients associated with a greater secretion of digestive enzymes. The physiological importance of these global transients has not been fully clarified. So far, it is difficult to explain how such complex patterns are generated. It is likely that they may be related to the pancreatic growth response induced by CCK [1].

Local Ca^{2+} Spiking in the Secretory Pole

Recent studies on the generation of elementary Ca^{2+} events provided evidence for cooperativity between the IP_3- and the ryanodine- receptors [3, 50, 51]. Elementary Ca^{2+} signals in cells are highly localized and correspond to small clusters of intracellular Ca^{2+} releasing channels [52-54]. These clusters contain either IP_3 or ryanodine receptors, but may also contain both receptors. In pancreatic acinar cells, studies using confocal line-scanning showed that the ACh and CCK-evoked elementary Ca^{2+}events are localized in the secretory pole, suggesting that ryanodine, IP_3 and NAADP receptors are present in the same clusters [3]. In support of this view, it has been shown that IP_3, cADPR and NAADP each can evoke Ca^{2+} release in the apical pole [11, 23, 43].

A model for Ca^{2+} spiking in the secretory pole of the pancreatic acinar cells has been proposed recently [3]. In this model, Ca^{2+} spikes originate from the concerted activity of IP_3- and ryanodine- receptors, irrespective of the agonist triggering them. ACh and CCK, but also IP_3, cADPR and NAADP, evoke Ca^{2+} spikes in the secretory pole by activation of both IP_3

and ryanodine receptors, which work as a common oscillator/amplifier unit (Fig 2). For example, in the cases of NAADP and CCK, the model suggests that the primary Ca^{2+} released by NAADP would then be amplified by the cADPR/ryanodine receptor, while the IP_3 receptor would be the final oscillator [2].

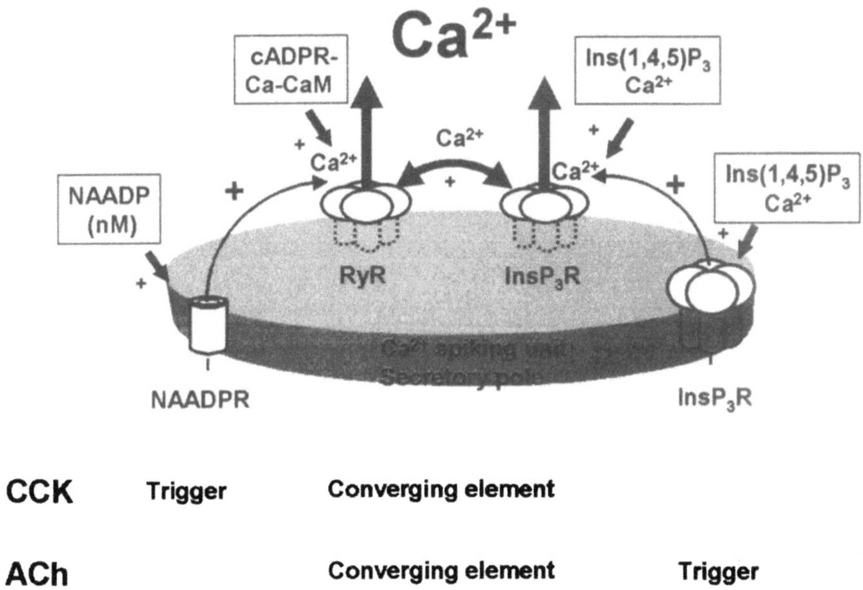

Figure 2. **Schematic model showing that CCK and ACh initiate Ca^{2+} spiking in the secretory pole from units of ryanodine and IP_3 receptors via separate intracellular pathways.** The peptide hormone CCK initiates Ca^{2+} release through a specific combination of intracellular Ca^{2+} release channels with separate functions, where the NAADP receptor is the trigger, the cADPR receptor via the ryanodine receptor (RYR) is an amplifier and the combination of IP_3-receptors and RYRs is the oscillator. The neurotransmitter ACh also acts through a combination of intracellular Ca^{2+} channels, where the IP_3 receptor is the trigger and the RYR/ IP_3 complex is the amplifier/oscillator. The RYR is most likely recruited by the action of Ca^{2+} released from the IP_3 receptor. Modified from Cancela *et al.* 2000 [3]. Reproduced with permission from The EMBO Journal [3].

Global Ca^{2+} Spiking in Pancreatic Acinar Cells

It is now established that the short-lasting Ca^{2+} spikes evoked by IP_3, cADPR and NAADP are specifically localized in the apical pole of the cells [11, 23, 43]. However, all three messengers may be involved in the globalization of the Ca^{2+} signals evoked by agonists. Interestingly, the ability of these messengers to elicit Ca^{2+} spikes in the apical pole may not be in agreement with the localization of their respective receptors [43]. The IP_3-receptors are mostly localized in the apical pole [23, 55-57]. This receptor partition is in agreement with the ability of IP_3 to evoke Ca^{2+} spikes in the apical pole. In contrast, the ryanodine receptors of type I, II, III have been found to be diffused in the whole cell [58-60]. Nothing is known about the NAADP receptor distribution in pancreatic acinar cells, or even in other cell types [2, 17, 41]. However, NAADP-evoked Ca^{2+} signals initiate at the apical pole and can spread occasionally toward the baso-lateral pole of the cells, which indicates that the NAADP receptors may be distributed over the whole cell, perhaps with different regional density.

We addressed the question of how local Ca^{2+} spikes, evoked by low concentrations of either ACh or CCK, can become global Ca^{2+} transients. In addition to the IP_3-receptors, CCK stimulates also the cADPR- and NAADP-receptors. The combined activation of these receptors may explain why Ca^{2+} signals evoked by CCK are initiated in the apical part and spread towards the basolateral part of the cells which is rich in ryanodine receptors [2, 3]. Since NAADP is proposed to be the trigger, the spatial distribution of its receptors may be crucial in explaining the ability of CCK to produce global responses. It is well established that ACh evokes Ca^{2+} release by stimulating the production of IP_3 [1]. It has recently been reported that ACh, at higher concentrations, also stimulates the activity of CD38, an ADP-ribosyl cyclase [45], indicating that in addition to IP_3, cADPR, and perhaps also NAADP, could be involved in the ACh-response. Interestingly, it would appear that synthesis of multiple Ca^{2+} releasing messengers may occur after agonist stimulation and we have therefore designed experiments to test their respective roles, if any, in shaping the Ca^{2+} signals.

Although each of the three Ca^{2+} releasing messengers tested elicited on their own repetitive local Ca^{2+} spikes in the granular pole, their effects on the responses induced by ACh and CCK were very different. In the presence of either cADPR or NAADP, the short-lasting repetitive local Ca^{2+} spikes normally evoked by ACh were transformed into a sustained global Ca^{2+} elevation. On the other hand, IP_3 only had a very small potentiating effect on the ACh response. In the case of CCK stimulation, it was the other way around. Internal IP_3 dramatically potentiated the CCK response, whereas cADPR or NAADP only had relatively small effects on the CCK response. The simplest explanation for these data is that only complementary messengers can effectively potentiate an agonist response. Thus the main messenger mediating the ACh response would appear to be IP_3, whereas the CCK response would seem to be primarily mediated by cADPR and NAADP. One important conclusion to be drawn from these data is that increasing the intensity of stimulation with any one of the agonists is likely to produce all three messengers and consequently generates substantial Ca^{2+} waves. To do so, ACh and CCK may recruit these messengers in a different sequence.

To further understand the roles of the Ca^{2+} releasing messengers in the globalization of the agonist-evoked Ca^{2+} response, cells were perfused with various combinations of Ca^{2+} releasing messengers [43]. The experiments with various combinations of messengers revealed that NAADP locally potentiates the IP_3-evoked Ca^{2+} spikes in the secretory pole, whereas no, or relatively minor, interaction between IP_3 and cADPR was observed. There was a somewhat stronger potentiating interaction between cADPR and NAADP, which gave rise to a global Ca^{2+} response. However, in order to obtain the type of global and sustained response, that can result from combined actions of CCK with IP_3, or ACh with cADPR or NAADP, it was necessary to use all three messengers together [43].

There would appear to be two separate aspects of the globalization process. The first aspect concerns Ca^{2+} diffusion. The cytosol appears to be compartmentalized by a major mitochondrial barrier placed on the border between the apical granular pole and the rest of the cell [22, 61]. This barrier has a role in buffering Ca^{2+} and confining the Ca^{2+} signals elicited in the apical region to this part of the cell [22, 61]. In order for a Ca^{2+} signal to become global, a substantial amount of Ca^{2+} must be released in the granular region, such that the mitochondrial barrier is overwhelmed. The second aspect in globalization concerns propagation of the Ca^{2+} wave throughout the cell, from the apical to the basal pole. This is likely to be via a Ca^{2+}-induced Ca^{2+} release mechanism through activation of the ryanodine receptors [47, 62, 63]. This means that during globalization, activation of Ca^{2+} release channels, such as the ryanodine receptors, also occurs in the basal pole. These data indicate that a combination of IP_3, cADPR and NAADP plays an important role in this process, since silent Ca^{2+} release channels in the basal pole become activated during globalization of Ca^{2+} signaling evoked by the coincidence presence of these messengers (Fig. 3).

A global Ca^{2+} wave can thus be generated by the concerted activity of elementary Ca^{2+} release units, which act as "building blocks" [52, 64, 65]. Increasing the frequency of elementary Ca^{2+} release events triggers a Ca^{2+} wave. Several mechanisms including Ca^{2+} diffusion, Ca^{2+}-induced Ca^{2+} release and also increase of IP_3 production are used to reach the threshold for the generation of a Ca^{2+} wave. In contrast, abortive Ca^{2+} waves or local Ca^{2+} signals are generated when the most sensitive Ca^{2+} release units are surrounded by poorly sensitive Ca^{2+} release units [52, 64, 65]. This is what may happen with the infusion of IP_3, cADPR or NAADP, which generates short-lasting Ca^{2+} spikes in the apical pole of the cell without triggering a Ca^{2+} wave. The basolateral part of the cell contains poorly sensitive Ca^{2+} release units that cannot support a regenerative Ca^{2+} wave in the sole presence of either IP_3, cADPR or NAADP (Fig. 3). In the case of the pancreatic acinar cell, the generation of a global Ca^{2+} wave is produced by a combination of potentiated Ca^{2+} release in the granular pole, overwhelming the mitochondrial barrier, and the sensitization of the cluster-containing Ca^{2+} release channels. The latter occurs by coincident activation of ryanodine, IP_3

and NAADP receptors by their respective messengers in the basal pole (Fig. 3).

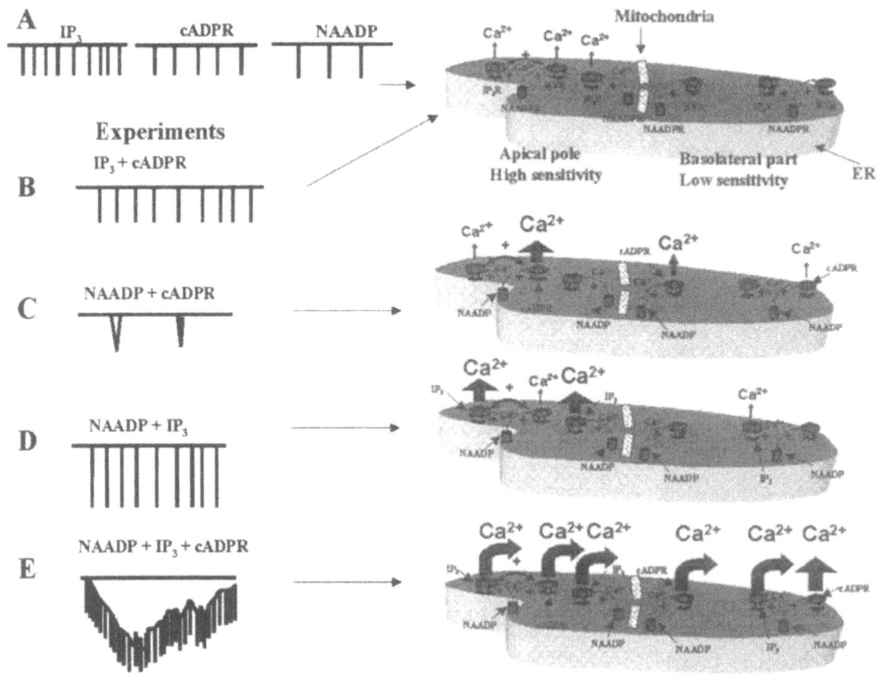

Figure 3. **Schematic model showing that every messenger on its own can initiate local Ca^{2+} spikes.** (**A**) There is little interaction between IP_3 and cADPR. (**B**) NAADP does potentiate the action of cADPR producing longlasting global spikes at long intervals. (**C**) In contrast NAADP only has a locally potentiating effect on the local IP_3-evoked Ca^{2+} spikes. (**D**) When all 3 messengers act together a large, sustained and global Ca^{2+} elevation is observed. (**E**) The apical pole is the most sensitive part of the cell. In the models shown to the right, the basolateral part of the cell contains poorly sensitive Ca^{2+} release units that cannot trigger a wave in the presence of either IP_3, cADPR or NAADP alone. To generate a Ca^{2+} wave across the cell, a combination of potentiated Ca^{2+} release in the apical pole, helping to overcome the mitochondrial barrier and sensitization of Ca^{2+} release channels by coincident activation of ryanodine, IP_3 and NAADP receptors by their respective messengers in the basal pole is necessary. Modified from Cancela *et al.* 2002 [43]. Reproduced with permission from The EMBO Journal [43].

From the work in the pancreatic acinar cell, NAADP is emerging as a key messenger in the globalization of Ca^{2+} signals. The Ca^{2+} signals induced by NAADP have a unique ability to interact with ryanodine- and IP_3- receptors, which allows a substantial increase in excitability that can be further enhanced by either cADPR or IP_3.

Other cell types have also been shown to possess IP_3, cADPR and NAADP sensitive stores [21, 24, 41], suggesting that this model may also be valid in these systems. However, cells may have one continuous Ca^{2+} store or multiple separate stores located in different regions, which may introduce further complexity in the generation of Ca^{2+} signals [15, 21, 41, 66-74]. In ascidian oocytes, starfish oocytes, T lymphocytes and sea urchin eggs, not only NAADP, but also cADPR and IP_3, release Ca^{2+} from internal stores in the same cell [66, 67, 75-78]. Interestingly, fertilization of sea urchin eggs triggers global Ca^{2+} waves, which can be markedly inhibited when both IP_3 and cADPR receptors are blocked [21, 79]. In addition, there is also evidence that NAADP receptors are involved in the fertilization process [80]. It has been proposed that NAADP releases Ca^{2+} from a pool separate from those sensitive to cADPR or IP_3 [41, 66, 67]. A two-pool model has been proposed to explain the Ca^{2+} oscillations evoked by NAADP. In this model, the Ca^{2+} released by NAADP is taken up into the Ca^{2+} pools that are sensitive to cADPR and IP_3, thereby generating further Ca^{2+} oscillations [41, 66, 67].

Finally, work on the pancreatic acinar cell has uncovered a complex relationship between IP_3, cADPR and NAADP. Information is still missing and further work is clearly needed to firmly establish NAADP as a second messenger. For example, the hypothesis that an extracellular agonist can control the NAADP synthesis remains to be tested. Importantly, isolation and characterization of the NAADP receptor is of crucial importance to further explore the physiological role of NAADP in various tissues.

ACKNOWLEDGEMENTS

I thank O.H. Petersen and A. Galione for their support. This work was supported by an MRC Programme Grant and by an EEC grant.

REFERENCES

1. Petersen OH, Petersen CCH and Kasai H. 1994. Ca^{2+} and hormone action. *Annu. Rev. Physiol.*, 56: 297-319.
2. Petersen OH and Cancela JM. 1999. New Ca^{2+}-releasing messengers: are they important in the nervous system? *Trends Neurosci.* 22: 488-494.
3. Cancela JM, Gerasimenko OV, Gerasimenko JV, Tepikin AV and Petersen, O.H. 2000. Two different but converging messenger pathways to intracellular Ca^{2+} release: the roles of NAADP, cADPR and IP_3. *EMBO J.* 19: 2549-2557.
4. Petersen CH, Toescu EC and Petersen OH. 1991. Different patterns of receptor-activated cytoplasmic Ca^{2+} oscillations in single pancreatic acinar cells: dependence on receptor type, agonist concentration and intracellular Ca^{2+} buffering. *EMBO J.* 10: 527-533.
5. Wakui M, Osipchuk YV and Petersen OH. 1990. Receptor-activated cytoplasmic Ca^{2+} spiking mediated by inositol trisphosphate is due to Ca^{2+}-induced Ca^{2+} release. *Cell* 63: 1025-1032.
6. Thorn P and Petersen OH. 1993. Ca^{2+} oscillations in pancreatic acinar cells, evoked by the cholecystokinin analogue JMV-180, depend on functional inositol 1,4,5-trisphosphate receptors. *J. Biol. Chem.* 268: 23219-23221.

7. Matozaki T, Goke B, Tsunoda Y, Rodriguez M, Martinez J and Williams JA. 1990. Two functionally distinct cholecystokinin receptors show different modes of action on Ca^{2+} mobilization and phospholipid hydrolysis in isolated rat pancreatic acini. Studies using a new cholecystokinin analog, JMV-180. *J. Biol. Chem.* 265: 6247-6254.
8. Xu X, Zeng W and Muallem S. 1996. Regulation of the inositol 1,4,5-trisphosphate-activated Ca^{2+} channel by activation of G-proteins. *J. Biol. Chem.* 271: 11737-11744.
9. Muallem S and Wilkie TM. 1999. G protein-dependent Ca^{2+} signaling complexes in polarized cells. *Cell Calcium.* 26: 173-180.
10. LeBeau AP, Yule DI, Groblewski GE and Sneyd J. 1999. Agonist-dependent phosphorylation of the inositol 1,4,5-trisphosphate receptor. *J. Gen. Physiol.* 113: 851-871.
11. Thorn P, Gerasimenko O and Petersen OH. 1994. Cyclic ADP-ribose regulation of ryanodine receptors involved in agonist evoked cytosolic Ca^{2+} oscillations in pancreatic acinar cells. *EMBO J.* 13: 2038-2043.
12. Schulz I, Krause E, Gonzalez A, Gobel A, Sternfeld L and Schmid A. 1999. Agonist-stimulated pathways of Ca^{2+} signaling in pancreatic acinar cells. *Biol. Chem.* 380: 903-908.
13. Galione A, Lee HC and Busa WB. 1991. Ca^{2+}-induced Ca^{2+} release in sea urchin egg homogenates and its modulation by cyclic ADP-ribose. *Science.* 253: 1143-1146.
14. Meszaros LG, Bak J and Chu A. 1993. Cyclic ADP-ribose as an endogeneous regulator of the non-skeletal type ryanodine receptor Ca^{2+} channel. *Nature.* 364: 76-79.
15. Lee HC. 1997. Mechanisms of Ca^{2+} signalling by cyclic ADP-ribose and NAADP. *Physiol. Rev.* 77: 1133-1164.
16. Sonnleitner A, Conti A, Bertocchini F, Schindler H and Sorrentino V. 1998. Functional properties of the ryanodine receptor type 3 (RYR3) Ca^{2+} release channel. *EMBO J.* 17: 2790-2798.
17. Lee HC. 2000. NAADP: an emerging Ca^{2+} signaling molecule. *J. Memb. Biol.* 173: 1-8.
18. Munshi C, Thiel DJ, Mathews II, Aarhus R, Walseth TF and Lee HC. 1999. Characterization of the active site of ADP-ribosyl cyclase. *J. Biol. Chem.* 274: 30770-30777.
19. Hirata Y, Kimura N, Sato K, Ohsugi Y, Takasawa S, *et al.* 1994. ADP ribosyl cyclase activity of a novel bone marrow stromal cell surface molecule, BST-1. *FEBS Lett.* 356: 244-248.
20. Guse AH. 1999. Cyclic ADP-ribose: A novel Ca^{2+}-mobilising second messenger. *Cell. Signal.* 11: 309-316.
21. Lee HC. 2001. Physiological functions of cADP-ribose and NAADP as Ca^{2+} messengers. *Annu. Rev. Pharmacol. Toxicol.* 41: 317-345.
22. Tinel H, Cancela JM, Mogami H, Gerasimenko JV, Gerasimenko OV, Tepikin AV and Petersen OH. 1999. Active mitochondria surrounding the pancreatic acinar granule region prevent spreading of inositol trisphosphate-evoked local cytosolic Ca^{2+} signals. *EMBO J.* 18: 4999-5008.
23. Thorn P, Lawrie AM, Smith PM, Gallacher DV and Petersen OH. 1993. Local and global cytosolic Ca^{2+} oscillations in exocrine cells evoked by agonists and inositol trisphosphate. *Cell* 74: 661-668.
24. Cancela JM. 2001. Specific Ca^{2+} signaling evoked by cholecystokinin and acetylcholine: the roles of NAADP, cADPR, and IP_3. *Annu. Rev. Physiol.* 63, 99-117.
25. Ziegler M, Jorcke D and Schweiger M. 1997. Identification of bovine liver mitochondrial NAD^+ glycohydrolase as ADP-ribosyl cyclase. *Biochem. J.* 326: 401-405.
26. Adebanjo OA, Anandatheerthavarada HK, Koval AP, Moonga BS, Biswast G, *et al.* 1999. A new function for CD38/ADP-ribosyl cyclase in nuclear Ca^{2+} homeostasis. *Nature Cell Biol.* 1: 409-414.
27. Liang M, Chini ED, Cheng J and Dousa TP. 1999. Synthesis of NAADP and cADPr in mitochondria. *Arch. Biochem. Biophys.* 15: 317-325.

28. Takasawa S, Nata S, Yonekura H and Okamoto H. 1993. Cyclic ADP-ribose in insulin secretion from pancreatic β cells. *Science*. 259: 370-373.
29. Kuemmerle JF and Makhlouf GM. 1995. Agonist-stimulated cyclic ADP ribose. Endogenous modulator of Ca^{2+}-induced Ca^{2+} release in intestinal longitudinal muscle. *J. Biol. Chem.* 270: 25488-25494.
30. Cancela JM and Petersen OH. 1998. The cyclic ADP-ribose antagonist 8-NH_2-cADP-ribose blocks cholecystokinin-evoked cytosolic Ca^{2+} spiking in pancreatic acinar cells. *Pflugers Arch.* 435: 746-48.
31. Cancela JM, Mogami H, Tepikin AV and Petersen OH. 1998. Intracellular glucose switches between cyclic ADP-ribose and inositol trisphosphate triggering of cytosolic Ca^{2+} spiking. *Curr. Biol.* 8: 865-68.
32. Walseth TF and Lee HC. 1993. Synthesis and characterization of antagonists of cyclic ADP-ribose. *Biochim. Biophys. Acta.* 1178: 235-242.
33. Gilon P, Arredouani A, Gailly, P, Gromada J and Henquin JC. 1999. Uptake and release of Ca^{2+} by the endoplasmic reticulum contribute to the oscillations of the cytosolic Ca^{2+} concentration triggered by Ca^{2+} influx in the excitable pancreatic β-cell. *J. Biol. Chem.* 274: 20197-20205.
34. Tengholm A, Hagman C, Gylfe E and Hellman B. 1998. In situ characterization of nonmitochondrial Ca^{2+} stores in individual pancreatic beta-cells. *Diabetes* 47: 1224-1230.
35. Korc M, Wiliams JA and Goldfine ID. 1979. Stimulation of the glucose transport system in isolated mouse pancreatic acini by Cholecystokinin and analogues. *J. Biol. Chem.* 254: 7624-7629.
36. Chini EN and Dousa TP. 1999. Differential effect of glycolytic intermediaries upon cyclic ADP-ribose, inositol 1,4,5-trisphosphate-, and nicotinate adenine dinucleotide phosphate-induced Ca^{2+} release systems. *Arch. Biochem. Biophys.* 370: 294-299.
37. Chini EN, Beers KW, Dousa TP. 1995. Nicotinate adenine dinucleotide phosphate (NAADP) triggers a specific Ca^{2+} release system in sea urchin eggs. *J. Biol. Chem.* 270: 3216-3223.
38. Lee HC and Aarhus R. 1995. A derivative of NADP mobilizes Ca^{2+} stores insensitive to inositol trisphophate and cyclic ADP-ribose. *J. Biol. Chem.* 270: 2152-2157.
39. Genazzani AA and Galione A. 1997. A Ca^{2+} release mechanism gated by the novel pyridine nucleotide, NAADP. *Trends Pharmacol. Sci.* 18: 108-110.
40. Cancela JM, Churchill GC and Galione A. 1999. Coordination of agonist-induced Ca^{2+}-signalling patterns by NAADP in pancreatic acinar cells. *Nature* 398: 74-76.
41. Patel S, Churchill GC and Galione A. 2001. Coordination of Ca^{2+} signalling by NAADP. *Trends Biol. Sci.* 26, 482-489.
42. Chini EN and Dousa TP. 1996. Nicotinate-adenine dinucleotide phosphate-induced Ca^{2+} release does not behave as a Ca^{2+}-induced Ca^{2+} release system. *Biochem. J.* 316: 709-711.
43. Cancela JM, Coppenolle FV, Galione A, Tepikin AV and Petersen OH. 2002. Transformation of local Ca^{2+} spikes to global Ca^{2+} transients: the combinatorial roles of multiple Ca^{2+} releasing messengers. *EMBO J* 21:909-19.
44. Burdakov D, Galione A. 2000. Two neuropeptides recruit different messenger pathways to evoke Ca^{2+} signals in the same cell. *Curr Biol* 10: 993-96.
45. Fukushi Y, Kato I, Takasawa S, Sasaki T, Ong BH, Sato M, Ohsaga A., Sato K, Shirato K, Okamoto H, Maruyama Y. 2001. Identification of Cyclic ADP-ribose-dependent Mechanisms inPancreatic Muscarinic Ca 21 Signaling Using CD38 Knockout Mice. *J. Biol. Chem.* 276: 649-655.
46. Ito,K., Miyashita,Y. and Kasai,H. 1999. Kinetic control of multiple forms of Ca^{2+} spikes by inositol trisphosphate in pancreatic acinar cells. *J. Cell Biol.* 146: 405-413.
47. Kasai H and Augustine GJ. 1990. Cytosolic Ca^{2+} gradients triggering unidirectional fluid secretion from exocrine pancreas. *Nature.* 348: 735-738.
48. Maruyama Y and Petersen OH. 1994. Delay in granular fusion evoked by repetitive cytosolic Ca^{2+} spikes in mouse pancreatic acinar cells. *Cell Calcium* 16: 419-430.

49. Park MK, Lomax RB, Tepikin AV and Petersen OH. 2001. Local uncaging of caged Ca^{2+} reveals distribution of Ca^{2+}-activated Cl^- channels in pancreatic acinar cells. *Proc. Natl. Acad. Sci USA* 98: 10948-10953.
50. Boittin FX, Coussin F, Macrez N, Mironneau C and Mironneau J. 1998. Inositol 1,4,5-trisphosphate- and ryanodine-sensitive Ca^{2+} release channel-dependent Ca^{2+} signalling in rat portal vein myocytes. *Cell Calcium* 23: 303-311.
51. Koizumi S, Bootman MD, Bobanovic LK, Schell MJ, Berridge MJ and Lipp P. 1999. Characteriztion of elementary Ca^{2+} release signals in NGF-differentiated PC12 cells and hippocampal neurons. *Neuron* 22: 125-37.
52. Parker I, Choi J and Yao Y. 1996. Elementary events of $InsP_3$-induced Ca^{2+} liberation in *Xenopus* oocytes: hot spots, puffs and blips. *Cell Calcium* 20: 105-121.
53. Berridge MJ. 1997. Elementary and global aspects of calcium signalling. *J. Physiol.* 499: 291-306.
54. Cannell MB and Soeller C. 1999. Mechanisms underlying Ca^{2+} sparks in cardiac muscle. *J. Gen. Physiol.* 113: 373-376.
55. Nathanson MH, Fallon MB, Padfield PJ and Maranto AR. 1994. Localization of the type 3 inositol 1,4,5-trisphosphate receptor in the Ca^{2+} wave trigger zone of pancreatic acinar cells. *J. Biol. Chem.* 269: 4693-4696.
56. Lee MG, Xu X, Zeng W, Diaz J, Wojcikiewicz JH, Kuo TH, Wuytack F, Racymaekers L and Muallem S. 1997. Polarized expression of Ca^{2+} channels in pancreatic and salivary gland cells. *J. Biol. Chem.* 272: 15765-15770.
57. Kasai H and Petersen OH. 1994. Spatial dynamics of second messengers: IP_3 and cAMP as long-range and associative messengers. *Trends Neurosci.* 17: 95-101.
58. Leite MF, Dranoff JA, Gao L and Nathanson MH. 1999. Expression and subcellular localization of the ryanodine receptor in rat pancreatic acinar cells. *Biochem. J.* 337: 305-309.
59. Fitzsimmons TJ, Gukovsky I, McRoberts JA, Rodriguez E, Lai FA and Pandol SJ. 2000 Multiple isoforms of the ryanodine receptor are expressed in rat pancreatic acinar cells. *Biochem. J.* 351: 265-271.
60. Straub SV, Giovannucci DR, Yule DI. 2000. Ca^{2+} wave propagation in pancreatic acinar cells: functional interaction of inositol 1,4,5-trisphosphate receptors, ryanodine receptors, and mitochondria. *J. Gen. Physiol.* 116: 547-559.
61. Park MK, Ashby MC, Erdemli G, Petersen OH and Tepikin AV. 2001. Perinuclear, perigranular and sub-plasmalemmal mitochondria have distinct functions in the regulation of cellular Ca^{2+} transport. *EMBO J.* 20: 1863-1874.
62. Toescu EC Lawrie AM, Petersen OH and Gallacher, DV. 1992. Spatial and temporal distribution of agonist-evoked cytoplasmic Ca^{2+} signals in exocrine acinar cells analysed by digital image microscopy. *EMBO J.* 11: 1623-1629.
63. Toescu EC, Gallacher DV and Petersen OH. 1994. Identical regional mechanisms of intracellular free Ca^{2+} concentration increase during polarized agonist-evoked Ca^{2+} response in pancreatic acinar cells. *Biochem J.* 304: 313-316.
64. Marchant J, Callamaras N and Parker I. 1999. Initiation of IP_3-mediated Ca^{2+} waves in Xenopus oocytes. *EMBO J.* 18: 5285-5299.
65. Berridge MJ, Lipp P and Bootman MD. 2000. The versality and universality of Ca^{2+} signaling. *Nature Rev.* 1: 11-21.
66. Churchill GC and Galione A. 2000. Spatial control of Ca^{2+} signaling by nicotinic acid adenine dinucleotide phosphate diffusion and gradients. *J. Biol. Chem.* 275: 38687-38692.
67. Churchill GC and Galione A. 2001. NAADP induces Ca^{2+} oscillations via a two-pool mechanism by priming IP_3 and cADPR-sensitive Ca^{2+} stores. *EMBO J.* 20: 2666-2671.
68. Malgaroli A, Fesce R and Meldolesi J. 1990. Spontaneous $[Ca^{2+}]_i$ fluctuations in rat chromaffin cells do not require inositol 1,4,5-trisphosphate elevations but are generated by a caffeine- and ryanodine-sensitive intracellular Ca^{2+} store. *J. Biol. Chem.* 265: 3005-3008.

69. Golovina VA and Blaustein MP. 1997. Spatially and functionally distinct Ca^{2+} stores in sarcoplasmic and endoplasmic reticulum. *Science.* 275: 1643-1648.
70. Hofer AM, Landolfi B, Debellis L, Pozzan, T and Curci S. 1998. Free $[Ca^{2+}]$ dynamics measured in agonist-sensitive stores of single living intact cells: A new look at the refilling process. *EMBO J.* 17: 1986-1995.
71. Mogami H, Nakano K, Tepikin AV and Petersen,O.H. 1997. Ca^{2+} flow via tunnels in polarized cells: recharging of apical Ca^{2+} stores by focal Ca^{2+} entry through basal membrane patch. *Cell* 88: 49-55.
72. Meldolesi J and Pozzan T. 1998. The heterogeneity of ER Ca^{2+} stores has a key role in non muscle cell signaling and function. *J. Cell Biol.* 142: 1395-1398.
73. Park MK, Petersen OH and Tepikin AV. 2000. The endoplasmic reticulum as one continuous Ca^{2+} pool: visualization of rapid Ca^{2+} movement and equilibration. *EMBO J.* 19. 5729-5739.
74. Petersen OH, Tepikin AV and Park MK. 2001. The endoplasmic reticulum: one continuous or several separate Ca^{2+} stores. *Trends Neurosci.* 24: 271-76.
75. Albrieux M, Lee HC and Villaz M. 1998. Ca^{2+} signaling by cyclic ADP-ribose, NAADP, and inositol trisphosphate are involved in distinct functions in ascidian oocytes. *J. Biol. Chem.* 273: 14566-14574.
76. Guse AH, da Silva CP, Berg I, Skapenko AL, Weber K, *et al.* 1999. Regulation of Ca^{2+} signalling in T lymphocytes by the second messenger cyclic ADP-ribose. *Nature.* 398: 70-73
77. Berg I, Potter BVL, Mayr GW and Guse AH. 2000. Nicotinic acid adenine dinuleotide phosphate ($NAADP^{+}$) is an essential regulator of T-lymphocyte Ca^{2+} signaling. *J. Cell Biol.* 150: 581-588.
78. Santella L, Kyozuka K, Genazzani AA, De Riso L and Carafoli E. 2000. Nicotinic acid adenine dinucleotide phosphate-induced Ca^{2+} release. *J. Biol. Chem.* 275: 8301-8306.
79. Galione A. 1994. Cyclic ADP-ribose, the ADP-ribosyl cyclase pathway and Ca^{2+} signalling. *Mol. Cell. Endocrinol.* 98: 125-131.
80. Perez-Tersic CM, Chini EN, Shen SS, Dousa TP and Clapham DE. 1995. Ca^{2+} release triggered by nicotinate adenine dinucleotide phosphate in intact sea urchin eggs. *Biochem. J.* 312: 955-959.

Jose Manuel Cancela, Affiliation Laboratoire de
Neurobiologie Cellulaire et Moléculaire. CNRS.
1,av de la terrasse. 91 198 Gif-sur-Yvette cédex. France.
e-mail: jose.cancela@nbcm.cnrs-gif.fr

Cyclic ADP-Ribose and NAADP. Structures, Metabolism and Functions. Lee HC (ed), pp. 381-396, Dordrecht: Kluwer.

Calcium and Calcium-Linked Second Messengers are Main Actors in the Maturation and Fertilization of Starfish Oocytes

Luigia Santella, Gilda A. Nusco and Dmitri Lim
Laboratory of Cell Biology, Stazione Zoologica "A. Dohrn" Villa Comunale I-80121, Naples, Italy.

INTRODUCTION

Since the early studies on sea urchin egg activation [1] and on starfish oocytes [2], to the more recent discoveries of the Ca^{2+}-mobilizing activities of cyclic ADP-ribose (cADPR) and nicotinic acid adenine dinucleotide phosphate (NAADP) in sea urchin egg homogenates [3-6], echinoderm gametes have remained a widely investigated system in the area of egg activation. In addition to fertilization, and at variance with sea urchin, starfish oocytes have also provided an exceptional model to investigate the re-initiation of the meiotic cycle (e.g., maturation) due to their synchrony, transparency and ease of handling. Maturation, which is induced by the hormone 1-methyladenine, takes these oocytes from the germinal vesicle stage (4n chromosomes, first prophase stage of meiosis) where they remain arrested to the spawning period at which they can be fertilized. During meiosis, re-initiated oocytes undergo a number of structural and biochemical changes, which prepare them for successful fertilization. Thus, oocytes have been a useful tool in investigations of the intracellular mechanisms regulating the prophase/metaphase transition. They are also a unique source of highly purified cell cycle control elements e.g., purified M-phase promoting factor [7, 8]. The latter protein has been invaluable in the study of the regulation of cyclin B phosphorylation [9] and for the discovery of specific inhibitors of cyclin-dependent kinases [10]. The advantages of starfish oocytes have been particularly evident in the study of the spatiotemporal aspects of calcium signalling during the re-initiation of the cell cycle, and will thus be discussed in some detail in the pages to follow. Attention will be also placed on the important findings these oocytes have allowed in the area of Ca^{2+} homeostasis at fertilization.

HORMONAL CONTROL OF OOCYTE MATURATION

Starfish oocytes develop in the ovary to the first prophase stage of meiosis and then become arrested. When fully grown within the ovary they possess a large nucleus (germinal vesicle) (Fig. 1). At this stage they are immature and fail to undergo normal fertilization if isolated and immediately inseminated. The re-initiation of the meiotic cycle is controlled by a peptide released from radial nerves [11]. It acts on the follicle cells surrounding the oocytes [12] and stimulates the production and release of 1-methyladenine (1-MA), which induces meiosis resumption *in vivo* or *in vitro* [13]. 1-MA acts on the plasma membrane since its direct injection into the oocytes fails to induce meiosis resumption [14]. The 1-MA receptor is coupled to a trimeric G protein: it has been claimed that its dissociated βγ subunits may mediate the 1-MA signal [15, 16]: A phosphatidylinositol 3-kinase (PI-3 kinase) activity may function downstream of the βγ subunits [17]. Early events following 1-MA-receptor interaction include a decrease in cAMP concentration and possibly involve tyrosine kinases, phospholipases 2A and C, proteases, and protein phosphatase 2A [18-20]. They also involve the activation of a M-phase promoting factor (MPF) in the cytoplasm [21] which correlates with increased protein phosphorylation, occurring initially in the cortical region of the oocyte [22]. The catalytic component of MPF is a serine-threonine kinase, which is associated to a cyclin B. The activity of the kinase, also termed Cdc2, in immature arrested oocytes is suppressed by Myt1, a member of the Wee1 family, which downregulates protein phosphatase Cdc25 [23]. Following 1-MA treatment MPF, partially activated in the cytoplasm, migrates to the germinal vesicle (nucleus) [24] together with phosphatase Cdc25 where it undergoes final activation [23]. It has been recently shown that the kinase Akt (protein kinase B, PKB) phosphorylates and inactivates Myt1 causing MPF activation [25]. Once activated, MPF phosphorylates a number of proteins involved in the disassembly of the nuclear lamina (lamin B receptors) and lamins [26], and in the breakdown of the germinal vesicle, (GVBD). The disassembly of the nuclear envelope is a process mediated by the cleavage of lamin (B type), which is one of the 5 proteins that have been identified by mass profile fingerprinting as one of the substrates of the Ca^{2+} dependent protease calpain [27]. The suggestion that calpain may act in the nucleus is supported by its presence in the mitotic spindle [28] and by the induction of GVBD in the absence of 1-MA by injecting calpain directly into the nucleus [27].

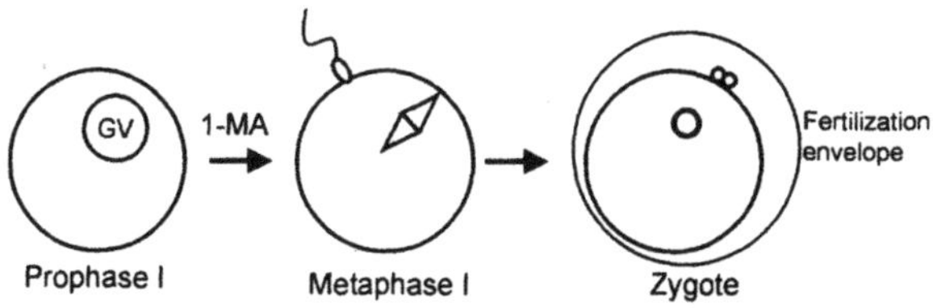

Figure 1. **Relationship between maturation and fertilization in starfish oocytes.** Fully grown oocytes within the ovary are arrested at the prophase of the first meiotic division and are characterized by a large nucleus (germinal vesicle, GV). 1-MA induces the re-initiation of the cell cycle (meiosis). Fertilization takes place at the metaphase I stage after the disappearance of the germinal vesicle and prior to the extrusion of the first polar body

CORTICAL, CYTOPLASMIC AND NUCLEAR CHANGES DURING OOCYTE MATURATION

Maturation involves a variety of electrical, biochemical and morphological changes that prepare the immature oocyte for fertilization. A temporal correlation between GVBD and fertilization in 1-MA treated oocytes has shown that the greatest percentage of oocytes become fertilizable following the intermixing of nucleoplasm with the cytoplasm [29], which is followed by the extrusion of the two polar bodies [30]. Relatively large potentials (-60 to -80mV) have been recorded in oocytes prior to the addition of the hormone: both an inward rectifying K^+ current and a L-type Ca^{2+} current have been detected. 10 min after 1-MA addition depolarization of the membrane is observed and the potential changes from -70 to about -10 mV [31-33]. In *Asterina pectinifera* oocytes the change in the resting potential induced by 1-MA depends on the decrease of the K^+ permeability and on the relative increase in Na^+ permeability [34].

1-MA also triggers a decrease in membrane surface area [33] in line with previous findings of cortical changes during the maturation process. The immature oocyte possesses an animal-vegetal polarity with the nucleus (germinal vesicle) under the animal pole surface. Secretory vesicles or cortical granules move from the center towards the oocyte surface where they localize and anchor to the cortex, until the activating Ca^{2+} signal induces a massive exocytosis [35, 36].

The important cortical changes that occur during oocyte maturation are linked to the restructuring of the cytoskeleton [37]. Cytoskeletal elements become more polarized during maturation and the polymerization/ depolymerization of cortical actin associated with surface microvilli induces structural and mechanical changes of the oocyte [38]. 10 min after 1-MA addition a rapid reorganization of the oocyte surface culminates in the formation of giant actin spikes that appear in places where the oocyte had established junctions with surrounding cells, and in the elimination of microvilli as spikes form [39]. Long after microvilli are eliminated, their number and shape are restored especially around the animal pole [40]. Cytoplasmic changes also occur, as indicated by the reorganization of the endoplasmic reticulum (ER) which becomes associated with yolk platelets, forming spherical shells [41]. In immature *Xenopus* oocytes a uniform three-dimensional network of tubules and cisternae has been described in the cortex of both the animal and vegetal hemispheres. In mature eggs, the ER appeared unchanged in the animal half but became organized in the vegetal side in clusters containing $InsP_3$ receptors [42]. Changes in the sensitivity to $InsP_3$ have been claimed to correlate with the changes that occur in the ER during maturation in hamster and mouse oocytes [43, 44]. In comparing immature and mature starfish oocytes, it was found that more $InsP_3$ was required to obtain a comparable rise in intracellular Ca^{2+} in the former [45]. $InsP_3$-dependent Ca^{2+} release (IICR) mediates Ca^{2+} release during fertilization. It becomes fully developed during the maturation process, but has been claimed not to be essential for the re-initiation of the cell cycle [46].

The factors necessary for the initial expansion of the incorporated sperm and for pronuclear development are absent in the cytoplasm of immature oocytes but develop following the dispersal of germinal vesicle contents into the cytoplasm. MPF-regulated changes in nuclear (chromatin) structures may be responsible for the sperm-induced nuclear transformation [47, 48]. In addition to MPF, MAP kinase activity also changes during the meiotic process. MAP kinase is inactive in immature oocytes and becomes activated just after GVBD and MPF activation. MAP kinase activity is dependent on *de novo* protein synthesis and it is maintained at elevated levels until the completion of meiosis I [49]. A direct link has been claimed between the fertilization signal transduction pathway and MAP kinase inactivation [50].

OLD AND NEW Ca^{2+}-LINKED MESSENGERS IN THE MATURATION OF STARFISH OOCYTES

The identification of 1-MA as the maturing hormone and the evidence for early MPF activation following the interaction of the oocytes with the hormone has triggered an impressive amount of work aimed at identifying the intracellular second messenger linking hormone/receptor interaction to MPF activation. Although the pioneering work of Dalcq [51] and Pasteels [52] indicated the importance of Ca^{2+} in re-initiating the meiotic process, the triggering role of Ca^{2+} in maturation has remained debated (reviewed in [53, 54, 55]). As early as 1978 starfish oocytes injected with the photoprotein aequorin were shown to respond to 1-MA with an increase in intracellular Ca^{2+}. Although this supported the suggestion that 1-MA induced Ca^{2+} release from intracellular stores [56] results in other laboratories failed to detect similar changes (the issue of Ca^{2+} involvement in meiosis resumption is discussed in depth in a recent review [55]).

In our laboratory, we routinely observe Ca^{2+} responses to 1-MA. Our earlier work had shown that 1-MA triggered three Ca^{2+} transients, two in the cytoplasm and one in the nucleus, the latter sandwiched in time between the two cytosolic transients. We found that the nuclear Ca^{2+} increase was essential for the continuation of the meiotic cycle, because the injection of BAPTA directly into the nucleus blocked GVBD [57, 58]. We have also investigated the source of Ca^{2+} liberated in response to the application of 1-MA. Although the activation of the oocytes by the hormone involves a complex series of events, it is eventually coupled to inositol 1,4,5-trisphosphate ($InsP_3$) production through the stimulation of phospholipase C (PLC). That the Ca^{2+} signal induced by the hormone is $InsP_3$-dependent is supported by the finding that heparin pre-injected into the oocytes prior to the addition of the hormone blocked the Ca^{2+} mobilization induced by the hormone (Lim *et al.*, in preparation).

Most recently, the spatio-temporal aspects of Ca^{2+} release have been explored by photoliberating $InsP_3$ at selected time intervals during the oocyte maturation process. The increased sensitivity to $InsP_3$ has been found to be inhomogeneous within the oocytes, i.e., it starts in the perinuclear area at the animal hemisphere and propagates to the entire oocyte along the animal-vegetal axis. This is in line with previous work, which had shown that the $InsP_3$-sensitive Ca^{2+} stores in starfish oocytes are fully replenished prior to the addition of 1-MA, and that Ca^{2+} in the ER lumen does not change during the transition from the immature to the mature stage.

A very sensitive CCD camera has permitted us to discover that the addition of 1-MA induces a Ca^{2+} transient lasting about 4 min in the same perinuclear region of the starfish oocytes where the increased sensitivity of the $InsP_3R$ is actually observed (Lim *et al.*, in preparation). $InsP_3$-linked Ca^{2+}-liberating mechanisms are present in the nucleus of these oocytes as well [59]: Ca^{2+} that appears in the nucleoplasm following the treatment with 1-MA is linked to the presence of $InsP_3$ receptors located in the inner membrane of the nuclear envelope. The very large size of the nucleus of the prophase arrested oocytes (approximately 70 μm across) permits the injection of Ca^{2+} indicators, activators, or inhibitors of Ca^{2+} signaling, making these oocytes ideal objects for the study of the thorny question of the permeability of the nuclear envelope to Ca^{2+}. For example, when caged $InsP_3$ was injected directly into the nucleus of the oocytes together with a large molecular mass Ca^{2+} indicator, it induced a nuclear Ca^{2+} transient which was abolished by the nuclear injection of the $InsP_3$ receptor inhibitor heparin [59].

Starfish oocytes also respond to cyclic ADP-ribose (cADPR). We have recently found that the Ca^{2+} response to cADPR also changes during the maturation process induced by 1-MA [60]. Immature *Astropecten auranciacus* oocytes injected with caged cADPR responded to the irradiation with a Ca^{2+} increase, which started at several cortical points. Ca^{2+} wave then propagated centripetally, decaying to the baseline after about 20 sec (Fig. 2A). As was the case for $InsP_3$, the Ca^{2+} response induced by the uncaging of cADPR also differed substantially in mature oocytes. The photoliberation of caged cADPR 70 min after 1-MA treatment induced Ca^{2+} release only at one (or at most two) circumscribed cortical regions and was then followed by a cortical Ca^{2+} flash. The uniform cortical Ca^{2+} increase failed to decay but spread instead centripetally to the entire oocyte in about 10 sec (Fig. 2B), eventually triggering the elevation of the fertilization envelope. The pre-injection of the specific antagonist of the cADPR/ryanodine receptors 8-NH_2cADPR completely blocked the Ca^{2+} response in both immature and mature oocytes. L-type Ca^{2+} channel inhibitors or removal of external Ca^{2+} instead only abolished the cortical flash, and induced a pattern of Ca^{2+} response more similar to that observed in oocytes not treated with 1-MA [60]. cADPR receptors (channels) are also present and operative in the nucleus of starfish oocytes, since caged cADPR injected into the nucleus together with a Ca^{2+} indicator produced a Ca^{2+} spike which persisted following the cytoplasmic injection of the antagonist ruthenium red prior to the nuclear introduction of caged cADPR [61].

Experiments were also performed to establish whether calmodulin (CaM) was involved in the nuclear Ca^{2+} elevation promoted by the hormone. Nuclear injection of antibodies against CaM strongly inhibited the nuclear Ca^{2+} signal promoted by the uncaging of cADPR (they inhibited the resumption of meiosis induced by 1-MA as well) [61]. This suggests that the mechanism by which cADPR liberated Ca^{2+} was dependent on CaM, in agreement with what observed in sea urchin eggs [5, 62]. The finding that cADPR receptors are present and functioning in the inner membrane of the envelope, is nicely in line with the recent finding that a functionally active CD38/ADP ribosyl cyclase, (the enzyme that catalyzes the cyclization of NAD to cADPR, and is also proposed to catalyze the synthesis of NAADP) is present in the inner membrane of the nuclear envelope with its catalytic site facing the nucleoplasm [63, 64].

The other derivative of pyridine nucleotide nicotinic acid adenine dinucleotide phosphate (NAADP) [6] which has recently been found to release Ca^{2+} from sea urchin eggs [65] and ascidian oocytes [66, 67] has also beeen explored in starfish oocytes: it elicited a rapid Ca^{2+} increase in both immature and mature oocytes [68]. Interestingly, the properties of the Ca^{2+} mobilization process also changed during the maturation process. However, the change did not involve the sensitivity of the putative NAADP receptors, as was the case for the $InsP_3$ receptors, but its relationship to the external Ca^{2+}: the removal of external Ca^{2+} inhibited very substantially the Ca^{2+} response induced by the photoliberation of NAADP in mature oocytes but affected it only marginally in immature oocytes. In both cell types, the Ca^{2+}

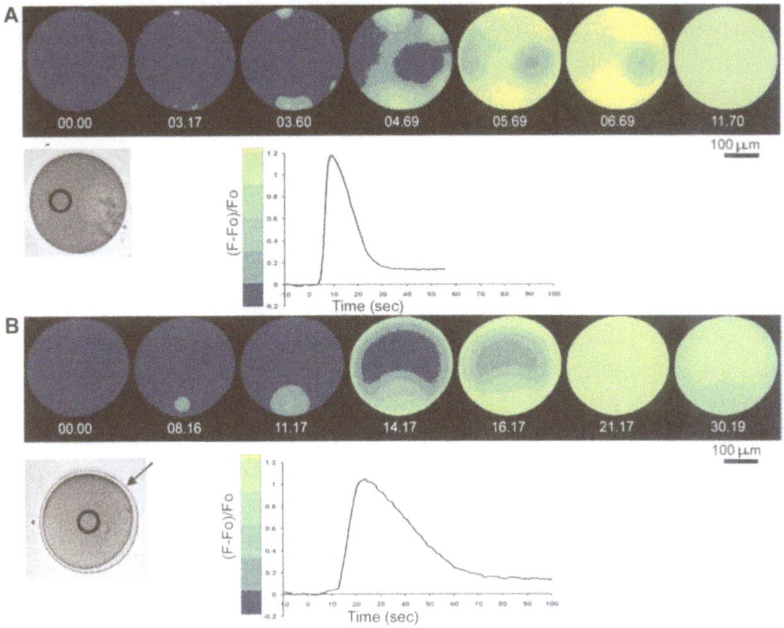

Figure 2. Ca^{2+} **increase induced by the photoliberation of injected caged cADPR visualized with a cooled CCD camera. A)** The uncaging of cADPR in an immature oocyte (no 1-MA addition) in the light microscopy image (lower panel on the left) induces a Ca^{2+} increase which starts simultaneously in several cortical regions. The graph shows the relative fluorescence following the photoliberation of cADPR. **B)** Ca^{2+} increase following the uncaging of caged cADPR in a mature oocyte 70 min after 1-MA addition. The Ca^{2+} increase starts at a circumscribed site in the cell cortex, expands to a cortical flash, and then spreads centripetally to the center of the oocyte. The elevation of the fertilization envelope following the Ca^{2+} increase induced by the uncaging of cADPR is indicated by the arrow in the light microscopy image in the lower panel on the left. The graph shows the relative fluorescence following the uncaging of cADPR measured in the center of the oocyte.

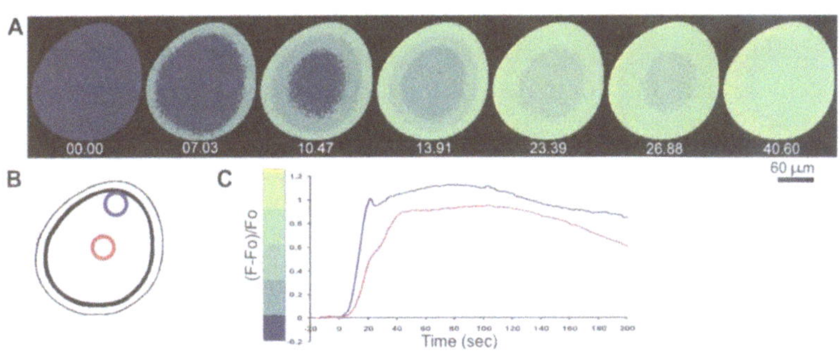

Figure 3. Ca^{2+} **release induced by the uncaging of injected NAADP in a mature oocyte of *Asterina pectinifera*.** The Ca^{2+} response occurs mainly in the cortical domain and then spreads to the center of the oocyte. The massive Ca^{2+} liberation induces the elevation of the fertilization envelope (lower panel on the left). The graph of the relative fluorescence of the Ca^{2+} increase after irradiation is measured in the cortex and in the center of the oocyte as indicated in the scheme.

spike was (partially) sensitive to the simultaneous injection of both heparin and 8-NH_2-cADPR, but was unaffected by the separate injection of the two antagonists. The interplay between the routes activated by $InsP_3$/cADPR and that promoted by NAADP appears possible.

Ca^{2+} SIGNALS AT FERTILIZATION

The activation of immature oocytes by sperm results in the asynchrony of the normal fertilization process since a necessary prerequisite for the occurrence of the cytoplasmic activation response is the dispersal of nuclear contents into the cytoplasm. Thus, the optimal period for fertilization is the metaphase I stage, after the disappearance of the germinal vesicle and before the extrusion of the first polar body. In the absence of fertilization meiotic maturation proceeds to the completion of the second meiosis (meiosis II), without the second metaphase arrest that occurs in ascidian (meiosis I) and mammalian oocytes (meiosis II). The process of fertilization in starfish oocytes occurs in the first metaphase (Fig. 1) when the spermatozoon binds to the outer surface of the jelly layer [69]. A long thin acrosomal filament is extruded from the anterior region of the sperm, which establishes contact with the oocyte surface [70]. The acrosomal process (more than 15 μm) is then incorporated by the growing fertilization cone followed by the engulfment of the sperm head by protrusions (microfilaments) from the fully grown fertilization cone [71].

In starfish, the sperm-oocyte interaction induces a Ca^{2+} transient in the form of a single wave, which starts from the point of sperm attachment and propagates to the antipode of the cell [72]. Following the early finding that the injection of $InsP_3$ in sea urchin eggs elicited a cortical granule exocytosis similar to that induced by the sperm [73], the pathways leading to the production of $InsP_3$ have been investigated. In sea urchin eggs and frog oocytes, the phosphoinositide lipid turnover and the generation of $InsP_3$ have been measured and found to coincide with the Ca^{2+} wave [74]. Experiments have been carried out to investigate whether the generation of $InsP_3$ at fertilization results from the activation of a specific isoform of phospholipase C (PLC). The components of a signaling pathway involving PLCβ, characteristic of G-protein coupled receptors, as well as components of a pathway involving PLCγ, characteristic of tyrosine kinase receptors, appear to be present in starfish oocytes. The evidence that both pathways lead to starfish activation came from experiments in which the expression of exogenous G-protein coupled receptors, such as that for serotonin 1c, and of exogenous tyrosine kinase coupled receptors such as that for the platelet-derived growth factor (PDGD) or the human fibroblast growth factor (EGF) induced responses similar to those seen at fertilization: i.e., a rise in intracellular Ca^{2+}, cortical granule exocytosis, and increased DNA synthesis

[75]. A requirement for PLCγ activation for the initiation of Ca^{2+} release at fertilization has been demonstrated by microinjecting the PLCγ SH2 domain in starfish oocytes and finding that it inhibited the cytoplasmic Ca^{2+} increase [76], whereas the injection of active Src-kinase initiated a Ca^{2+} wave in the absence of fertilization and caused resumption of the cell cycle and replication, as well as condensation of DNA [77]. All these results support the generally accepted proposal of the involvement of $InsP_3$ in triggering the Ca^{2+} rise at fertilization, as suggested for sea urchin eggs [73, 74, 78].

However, the Ca^{2+} rise seen in starfish oocytes at fertilization consists of two phases. The initial response is a Ca^{2+} increase restricted to the cortical layer (cortical flash), the second is the globalization of the signal [76]. The initial response has been claimed to result from Ca^{2+} entry through voltage-gated channels in the plasma membrane [76]. In sea urchin and starfish oocytes the initial and global increases in intracellular Ca^{2+} may also be linked to the activation of ryanodine/cyclic ADP-ribose and NAADP receptors, since the injection of either cADPR or NAADP induces a massive Ca^{2+} increase, which leads to cortical granules exocytosis and to the elevation of the fertilization envelope [79, 65, 59, 68].

A role in the Ca^{2+} swings at fertilization has also been suggested for the second messenger NO. The injection of cGMP in sea urchin eggs [80] or the direct application of NO induces intracellular Ca^{2+} mobilization [81]. It has also been recently suggested that nitric oxide synthase may be the factor from sea urchin sperm that causes Ca^{2+} release at fertilization [82].

However, NAADP also appears to be important: One peculiar aspect of NAADP which has been first documented in starfish oocytes is the linkage of the Ca^{2+} signal mediated by it to the influx of external Ca^{2+}, which is responsible for the cortical flash that follows the photoliberation of injected caged NAADP [83]. Interestingly, species differences among starfish exist: in *Asterina pectinifera* the cortical flash induced by the uncaging of NAADP is followed by the globalization of the signal to the oocyte (Fig. 3) whereas the Ca^{2+} response remains circumscribed to the cortical layer in *Astropecten auranciacus* (Fig. 4) [60]. In both cases, however, the activation of cortical NAADP "receptors" is preliminar to the propagation of the Ca^{2+} wave to the remainder of the cell, which is later induced by the activation of $InsP_3$ receptors [83].

The respective roles of $InsP_3$ and NAADP in the Ca^{2+} responses at fertilization have been recently dissected using enucleated oocytes. The rationale for this kind of experiment came from the knowledge that the intermixing of the nucleoplasm with the cytoplasm is a necessary requisite for a successful fertilization (see above). We have found that the uncaging of $InsP_3$ in a mature oocyte from which the nucleus had been removed prior to the addition of 1-MA to induce maturation was followed by a Ca^{2+} response in which the time necessary to reach the plateau was longer than that of the

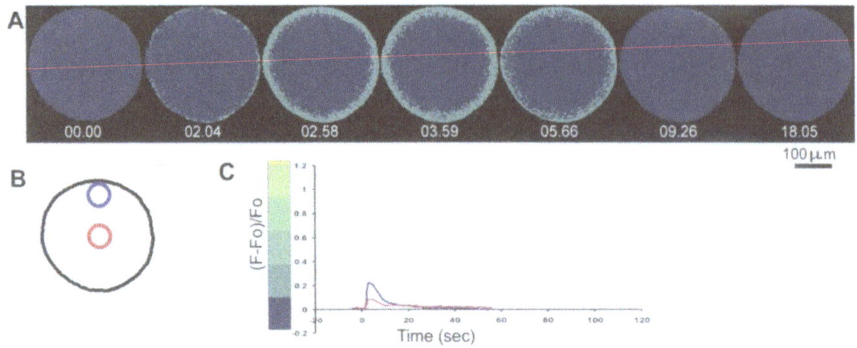

Figure 4. **Ca^{2+} response following the photoliberation of injected caged NAADP in a mature oocyte of *Astropecten auranciacus*.** The Ca^{2+} increase is limited to the cortical region of the oocyte and decays to the baseline level without spreading to the center of the oocyte. The graph shows the relative fluorescence measured in the cortex and in the center of the oocyte (see blue and red circles in the scheme on lower left).

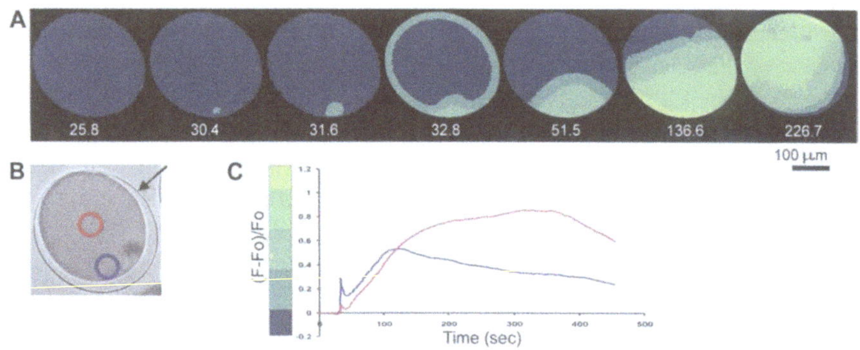

Figure 5. **Ca^{2+} responses at fertilization.** The sperm induces a limited, circumscribed Ca^{2+} elevation in the cortical region of the oocyte. This circumscribed hotspot develops into a uniform Ca^{2+} flash that disappears in a few seconds. The wave then spreads to the entire cell. The graph shows the relative fluorescence measured in the cortex and in the center of the oocyte (see blue and red circles in the light microscopy image on lower left).

control experiment in nucleated oocytes. After the peak was eventually reached, Ca^{2+} stayed at the peak level for a longer time. In these oocytes the intracellular Ca^{2+} liberation eventually became massive, but it failed to elevate the fertilization envelope.

In contrast to $InsP_3$, the response to the uncaging of NAADP in the cortical layer of the enucleated oocytes was the same as in the controls. In the case of NAADP, the removal of the nucleus failed to affect the Ca^{2+} response and the cortical exocytotic process. This was also seen at fertilization, except that in that case the removal of the nucleus strongly delayed the globalization of the signal. Apparently, then, the Ca^{2+} release at fertilization is initiated by NAADP "receptors" and is then propagated by those sensitive to $InsP_3$ [83]. We have also found that the sperm induces an initial limited elevation of Ca^{2+} at the site of interaction, which expands to a uniform cortical flash and then spreads to the entire oocyte (Fig. 5). The locally restricted Ca^{2+} release upon sperm interaction was not easy to demonstrate. We missed it using the confocal microscope due to the difficulty of detecting a Ca^{2+} increase on a very limited cell area out of the confocal plane, which explains the failure of previous reports in the literature to detect a distinct point-source pattern of propagation for the Ca^{2+} signal. To observe the initial Ca^{2+} hotspot it was necessary to use a cooled, sensitive CCD camera [60].

A set of interesting recent observations on starfish oocytes suggest that the actin cytoskeleton may have a role in the Ca^{2+} response at fertilization. The actin depolymerizing drug latrunculin A has been found to mimic the cascade of Ca^{2+} events seen at fertilization indicating that actin could play an important role in the initial liberation of Ca^{2+} and in the production of the cortical flash [84]. The sperm, or rather factors brought into the cell by it, according to the prevailing sperm factor hypotheses (see [85] for a recent discussion of the problem) could somehow locally depolymerize actin, producing a restricted Ca^{2+} hotspot, and would then initiate the cortical flash by promoting the entry of Ca^{2+} into the cell. This would be in line with the notion that the cytoskeleton regulates the activity of plasma membrane Ca^{2+} channels [86].

CONCLUDING REMARKS

Starfish oocytes are a convenient tool for the study of the complex set of Ca^{2+} responses at maturation and fertilization. Recent work has shown that the Ca^{2+}-linked second messengers $InsP_3$, cADPR, and NAADP have distinct roles in the two processes. It has also revealed a hitherto unknown spatio-temporal pattern of development of the Ca^{2+} liberation events in which an initial circumscribed cortical Ca^{2+} hotspot is the first response to the interaction of the sperm. It now begins to emerge that the actin cytoskeleton may have an important role in the initiation of the Ca^{2+} response.

REFERENCES

1. Hertwig O. 1876. Beiträge zur Kenntniss der Bildung, Befruchtung und theilung des thierischen Eies. *Gegenbaurs Morph. Jb.* 1: 347-434.
2. Fol H. 1877. Sur le commencement de l'hénogenie chez divers animaux. *Arch. Sci. Gen.* 58: 439-472.
3. Clapper DL, Walseth TF, Dargie PJ and Lee HC. 1987. Pyridine nucleotide metabolites stimulate Ca^{2+} release from sea urchin egg microsomes desensitized to inositol trisphosphate. *J. Biol. Chem.* 262: 9561-9568.

4. Lee HC, Walseth TF, Bratt GT, Hayes RN and Clapper DL. 1989. Structural determination of a cyclic metabolite of NAD^+ with intracellular Ca^{2+}-mobilizing activity. *J. Biol. Chem.* 264: 1608-1615.
5. Lee HC, Aarhus R and Levitt D. 1994. The cristal structure of cADP-ribose. *Nature Struct. Biol.* 1: 143-144.
6. Lee HC and Aarhus R. 1995. A derivative of NADP mobilizes Ca^{2+} stores insensitive to inositol trisphosphate and cyclic ADP-ribose. *J. Biol. Chem.* 270: 2152-2157.
7. Labbée JC, Capony JP, Caput D, Cavadore JC, Derancourt J, Kaghad M, Lelias JM, Picard A and Dorée M. 1989. MPF from starfish oocytes at first meiotic metaphase is a heterodimer containing one molecule of cdc2 and one molecule of cyclin B. *EMBO J.* 8: 3053-3058.
8. Pondaven P, Meijer L and Beach D. 1990. Activation of M-phase-specific histone H1 kinase by modification of the phosphorylation of its p34cdc2 and cyclin components. *Genes Dev.* 4: 9-17.
9. Borgne A, Ostvold AC, Flament S and Meijer L. 1999. Intra-M phase-promoting factor phosphorylation of cyclin B at the prophase/metaphase transition. *J. Biol. Chem.* 23: 11977-11986.
10. Meijer L, Borgne A, Mulner O, Chong JP, Blow JJ, Inagaki N, Inagaki M, Delcros JG and Moulinoux JP. 1997. Biochemical and cellular effects of roscovitine, a potent and selective inhibitor of the cyclin-dependent kinases cdc2, cdk2 and cdk5. *Eur. J. Biochem.* 243: 527-536.
11. Kanatani H, Ikegami S, Shirai H, Oide H and Tamura S. 1971. Purification of gonad-stimulating substance obtained from radial nerves of the starfish, *Asterias amurensis*. *Dev. Growth Differ.* 13: 151-164.
12. Schroeder TE. 1981. Microfilament-mediated surface change in starfish oocytes in response to 1-methyladenine: implications for identifying the pathway and receptor sites for maturation-inducing hormones. *J. Cell Biol.* 90: 362-371.
13. Kanatani H, Shirai H, Nakanishi K and Kurosawa T. 1969. Isolation and identification of meiosis-inducing substance in starfish, *Asterias amurensis*. *Nature* 221: 273-274.
14. Kanatani H and Hiramoto Y. 1970. Site of action of 1-methyladenine in inducing oocyte maturation in starfish. *Exp. Cell Res.* 61: 280-284.
15. Jaffe LA, Gallo CJ, Lee RH, Ho Y-K and Jones TLZ. 1993. Oocyte maturation in starfish is mediated by the βγ subunit complex of a G protein. *J. Cell Biol.* 121: 755-783.
16. Chiba K, Kontani K, Tadenuma H, Katada T and Hoshi M. 1993. Induction of starfish oocyte maturation by the βγ subunit of starfish G protein and possible existence of the subsequent effector in cytoplasm. *Mol. Biol. Cell.* 4: 1027-1034.
17. Sadler KC and Ruderman JV. 1998. Components of the signaling pathway linking the 1-methyladenine receptor to MPF activation and maturation in starfish oocytes. *Dev. Biol.* 197: 25-38.
18. Meijer L and Zarutskie P. 1987. Starfish oocyte maturation: 1-methyladenine triggers a drop of cAMP concentration related to the hormone-dependent period. *Dev. Biol.* 121: 306-315.
19. Meijer L and Guerrier P. 1984. Maturation fertilization in starfish oocytes. *Int. Rev. Cytol.* 86: 129-196.
20. Meijer L and Mordret G. 1994. Starfish oocyte maturation: from prophase to metaphase. *sem in Dev. Biol.* 5: 165-171.
21. Kishimoto T, Kanatani H. 1976 Cytoplasmic factor responsible for germinal vesicle breakdown and meiotic maturation in starfish oocyte. *Nature* 260: 321-322.
22. Guerrier P, Moreau M and Dorée M. 1977. Hormonal control of meiosis in starfish: stimulation of protein phosphorylation induced by 1-methyladenine. *Mol. Cell Endocrinol.* 7: 137-150.
23. Kishimoto T. 1999. Activation of MPF at meiosis reinitiation in starfish oocytes. *Dev. Biol.* 214: 1-8.

24. Ookata K, Hisanaga S, Okano T, Tachibana K and Kishimoto T. 1992. Relocation and distinct subcellular localization of p34cdc2-cyclin B complex at meiosis reinitiation in starfish oocytes. *EMBO J.* 11: 1763-1767.
25. Okumura E, Fukuhara T, Yoshida H, Hanada S, Kozutsumi R, Mori M, Tachibana K and Kishimoto T. 2002. Akt inhibits Myt1 in the signalling pathway that leads to meiotic G2/M-phase transition. *Nature Cell Biol.* 4: 111-116.
26. Nigg EA. 1993. Cellular substrates of p34cdc2 and its companion cyclin-dependent kinases. *Trends Cell Biol.* 3: 296-301.
27. Santella L, Kyozuka K, Hoving S, Munchbach M, Quadroni M, Dainese P, Zamparelli C, James P and Carafoli E. 2000. Breakdown of cytoskeletal proteins during meiosis of starfish oocytes and proteolysis induced by calpain. *Exp. Cell. Res.* 259: 117-126.
28. Schollmeyer JE. 1988 Calpain-II involvement in mitosis. *Science* 240: 911-913.
29. Stevens M. 1970. Procedures for induction of spawning and meiotic maturation of starfish oocytes by treatment with 1-methyladenine. *Exp. Cell Res.* 59: 482-484.
30. Kanatani H. 1985. Oocyte growth and maturation in starfish. In *Biology of fertilization*, eds. CB Metz and A Monroy, vol. I, pp. 119-140. Academic Press, Inc Orlando, Florida.
31. Dale B, de Santis A and Hoshi M. 1979. Membrane response to 1-methyladenine requires the presence of the nucleus. *Nature* 282: 89-90.
32. Moody WJ and Bosma MM. 1985. Hormone-induced loss of surface membrane during maturation of starfish oocytes: differential effects on potassium and Ca^{2+} channels. *Dev. Biol.* 112: 396-404.
33. Moody WJ and Lansman JB. 1983. Developmental regulation of Ca^{2+} and K^{+} currents during hormone-induced maturation of starfish oocytes. *Proc. Natl. Acad. Sci. USA.* 80: 3096-3100.
34. Miyazaki SI, Ohmori H and Sasaki S. 1975. Potassium rectifications of the starfish oocyte membrane and their changes during oocyte maturation. *J. Physiol.* 246: 55-78.
35. Longo FJ, Woerner M, Chiba K and Hoshi M. 1995. Cortical changes in starfish (*Asterina pectinifera*) oocytes during 1-methyladenine-induced maturation and fertilisation/activation. *Zygote* 3: 225-239.
36. Santella L, De Riso L, Gragnaniello G and Kyozuka K. 1999. Cortical granule translocation during maturation of starfish oocytes requires cytoskeletal rearrangement triggered by $InsP_3$-mediated Ca^{2+} release. *Exp. Cell Res.* 248: 567-574.
37. Sardet C, Prodon F, Dumollard R, Chang P and Chênevert J. 2002. Structure and function of the egg cortex from oogenesis through fertilization. *Dev. Biol.* 241: 1-23.
38. Heil-Chapdelaine RA and Otto JJ. 1996. Characterization of changes in F-actin during maturation of starfish oocytes. *Dev. Biol.* 177: 204-216.
39. Schroeder TE and Stricker SA. 1983. Morphological changes during maturation of starfish oocytes: surface ultrastructure and cortical actin. *Dev. Biol.* 98: 373-384.
40. Hirai H and Shida H. 1979. Shortening of microvilli during the maturation of starfish oocyte from which vitelline coat was removed. *Bull. Mar. Biol. Sta. Asamushi, Tohoku, Univ.* 16: 161-167.
41. Jaffe LA and Terasaki M. 1994. Structural changes in the endoplasmic reticulum of starfish oocytes during meiotic maturation and fertilization. *Dev. Biol.* 164: 579-587.
42. Terasaki M, Runft LL and Hand AR. 2001. Changes in organization of the endoplasmic reticulum during *Xenopus* oocyte maturation and activation. *Mol. Biol. Cell* 12: 1103-1116.
43. Fujiwara T, Nakada K, Shirakawa H and Miyazaki S. 1993. Development of inositol trisphosphate-induced Ca^{2+} release mechanism during maturation of hamster oocytes. *Dev. Biol.* 156: 69-79.

44. Mehlmann LM and Kline D. 1994. Regulation of intracellular Ca^{2+} in the mouse egg: Ca^{2+} release in response to sperm or inositol trisphosphate is enhanced after meiotic maturation. *Biol. Reprod.* 51: 1088-1098.
45. Chiba K, Kado RT and Jaffe LA. 1990. Development of Ca^{2+} release mechanisms during starfish oocyte maturation. *Dev. Biol.* 140: 300-306.
46. Iwasaki H, Chiba K, Uchiyama T, Suzuki F, Ikeda M, Furuichi T and Mikoshiba K. 2002. Molecular characterization of the starfish $InsP_3$ receptors and its role during oocyte maturation and fertilization. *J. Biol. Chem.* 277: 2763-2772.
47. Schuetz AW and Longo FJ. 1981. Hormone-cytoplasmic interactions controlling sperm nuclear decondensation and male pronuclear development in starfish oocytes. *J. Exp. Zool.* 215: 107-111.
48. Longo FJ, Cook S and Mathews L. 1991. Pronuclear formation in starfish eggs inseminated at different stages of meiotic maturation: correlation of sperm nuclear transformation and activity of the maternal chromatin. *Dev. Biol.* 147: 62-72.
49. Kishimoto T. 1998. Cell cycle arrest and release in starfish oocytes and eggs. *Sem. Dev. Biol.* 9: 549-557.
50. Tachibana K, Machida T, Nomura Y and Kishimoto T. 1997. MAP kinase links the fertilization signal transduction pathway to the G1/S-phase transition in starfish eggs. *EMBO J.* 16: 4333-4339.
51. Dalcq A. 1925. Recherches expérimentales et cytologiques sur la maturation et l'activation de l'œuf d'Asterias glacialis. *Arch. Biol.* 34: 507-674.
52. Pasteels J. 1935. Recherches sur le déterminisme de l'entrée en maturation de l'œuf chez divers Invertébrés marins. *Arch. Biol.* 46: 229-262.
53. Whitaker M and Patel R. 1990. Ca^{2+} and cell cycle control. *Development* 108: 525-542.
54. Means AR. 1994. Calcium, calmodulin and cell cycle regulation. *FEBS Lett.* 347: 1-4.
55. Santella L. 1998. The role of calcioum in the cell cycle: facts and hypothesis. *Biochem. Biophys. Res. Commun.* 244: 317-324.
56. Moreau M, Guerrier P, Dorée M and Ashley CC. 1978. 1-methyladenine induced release of intracellular Ca^{2+} triggers meiosis in starfish oocytes. *Nature* 272: 251-253.
57. Santella L and Kyozuka K. 1994. Reinitiation of meiosis in starfish oocytes requires an increase in nuclear Ca^{2+}. *Biochem. Biophys. Res. Commun.* 203: 674-680.
58. Santella L, De Riso L, Gragnaniello G and Kyozuka K. 1998. Separate activation of the cytoplasmic and nuclear Ca^{2+} pools in maturing starfish oocytes. *Biochem. Biophys. Res. Commun.* 252: 1-4.
59. Santella L and Kyozuka K. 1997. Effects of 1-methyladenine on nuclear Ca^{2+} transients and meiosis resumption in starfish oocytes are mimicked by the nuclear injection of inositol 1,4,5-trisphosphate and cADP-ribose. *Cell Calcium* 22: 11-20.
60. Nusco GA, Lim D, Sabala P and Santella L. 2002. Ca^{2+} response to cADPR during maturation and fertilization of starfish oocytes. *Biochem. Biophys. Res. Commun.* 290: 1015-1021.
61. Santella L and Kyozuka K. 1997. Association of calmodulin with nuclear structures in starfish oocytes and its role in the resumption of meiosis. *Eur. J. Biochem.* 246: 602-610.
62. Lee HC, Aarhus R and Graeff RM. 1995. Sensitization of Ca^{2+}-induced Ca^{2+} release by cyclic ADP-ribose and calmodulin. *J. Biol. Chem.* 270: 9060-9066.
63. Adebanjo OA, Anandatheerthavarada HK, Koval AP, Moonga BS, Biswas G, Sun L, Sodam BR, Bevis PJ, Huang CL, Epstein S, Lai FA, Avadhani NG and Zaidi M. 1999. A new function for CD38/ADP-ribosyl cyclase in nuclear Ca^{2+} homeostasis. *Nature Cell Biol.* 17: 409-414.

64. Khoo KM, Han MK, Park JB, Chae SW, Kim UH, Lee HC, Bay BH and Chang CF. 2000. Localization of the cyclic ADP-ribose-dependent Ca^{2+} signaling pathway in hepatocyte nucleus. *J. Biol. Chem.* 275: 24807-24817.
65. Galione A, Patel S and Churchill GC. 2000. $NAADP^+$-induced Ca^{2+} release in sea urchin eggs. *Biol. Cell.* 92: 197-204.
66. Albrieux M. Sardet C and Villaz M. 1997 The two intracellular Ca^{2+} release channels, ryanodine receptors and inositol 1,4,5-trisphosphate receptor, play different roles during fertilization in Ascidians. *Dev. Biol.* 189: 174-185.
67. Albrieux M, Lee HC and Villaz M. 1998 Calcium signaling by cyclic ADP-ribose, NAADP, and inositol trisphosphate are involved in distinct functions in Ascidian oocytes. *J. Biol. Chem.* 273: 14566-14574.
68. Santella L, Kyozuka K, Genazzani AA, De Riso L and Carafoli E. 2000b. Nicotinic acid adenine dinucleotide phosphate-induced Ca^{2+} release. *FASEB J.* 275: 8301-8306.
69. Hoshi M, Nishigaki T, Ushiyama A, Okinaga T, Chiba K and Matsumoto M. 1994. Egg-jelly signal molecules for triggering the acrosome reaction in starfish spermatozoa. *Int. J. Dev. Biol.* 38 : 167-174.
70. Dan JC. 1954. Studies on the acrosome. II. Acrosome reaction in starfish spermatozoa. *Biol. Bull.* 107: 203-218.
71. Kyozuka K and Osanai K. 1988. Fertilization cone formation in starfish oocytes: the role of the egg cortex actin microfilaments in sperm incorporation. *Gam. Res.* 20: 275-285.
72. Stricker SA. 1999. Comparative biology of calcium signaling during fertilization and egg activation in animals. *Dev. Biol.* 15: 157-176.
73. Swann K, Whitaker M. 1986. The part played by inositol trisphosphate and calcium in the propagation of the fertilization wave in sea urchin eggs. *J. Cell. Biol.* 103: 2333-42
74. Ciapa B, Borg B, Whitaker M. 1992. Polyphosphoinositide metabolism during the fertilization wave in sea urchin eggs. *Development* 115: 187-95.
75. Shilling FM, Carroll DJ, Muslin AJ, Escobedo JA, Williams LT, Jaffe LA. 1994. Evidence for both tyrosine kinase and G-protein-coupled pathways leading to starfish egg activation. *Dev. Biol.* 162: 590-599.
76. Carroll DJ, Ramarao CS, Mehlmann LM, Roche S, Terasaki M and Jaffe LA. 1997. Calcium release at fertilization in starfish eggs is mediated by phospholipase Cγ. *J. Cell Biol.* 138: 1303-1311.
77. Giusti AF, Carroll DJ, Abassi YA and Foltz KR. 1999. Evidence that a starfish egg Src family tyrosine kinase associates with PLC-γ1 SH2 domains at fertilization. *Dev. Biol.* 208: 189-199.
78. Abassi YA, Carroll DJ, Giusti AF, Belton RJ Jr and Foltz KR. 2000. Evidence that Src-type tyrosine kinase activity is necessary for initiation of calcium release at fertilization in sea urchin eggs. *Dev. Biol.* 15: 206-219.
79. Galione A, Lee HC and Busa WB. 1991. Ca^{2+}-induced Ca^{2+} release in sea urchin egg homogenates: modulation by cyclic ADP-ribose. *Science* 253: 1143-1146.
80. Whalley T, McDougall A, Crossley I, Swann K and Whitaker M. 1992. Internal Ca^{2+} release and activation of sea urchin eggs by cGMP are independent of the phosphoinositide signaling pathway. *Mol. Biol. Cell.* 3: 373-383.
81. Willmott N, Sethi JK, Walseth TF, Lee HC, White AM and Galione A. 1996. Nitric oxide-induced mobilization of intracellular calcium via the cyclic ADP-ribose signaling pathway. *J. Biol. Chem.* 271: 3699-3705.
82. Kuo RC, Baxter GT, Thompson SH, Stricker SA, Patton C, Bonaventura J and Epel D. 2000. NO is necessary and sufficient for egg activation at fertilization. *Nature* 406: 633-636.

83. Lim D, Kyozuka K, Gragnaniello G, Carafoli E and Santella L. 2001. NAADP$^+$ initiates the Ca^{2+} response during fertilization of starfish oocytes. *FASEB J.* 15: 2257-2267.
84. Lim D, Lange K and Santella L. 2002. Activation of oocytes by latrunculin A. *FASEB J.* (in press)
85. Carafoli E, Santella L, Branca D and Brini M. 2001. Generation, control, and processing of cellular calcium signals. *Crit. Rev. Biochem. Mol. Biol.* 36: 107-260.
86. Nilius B and Droogmans G. 2001. Ion channels and their functional role in vascular endothelium. *Physiol. Rev.* 81: 1415-1459.

Luigia Santella, Laboratory of Cell Biology,
Stazione Zoologica "A. Dohrn"
Villa Comunale I-80121, Naples, Italy.
e-mail: santella@alpha.szn.it

Cyclic ADP-Ribose and NAADP. Structures, Metabolism and Functions. Lee HC (ed), pp. 397-407, Dordrecht: Kluwer.

ADP-Ribosylation and CD38 Signaling

Uh-Hyun Kim[1,3], Myung-Kwan Han[1,3] and Chang-Yeol Yim[2,3]
[1]Department of Biochemistry and [2]Department of Internal Medicine, [3]Institute for Medical Sciences, Chonbuk National University Medical School, Chonju, 561-182 Korea.

INTRODUCTION

Mono-ADP-ribosylation is a posttranslational protein modification reaction that was originally discovered as a mechanism by which bacterial toxins interfere with signal transduction in human host cells [1, 2]. Mono-ADP-ribosylation is also used as a mechanism to regulate endogenous metabolism, as clearly demonstrated in photosynthetic bacteria [3]. Mammalian endogenous mono-ADP-ribosylation has also been studied and the responsible enzymes have been purified and defined at the molecular level [4-9].

Lymphocyte surface molecules including ecto-enzymes are known to play an important regulatory role by modulating their expression, binding activity and/or signal transduction. Mono-ADP-ribosylation of surface proteins on T cells by glycosylphosphatidylinositol (GPI)-anchored mono-ADP ribosyltransferase (ADPRT) results in the inhibition of cytotoxic T cell functions, such as cell proliferation, cytotoxicity, and cytokine secretion [10], supporting the previous finding of the link between defective expression of RT6, a member of the ADPRT family and increased susceptibility to autoimmune disease [11]. It is of great interest to identify surface target protein(s) in order to clarify the function of mono-ADP-ribosylation on T cells by ADPRT.

The type II transmembrane glycoprotein CD38 is the prototypical member of the class of adenosine diphosphate ribose (ADPR) transfer enzymes known as nicotinamide adenine dinucleotide (NAD) glycohydrolases (NADases) (for review, [12]). CD38 is a bifunctional ectoenzyme that catalyzes the formation and hydrolysis of cyclic adenosine diphosphate ribose (cADPR) [13-15]. Thus, this enzyme possesses both ADP-ribosyl cyclase and cADPR hydrolase, the net reaction of which is identical to that catalyzed by NADase. cADPR is a potent Ca^{2+} mobilizing agent believed to be involved in Ca^{2+}-induced Ca^{2+} release in a variety of cells ranging from plants to humans [13-17], and it has also been shown to augment the proliferative response of activated murine B lymphocytes [14], suggesting that cADPR may function as a signaling second messenger in multiple hematopoietic cell types. Glucose raises cADPR concentration in pancreatic islets and cADPR has been shown to induce insulin secretion from permeabilized islets [18], indicating that cADPR is a physiologically relevant second messenger for insulin secretion in islets.

Most ecto-NADases are known to be inhibited by their substrate, NAD [19-21]. Therefore, we explored whether CD38, a member of the NADase family, is also inactivated by NAD. Our results show that ADP-ribosylation of CD38 by ADPRT in activated T cells results in apoptosis as well as CD38 inactivation, suggesting that ADP ribosylation of CD38 plays an important role in immune regulation. CD38 in the pancreatic islet cells was also found to be ADP ribosylated by ADPRT, resulting in blockade of islet's insulin secretion.

ADP-RIBOSYLATION OF CD38 RESULTS IN INACTIVATION OF ITS ADP-RIBOSYL CYCLASE AND cADPR HYDROLASE ACTIVITIES

Murine resting T cells showed low CD38 expression whereas its expression was induced by activation with anti-CD3 antibodies [22]. When the activated T cells were incubated with NAD, the ADP-ribosyl cyclase and cADPR hydrolase activities of CD38 decreased in a dose-dependent manner, as the NAD concentrations increased (Fig 1). This result indicated that both enzyme activities of CD38 were affected by the addition of NAD. Furthermore, when the effects of metabolites such as adenosine, AMP, ADP, ADP-ribose and cADPR were tested to exclude the possible argument that NAD metabolites might have inactivated CD38 activities, no significant changes in CD38 activities were found [22].

The mechanism responsible for the NAD-dependent inhibition of CD38 enzyme activities was found to be the transfer of ADPR from NAD to the CD38 molecule, evidenced by the observation that CD38 immunoprecipitated with anti-CD38 antibodies from intact T cells which had been incubated with ^{32}P-adenylate-NAD contained ^{32}P-radioactivity (inset in Fig. 1). Since ADPRT had been reported to be anchored on the cell membrane via a glycosylphosphatidylinositol (GPI)-linkage [8], we examined whether treatment of T cells with phatidylinositol-specific phospholipase C (PI-PLC) would block ADP-ribosylation of CD38. It was found that tretreatment of T cells with PI-PLC prevented NAD-dependent

ADP-ribosylation of CD38, however it did not affect either the amount of membrane associated CD38 or the activities of CD38 [22]. This result indicated that the ADP-ribosylation of CD38 was mediated by a GPI-anchored cell surface ADPRT but not due to automodification. Next, the ADPR acceptor site(s) on CD38 was determined. Thus, ^{32}P-ADP-ribosylated CD38 was treated with water (control), hydroxylamine (which specifically cleaves arginine-linked ADPR) or mercuric chloride (which specifically cleaves cysteine-linked ADPR). The ^{32}P-ADPR on CD38 was removed completely only by treatment with both hydroxylamine and mercuric chloride. Treatment with either hydroxylamine or mercuric chloride alone were found to release only partially the ^{32}P-ADPR [22]. This suggested that both arginine- and cysteine-specific ADPRTs are involved in the ADP-ribosylation of CD38 in the activated T cells.

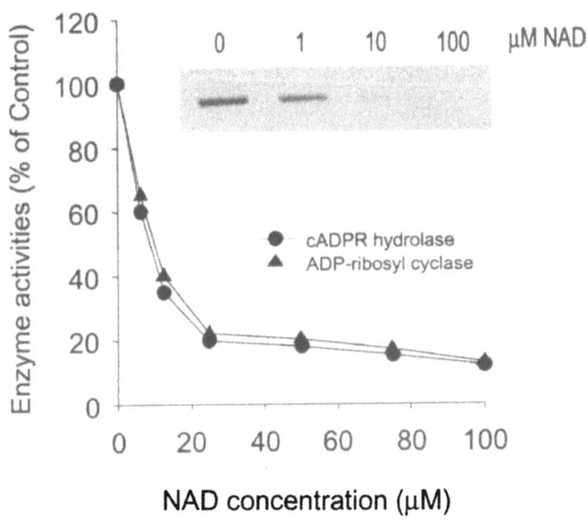

Figure 1. **Inactivation of CD38 activities by ADPRT-mediated ADP-ribosylation.** Effect of NAD on enzyme activities of CD38. Intact T cells (1 X 10^5 cells/ml) were incubated at 37°C for 1 hour in the presence of various concentrations of NAD, and the enzyme activities were measured. *Inset*, After incubation of activated T cells (1 X 10^5 cells/ml) with the indicated concentrations of unlabeled NAD for 1 hour, the cells were further incubated for additional 1 hour in the presence of 10 μM [^{32}P]NAD. CD38 was immunoprecipitated, separated on a SDS-polyacrylamide gel and subjected to autoradiography.

APOPTOSIS OF ACTIVATED T CELLS BY ADP-RIBOSYLATION OF CD38

When the activated T cells were incubated with NAD, the percentage of viable T cells was significantly decreased, compared with control incubated under the same conditions without NAD [22]. A question of whether the reduction of viable T cells was due to apoptosis was examined and it was found that, as shown in Fig. 2, an apoptotic signature, DNA ladder pattern, was significantly increased in the activated T cells when treated with NAD. ADP-ribosylation of CD38 could induce apoptosis in CD38-transfected cell lines; Jurkat cells do not normally express CD38 but have both arginine- and cysteine-specific ADPRTs. When the CD38-transfected and non-transfected Jurkat cells were exposed to NAD, only the CD38-transfected Jurkat cells were found to undergo apoptosis (Fig 3). These results strongly suggested that CD38 was indeed a target molecule for ADP-ribosylation by ADPRTs leading to programmed cell death and that the ADP-ribosylation of CD38 could induce apoptosis in CD38-expressing cells.

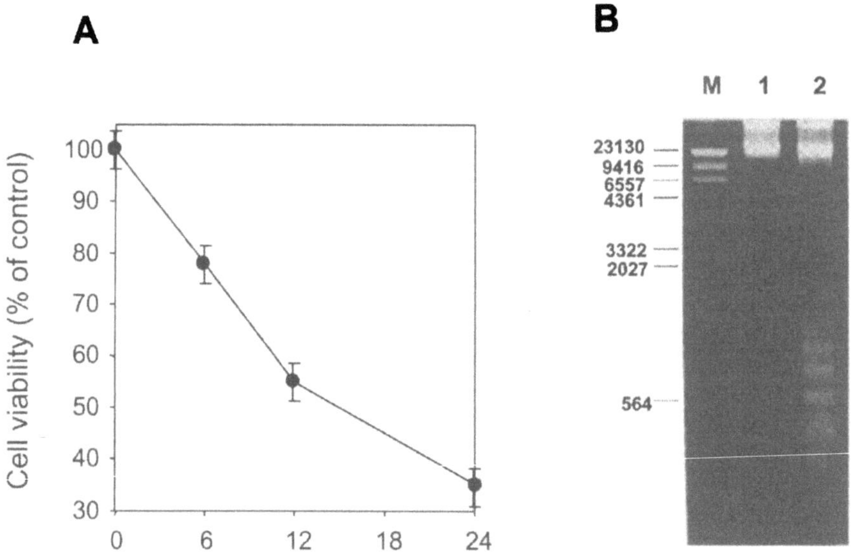

Figure 2. **Apoptosis of activated T cells by ADP-ribosylation of CD38. A.** Effect of NAD on T cell viability. The number of viable cells was determined by trypan blue exclusion after T cells were cultured for the indicated times following their incubation for 1 hour in the presence of 100 μM NAD. The data shown are the mean ± SD of three independent experiments. **B.** DNA fragmentation in NAD-treated activated T cells. DNA was extracted using a DNA extraction kit (Promega) of T cells that had been incubated for 12 hours in the presence (lane 2) or absence (lane 1) of 100 μM NAD. DNA fragments were visualized after agarose gel electrophoresis by ethidium bromide staining. Lane M, DNA molecular markers (λ DNA cleaved with HindIII).

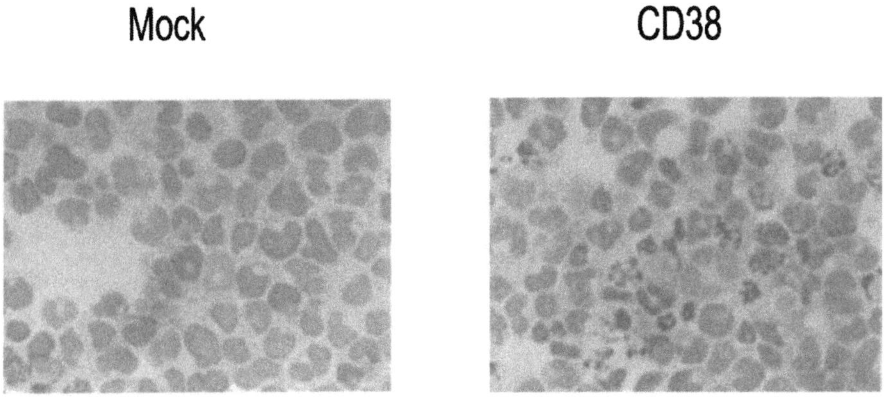

Figure 3. **NAD-dependent apoptosis of CD38-transfected cells.** Wild-type (Mock) and CD38-transfected Jurkat were incubated for 16 hours with 100 μM NAD in RPMI medium containing 10% fetal bovine serum, following serum deprivation for 24 hours [22]. The cells were centrifuged onto slides and stained with hematoxylin-eosin to identify cells that lost nuclear structure.

EXOGENOUS cADPR PREVENTS T CELLS FROM NAD-INDUCED APOPTOSIS

Franco *et al.* [23, 24] demonstrated that transport of cADPR could be carried out through two different systems, transmembrane CD38 itself and a non-CD38 transporter. Consistent with the above, our data showed that CD38, once dimerized, could be internalized through endocytic vesicle formation [25]. In the intracellular vesicle, cytoplasmic NAD would be accessible to the catalytic site of CD38 by an equilibratory NAD transport system, thus producing cADPR within the endocytic vesicle [26]. cADPR produced in the endocytic vesicle crosses the plasma membrane after CD38 dimerization [24].

Intact T cells did not show any increase in intracellular cADPR concentration ($[cADPR]_i$) by the addition of exogenous cADPR, while NAD-treated cells with inactivated cADPR hydrolase of CD38 showed an increase in $[cADPR]_i$ by the addition of cADPR (Fig 4). Simultaneously, exogenous cADPR could protect NAD-dependent apoptosis of the T cells by increasing the cytosolic Ca^{2+} level, indicating that cADPR has a role in Ca^{2+} homeostasis that is essential for the survival of activated T lymphocytes. This finding also supported the existence of a non-CD38 cADPR transporter in T cells, because CD38 itself was catalytically inactivated by the ADP-ribosylation.

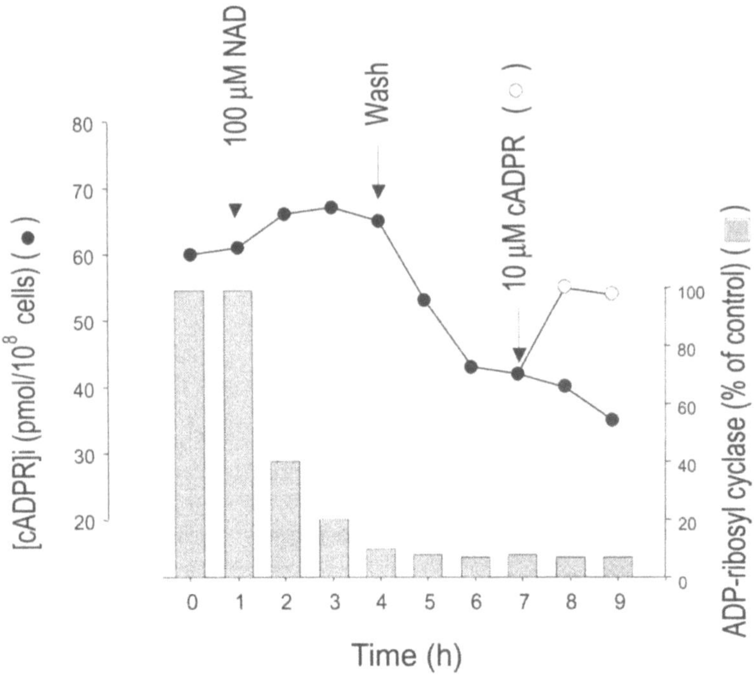

Figure 4. **Effect of NAD and cADPR on $[cADPR]_i$ and cyclase activity of CD38 in activated T cells.** $[cADPR]_i$ and CD38 cyclase activity were measured every hour as described [22]. 100 μM NAD and 10 μM cADPR were added at the indicated times. After 3 hours of NAD treatment, the cells were washed with RPMI containing 10% FBS.

ARGININE- VERSUS CYSTEINE- ADP-RIBOSYLATION OF CD38

When determined by reverse transcriptase-polymerase chain reaction and southern blot, B cells were found not to express arginine-specific mono-ADPRT, whereas T cells expressed this enzyme [22]. The finding was consistent with data that the ADP-ribosylation of CD38 in B cells occurred only on the cysteine residue, in contrast to the ADP-ribosylation of CD38 in T cells which occurred on both arginine and cysteine residues, as discussed above.

B cells showed no detectable apoptosis and no inhibition of proliferation when they were incubated with NAD, although CD38 was ADP-ribosylated. Arginine-ADP-ribosylation of CD38 resulted in the inactivation of both the cyclase and hydrolase activities of CD38, while cysteine-ADP-ribosylation resulted in the inactivation of hydrolase activity but not the cyclase activity [22]. To test whether arginine-specific ADRPT played an essential role in the induction of apoptosis, B cells were incubated with NAD and arginine-specific ADPRT purified from splenocytes, and the results showed that the cells underwent apoptosis after incubation with NAD, similar to T cells. This result suggested that the ADP-ribosylation of an arginine residue(s) on CD38 played a critical role in the induction of apoptosis, and that this was due to reduced cyclase activity, which most likely resulted in reduced intracellular $[cADPR]_i$. As expected, prior incubation of T cells with NAD significantly reduced $[cADPR]_i$, compared to the cells not treated with NAD. Taken together, these results suggest that arginine-specific ADP-ribosylation is responsible for apoptosis through the inhibition of the cyclase activity of CD38 and blockade of intracellular cADPR production, consequently resulting in the decrease of intracellular free Ca^{2+}.

SITE OF ADP-RIBOSYLATION OF CD38

Site-specific mutagenesis of CD38 showed that Arg 269 was the only acceptor of ADPR [22]. Furthermore, R269K mutant-transfected Hela cells retained WT levels of ADP-ribosyl cyclase; however, they could not serve as a substrate for the Arg-ADPRT and also did not undergo any apoptosis by NAD treatment. This result indicated that ADP-ribosylation of CD38 was an important signal leading to apoptosis.

AVAILABILITY OF EXTRACELLULAR NAD FOR ADP-RIBOSYLATION OF CD38

It is imperative that NAD, an intracellular metabolite, should be available for extracellularly occurring ADP-ribosylation of CD38 to have any physiological relevance, since extracellular NAD concentrations have been reported to be submicromolar in the plasma of mammals including mice [27], whereas micromolar concentrations are likely to be needed for effective ADP-ribosylation of CD38 in the extracellular compartment. However, Bruzzone *et al.* [26] recently reported that cytosolic NAD can be released from fibroblasts through hexameric connexin 43 hemichannels, which supports the relevance of our observations. Furthermore, conditions such as massive cell lysis during inflammatory immune reactions also increase plasma NAD concentration, consequently enhancing the ADP-ribosylation of CD38.

EFFECT OF ADP-RIBOSYLATION ON CD38 SIGNALING

The ligation of CD38 by agonistic CD38 antibodies has earlier been shown to elicit cellular responses, including elevation in cytoplasmic Ca^{2+} due to Ca^{2+} influx [28]. Also, Guse *et al.* [29] recently demonstrated sustained increases in $[cADPR]_i$ and Ca^{2+} entry by the stimulation of the TCR/CD3 complex in T lymphocytes. The mechanism by which cADPR regulates Ca^{2+} entry has been suggested to be similar to that of inositol 1,4,5-trisphosphate in the capacitative Ca^{2+} model, because cADPR mediated depletion of the Ca^{2+} pool by releasing Ca^{2+} from its target store [29]. Indeed, preincubation of WT-transfected HeLa cells with NAD completely abolished the CD38 ligation-induced increase of cADPR production and $[Ca^{2+}]_i$, whereas cells without NAD treatment had increased $[cADPR]_i$ and $[Ca^{2+}]_i$ [22]. The increase in the $[Ca^{2+}]_i$ by CD38 ligation was most likely due to Ca^{2+} influx, since the increase was completely abrogated by EGTA, a Ca^{2+} chelator. In contrast, pretreatment of R269K mutant-transfected cells with NAD could not counteract the CD38 ligation-induced increases in cADPR production and Ca^{2+} influx [22]. These findings were also observed in the WT- and R269K mutant-transfected Jurkat T lymphocytes. Taken together, these results indicated that ADP-ribosylation of CD38 blocked intracellular cADPR production, which subsequently resulted in the decrease of intracellular free Ca^{2+} and apoptosis.

ADP-RIBOSYLATION OF CD38 IN PANCREATIC ISLET CELLS

Similar to T cells, pancreatic islets showed that ADP-ribosylation of CD38 by GPI-anchored cell surface ADPRT resulted in inactivation of both the cyclase and hydrolase activities of CD38 [30]. In order to exclude the possibility that even such a small number of, if any, infiltrated T cells might have been responsible for the ADP-ribosylation of CD38, we removed T cells in the single cell suspensions of islets, and ADPRT activities in the islet cells were measured before and after the removal of T cells. The findings [30] that specific activities of ADPRT in the suspensions before and after the depletion of T cells were the same and that the T cell-depleted islet cells were CD38-positive suggested that ADP-ribosylation of CD38 was not due to ADPRT of infiltrated T cells but rather due to the islets' co-expressed ADPRT.

In addition to the above, glucose–induced insulin secretion was completely blocked by the treatment of the islet cells with NAD [30]. This suggested that ADP-ribosylation of CD38 played a critical role in the suppression of insulin secretion and that this was possibly due to reduced cyclase activity, resulting in the reduction of $[cADPR]_i$ and $[Ca^{2+}]_i$. Preincubation of the cells with NAD significantly reduced $[cADPR]_i$ and $[Ca^{2+}]_i$, compared to the cells not treated with NAD [30]. Taken together, these results suggested that the inhibition of the CD38 cyclase activity through ADPRT–mediated ADP-ribosylation decreased intracellular cADPR production, consequently resulting in the decrease of intracellular free Ca^{2+} in the islet cells and the blockade of glucose–induced insulin secretion.

COMPARISON OF THE EFFECTS OF ADP-RIBOSYLATION OF CD38 BETWEEN T CELLS AND PANCREATIC ISLET CELLS

Even though ADP-ribosylation of CD38 resulted in apoptosis in activated T cells, it was not likely that ADP-ribosylation of CD38 in islet cells resulted in apoptosis, since these two cell systems are different in their CD38 expression patterns and roles; CD38 expression in T cells can be induced by activation triggered through specific antigen receptors, whereas CD38 in islet cells is constitutively expressed. The role of CD38 in T cells is for proliferation when antigen receptors are triggered to express CD38 and for death when CD38 is ADP-ribosylated and inactivated. In contrast, however, the role of CD38 in islet cells may be to produce intracellular cADPR to mobilize Ca^{2+} for insulin secretion. Therefore, the CD38 modification down-regulates insulin secretion. Consistent with our data, some patients with type 2 diabetes were found to have a point mutation on the CD38 gene or circulating auto-antibodies against CD38, both of which resulted in reduction of the enzyme activity [31, 32]. This indicated the essential role of CD38 in insulin secretion of islet cells. Our results that an inactivation of ecto-enzyme activities of CD38 by its covalent modification results in down regulation of insulin secretion suggest a possibility that a permanent inactivation of CD38 can be a contributory factor to the cause of diabetes. From the above consideration, it is highly possible that accelerated cell destruction in such states as massive pancreatitis induces diabetes.

ACKNOWLEDGMENTS

We thank Dr. W.K. Paik, Department of Biochemistry, Ajou University, Korea, for his critical reading of the manuscript. This work was supported in part by grant [2000-2-20900-004-3] from Korea Science and Engineering Foundation and the Chonbuk National University Postdoctoral Fellow Program (to M.K.H.).

REFERENCES

1. Ueda K and Hayaishi O. 1985. ADP-ribosylation. *Annu. Rev. Biochem.* 54: 73-100.
2. Wreggett KA. 1986. Bacterial toxins and the role of ADP-ribosylation. *J. Recept. Res.* 6: 95-126.
3. Ludden PW. 1994. Reversible ADP-ribosylation as a mechanism of enzyme regulation in procaryotes. *Mol. Cell. Biochem.* 138: 123-129.
4. Yost DA and Moss J. 1983. Amino acid-specific ADP-ribosylation. Evidence for two distinct NAD:arginine ADP-ribosyltransferases in turkey erythrocytes. *J. Biol. Chem.* 258: 4926-4929.
5. Godeau F, Belin D, and Koide SS. 1984. Mono(adenosine diphosphate ribosyl) transferase in Xenopus tissues. Direct demonstration by a zymographic localization in sodium dodecyl sulfate--polyacrylamide gels. *Anal. Biochem.* 137: 287-296.
6. Peterson JE, Larew JS and Graves DJ. 1990. Purification and partial characterization of arginine-specific ADP- ribosyltransferase from skeletal muscle microsomal membranes. *J. Biol. Chem.* 265: 17062-17069.

7. Maehama T, Takahashi K, Ohoka Y, Ohtsuka T, Ui M and Katada T. 1991. Identification of a botulinum C3-like enzyme in bovine brain that catalyzes ADP-ribosylation of GTP-binding proteins. *J. Biol. Chem.* 266: 10062-10065.
8. Zolkiewska A, Nightingale MS and Moss J. 1992. Molecular characterization of NAD:arginine ADP-ribosyltransferase from rabbit skeletal muscle. *Proc. Natl. Acad. Sci. USA* 89: 11352-11356.
9. Tsuchiya M, Hara N, Yamada K, Osago H and Shimoyama M. 1994. Cloning and expression of cDNA for arginine-specific ADP- ribosyltransferase from chicken bone marrow cells. *J. Biol. Chem.* 269: 27451-27457.
10. Wang J, Nemoto E, Kots AY, Kaslow HR and Dennert G. 1994. Regulation of cytotoxic T cells by ecto-nicotinamide adenine dinucleotide (NAD) correlates with cell surface GPI-anchored/arginine ADP-ribosyltransferase. *J. Immunol.* 153: 4048-4058.
11. Greiner DL, Mordes JP, Handler ES, Angelillo M, Nakamura N and Rossini AA. 1987. Depletion of $RT6.1^+$ T lymphocytes induces diabetes in resistant biobreeding/Worcester (BB/W) rats. *J. Exp. Med.* 166: 461-475.
12. Mehta K and Malavasi F. 2000. Human CD38 and related molecules, *Chem. Immunol.* 75, Karger, Basel, Switzerland.
13. Kim H, Jacobson EL and Jacobson MK. 1993. Synthesis and degradation of cyclic ADP-ribose by NAD glycohydrolases. *Science* 261, 1330-1333.
14. Howard M, Grimaldi JC, Bazan JF, Lund FE, Santos-Argumedo L, Parkhouse RM, Walseth TF and Lee HC. 1993. Formation and hydrolysis of cyclic ADP-ribose catalyzed by lymphocyte antigen CD38. *Science* 262: 1056-1059.
15. Zocchi E, Franco L, Guida L, Benatti U, Bargellesi A, Malavasi F, Lee HC and De Flora A. 1993. A single protein immunologically identified as CD38 displays NAD^+ glycohydrolase, ADP-ribosyl cyclase and cyclic ADP-ribose hydrolase activities at the outer surface of human erythrocytes. *Biochem. Biophys. Res. Commun.* 196: 1459-1465.
16. Lee HC. 1997. Mechanisms of calcium signaling by cyclic ADP-ribose and NAADP. *Physiol. Rev.* 77: 1133-1164.
17. Wu Y, Kuzma J, Marechal E, Graeff R, Lee HC, Foster R and Chua NH. 1997. Abscisic acid signaling through cyclic ADP-ribose in plants. *Science* 278: 2126-2130.
18. Takasawa S, Nata K, Yonekura H and Okamoto H. 1993. Cyclic ADP-ribose in insulin secretion from pancreatic beta cells. *Science* 259: 370-373.
19. Liberman 1957. The mechanism of the specific depression of an enzyme activity in cells in tissue culture. *J. Biol. Chem.* 225: 883-898.
20. Green S and Dobrjansky A. 1971. pH-dependent inactivation of nicotinamide--adenine dinucleotide glycohydrolase by its substrate, oxidized nicotinamide--adenine dinucleotide. *Biochemistry* 10: 2496-2500.
21. Han MK, Lee JY, Cho YS, Song YM, An NH, Kim HR and Kim UH. 1996. Regulation of NAD^+ glycohydrolase activity by NAD^+-dependent auto-ADP- ribosylation. *Biochem. J.* 318: 903-908.
22. Han MK, Cho YS, Kim YS, Yim CY and Kim UH. 2000. Interaction of two classes of ADP-ribose transfer reactions in immune signaling. *J. Biol. Chem.* 275: 20799-20805.
23. Franco L, Guida L, Bruzzone S, Zocchi E, Usai C and De Flora A. 1998. The transmembrane glycoprotein CD38 is a catalytically active transporter responsible for generation and influx of the second messenger cyclic ADP-ribose across membranes. *FASEB J.* 12: 1507-1520.
24. Franco L, Zocchi E, Usai C, Guida L, Bruzzone S, Costa A and De Flora A. 2001. Paracrine roles of NAD^+ and cyclic ADP-ribose in increasing intracellular calcium and enhancing cell proliferation of 3T3 fibroblasts. *J. Biol. Chem.* 276: 21642-21648.
25. Han MK, Kim SJ, Park YR, Shin YM, Park HJ, Park KJ, Park KH, Kim HK, Jang SI, An NH and Kim UH. 2002. Antidiabetic Effect of a Prodrug of Cysteine, L-2-Oxothiazolidine-4- carboxylic Acid, through CD38 Dimerization and Internalization. *J. Biol. Chem.* 277: 5315-5321.

26. Bruzzone S, Guida L, Zocchi E, Franco L and De Flora A. 2001. Connexin 43 hemi channels mediate Ca^{2+}-regulated transmembrane NAD^+ fluxes in intact cells. *FASEB J.* 15: 10-12.
27. Kim UH, Kim MK, Kim JS, Han MK, Park BH and Kim HR. 1993. Purification and characterization of NAD glycohydrolase from rabbit erythrocytes. *Arch. Biochem. Biophys.* 305: 147-152.
28. Santos-Argumedo L, Teixeira C, Preece G, Kirkham PA and Parkhouse RM. 1993. A B lymphocyte surface molecule mediating activation and protection from apoptosis via calcium channels. *J. Immunol.* 151: 3119-3130.
29. Guse AH, da Silva CP, Berg I, Skapenko AL, Weber K, Heyer P, Hohenegger M, Ashamu GA, Schulze-Koops H, Potter BV and Mayr GW. 1999. Regulation of calcium signalling in T lymphocytes by the second messenger cyclic ADP-ribose. *Nature* 398: 70-73.
30. An NH, Han MK, Um C, Park BH, Park BJ, Kim HK and Kim UH. 2001. Significance of ecto-cyclase activity of CD38 in insulin secretion of mouse pancreatic islet cells. *Biochem. Biophys. Res. Commun.* 282: 781-786.
31. Ikehata F, Satoh J, Nata K, Tohgo A, Nakazawa T, Kato I, Kobayashi S, Akiyama T, Takasawa S, Toyota T and Okamoto H. 1998. Autoantibodies against CD38 (ADP-ribosyl cyclase/cyclic ADP-ribose hydrolase) that impair glucose-induced insulin secretion in noninsulin- dependent diabetes patients. *J. Clin. Invest.* 102: 395-401.
32. Yagui K, Shimada F, Mimura M, Hashimoto N, Suzuki Y, Tokuyama Y, Nata K, Tohgo A, Ikehata F, Takasawa S, Okamoto H, Makino H, Saito Y and Kanatsuka A. 1998. A missense mutation in the CD38 gene, a novel factor for insulin secretion: association with Type II diabetes mellitus in Japanese subjects and evidence of abnormal function when expressed in vitro. *Diabetologia* 41: 1024-1028.

Uh-Hyun Kim, Department of Biochemistry,
Chonbuk National University Medical School,
Chonju, 561-182 Korea.
e-mail: uhkim@moak.chonbuk.ac.kr

Cyclic ADP-Ribose and NAADP. Structures, Metabolism and Functions. Lee HC (ed), pp. 409-425, Dordrecht: Kluwer.

Retinoid-Mediated Signaling and CD38 Expression

Kapil Mehta
Department of Bioimmunotherapy, The University of Texas M.D. Anderson Cancer Center, 1515 Holcombe Blvd., Houston, TX 77030, USA.

INTRODUCTION

CD38 was initially identified more than two decades back when murine monoclonal antibodies (mAbs) were being employed to define newer surface molecules expressed by the lymphocytes during various developmental stages. One such mAb (T10) was successfully employed in the analysis of distribution [1] and biochemical characterization of CD38 antigen [2]. The conclusions derived from the biochemical analysis of the molecule were independently confirmed by cloning of a cDNA for human CD38 in 1990 [3]. On the basis of its expression and allocation among immune cells, CD38 was referred to as an "activation marker" and till recently was used for phenotyping the differentiation state of lymphocytes and classifying leukemias.

In late 1992 a report described significant molecular and functional homology of CD38 to an enzyme purified from the mollusk, *Aplysia californica* [4], which rejuvenated the interest of the scientific community in this molecule. CD38 was then found to be capable of functioning as a bi-functional ectoenzyme, catalyzing the synthesis and hydrolysis of compounds involved in the regulation of intracellular calcium [5]. These findings were further extended to cells such as erythrocytes [6] and platelets [7] that were previously considered to be non-CD38 expressers, and suggested that the molecule may be ubiquitously present albeit at low levels in certain cell types or under certain conditions. The molecule was also found in mouse and in rat, where it maintained high level of homology among the species and with the *Aplysia* enzyme [5]. Because CD38 turned out to be an ectoenzyme with multiple catalytic activities (NAD^+ glycohydrolase, ADP-ribosyl cyclase and cyclic ADP-ribose hydrolase), the biochemists working in the field raised the issue of topological paradox: how can an ectoenzyme operate in an environment where its substrate (NAD^+) is present only in trace amounts and whose product (cADPR) is used inside the cell? Immunologists, on the other hand, viewed CD38 as a cell-surface receptor with an intent that it may play a role in cell-to-cell communication by establishing contacts and communicating with neighboring cells.

The first evidence that CD38 antigen might play a role as a signaling molecule came from the observations that ligation of CD38 on peripheral blood mononuclear cells and T cell lines induced activation and proliferation signals [8]. Subsequent studies revealed that ligation of human CD38 with specific moAbs induces transcription of the cytokines IL-1, TNFα, interferon-gamma (IFNγ), and GM-CSF [9]. Studies with freshly isolated leukemia cells suggested that CD38 ligation could induce signals leading to the proliferation of leukemic cell clones [10]. These initial findings prompted the search for a natural ligand of CD38 and demonstration that CD31 (PE-CAM-1) can serve a counter-receptor for CD38 marked another important landmark in the field [11]. Proof for the receptorial nature of CD31 was revealed by its ability to reproduce the majority of the signaling events similar to those induced by agonistic anti-CD38 mAbs [12]. Thus, cross-talk between CD38 and CD31 induced increased cytosolic calcium levels, activation of the transcription factor NFκB, and release of cytokines [12].

The enzymatic and receptorial functions of CD38 are likely to influence certain biological processes by affecting cell growth, differentiation and apoptotic events. For example, the expression of CD38 during the life cycle of hematopoietic cells is considerably altered; immature precursor cells in the bone marrow have high expression, which declines as the cells differentiate to mature phenotype, and increases again on activated lymphocytes [13]. CD38 expression is also high on malignant hematopoietic cells. Moreover, the lymphocytes in neonates display much higher CD38 than do lymphocytes in adults. This distinct pattern of CD38 expression on hematopoietic cells suggests that this antigen may play a role in proliferation of immature cells and in the proliferation of certain leukemia clones [10, 14, 15]. Moreover, aberrant expression of CD38 has been linked to certain pathological conditions, such as diabetes [15-17], cancer [18-20], X-linked agammaglobulinemia, and human immunodeficiency virus infection [21]. In view of CD38's dual role in health and disease, it is imperative that we understand the molecular mechanisms that underlie the regulation of CD38 gene expression. The intent of this chapter is to inform the reader of our current understanding on transcriptional regulation of CD38. Special reference is made to retinoids (vitamin A and analogues), since they are the most potent inducers of CD38 expression, and their mechanism of action is relatively well understood.

RETINOIDS AS MODULATORS OF TRANSCRIPTION

Table 1. tRA-induced expression of genes during granulocytic differentiation.

Target gene	RARE in the promoter	Function
CD38	Yes	Inflammation, cell growth
C/EBPα	?	Differentiation
CIP/WAF/p21	Yes	Cell cycle inhibitor
Transglutaminase	Yes	Apoptosis
CD11b	Yes	Adhesion
Myeloperoxidase	Yes	Pro-inflammatory
TRAIL	?	Apoptosis

Retinoids in general and all trans-retinoic acid (tRA) in particular, play important physiological functions by regulating organogenesis, organ homeostasis, and cell growth, differentiation and death [22]. The ability of retinoids to influence these complex biological processes resides in their ability to modulate the expression of multiple genes (Table 1). Some of these genes are expressed immediately following treatment with tRA while the others are delayed. The net effect of tRA-induced gene expression is the inhibition of cell growth and induction of differentiation or apoptosis. tRA-induced expression of the target genes is mediated through the nuclear receptors that belong to a superfamily of ligand-inducible transcription factors including steroid, vitamin D, thyroid hormone, peroxisome proliferator-activated receptor and orphan receptors, whose functions are still not known [23]. Specifically, retinoid ligands can bind and activate two major types of nuclear receptors, the retinoic acid receptors (RARs) and retinoid X receptors (RXRs). Both, the RARs and RXRs consist of three subtyptypes (α-, β- and γ-) that are encoded by separate genes and have been discussed in great detail in a recent review [24].

Each of the RAR and RXR subtypes contains six distinct domains referred to as, A through F, and each of these domains has diverse functions (Fig. 1A). A and B domains are located at the amino terminal and contain isoform specific, ligand-independent transactivation functions, AF-1. The DNA-binding domain (domain C) is highly conserved among different retinoid receptor isoforms and binds to specific upstream nucleotide sequences (referred to as retinoic acid response element or RARE), located in the promoter region of the target genes. (Fig. 1). Ligands (retinoids) bind to a ligand-binding domain (LBD) or E domain at the C-terminus of the receptors (Fig, 1C) that contain sequences involved in dimerization of the receptors, ligand dependent transactivation (AF-2), and translocation to the nucleus. The functions for D and F domains have not been clearly defined. RARs and RXRs mainly act as heterodimers to affect gene transcription after binding to RAREs in the promoter region of the target genes (Fig. 1B). The RARs can be transcriptionally activated by binding to either tRA or 9-cis-RA, however RXRs can be activated only by 9-cis-RA (FIG. 1C). Upon ligand binding, activated nuclear receptors that bind to RAREs found in the upstream sequences (promoters) of RA responsive genes lead to transcription of the target genes.

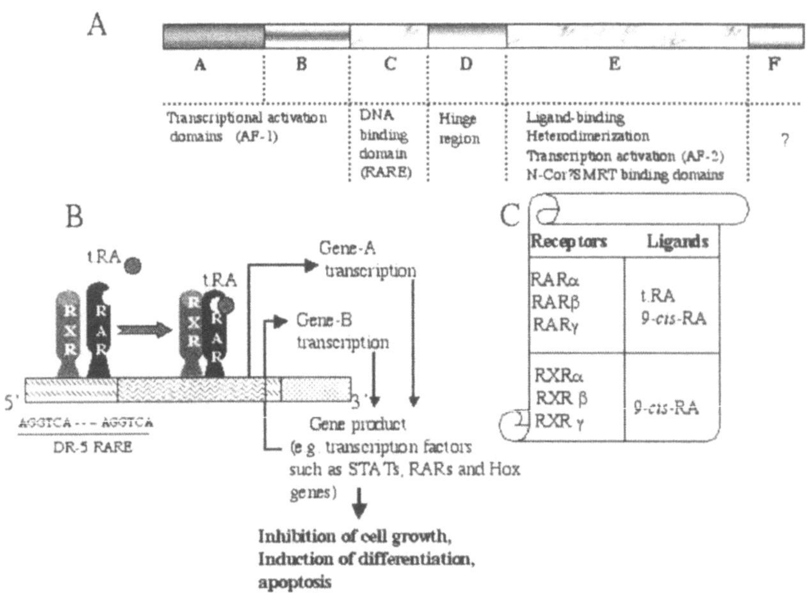

Figure 1. **Structure and functions of retinoid receptors. A**). Schematic representation of retinoid receptor depicting various functional domains. **B**). A molecular model for retinoid action. The ligated retinoid receptor forms heterodimer with RXR, binds to specific regulatory sequences (RARE) in the promoter region of a target gene A. Transactivation of gene A is a primary event of retinoid action. In an another scenario, the product of gene A can itself activate the transcription of a second gene B. Transactivation of gene B represents a secondary action of retinoid since its transcription requires the synthesis of protein A. This cascade of gene events leads to secondary and tertiary events that eventually produce a phenotype that is characteristic of retinoid action. **C**). Natural isoforms of retinoid receptors (RARs and RXRs) and their physiological ligands are listed.

The complex diversity and pleiotropic effects in tRA signaling pathway is provided not only due to existence of multiple isoforms of RA receptors but also as a result of combinations of RAR-RXR heterodimers and presence of different ligands [24]. tRA treatment, for example, can induce the expression of RARα and RARβ genes, suggesting that tRA can modulate its own receptors, in addition to differentiation-related genes [25]. The availability of the retinoid ligands to its cognate receptors can be regulated by the level of certain non-receptor proteins, such as cytoplasmic retinoic acid binding proteins (CRABP) and heat shock proteins [26]. The formation of a RXR/RAR heterodimer is required for the high-affinity binding to RAREs that is critical for subsequent retinoid-induced transcription of target genes, such as CD38. Most RAREs have been identified in the regulatory regions of genes whose transcription is induced by retinoids. These cognate response elements consist of direct repeats (DR) of 2 core motifs in the sequence AGGTCA(X)nAGGTCA. The most common spacing observed in RAREs is 5 bp (DR5); however, RAREs containing these or similar motifs separated by 1 (DR1) or 2 bp (DR2) are also common.

As noted above, the RAR-RXR heterodimer constitutes a high affinity receptor that can bind to the specific regulatory elements of target genes. However, in the absence of ligands for RAR/RXR dimer, or in the presence of some antagonists, the receptor's target genes are repressed. This is due to the recruitment of histone deacetylase (HDAC) containing complexes that

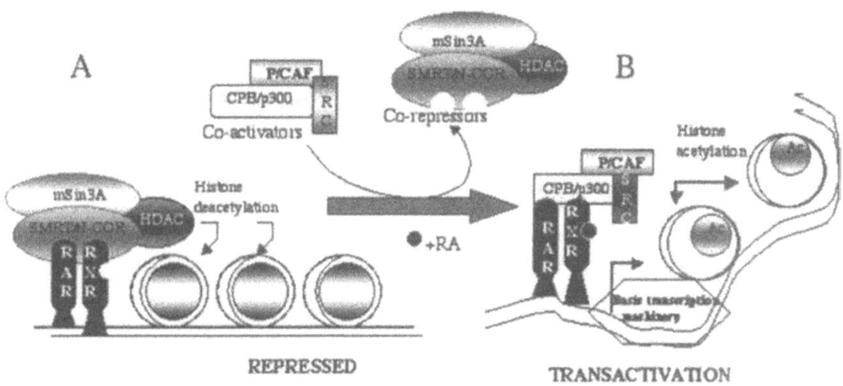

Figure 2. **Schematic representation of the basal repression and retinoid-induced transcriptional regulation of the target gene(s).** (**A**) The RAR/RXR heterodimer binds to a specific direct-repeat regulatory element that is separated by 5 bp (DR5-RARE) in the regulatory site of the target gene. In the absence of ligand (tRA), the RAR/RXR heterodimer recruits a transcriptional repressor complex that comprises N-CoR, mSin3a, SMRT, and a histone deacetylase (HDAC-1). The HDAC-1 activity helps suppress transcriptional activity. (**B**) Addition of tRA results in a conformational change in the RAR/RXR heterodimer that leads to release of the repressor complex and recruitment of a transcriptional-activator complex that has histone acetyltransferase (HAT) activity. That activity destabilizes nucleosomes and creates a permissive state for promoter activation.

are tethered through corepressors, such as the silencing mediator for retinoid and thyroid hormone receptors (SMRT) or the nuclear receptor co-repressor (NCoR) to a non-liganded RAR-RXR dimmer (Fig. 2). This process results in histone deacetylation, chromatin compaction and silencing of target-gene promoter regions. Current understanding of RAR-mediated gene transactivation is illustrated in Figure 2. The binding of RAR ligand (e.g. tRA) destabilizes the corepressor binding and induces allosteric changes in the ligand binding domain. This results in the release of the repressor-protein complex and the subsequent recruitment of transcriptional coactivator complexes (containing CREB-binding protein [CBP]/p300 or other co-activators) that contain or recruit histone acetyltransferase (HAT) [22, 27-29]. Thus, the binding of ligand to the RAR/RXR dimer converts it from a transcriptional repressor (Fig. 2A) to a transcriptional activator (Fig. 2B). Evidently, modulation of transcriptional activity involves close links between acetylation/deacetylation and stabilization/destabilization of repressive chromosomes.

Retinoid-Mediated Regulation of CD38 Expression

The expression of the CD38 gene is physiologically regulated during hematopoiesis. A significant proportion of $CD34^+$ normal bone marrow precursor cells express CD38 antigen, whereas the expression is downmodulated or even lost once these cells differentiate into mature polymorphonuclear cells [30]. The human myeloblast cell line HL-60 offers a good model for studying the regulatory mechanisms of myeloid cell differentiation. This cell line can be terminally differentiated into mature neutrophils following their treatment with tRA. tRA-differentiated HL-60 cells, like normal neutrophils, have a limited in vitro life span and can be further induced to undergo apoptosis in response to certain stimuli [31]. tRA-induced differentiation of HL-60 cells is associated with a large increase in CD38 expression. In addition to HL-60 cells, other myeloid leukemia cell lines exhibit a similar increase in CD38 expression in response to tRA treatment [32]. As early as 1982, Hemmi and Brietman [33] observed that tRA-induced differentiation of HL-60 cells was associated with an increased ability of these cells to hydrolyze NAD^+, which was later shown to result from an increase in membrane enzyme that could prevent ADP-ribosylation reactions. More recent work on these lines suggested that tRA-induced increase in NAD-hydrolase activity was due to the induction of CD38 [34-37]. tRA is quite unique and selective in its ability to induce the expression of CD38, in that at picomolar concentrations an appreciable accumulation of mRNA and protein can be observed [32]. Moreover, t-RA-induced increase in CD38 expression is rapid and was not observed in response to other differentiation-inducing agents such as dimethylsulfoxide

(DMSO), G-CSF, GM-CSF, IFN-γ, and 12-O-tetradecanoylphorbol-13-acetate (TPA) [35]. Also, tRA could modify the catalytic functions of CD38 as a result of posttranslational cross-linking of CD38 into a high-molecular-weight form, p190 [36], which was determined to be a transglutaminase-catalyzed cross-linked product of native membrane-bound CD38. The expression of transglutaminase in myeloid cells has also been suggested to be under the control of retinoids [38, 39]. These results imply that retinoids can modulate CD38 antigen expression by functioning at the transcriptional as well as posttranslational level.

Since retinoids can intercede CD38 gene transactivation via 6 different subtypes of nuclear receptors (RARα, -β, -γ or RXRα, -β, -γ), next we delineated the nature of retinoid receptor(s) involved in CD38 regulation. We used an HL-60 cell line variant (HL-60R) in which the retinoid receptor function has been abrogated as a result of trans-dominant negative mutation. HL-60R cells were resistant to tRA-induced CD38 expression. However, retrovirally mediated transduction of cDNA coding for full-length functional RARα in HL-60R cells, fully restored this ability [35]. In contrast, transduction with cDNA coding for RARβ, RARγ, or RXRα failed to restore their responsiveness to tRA. These observations suggested that RARα may play a critical role in tRA-induced expression of CD38 in HL-60 cells. To further confirm the involvement of RARα, we used several synthetic

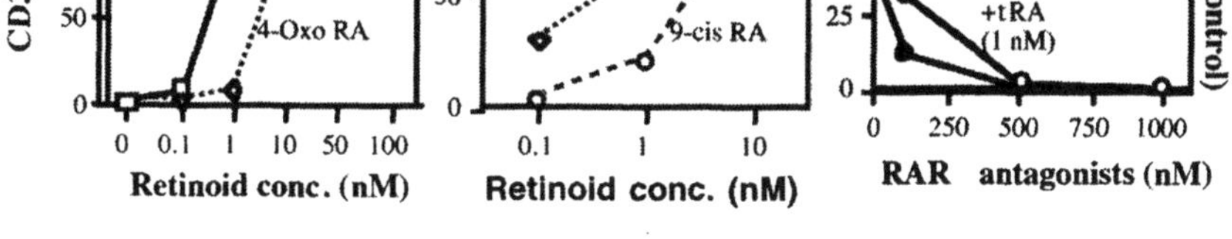

Figure 3. **RARα-selective driven expression of CD38 in myeloid leukemia HL-60 cells.** The ability of a particular retinoid to induce the expression of CD38 is dependent on its affinity for RARα. For example, tRA is about 10 times more active than 4-oxo RA (A). Similarly, the synthetic retinoid TTNB that has much higher affinity for RARs than for RXRs is more potent inducer of CD38 expression (B). Conversely, 3-methyl-TTNB has higher affinity towards RXRs than RARs and accordingly is less active in inducing CD38 expression (B). Similarly, 9-cis RA, a panagonist that can bind and activate all the RARs and RXRs, was least active (B). Interestingly, tRA and 4-oxo-RA that bind all three isoforms of RARs (α, β, and γ), were completely inactive when co-cultured in the presence of RARα selective antagonist (C), suggesting that retinoid-mediated expression of CD38 is completely dependent on RARα nuclear receptors.

retinoid analogues that selectively bind to and activate one of the six retinoid receptors. Indeed, the RARα-selective retinoids were highly potent inducers of CD38 expression (Figs. 3A and B; [40]); RARβ, -γ or RXRs-selective retinoids were relatively inactive in this regard. Moreover, RARα-specific antagonists, which bind but do not transcriptionally activate the receptor, completely blocked tRA-induced expression of CD38 in HL-60 cells (Fig. 3C). The significance of RARα in regulating CD38 gene expression was further confirmed in a study of RARα-antisense transgenic mice [37]. In these mice, tissues that expressed an antisense construct of RARα produced little or no CD38 and expressed 50-80% less RARα protein than did the same tissues from normal control mice. These results suggested that regulation of CD38 antigen, both in vitro and in vivo, is under the direct control of RARα subtype of nuclear receptors.

The low basal expression of CD38 antigen on APL blasts provides further evidence for the involvement of RARα in regulation of CD38 expression. In the t(15;17) variant of APL, the RARα locus is fused to that of the PML gene [25, 27], which results in the production of polymorphonuclear cells with abnormal granulation and CD38 expression. The basal expression of CD38 antigen on blast cells from patients with APL is very low (signal to noise ratio, 2.8 + 0.4) compared to the blasts cells from non-APL patients (S/N ratio, 58.6 + 8.2), which harbor full-length functional RARα. Treatment of APL cells with tRA, in vitro or in vivo, however, strongly induces CD38 expression [32, 35], suggesting that defective RARα-signaling pathways may attenuate basal CD38 expression in APL cells.

Retinoid-induced expression of CD38 protein depends on transcriptional regulation of the gene. Thus, accumulation of CD38 mRNA transcript in tRA-treated HL-60 cells could be blocked by actinomycin D but not by cyclohexamide. In an attempt to locate the RARE, Katada's group in Japan isolated and sequenced the genomic DNA ranging from the 5'-flanking region of CD38 gene to exon 1 and partial intron 1. Transient transfection experiments revealed that the responsiveness to tRA was conferred through a DR5 element consisting of 2 direct-repeat TGACCT-like hexamer motifs located at base pairs 437 to 453 of the first intron rather than in the 5'-upstream region of the CD38 gene [41]. On gel shift assays, the DR5 element interacted with a heterodimer composed of RAR and RXR. These results provided direct evidence that tRA-induced expression of CD38 is mediated by direct transcriptional regulation via activation of a RAR/RXR heterodimer through a RARE located in the first intron of the CD38 gene.

As discussed earlier in this chapter, retinoid receptors like other nuclear receptors, can regulate the basal and ligand-induced gene expression by interacting with a select group of proteins called corepressors and coactivators (Fig. 3). The 2 transcriptional corepressors that interact with the

RAR/RXR dimer include the 270-kDa N-CoR protein and the 168-kDa SMRT protein. Both of these corepressors can interact directly with retinoid receptors as well as with another corepressor, Sin3A. Furthermore, histone deacetylase (HDAC1) has been found to interact with Sin3A and SMRT to form a multisubunit repressor complex at the regulatory site of the target gene, which suppress basal transcriptional activity (Fig. 2A). Binding with the tRA ligand produces conformational changes in the RAR/RXR complex that result in the release of the repressor complex and the recruitment of another set of auxiliary proteins called transcriptional coactivators (Fig. 2B). Coactivators such as CBP (CREB-binding protein) or its homologue p300 either have intrinsic HAT activity or can associate with other HATs such as p/CAF. Recruitment of HATs to the DNA thus locally destabilizes nucleosomes, creating a permissive state for promoter activation. To gain insight into the possible role of HDAC in retinoid-dependent transcription of the CD38 gene, we studied the effect of trichostatin A (TSA), a specific inhibitor of HDAC, on HL-60 cells. TSA induced the expression of CD38 transcripts in these cells [40], suggesting that the CD38 gene in its basal state can stay silent due to compaction of the DNA caused by HDAC-catalyzed deacetylation.

Moreover, the 5' flanking region of the CD38 gene contains GC-rich sequences without classic TATA or CCAAT boxes [42], a feature that is usually shared by noninducible housekeeping genes [43]. Furthermore, in the upstream region (promoter) of the CD38 gene, potential binding sites for some common transcription factors, such as PEA-3, CP-2, and PuF have been identified [42]. Sites for T-cell transcription factor 1-alpha (TCF-1α), nuclear factor interleukin-6 (NFIL-6), and interferon responsive factor 1 (IRF-1) have also been identified [42]. The presence of an estrogen response half-palindromic element and a glucocorticoid response element suggests that the CD38 gene may also be hormonally regulated. Similarly, the 5' end of the human CD38 gene consists of a CpG-rich island. This island is approximately 900 bp long and encompasses exon 1 and the 5' end of intron 1. Within the intronic portion of the island, there are two binding sites for the SP1 transcription factor [42]. The functional significance of CpG island is that hypermethylation of this island may silence the expression of the gene. CD38 is likely to depend on the methylation or demethylation status of these regulatory sequences. Methylation of the CpG island generally represses gene expression, whereas demethylation induces it [34]. All these observations suggest the complex nature of CD38 gene regulation that may affect its expression during various physiological and pathological conditions.

Other Modulators of CD38 Expression

Table 2 lists some physiological and pharmacological agents that have been shown to regulate the expression of CD38. For example, cytokines, such as interferon-γ (IFN-γ) and IFN-α are known to induce CD38 expression in resting mature B cells and myeloma cells. In one study, IFN-α could upregulate CD38 expression on activated B cells, monocytes and the monocytic leukemia cell line THP-1 [44], but other cytokines (IL-1, IL-2, IL-3, IL-4, IL-5, IL-6, IL-7, IL-10, IL-12, IL-13, granulocyte-macrophage colony stimulating factor (GM-CSF), and tumor necrosis factor alpha [TNF-α]) were completely ineffective. Similarly, up-regulation of CD38 was observed in monocytes and monocytic cell lines in response to IFN-α treatment whereas TNF-α, GM-CSG, and lipopolysaccharide were completely ineffective [45]. In support to these findings is the observation that the regulatory region of CD38 gene contains an interferon-responsive element [42].

Table 2. Agents known to induce CD38 antigen expression

Retinoids
HL-60 cell line
KG-1 cell line
$CD34^+$ progenitors
Myeloid leukemia cells (APL, AML)
Myeloma cell line
Lymphoma cell line
T- and B-cells
Vitamin D3
T- and B-cells
HL-60 cell line
Interferons
Activated T- and B-cells
Myeloma cell lines
$CD34^+$ progenitors
Monocytic cell lines and monocytes
Cyclic-AMP
T-cells
Promonocytic cell lines
Rotenone and arsenic tetraoxide
HL-60 cell line

A metabolite of vitamin D, vitamin D3 (1-25 α-dihydroxyvitamin D3) is known to induce monocytic differentiation in the human myeloblastic leukemia cell line HL-60. Like retinoids, vitamin D also exerts its biological effects by interacting and activating a distinct class of nuclear receptor, called vitamin D receptor (VDR). In one study, vitamin D3 induced CD38 expression in HL-60 cells as well as in mature tonsillar B cells and peripheral T cells [44]. Similarly, rotenone, a mitochondrial NADH dehydrogenase inhibitor, can also induce expression of CD38 in HL-60 cells [46]. β-estradiol, an archetypal steroid hormone, has been shown to induce ADP ribosyl-cyclase activity in its target organ, the uterus, and thus to enhance the synthesis of cADPR from NAD^+ [47]. However, it is not known whether the β-estradiol-induced increase in cyclase activity is due to an increase in CD38 expression. Nevertheless, the genomic action of estrogens could be augmented by the release of calcium triggered by cADPR, which would enhance calcium-dependent contractile responses of myometrial smooth muscle cells as well as activate the calcium- dependent phosphorylation of nuclear transcription factors. Other ligands that act on nuclear receptors, namely tRA and triiodothyronine, could also induce ADPR-cyclase activity in renal cell carcinoma cell lines without an increase in CD38 protein expression.

CLINICAL IMPLICATIONS OF RETINOID-INDUCED CD38 EXPRESSION

Retinoid-Induced CD38 Expression and Retinoic Acid Syndrome

The clinical outcome of acute promyelocytic leukemia (APL) has been dramatically altered by the discovery of the striking differentiating potential of tRA. tRA treatment induces complete remission in >90% of APL patients. Although tRA therapy is well tolerated by patients, two problems frequently occur. The first problem is leukocytosis, reflecting an increase in peripheral leukocytes to >20,000 cells/mm3 that occurs in about half of the APL patients treated with tRA. The second, more serious problem is a condition called, retinoic acid syndrome (RA syndrome), which develops in about 20 - 30% APL patients treated with tRA. Early signs of this problem are characterized by fluid retention, weight gain, hectic fever and musculoskeletal pain. Later signs include progressive respiratory distress, pulmonary infiltrates, pleural effusions, renal insufficiency, skin infiltrates, hypotension and death [25]. It is postulated that RA syndrome is secondary to the aberrant interactions taking place between differentiating APL cells and host tissues. Indeed, the autopsies of patients who died of the syndrome have revealed extensive organ infiltration by differentiating APL cells.

Because the clinical signs of RA syndrome closely resemble those of endotoxin shock and acute respiratory distress syndrome (ARDS), it has been suggested that tRA treatment may affect in some way the cytokine expression by differentiating granulocytes. Although the pathobiology of ARDS is poorly understood, several observations suggest an important role for interaction between polymorphonuclear neutrophils (PMN) and cytokines in this process. High levels of IL-1, TNF-α, IL-8, IL-6, and macrophage inflammatory protein-2 (MIP-2) have been found in plasma and in bronchoalveolar lavage supernatant of patients with ARDS [48]. Induction of IL-1 and granulocyte-colony stimulating factor (G-CSF) like molecules under tRA influence may contribute towards hyperleukocytosis. Whereas, secretion of IL-6, IL-8, and TNF-α like molecules, all of which can activate leukocytes and are implicated in the development of ARDS, can induce lung injury observed in RA syndrome-related pathogenesis. Since RA syndrome occurs only in patients with APL who have been treated with tRA, this syndrome is considered to reflect an aberrant interaction between maturing granulocytes and host tissues.

As discussed in the previous section, tRA-treated APL cells express large amounts of CD38 antigen; it is therefore likely that maturing APL cells in

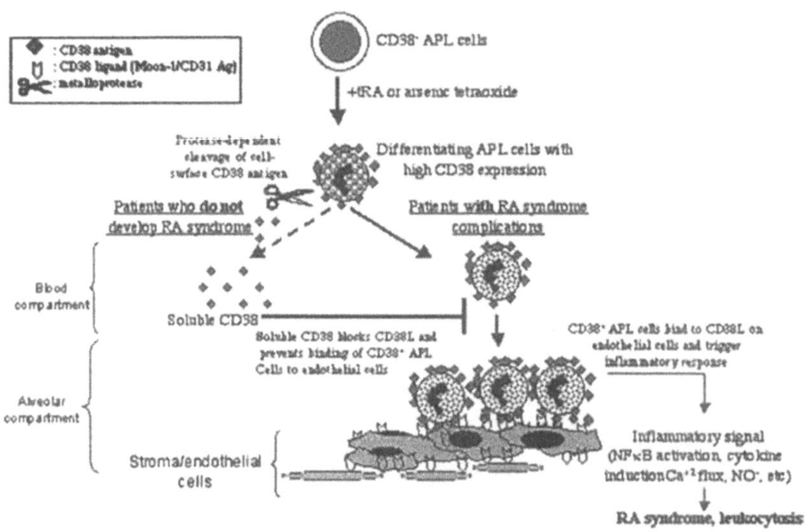

Figure 4: **Hypothetical model for RA-induced CD38 antigen's involvement in RA syndrome.**

tRA treated patients interact with CD31 antigen on the lung endothelial cells and trigger inflammatory signals. Human lung endothelial cells express high basal levels of CD31, a cell-surface protein that serves as a ligand for CD38

[11]. In support to this contention, our results suggested that tRA-treated leukemia cells accumulate large amounts of transcripts for inflammatory cytokines such as TNFα, IL-1β, IL-6, and IL-8 after co-culture with human lung endothelial cells. These cytokines and possibly other mediators produced in the alveolar interstitium may contribute to the pathogenesis of RAS. Possible CD38-mediated events that might lead to the development of RA syndrome are illustrated in Figure 4. Further delineation of these pathways may yield important benefits in designing effective strategies to control not only RA syndrome pathogenesis but also for evaluating the therapeutic superiority of tRA in combination with other chemotherapeutic agents, given tRA's ability to downregulate Bcl2 and perhaps other proteins that inhibit chemotherapy-induced apoptosis.

Retinoid-Induced CD38 as Target for Immunotherapy

The use of monoclonal antibodies for delivering drugs or toxins to molecular structures expressed on the surfaces of tumor cells represents an attractive and potentially useful strategy. Theoretically, such a targeted approach to cancer therapy could selectively eliminate tumor cells while reducing the toxicity of the treatment to normal, non-target tissues. Nevertheless, in practice many problems exist that need to be addressed before immunotoxin or antibody-drug therapies become practical. One limitation to the success of such targeted therapy is the heterogeneity of target antigen expression within a population of tumor cells. If a small number of cells within a tumor were negative for the target antigen or expressed the antigen only weakly, then these cells would escape destruction from the antibody-mediated delivery of an agent that is cytotoxic to those cells. One possible means of overcoming this problem would be to identify agents that induce high levels of cell surface target molecules, in the expectation that target tumor cells that were antigen-negative could be induced to express those target molecules in abundance. We have exploited the ability of tRA to induce high levels of CD38 antigen on leukemia cells and then used this antigen as a target for delivering the anti-CD38-conjugated plant toxin gelonin. Our preliminary results indicate that tRA treatment of leukemia cells, even at sub-nanomolar concentrations made those cells exquisitely sensitive to immunotoxin-induced killing. Co-culture of leukemia cells with as little as 0.1 nM tRA in the presence of picomolar concentrations of the immunotoxin killed more than 90% of the leukemia cells. In contrast, similar concentrations of the immunotoxin without tRA killed only 40-50% of the leukemia cells (Fig. 5). Since tRA is currently being used in the clinic for treating APL patients, its combination with anti-CD38-based therapies may offer new exciting opportunities for the treatment of certain leukemias.

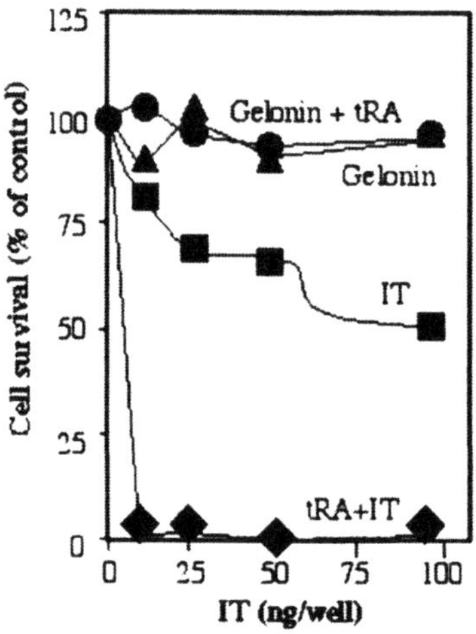

Figure 5. **tRA-induced CD38 expression augments anti-CD38-based delivery of the plant toxin (gelonin) to leukemia (HL-60) cells.** HL-60 cells were incubated with gelonin alone, gelonin plus tRA (5 nM), anti-CD38-conjugated gelonin (IT) or IT plus tRA for 48 h at 37^0 C. The cells surviving under various culture conditions were determined by MTT assay. As a control, the simultaneous presence of unconjugated anti-CD38 antibody (100-fold excess) completely abrogated IT-induced cytotoxicity in the presence of tRA (not shown).

CONCLUSION

CD38 has emerged out as a multifunctional protein from a simple cell-surface differentiation antigen title. Originally identified as a differentiation marker on hematopoietic cells, CD38 is now known to be present on nonhematopoietic tissues as well. The best-characterized feature of the CD38 antigen is its ability to catalyze the synthesis of cADPR, which serves as an important second messenger in regulating calcium release from intracellular pools. By virtue of its ability to synthesize cADPR as well as to serve as a signal transduction molecule, CD38 may participate in physiological processes as diverse as cell growth, apoptosis, differentiation, and inflammation. Equally interesting is the decline in CD38 expression on hematopoeitic cells as they mature and on normal prostate epithelial cells as they become cancerous. Little information is available on the factors that regulate the expression of CD38. In this chapter, I have attempted to summarize current knowledge with special reference on retinoic acid, which to my knowledge is the most well-characterized and potent positive regulators of CD38 expression. The retinoid-induced expression of CD38 is mediated via selective activation of RARα. The tRA-induced upregulation of CD38 in APL cells can be considered pathologic and may contribute to the development of RA syndrome and leukocytosis like conditions that frequently occur and are often fatal in patients treated with tRA. However, tRA-induced CD38 expression may offer new means for treating certain leukemias in combination with anti-CD38-based therapies.

ACKNOWLEDGMENTS

The work in author's laboratory was supported in part by grants from the U.S. Food and Drug Administration (FDR-000923) and Department of Defense (DAMD 17-1-1-0645). I wish to thank all the research colleagues working in my laboratory whose dedicated efforts made this publication possible. Also, I would like to acknowledge the invaluable help of several collaborators who made the reagents available for these studies.

REFRENCES

1. Reinherz EL, Kung PC, Goldstein G, Levey RH and Schlossman SF. 1980. Discrete stages of intrathymic differentiation: analysis of normal thymocytes and leukemic lymphoblasts of T lineage. *Proc. Natl. Acad. Sci. USA* 77: 1588-1592.
2. Terhorst C, van Agthoven A, Le Clair K, Snow P, Reinherz EL and Schlossman SF. 1981. Biochemical studies of the human thymocyte cell-surface antigens T6, T9 and T10. *Cell* 23: 771-778.
3. Jackson D and Bell JI. 1990. Isolation of a cDNA encoding the human CD38 (T10) molecule, a cell surface glycoprotein with unusual discontinuous pattern of expression during lymphocyte differentiation. *J. Immunol.* 144: 2811-2817.
4. States DJ, Walseth TF and Lee HC. 1992. Similarities in amino acid sequences of *Aplysia* ADP ribosyl cyclase and human lymphocyte antigen CD38. *Trends Biochem. Sci.* 17: 495.
5. Howard M, Grimaldi JC, Bazan JF, Lund FE, Santos-Argumedo L, Parkhouse RM, Walseth TF and Lee HC. 1993. Formation and hydrolysis of cyclic ADP ribose by lymphocyte antigen CD38. *Science* 262: 1056-1059.
6. Zocchi E, Franco L, Guida L, Benatti U, Bargellesi A, Malavasi F, Lee HC and De Flora F. A single protein immunologically identified as CD38 displays NAD^+-glycohydrolase, ADP ribosyl cyclase and cyclic ADP-ribose hydrolase activities in the auter surface of human erythrocytes. *Biochem. Biophys. Res. Commun.* 196: 1459-1465.
7. Ramaschi G, Torti M, Tolnai Festetics E, Sinigaglia F, Malavasi F and Balduini C. 1996. Expression of cyclic ADP-ribose-synthetizing CD38 molecule of human platelet membrane. *Blood* 87: 2308-2313.
8. Dianzani U, Funaro A, DiFranco D, Garbarino G, Bragardo M, Redoglia V, Buonfiglio D, DeMonte LB, Pileri A and Malavasi F. 1994. Interaction between endothelium and $CD4^+CD45RA^+$ lymphocytes. *J. Immunol.* 22: 952-959.
9. Ausiello CM, Urbani F, LaSala A, Funaro A and Malavasi F. 1995. CD38 ligation induces discrete cytokine mRNA expression in human cultured lymphocytes. *Eur. J. Immunol.* 25: 1477-1480.
10. Konopleva M, Estrov Z, Zhao S, Andreeff M and Mehta K. 1998. Ligation of cell surface CD38 protein with agonistic monoclonal antibody induces a cell growth signal in myeloid leukemia cells. *J. Immunol.* 161: 4702-4708.
11. Deaglio S, Morra M, Mallone R, Ausiello CM, Prager E, Garbarino G, Dianzani U, Stockinger H and Malavasi F. 1998. Human CD38 (ADP ribosyl cyclase) is a counter-receptor of CD31, an Ig superfamily member. *J. Immunol.* 160: 395-402.
12. Deaglio S, Mehta K and Malavasi F. 2001. Human CD38: a (r)evolutionary story of enzyme and receptor. *Leuk. Res.* 25: 1-12.
13. Funaro A, Ferrero E, Mehta K and Malavasi F. 2000. Schematic portrait of CD38 and related molecules. *Chem. Immunol.* 75: 256-273.
14. Podesta M, Zocchi E, Pitto A, Usai C, Franco L, Bruzzone S, Guida L, Bacigalupo A, Scadden DT, Walseth TF and De Flora A, Daga A. 2000. Extracellular cyclic ADP-ribose increases intracellular free calcium and stimulates proliferation of human hematopoietic progenitors. *FASEB Lett.* 14: 680-690.

15. Okamoto H, Takasawa S, Nata K, Tohgo A and Noguchi N. 2000. Physiological and pathological significance of the CD38-cyclic ADP-ribose signaling system. *Chem. Immunol.* 75: 122-145.
16. Antonelli A, Baj G, Marchetti P, Fallahi P, Surico N, Pupilli C, Malavasi F and Ferrannini E. 2001. Human anti-CD38 autoantibodies raise intracellular calcium and stimulate insulin release in human pancreatic islets. *Diabetes* 50: 985-995.
17. An NH, Han MK, Um C, Park BH, Park BJ, Kim HK and Kim UH. 2001. Significance of ecto-cyclase activity of CD38 in insulin secretion of mouse pancreatic islet cells. *Biochem. Biophys. Res. Commun.* 282: 781-786.
18. Ibrahim S, Keating M, Do KA, O'Brien S, Huh YO, Jilani I, Lerner S, Kantarjian HM and Albitar M. 2001. CD38 expression as an important prognostic factor in B-cell chronic lymphocytic leukemia. *Blood* 98: 181-186.
19. Damle RN, Wasil T, Fais F, Ghiotto F, Valetto A, Allen SL, Buchbinder A, Budman D, Dittmar K, Kolitz J, Litchman SM, Schulman P, Vinciguerra VP, Rai KR, Ferrarini, M and Chiorazzi N. 1999. Ig V gene mutation status and CD38 expression as novel prognostic indicators in chronic lymphocytic leukemia. *Blood* 97: 1840-1847.
20. Kramer G, Steiner G, Fodinger D, Fiebiger E, Rappersberger C, Binder S, Hofbauer J and Marberger M. 1995. High expression of a CD38-like molecule in normal prostatic epithelium and its differential loss in benign and malignant disease. *J. Urol.* 154: 1636-1641.
21. Bofill M and Borthwick NJ. 2000. CD38 in health and disease. *Chem. Immunol.* 75: 218-234.
22. Altucci L and Gronemeyer H. 2001. The promise of retinoids to fight against cancer. *Nature Rev.* 1: 181-193.
23. Collinwood TN, Urnov FD and Wolffe AP. 1999. Nuclear receptors: coactivators, corepressors and chromatin remodeling in the control of transcription. *J. Mol. Edocrinol.* 23: 255-275.
24. Chambon P. 1996. A decade of molecular biology of retinoic acid receptors. *FASEB J.* 10: 940-954.
25. Melnick A and Licht JD. 1999. Deconstructing a disease. RARalpha, its fusion partners, and their roles in the pathogenesis of acute promyelocytic leukemia. *Blood* 93: 3167-3215.
26. Boylan JF and Gudas LJ. 1991. Overexpression of the CRABP-1 results in a reduction in differentiation specific gene expression in F9 teratocarcinoma cells. *J. Cell Biol.* 112:965-979.
27. Ozpolat B, Lopez-Berestein G and Mehta K. 2001. ATRA(ouble) in the treatment of acute promyelocytic leukemia cells. *J. Biol. Regul. Homeost. Agents* 15: 107-122.
28. Glass CK, Rose DW and Rosenfeld MG. 1997. Nucelar receptors coactivators. *Curr. Opin. Cell Biol.* 9: 222-232.
29. Collingwood TN, Urnov FD and Wolffe AP. 1999. Nucelar receptors: coactivators, corepressors and chromatin remodeling in the control of transcription. *J. Mol. Endorinol.* 23: 255-275.
30. Malavasi F, Funaro A, Rogerro S, Horenstein A, Calosso L and Mehta K. 1994. CD38: A glycoprotein in search of a function. *Immunol. Today* 15: 95-97.
31. Mehta K, McQueen T, Neamati N, Collins S and Andreeff M. 1996. Activation of retinoid receptors RAR alpha and RXR alpha induces differentiation and apoptosis, respectively, in HL-60 cells. *Cell Growth Diffren.* 7: 179-186.
32. Drach J, Zhao S, Malavasi F and Mehta K. 1993. Rapid induction of CD38 antigen on myeloid leukemia cells by all trans-retinoic acid. *Biochem. Biophys. Res. Commun.* 195: 545-550.
33. Hemmi H and Brietman TR. 1982. Induction by retinoic acid of NAD-glycohydrolase activity of myelomonocytic cell lines HL-60, THP-1 and U937, and fresh human acute promyelocytic leukemia cells in primary culture. *Biochem. Biophys. Res. Commun.* 109: 669-674.

34. Kontani K, Nishina H, Ohoka Y, Takahashi K and Katada T. 1993. NAD-glycohydrolase specifically induced by retinoic acid in human leukemic HL-60 cells. Identification of the NAD-glycohydrolase as leukocyte cell surface antigen CD38. *J. Biol. Chem.* 268: 16895-16898.
35. Drach J, McQueen T, Engel H, Andreeff M, Robertson K, Collins SJ, Malavasi F and Mehta K. 1994. Retinoic acid-induced expression of CD38 antigen in myeloid cells is mediated through retinoic acid receptor-alpha. *Cancer Res.* 54: 1746-1752.
36. Umar S, Malavasi F and Mehta K. 1996. Post-translational modification of CD38 protein into a high molecular weight form alters its catalytic properties. *J. Biol. Chem.* 271: 15922-15927.
37. Mehta K, McQueen T, Manshouri T, Andreeff M, Collins S and Albitar M. 1997. Involvement of retinoic acid receptor-alpha-mediated signaling pathway in induction of CD38 cell-surface antigen. *Blood* 89: 3607-3614.
38. Mehta K and Berestein GL. 1986. Expression of tissue transglutaminase in cultured monocytic leukemia (THP-1) cells during differentiation. *Cancer Res.* 46:1388-1394.
39. Davies PJA, Murtaugh MP, Moore WT, Johnson GS and Lucas D. 1985. Retinoic acid-induced expression of tissue transglutaminase in human promyelocytic leukemia (HL-60) cells. *J. Biol. Chem.* 260: 5166-5174.
40. Mehta K. 2000. Retinoid-mediated signaling in CD38 antigen expression. *Chem. Immunol.* 75:20-38.
41. Kishimoto H, Hoshino S, Ohori M, Kontani K, Nishina H, Suzawa M, Kato S and Katada T. 1997. Molecular mechanism of human CD38 gene expression by retinoic acid. *J. Biol. Chem.* 273:15429-14434.
42. Ferrero E, Sccucci F and Malavasi F. 2000. The making of leukocyte receptor: origin, genes and regulation of human CD38 and related molecules. *Chem. Immunol.* 75:1-19.
43. Ernst P and Smale ST. 1995. Combinatorila regulation of transcription: general aspects of transcriptional control. *Immunity* 2: 311-319.
44. Stoeckler JD, Stoeckler HA, Kouttab N and Maizel AL. 1996. Dihydroxyvitamin D3 modulates CD38 expression on human lymphocytes. *J. Immunol.* 157: 4908-4917.
45. Musso T, Deaglio S, Franco L, Calosso L, Badolato R, Garbarino G, Dianzani U and Malavasi F. 2001. CD38 expression and functional activities are upregulated by IFN-γ on human monocytes and monocytic cell lines. *Leuk. Biol.* 69: 605-612.
46. Matsunaga T, Kudo J, Takahashi K, Dohmen K, Hayashida K, Okamura S, Ishibashi H and Niho Y. 1996. Rotenone, a mitochondrial NADH dehydrogenase inhibitor, induces cell surface expression of CD13 and CD38 and apoptosis in HL-60 cells. *Leuk. Lympho.* 20: 487-494.
47. Chini EN, de Toledo FG, Thompson MA and Dousa TP. 1997. Effect of estrogen upon cyclic ADP ribose metabolism: β-estradiol stimulates ADP ribosyl cyclase in rat uterus. *Proc. Natl. Acad. Sci. USA* 94: 5872-5876.
48. Abraham E, Carmody A, Shenkar R, Arcaroli J. 2000. Neutrophils as early immunologic factors in hemorrhage-induced acute lung injury. *Am. J. Physiol.* 279: 1137-1145.

Kapil Mehta, Department of Bioimmunotherapy, The University of Texas M.D. Anderson Cancer Center, 1515 Holcombe Blvd., Houston, TX 77030, USA.
e-mail: kmehta@mdanderson.org

Cyclic ADP-Ribose and NAADP. Structures, Metabolism and Functions. Lee HC (ed), pp. 427-449, Dordrecht: Kluwer.

Calcium Regulation in Smooth Muscle Through the CD38/Cyclic ADP-Ribose Pathway

Thomas A. White[1], Deepak A. Deshpande[1], Soner Dogan[1], Reynold A. Panettieri[3], Timothy F. Walseth[2] and Mathur S. Kannan[1]

[1]Department of Veterinary PathoBiology, College of Veterinary Medicine, University of Minnesota, St. Paul, Minnesota; [2]Department of Pharmacology, University of Minnesota, Minneapolis, Minnesota; [3]Pulmonary and Critical Care Division, Department of Medicine, University of Pennsylvania School of Medicine, Hospital of the University of Pennsylvania, Philadelphia, Pennsylvania.

INTRODUCTION

An elevation of intracellular Ca^{2+} is required for smooth muscle contraction [1,2]. Smooth muscle cells maintain a relatively low intracellular Ca^{2+} ($[Ca^{2+}]_i$) concentration under unstimulated conditions. Upon stimulation with hormones, neurotransmitters or other agents, there is a rapid elevation of $[Ca^{2+}]_i$ to a peak level, which then declines to a lower steady-state value [3,4]. The mechanisms underlying this "biphasic" pattern of $[Ca^{2+}]_i$ elevation have been the subject of intense investigation in different laboratories. There are two principal pathways by which elevation of $[Ca^{2+}]_i$ in smooth muscle cells occurs. Ca^{2+} influx from the extracellular space and Ca^{2+} release from the sarcoplasmic reticulum (SR) can both account for the elevation of $[Ca^{2+}]_i$ [5,6]. The extent to which either or both of these processes contribute to the elevation of Ca^{2+} varies depending on the activating mechanism and/or the smooth muscle cell type.

In most smooth muscle cells, Ca^{2+} release from the SR stores plays a key role in $[Ca^{2+}]_i$ regulation and excitation-contraction coupling [1,2,7]. In this regard, two distinct receptors and their associated Ca^{2+} release channels have been identified. One of these release channels is coupled to receptors in the SR membrane activated by the second messenger inositol 1,4,5-trisphosphate

(IP_3) [8] and the other to ryanodine receptors (RyR)[9-12]. Activation of most G-protein-coupled receptors in smooth muscle cells results in stimulation of phospholipase C (PLC), which catalyzes the hydrolysis of phosphatidyl inositol bisphosphate leading to formation of IP_3 [1,2,7]. IP_3 binds to the IP_3 receptors in the SR to cause Ca^{2-} release [13,14]. The mechanisms underlying IP_3-induced SR Ca^{2+} release and their regulation have been investigated in smooth muscle cells [15]. IP_3-induced SR Ca^{2+} release is regulated by Ca^{2+} and phosphorylation of the IP_3 receptor. Activation of the IP_3 receptor exhibits a bell-shaped dependence on Ca^{2+}, with lower concentrations of Ca^{2+} activating and higher concentrations inhibiting the channel [16]. Both cyclic AMP- and cyclic GMP-dependent protein kinases have been shown to regulate Ca^{2+} release through IP_3 receptor channels [17,18]. However, the mechanisms by which Ca^{2+} release occurs through RyR channels in smooth muscle are not entirely understood.

The investigations in our laboratory have centered largely on Ca^{2+} regulation in airway smooth muscle cells. In airway smooth muscle cells, previous investigations in our laboratory have shown that stimulation of muscarinic receptors by acetylcholine (ACh) causes a "biphasic" pattern of Ca^{2+} oscillations, when intracellular Ca^{2+} is measured within localized regions of the cell using the enhanced spatial and temporal resolution features of confocal microscopy [6]. These Ca^{2+} oscillations require Ca^{2+} release from the SR. IP_3-induced Ca^{2+} release is critical for the initiation of these Ca^{2+} oscillations, while Ca^{2+} release from the SR through RyR channels is critical for their maintenance. In airway as well as in other smooth muscle cells, there is now evidence that cADPR may mediate Ca^{2+} release through RyR channels [19-21]. This chapter will review existing literature regarding the role of the CD38/cyclic ADP-ribose (cADPR) pathway of Ca^{2+} regulation in airway, vascular, uterine, and intestinal smooth muscles. We will discuss evidence for regulation of CD38 expression in smooth muscles by hormones and inflammatory mediators. Finally, we will discuss the mechanisms by which cADPR causes Ca^{2+} release from the SR in different smooth muscles, based on available evidence.

cADPR METABOLISM IN SMOOTH MUSCLE

cADPR is a cyclized derivative of β-NAD. Conversion of β-NAD to cADPR is catalyzed by ADP-ribosyl cyclase and cADPR is degraded to the biologically inactive ADPR by cADPR hydrolase [22-25]. Both these enzyme activities are associated with a single bifunctional protein, CD38 in many mammalian cells [26,27]. CD38 is a type II transmembrane glycoprotein with an amino-terminal cytoplasmic tail of 23 amino acids, a transmembrane region of 21 amino acids and an extracellular region of 256 amino acids [28]. The membrane monomeric forms of CD38 are glycosylated proteins with a molecular weight of ~45-kDa. Unlike the CD38 of mammalian species, the ADP-ribosyl cyclase of the *Aplysia* ovotestis is a 29-kDa cytosolic protein that has no cADPR hydrolase activity [29,30]. To date, very few studies have specifically examined CD38 distribution, and ADP-ribosyl cyclase and cADPR hydrolase activities in smooth muscle. In this context, we have investigated the subcellular localization of CD38 in airway smooth muscle and measured ADP-ribosyl cyclase and cADPR hydrolase activities in different subcellular fractions [31].

The first evidence that cADPR has a role as a Ca^{2+}-mobilizing messenger in smooth muscle came from studies by Kuemmerle and Makhlouf [20]. They demonstrated that ryanodine-sensitive Ca^{2+} channels mediate agonist-induced Ca^{2+} mobilization in rabbit longitudinal smooth muscle of the intestine and that cADPR was capable of stimulating this release. In muscle extracts incubated with β-NAD^+, cADPR was produced in a time-dependent fashion, demonstrated by the ability of the extracts to cause Ca^{2+} release in permeabilized longitudinal smooth muscle cells. Identification of the agent in the extracts as authentic cADPR was provided by the ability of the cADPR antagonist, 8-amino-cADPR, to block this release. This was the first report suggesting ADP-ribosyl cyclase activity being present in a smooth muscle. In crude membrane preparations from rat uterine smooth muscle, Chini *et al* demonstrated the presence of ADP-ribosyl cyclase and cADPR hydrolase activities [32]. Conversion of β-NAD to cADPR was measured using a commonly used bio-assay for cADPR-induced Ca^{2+} release using sea urchin microsomes as well as directly by following the formation of the fluorescent product, cyclic GDP-ribose (cGDPR), from the NAD analog NGD [33]. Hydrolysis of cADPR to ADPR was followed using the sea urchin microsome bioassay. Later investigation from the same laboratory provided evidence for the presence of ADP-ribosyl cyclase activity in rat vascular smooth muscle cell membranes [34]. We have determined the presence as well as the subcellular localization of ADP-ribosyl cyclase and cADPR hydrolase activities in porcine tracheal smooth muscle [31]. In subcellular fractions obtained by differential and sucrose density gradient centrifugation of smooth muscle homogenates, the ADP-ribosyl cyclase and cADPR hydrolase activities were associated with a membrane fraction enriched in plasma membrane marker enzymes. There was no evidence for the presence of these activities in the cytosol or membrane fractions enriched in SR marker proteins. Plasma membrane enriched fractions incubated with NGD, produced time- and concentration-dependent increases in fluorescence due to formation of cGDPR (Figure 1). NGD was used as a substrate to measure ADP-ribosyl cyclase activity because the product, cGDPR, can be measured fluorometrically and the resistance of cGDPR to hydrolysis by cADPR hydrolase eliminates the confounding effects of cADPR hydrolysis on the assay [33]. cADPR hydrolase activities were determined in the same fractions by following the conversion of ^{32}P-cADPR to ^{32}P-ADPR using thin layer chromatographic separation and quantitation by phosphorimaging (Figure 2) [35]. Investigation of the kinetics revealed K_m values of 30.4 and 695.3 μM and V_{max} values of 330.4 and 102.8 nmol/mg/h for ADP-ribosyl cyclase and cADPR hydrolase activities respectively. Many other investigators have provided evidence for the enzymatic machinery responsible for the synthesis and degradation of cADPR in other smooth muscle types [20,32,34,36,37].

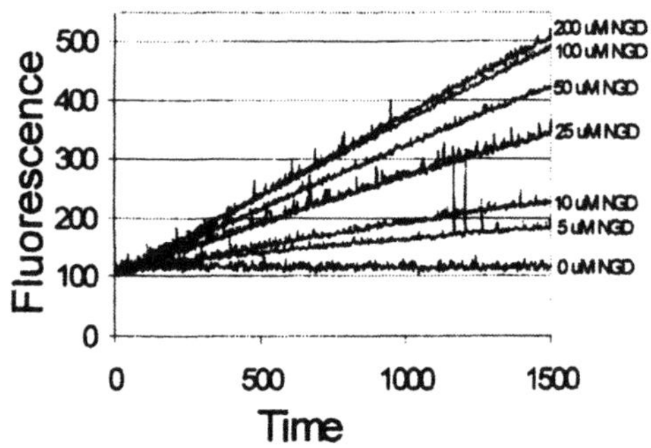

Figure 1. **The NGD assay.** Representative traces showing conversion of NGD to cGDPR measured fluorometrically at an excitation wavelength of 305 nm and emission wavelength of 410 nm. Note the increased fluorescence with time of incubation of membranes from tracheal smooth muscle (150 μg protein) as well as a concentration dependent increase in ADP-ribosyl cyclase activity.

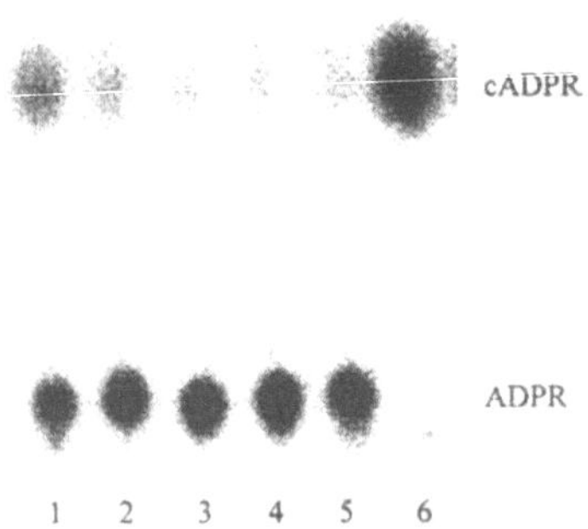

Figure 2. Phosphorimage of a TLC plate showing hydrolysis of ^{32}P-cADPR (1200 cpm/μl) to ^{32}P-ADPR with a 60 minute incubation of tracheal smooth muscle membranes (40 μg protein). Lanes 1-5 contain increasing amounts of unlabelled cADPR; 1, 30, 100, 300 and 490 μM. Lane 6 is a control lane run in the absence of membranes. Note increased hydrolysis of cADPR with increasing concentrations of unlabelled cADPR.

In most mammalian systems examined thus far, ADP-ribosyl cyclase and cADPR hydrolase activities are associated with a single protein, CD38 [27,38-41]. In porcine tracheal and rat uterine smooth muscles, we examined whether the ADP-ribosyl cyclase and cADPR hydrolase activities are also associated with a single bifunctional protein, presumably CD38 [31,37]. We assayed protein fractions separated by SDS-PAGE for ADP-ribosyl cyclase and cADPR hydrolase activities, after SDS removal. The activities were associated with a single protein fraction with an apparent molecular weight of 40 and 45 kDa in tracheal and uterine smooth muscle respectively. The co-migration of these two enzyme activities with a single protein fraction was further evaluated by separation of detergent solubilized membrane proteins by gel filtration chromatography (Figures 3 and 4). Western blot analysis using an anti-CD38 polyclonal antibody, confirmed the presence of CD38 in the column fractions which contained both ADP-ribosyl cyclase and cADPR hydrolase activities. Combined, these results provide strong evidence for the presence of CD38 in airway and uterine smooth muscles. Further evidence for the presence of CD38 is provided by reverse transcriptase polymerase

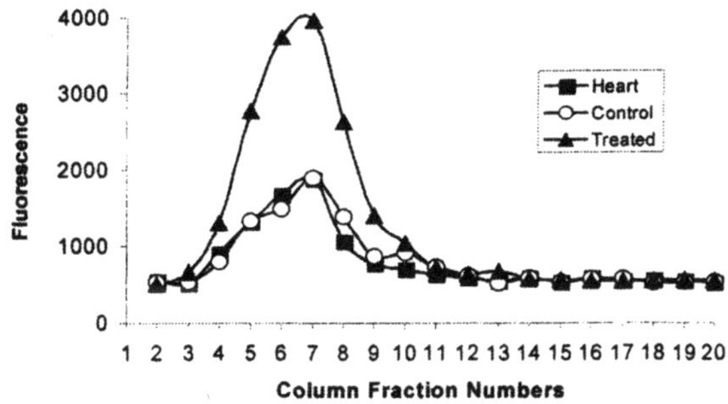

Figure 3. Proteins from rat uterine smooth muscle (ovariectomized and control or estrogen treated) and heart membranes were solubilized with lubrol PX and separated using gel filtration chromatography. Fractions were analyzed for ADP-ribosyl cyclase activity using NGD as substrate. Note that ADP-ribosyl cyclase activity is present in similar fractions from rat heart and from control and estrogen treated rat uterine smooth muscle.

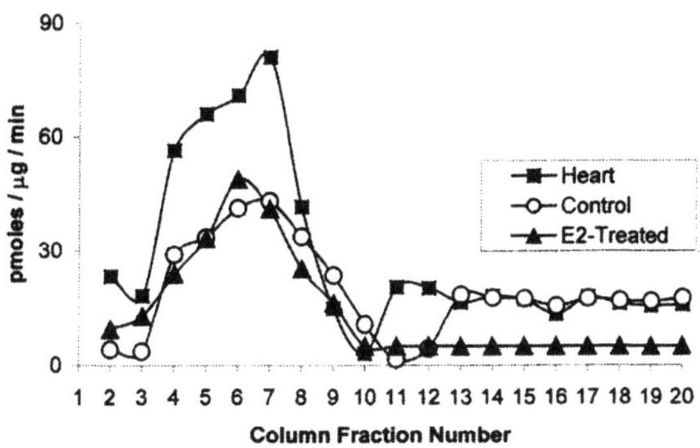

Figure 4. Fractions as described in figure 3 were assayed for cADPR hydrolase activities. Once again activity is found in similar fractions from the different tissues. Furthermore, both ADP-ribosyl cyclase and cADPR hydrolase activities co-migrate with the same column fractions suggesting that the activities are associated with a single bifunctional protein.

chain reaction (RT-PCR) of mRNA extracted from rat uterine, and porcine and human airway smooth muscles using primers specific for human CD38. Nucleotide sequencing of the cloned PCR product from porcine airway smooth muscle revealed sequence homology to human/mouse/rat CD38.

These results provide evidence that the metabolic pathways necessary for the synthesis and degradation of cADPR exist in smooth muscle. Furthermore, these results suggest that the enzymatic activities are associated with the cell surface antigen CD38. The presence of the enzymes involved in the synthesis and hydrolysis of cADPR appear to be wide spread in smooth muscles of different tissues [20,31,32,34,36,37]. Recent reports have described these activities in renal, cardiac and pulmonary vascular smooth muscles [42-44]. This provides the basis for the hypothesis that cADPR is an endogenous mediator of Ca^{2+} regulation in smooth muscle cells. This underscores the potential importance of the CD38/cADPR pathway in Ca^{2+} regulation in all smooth muscles, which may have implications in disease states in which altered Ca^{2+} regulation plays a role.

ROLE OF cADPR IN Ca^{2+} REGULATION IN SMOOTH MUSCLE

The mechanisms underlying cADPR-mediated Ca^{2+} regulation have been investigated in a variety of smooth muscle types. Previous work in other cell types has indicated that cADPR induces Ca^{2+} release from SR/endoplasmic reticulum [45]. This release occurs when IP_3 receptors are either blocked or IP_3 sensitive Ca^{2+} stores are depleted, suggesting that cADPR causes SR Ca^{2+} release through an IP_3 insensitive mechanism [19,24,46]. Ca^{2+} release induced by cADPR in many of these systems is sensitive to agents that are known to modulate RyR channels [47,48]. cADPR-induced Ca^{2+} release is inhibited by ruthenium red, ryanodine and dantrolene, agents known to inhibit RyR channels. If intracellular Ca^{2+} stores are depleted by repeated exposures to caffeine, which activates RyR channels, cADPR cannot elicit a subsequent Ca^{2+} response. Caffeine, at low concentrations, can augment cADPR mediated Ca^{2+} release. These results have been interpreted as evidence that the actions of cADPR are due to activation, directly and/or indirectly, of RyR channels in the SR.

We have investigated the role of cADPR in $[Ca^{2+}]_i$ regulation in airway and coronary artery smooth muscles [19,49]. In airway smooth muscle cells, stimulation with ACh results in a Ca^{2+} response with a biphasic pattern. This $[Ca^{2+}]_i$ response involves SR Ca^{2+} release as well as Ca^{2+} influx through receptor-operated channels. The initial Ca^{2+} response to ACh is attributable primarily to activation of M_3 muscarinic receptors that are coupled to phospholipase C (PLC), resulting in IP_3 production and subsequent Ca^{2+} release through IP_3R channels [2,3,5]. This initial IP_3 induced rise in $[Ca^{2+}]_i$ results in subsequent Ca^{2+} release through a process termed Ca^{2+} induced Ca^{2+} release (CICR). The steady state phase of ACh-induced $[Ca^{2+}]_i$ response reflects CICR and the maintenance of the steady state phase requires Ca^{2+} influx [3,4,6,50]. We have further demonstrated that the steady state phase can be maintained in the absence of extracellular Ca^{2+}, if efflux is blocked by the addition of lanthanum to the medium [6]. These results suggest that influx is necessary to maintain the $[Ca^{2+}]_i$, however the steady state phase of the $[Ca^{2+}]_i$ response to ACh *per se* is due to Ca^{2+} release and reuptake from the SR. During the steady state phase of the response, exposure to caffeine results in a Ca^{2+} transient, which eliminates the ongoing steady state Ca^{2+} response. This finding suggests that the steady state response may involve Ca^{2+} release through RyR channels. Studies in permeabilized tracheal smooth muscle cells provide direct evidence for RyR channel activation during the ACh-induced Ca^{2+} response [11]. In cells permeabilized with β-escin, ACh induces Ca^{2+} release, which is inhibited by ruthenium red. These findings clearly demonstrate activation of RyR in the ACh-induced Ca^{2+} response. We have investigated the mechanisms underlying the ACh- induced Ca^{2+} response, specifically the role of cADPR in RyR channel activation in airway smooth muscle cells [19].

In order to elucidate the role of cADPR in $[Ca^{2+}]_i$ regulation, we again used a permeabilized cell preparation from tracheal smooth muscle. In permeabilized cells maintained at a resting $[Ca^{2+}]_i$ of approximately 100-300 nM, cADPR induced a concentration-dependent increase in $[Ca^{2+}]_i$, which was transient in nature similar to that induced by caffeine. The IP_3R antagonist heparin was ineffective in blocking this response. However, heparin prevented IP_3 induced Ca^{2+} release in these cells. In the presence of ruthenium red or ryanodine, IP_3, but not cADPR, induced a Ca^{2+} response, indicating that cADPR was acting through RyR channel activation. Furthermore, the cADPR "receptor" antagonist, 8-amino-cADPR, prevented the cADPR-induced Ca^{2+} response, but not the release induced by caffeine. 8-amino-cADPR inhibited the steady state phase of the ACh induced Ca^{2+} response in these cells as well. These results indicated that cADPR mediates Ca^{2+} release in tracheal smooth muscle cells via activation of RyR channels. Although caffeine and cADPR both release Ca^{2+} through RyR channel activation, the mechanism of release appears to differ, as can be seen by the differential sensitivity to inhibition by 8-amino-cADPR of the responses to these two agents. During ACh stimulation, cADPR mediated Ca^{2+} release clearly plays a role in the Ca^{2+} response. In this regard, we have investigated the effects of a membrane permeant cADPR "receptor" antagonist, 8-bromo-cADPR, on agonist-induced $[Ca^{2+}]_i$ responses in tracheal smooth muscle cells. cADPR antagonism resulted in significant attenuation of the $[Ca^{2+}]_i$ response to ACh and endothelin-1, but not to histamine (Figures 5 and 6). This further confirms the role of cADPR in agonist-induced $[Ca^{2+}]_i$ mobilization in porcine airway smooth muscle cells. Furthermore, the agonist-specific effects of cADPR antagonism demonstrate divergence of signaling pathways involving cADPR in airway smooth muscle cells. These findings would have implications in excitation-contraction coupling. Contradictory evidence for the role of cADPR in agonist responses of smooth muscle has been reported [51]. In β-escin permeabilized rabbit tracheal smooth muscle strips, these authors failed to demonstrate contractile responses to a range of cADPR concentrations, while eliciting responses to IP_3, caffeine, or carbachol. These findings may be attributable to species differences or sensitivity to cADPR.

The role of cADPR induced Ca^{2+} mobilization has been investigated in coronary artery smooth muscle cells [49]. In β-escin permeabilized porcine coronary artery smooth muscle cells, cADPR stimulates SR Ca^{2+} release, which is not inhibited by heparin, depletion of caffeine-sensitive Ca^{2+} stores, or by a high concentration of ryanodine. In fact, in the presence of ryanodine, we observed potentiation of cADPR-induced $[Ca^{2+}]_i$ response. At the concentration used, ryanodine inhibited caffeine-induced $[Ca^{2+}]_i$ response suggesting that a Ca^{2+} store exists which is cADPR sensitive and distinct from the ryanodine or IP_3 sensitive stores. These results provide evidence for

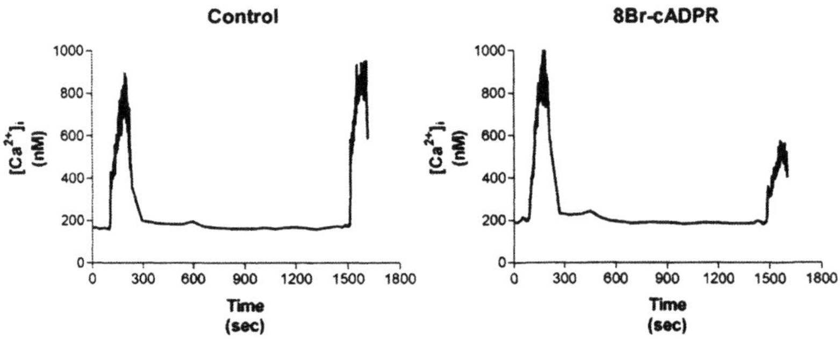

Figure 5. Intracellular Ca^{2+} responses in porcine tracheal smooth muscle cells elicited by 100 nM ACh in the absence of extracellular Ca^{2+} and the presence of 1 mM lanthanum. Cells were stimulated with 100 nM ACh followed by a 5 minute "wash", then incubated for 15 minutes with either Ca^{2+} containing medium (control; left panel) or 100 μM 8-bromo-cADPR in the same medium (right panel). Cells were then re-exposed to 100 nM ACh. Note the second response in the control group is similar in magnitude to the first. In the cells incubated with 8-bromo-cADPR the second response is significantly attenuated.

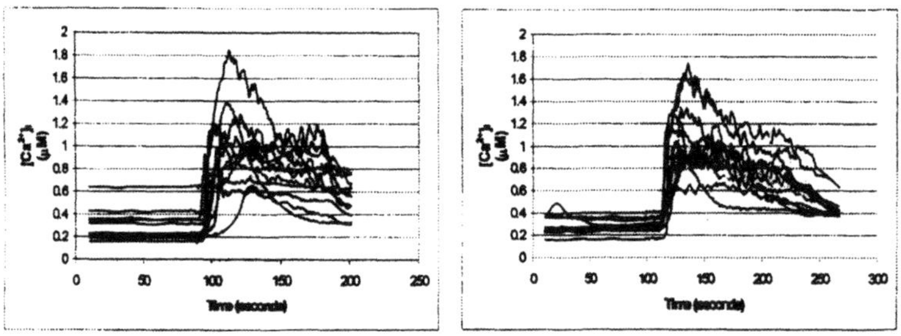

Figure 6. Intracellular Ca^{2+} responses in porcine tracheal smooth muscle cells elicited by 50μM histamine in the absence of extracellular Ca^{2+} and the presence of 1mM lanthanum. In the left panel are responses in control and in the right panel are responses in the presence of 100 μM 8-bromo-cADPR. Note that there is no significant attenuation of the responses in the presence of 8-bromo-cADPR.

a cADPR-induced Ca^{2+} response in coronary artery smooth muscle cells, the mechanism of which appears to differ from that of airway smooth muscle cells.

In rabbit longitudinal smooth muscle of the intestine, Makhlouf's laboratory has obtained evidence for a role of cADPR in $[Ca^{2+}]_i$ mobilization [20]. In permeabilized cells, cADPR induced Ca^{2+} release, increased $[Ca^{2+}]_i$, and caused contraction of the cells in a concentration-dependent fashion. The cADPR-induced Ca^{2+} release was dependent on basal Ca^{2+} concentration and inhibited by 8-amino-cADPR, ruthenium red and dantrolene, while unaffected by heparin. In this regard, the Ca^{2+} release mechanisms in rabbit longitudinal smooth muscle cells appear to be similar to those described in other cell types including that of airways. Furthermore, they demonstrated high affinity cADPR binding in these cells, suggesting that the actions of cADPR may be mediated through a receptor. This is the first such report for the presence of high affinity cADPR binding sites in smooth muscle cells. In these studies, the authors reported an absence of cADPR effects in the circular smooth muscle cells. This finding demonstrates that the cADPR pathway of $[Ca^{2+}]_i$ regulation in smooth muscle cells can be cell-type specific even within the same species.

Most evidence reported on the effects of cADPR on RyR channels in smooth muscle cells comes from use of pharmacological agents that are known to regulate these channels. In a recent study, Campbell and colleagues have described the effects of cADPR on reconstituted RyR from coronary artery smooth muscle cells [21]. Reconstitution of purified RyR from the smooth muscle cells into planar lipid bilayers revealed ryanodine sensitive Ca^{2+} currents. The open probability of these RyR channels was significantly attenuated by ruthenium red. Ryanodine at relatively low concentrations increased the open probability of these channels, with significant inhibition at high concentrations. Caffeine increased channel activity over a wide range of concentrations, as did cADPR. The effects of cADPR were attenuated at high ryanodine concentrations. These results provide the first direct evidence for cADPR activation of the RyR channels in smooth muscle. Effects of cADPR not directly related to RyR channel activation have also been reported. In one such study, cADPR was shown to inhibit Ca^{2+}-activated K^+ channels in inside-out membrane patches of single coronary artery smooth muscle cells [42]. An interesting observation was that ADPR, the hydrolysis product of cADPR, had the opposite effect on these channels. The authors speculated that the inhibition of these channels by cADPR would result in depolarization and Ca^{2+} mobilization, resulting in vasoconstriction. On the other hand, ADPR activation of these channels would lead to hyperpolarization and decreased vascular tone. Along the same lines, vasodilatation of pre-contracted coronary artery smooth muscle is attained in the presence of nicotinamide, which inhibits the formation of cADPR [52]. Pre-exposure to nicotinamide also resulted in attenuation of KCl- or Bay K8644- (a Ca^{2+}

channel activator) induced vasoconstriction. In a study using smooth muscle cell lines, cADPR was shown to inhibit IP_3-induced Ca^{2+} release, suggesting yet another mechanism of cADPR action. In a recent study involving bovine tracheal smooth muscle, cADPR as well as a non-hydrolyzable cADPR analog, 3-deaza-cADPR, applied extracellularly were shown to result in elevation of basal $[Ca^{2+}]_i$, augmentation of ACh-induced increase in $[Ca^{2+}]_i$, and ACh-induced contractility [53]. Exposure to 8-amino-cADPR decreased both the $[Ca^{2+}]_i$ as well as the contractile responses to ACh. These findings indicate that the CD38/cADPR signaling pathway has a physiological role in smooth muscle contractility through its effects on $[Ca^{2+}]_i$ regulation and K^+ conductance.

Other investigations have provided insights into the role of cADPR in excitation-contraction coupling in smooth muscle. In a model of hypoxia-induced pulmonary vasoconstriction, the effects of 8-bromo-cADPR were evaluated [54]. In isolated pulmonary arteries, constriction induced by hypoxia was attenuated by 8-bromo-cADPR, suggesting that cADPR mediates the response. In isolated perfused lung, these investigators showed that the effects of 8-bromo-cADPR were highly selective. Vasoconstriction induced by hypoxia in these lungs was inhibited by cADPR antagonism, without any significant effects on vasoconstriction induced by KCl depolarization. Vasoconstriction induced by $PGF_{2\alpha}$, which acts through an endothelium-dependent mechanism, was unaffected by 8-bromo-cADPR. These selective effects of cADPR antagonism demonstrate that different mechanisms of Ca^{2+} mobilization are involved in contractility and that the CD38/cADPR signaling pathway is recruited differentially depending on the agonist used. These results are also consistent with our findings in tracheal smooth muscle cells, where utilization of the CD38/cADPR signaling pathway in Ca^{2+} mobilization is agonist specific.

REGULATION OF CD38/cADPR PATHWAY IN SMOOTH MUSCLE

How the CD38/cADPR pathway is activated has been the focus of intense investigation. The mechanism through which cADPR production is regulated has been described in various cell types. In sea urchin eggs, earlier investigations have provided evidence that ADP-ribosyl cyclase is regulated by cGMP, through a mechanism involving activation of cGMP-dependent kinase [55-57]. In several mammalian cell types evidence has been provided for the stimulation of ADP-ribosyl cyclase and cADPR production in response to various extracellular stimuli. In pancreatic β-cells, glucose stimulates cADPR production and insulin release [58]. There is also evidence in several cell types for stimulation of ADP-ribosyl cyclase by agonists that activate G-protein-coupled receptors [59-64]. There is a suggestion by these studies that CD38 may be directly coupled to G-proteins. Collectively, there appears to be a variety of mechanisms underlying regulation of CD38 activity. Several recent studies have examined CD38 regulation in smooth muscle in response to hormones, agonists, and nitric oxide (NO).

We and others have obtained evidence that in uterine smooth muscle, CD38 expression is regulated by estrogen [32,37]. In ovariectomized rats treated with estrogen, ADP-ribosyl cyclase activity was increased when compared to untreated controls and this effect was prevented by pre-treatment of the animals with the estrogen antagonist tamoxifen [32]. Dousa and colleagues have also demonstrated increased ADP-ribosyl cyclase activity in vascular smooth muscle cells stimulated with retinoic acid and triiodothyronine [34]. We predicted that the increased CD38 expression in the uterus of estrogen-treated rats would result in increased ADP-ribosyl cyclase and cADPR hydrolase activities. An interesting observation of the work in our laboratory demonstrated a differential regulation of CD38 activity in rat uterine smooth muscle by estrogen [37]. Estrogen induced increased ADP-ribosyl cyclase activity, with no change in cADPR hydrolase activity (Figures 7 and 8). These results indicated that such a differential effect of estrogen treatment on a bifunctional protein must arise from a post-translational modification of CD38. Although the nature of this post-translational modification is not understood, there are some potential mechanisms that can account for this. For example, specific amino acid residues are important for the catalytic activities in CD38 [65]. Therefore, modification of these specific

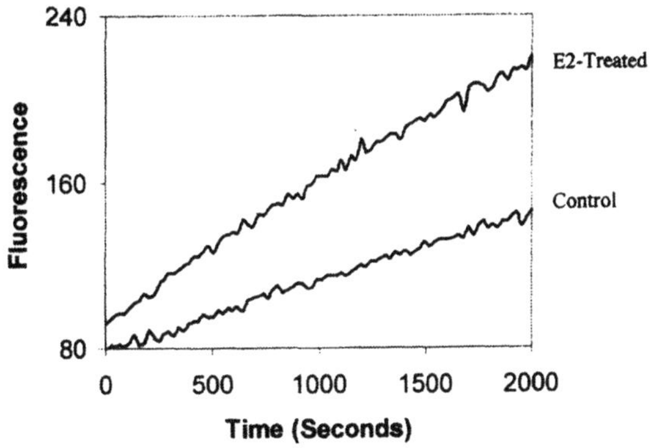

Figure 7. Uterine smooth muscle membranes from ovariectomized control and estrogen-treated rats were assayed for ADP-ribosyl cyclase activity. Using similar amounts of protein, activity was significantly higher in estrogen treated rats when compared to controls. A representative trace of fluorescence versus time is shown.

residues may contribute to the observed differential regulation. Furthermore, ADP-ribosylation of specific amino acid residues in CD38 is known to differentially modulate ADP-ribosyl cyclase and cADPR hydrolase activities [66]. The differential regulation of CD38 by estrogen may reflect ADP-ribosylation of residues important for the cADPR hydrolase activity. Another

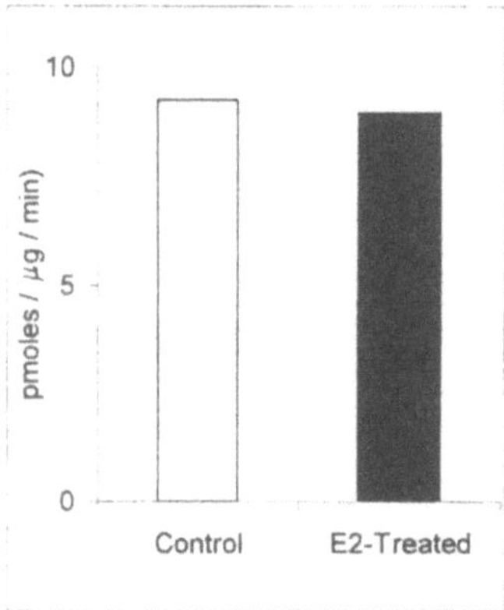

Figure 8. Uterine smooth muscle membranes from ovariectomized control and estrogen-treated rats were assayed for cADPR hydrolase activity. Using similar amounts of protein, activity was not significantly different in estrogen-treated rats when compared to controls.

potential mechanism includes phosphorylation. Sequence analysis of cloned CD38 reveals several potential phosphorylation sites. The stimulatory effects of cGMP and cAMP on ADP-ribosyl cyclase activity appear to support the hypothesis that phosphorylation by their respective kinases may be responsible for this activation [35,67-70]. In studies in our laboratory, we have demonstrated phosphorylation of a protein in tracheal smooth muscle microsomes with a molecular weight similar to that of CD38 by calmodulin-dependent protein kinase (Figure 9). The calmodulin antagonists W-7 and calmodulin binding peptide inhibit this phosphorylation. It is not known whether this calmodulin-dependent phosphorylation of CD38 results in changes in ADP-ribosyl cyclase and/or cADPR hydrolase activities in smooth muscle.

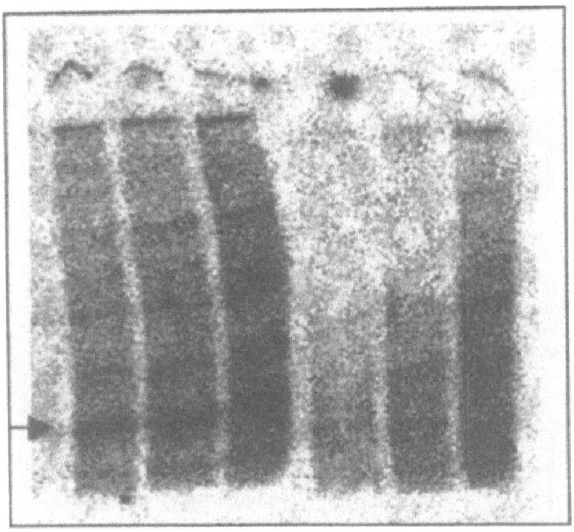

Figure 9. Phosphorimage of ^{32}P-labeled proteins in ASM microsomes in the presence of calmodulin. Reactions without (lanes 1-3) or with (lanes 4-6) W-7 at 1,2,5 min. ~40-kDa band (arrow) has both cyclase and hydrolase activities. Western blot of SDS-PAGE separated proteins reveals CD38 in the same band.

We and others have obtained evidence for regulation of CD38 in smooth muscle by NO [36]. In smooth muscle, NO is known to inhibit the $[Ca^{2+}]_i$ response to agonists [36,45,71] and therefore contribute to relaxation [72,73]. NO is also known to cause stimulation of Ca^{2+} activated K^+ currents [72,74,75]. These effects of NO are mediated through both cGMP-dependent and –independent mechanisms [36,45,76]. Whether NO has an effect on Ca^{2+} mobilization in smooth muscle by modulation of the CD38/cADPR signaling pathway has been investigated. In coronary artery smooth muscle, Campbell's laboratory demonstrated regulation of ADP-ribosyl cyclase activity by a NO donor [36]. The NO donor SNP inhibited ADP-ribosyl cyclase activity in a concentration-dependent fashion with no significant effect on cADPR hydrolase activity. We have also demonstrated inhibition of ADP-ribosyl cyclase activity in tracheal smooth muscle microsomes by NO donors. As in the case of coronary artery smooth muscle, there is a differential effect of NO donors on ADP-ribosyl cyclase and cADPR hydrolase activities. We examined the mechanism of this regulation of CD38 in tracheal smooth muscle membranes. The results from these studies on the effects of NO donors support a mechanism involving S-nitrosylation of CD38, independent of activation of cGMP-dependent protein kinase. This differential effect of NO donors on ADP-ribosyl cyclase and cADPR hydrolase activities would have an inhibitory effect on cADPR-mediated Ca^{2+} release and contractility. Since NO is released from intrinsic nerves in airway smooth muscle [73] as well as from endothelial cells adjacent to vascular smooth muscle [77], this mechanism of Ca^{2+} regulation by NO should have a physiological role on smooth muscle tone. These results collectively point to regulation of $[Ca^{2+}]_i$ in smooth muscle by NO through both cGMP-dependent and –independent mechanisms. The cGMP-dependent mechanism of regulation involves inhibition of Ca^{2+} release and contractility through activation of cGMP-dependent kinase. The cGMP-independent mechanism of $[Ca^{2+}]_i$ regulation may arise from inhibition of ADP-ribosyl cyclase activity and decreased cADPR production during agonist stimulation. There is no evidence for CD38 regulation in smooth muscle through a cGMP-dependent mechanism.

Regulation of CD38 in smooth muscle through a rise in $[Ca^{2+}]_i$ has been described. In rabbit intestinal longitudinal smooth muscle cells, Makhlouf's laboratory provided evidence that stimulation with cholecystokinin octapeptide results in Ca^{2+} influx through voltage-gated channels, which leads to the activation of ADP-ribosyl cyclase and production of cADPR [20,78,79]. This was the first evidence for agonist stimulation of cADPR production and subsequent enhancement of CICR in a mammalian smooth muscle. From this study, it is obvious that Ca^{2+} influx is required for the activation of ADP-ribosyl cyclase. However, whether Ca^{2+} directly activates the ADP-ribosyl cyclase or if some other Ca^{2+} dependent process mediates this effect is not clear. In our laboratory, we have investigated agonist recruitment of the CD38/cADPR signaling pathway in airway smooth muscle. We chose agonists which cause elevation of $[Ca^{2+}]_i$ through activation of G proteins. We hypothesized that this agonist stimulated rise in $[Ca^{2+}]_i$ would lead to activation of the CD38/cADPR pathway based on the model proposed for longitudinal smooth muscle of the intestine. The results of these studies provided evidence for agonist specificity of the CD38/cADPR signaling pathway that was not directly related to a rise in $[Ca^{2+}]_i$. Activation of $G_{\alpha q}$ type G proteins by itself appears to be insufficient for the recruitment of the CD38/cADPR signaling, since all three agonists chosen in these studies activate this G protein subtype. Only the responses to ACh and endothelin-1, but not histamine, are inhibited by the cell permeant cADPR antagonist 8-bromo-cADPR. These studies underscore the complexity of the signaling mechanisms involved in the regulation of the CD38/cADPR signaling pathway in different smooth muscle cells.

Smooth muscle cells are known to play an immunomodulatory role in a variety of diseases. Asthma is considered as a chronic inflammatory disease and proinflammatory cytokines released from a variety of cells within the lungs are important in the initiation and maintenance of airway inflammation [80-86]. Levels of inflammatory mediators such as TNF-α, IL1-β and IFN-γ are elevated within the airways of asthmatics [80]. Previous investigations have shown that exposure to TNF-α results in altered Ca^{2+} homeostasis in cultured human airway smooth muscle cells [87-89]. Stimulation of these cells with agonists results in increased $[Ca^{2+}]_i$ responses and production of inositol phosphates [90]. We have extended these studies to determine whether the CD38/cADPR pathway of Ca^{2+} regulation contributes to the altered Ca^{2+} homeostasis. Human airway smooth muscle cells maintained in culture express CD38 and its expression is significantly augmented by exposure to the inflammatory cytokines (Figure 10). Furthermore, the $[Ca^{2+}]_i$

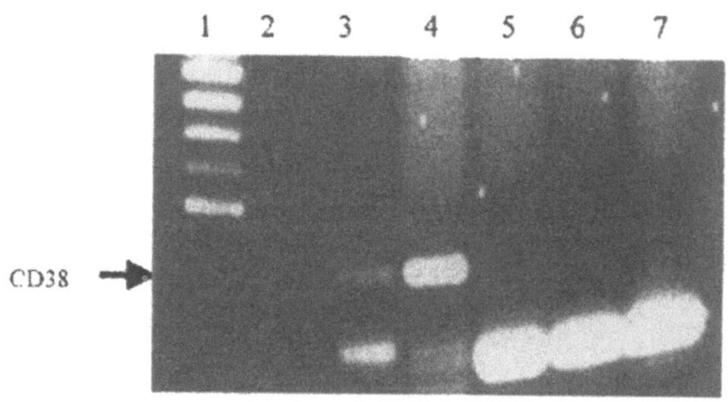

Figure 10. RT-PCR of RNA from cultured human airway smooth muscle cells showing CD38 mRNA expression in control and TNF-α treated cells. Note high expression in both TNF-α treated samples: lanes 3&4.

responses elicited by agonists such as bradykinin, ACh and thrombin are also significantly higher than in controls (Responses to bradykinin in control and TNF-α treated cells are shown in Figure 11). Pretreatment with 8-bromo-cADPR results in significant attenuation of these inflammatory cytokine-induced increases in $[Ca^{2+}]_i$ responses. These results provide evidence for a pathophysiological role for the CD38/cADPR signaling pathway in airway smooth muscle. It is known that the CD38 promoter region has IFN responsive elements regulating its expression in immune cells [28]. The increased CD38 expression, which we observe in airway smooth muscle cells, may result from such regulation. Whether this cytokine induced alterations in CD38/cADPR signaling have a role in inflammatory diseases involving other tissues remains to be determined.

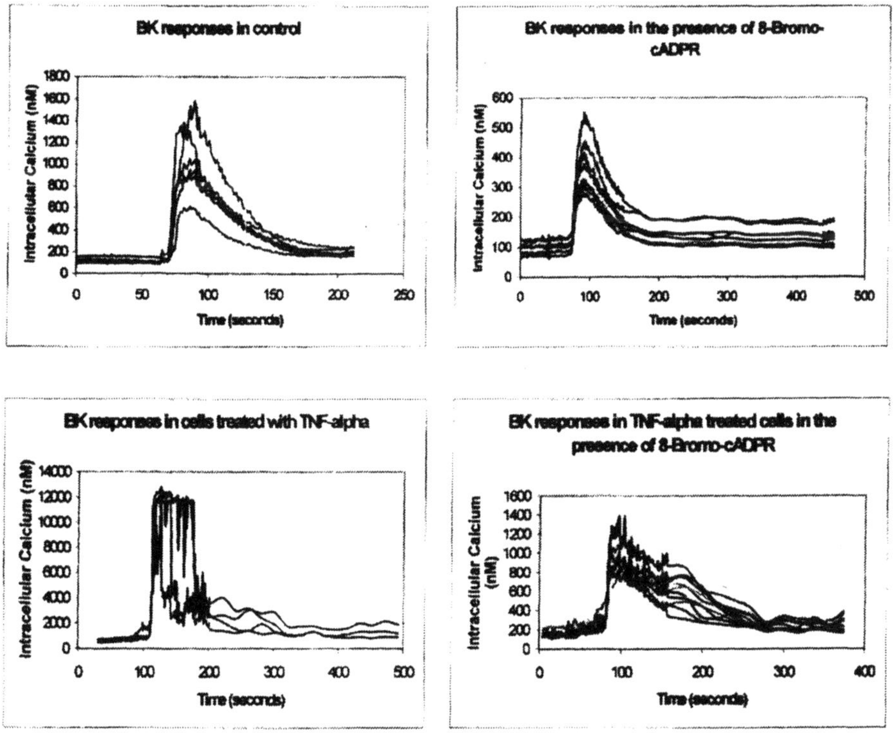

Figure 11. Human airway smooth muscle cells were cultured in medium containing TNF-α (20 ng/ml) or vehicle (control) for 22 hours, loaded with 5μM Fura-2/AM and intracellular Ca^{2+} responses to 100 nM bradykinin were measured in the presence or absence of 100 μM 8-bromo-cADPR. Note significant augmentation of responses to bradykinin in cells treated with TNF-α compared to control. The responses in the presence of 8-bromo-cADPR were significantly attenuated in both control and TNF-α treated cells.

REGULATION OF cADPR INDUCED $[Ca^{2+}]_i$ RELEASE

The evidence to date suggests that cADPR induces Ca^{2+} release from RyR sensitive Ca^{2+} stores in smooth muscle. However, it is not clear as to the exact mechanisms involved in the regulation of the cADPR induced Ca^{2+} responses. In this regard, there are at least two distinct hypotheses to be considered. First, cADPR acts directly on the RyR channels to increase the open probability of these channels resulting in increased Ca^{2+} release. The evidence supporting this hypothesis has been provided in studies of reconstituted RyR channels in planar lipid bilayers [21,91-93]. In studies of coronary artery smooth muscle, cADPR induces an increase in RyR channel activity, which is inhibited by high concentrations of ryanodine [21]. Also Ca^{2+} induced activation of these channels is enhanced by cADPR. This appears to be the only evidence for activation of RyR channels directly by cADPR in smooth muscle, although such evidence has been provided in other systems. Second, cADPR modulates RyR channel activity indirectly through a mechanism requiring other modulators such as Ca^{2+}, calmodulin, FK binding proteins, etc. In this respect, cADPR sensitizes the CICR mechanism involving RyR channel activity. In the longitudinal smooth muscle of rabbit intestine, Makhlouf's laboratory describes potentiation of CICR by cADPR [20]. They also provide evidence that cADPR binding in these cells exhibits a "bell-shaped" dependence on Ca^{2+} concentration. These results provide circumstantial evidence for cADPR sensitization of RyR channel activity. It is important to point out that there is no evidence that cADPR binds directly to RyR channels, although the effects of cADPR require RyR channel activation.

The role of calmodulin in cADPR-induced Ca^{2+} release in sea urchin eggs has been described. In preparations of sea urchin egg microsomes, cADPR-induced Ca^{2+} release requires the presence of calmodulin [94,95]. The mechanism by which calmodulin confers cADPR sensitivity is not clear. RyR channel activity is also regulated by FK binding proteins [96-99]. In cardiac myocytes, FK binding protein 12.6 (FKBP12.6) associates with the RyR2 channel [96,97]. In the bound state, FKBP12.6 inhibits RyR channel activity. In pancreatic islet microsomes, cADPR binds to FKBP12.6 resulting in the appearance of FKBP12.6 in the soluble fraction, suggesting dissociation of the FKBP/RyR complex upon cADPR binding [100]. Furthermore, in the presence of FK506, which binds and dissociates FKBP/RyR complex, cADPR no longer elicits Ca^{2+} release. These results suggest that cADPR mediates Ca^{2+} release by binding to FKBP12.6, dissociating the FKBP/RyR complex and thereby causing RyR channel activation. The lack of a cADPR effect on Ca^{2+} release in the presence of FK506 further supports this hypothesis. In porcine airway smooth muscle, we have determined expression of FKBP12.6 by RT-PCR of RNA using primers based on the sequence of human FKBP12.6. Sequence analysis of the PCR product reveals considerable homology. In porcine airway smooth muscle, the predominant RyR isoform is RyR2. We predict that FKBP12.6 has a similar role in mediating cADPR effects in smooth muscle, although this remains to be determined.

In summary, there appears to be sufficient evidence for a role of cADPR as an endogenous modulator of intracellular Ca^{2+} in airway, vascular and intestinal smooth muscles. The enzymatic machinery for the synthesis and degradation of cADPR is present in these tissues. cADPR has been found to induce Ca^{2+} release from the SR most likely reflecting activation of RyR channels. We and others have provided evidence for the regulation of CD38 activity in response to extracellular signals. The exact mechanism(s) by which cADPR induces intracellular Ca^{2+} release and the regulation of this process remain unclear. Based on the information gathered regarding the role of the CD38/cADPR pathway in smooth muscle, we provide a model depicting the different components (Figure 12).

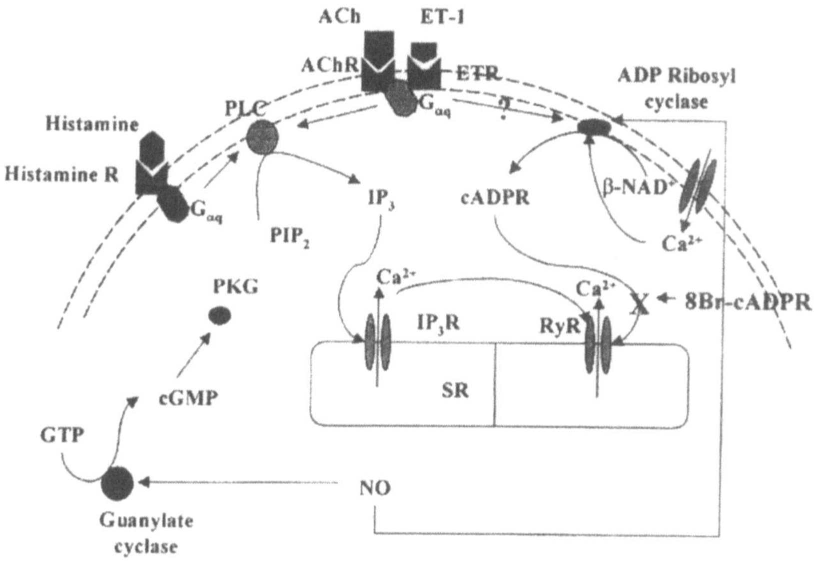

Figure 12. Model depicting the CD38/cADPR signaling pathway of intracellular Ca^{2+} regulation in smooth muscle. In longitudinal smooth muscle of the intestine, cholecystokinin elicits Ca^{2+} influx, which results in activation of ADP-ribosyl cyclase and cADPR mediated SR Ca^{2+} release. In airway as well as vascular smooth muscles, NO inhibits ADP-ribosyl cyclase through a cGMP independent mechanism. In airway smooth muscle recruitment of the CD38/cADPR signalling pathway is agonist specific. The Ca^{2+} responses to ACh and endothelin-1 are partially sensitive to inhibition by the cADPR antagonist 8-bromo-cADPR, while the response to histamine is not. Ca^{2+} release induced by cADPR involves activation of RyR channels in all smooth muscles, which may reflect sensitization of the CICR mechanism.

ACKNOWLEDGEMENTS

We would like to acknowledge Jill Johnson for her involvement in the studies related to phosphorylation. We acknowledge grant support from the following sources: National Institutes of Health HL57498 (to MSK), HL55301, HL64063 and HL67663 (to RAP), National Institutes of Drug Abuse DA11806 (to TFW), and an University of Minnesota Academic Health Center Faculty Development Grant (to MSK and TFW). Direct correspondence to Dr. M. S. Kannan: 1971 Commonwealth Avenue, St. Paul, MN 55108.

REFERENCES

1. Somlyo AP and Somlyo AV. 1994. Signal transduction and regulation in smooth muscle. *Nature*. 372: 231-236.
2. Coburn RF and Baron CB. 1990. Coupling mechanisms in airway smooth muscle. *Am. J. Physiol.* 258: L119-L133.
3. Shieh CC, Petrini MF, Dwyer TM and Farley JM. 1991. Concentration-dependence of acetylcholine-induced changes in calcium and tension in swine trachealis. *J. Pharmacol. Exp. Ther.* 256: 141-148.
4. Murray RK and Kotlikoff MI. 1991. Receptor-activated calcium influx in human airway smooth muscle cells. *J. Physiol.* 435: 123-144.
5. Sims SM, Jiao Y and Zheng ZG. 1996. Intracellular calcium stores in isolated tracheal smooth muscle cells. *Am J Physiol.* 271: L300-L309.
6. Prakash YS, Kannan MS and Sieck GC. 1997. Regulation of intracellular calcium oscillations in porcine tracheal smooth muscle cells. *Am. J. Physiol.*. 272: C966-C975.
7. Baron CB, Cunningham M, Strauss JF 3rd and Coburn RF. 1984. Pharmacomechanical coupling in smooth muscle may involve phosphatidylinositol metabolism. *Proc. Natl. Acad. Sci. USA*. 81: 6899-6903.
8. Berridge MJ and Irvine RF. 1984. Inositol trisphosphate, a novel second messenger in cellular signal transduction. *Nature*. 312: 315-321.
9. Wagner-Mann C, Hu Q and Sturek M. 1992. Multiple effects of ryanodine on intracellular free Ca^{2+} in smooth muscle cells from bovine and porcine coronary artery: modulation of sarcoplasmic reticulum function. *Br. J. Pharmacol.* 105: 903-911.
10. Lynn S, Morgan JM, Lamb HK, Meissner G and Gillespie JI. 1995. Isolation and partial cloning of ryanodine-sensitive Ca^{2+} release channel protein isoforms from human myometrial smooth muscle. *FEBS Lett.* 372: 6-12.
11. Kannan MS, Prakash YS, Brenner T, Mickelson JR and Sieck GC. 1997. Role of ryanodine receptor channels in Ca^{2+} oscillations of porcine tracheal smooth muscle. *Am. J. Physiol.* 272 : L659-L664.
12. Lynn S and Gillespie JI. 1995. Basic properties of a novel ryanodine-sensitive, caffeine-insensitive calcium-induced calcium release mechanism in permeabilised human vascular smooth muscle cells. *FEBS Lett.* 367: 23-27.
13. Morgan JM, De Smedt H and Gillespie JI. 1996. Identification of three isoforms of the InsP3 receptor in human myometrial smooth muscle. *Pflugers Arch.* 431: 697-705.
14. Schramm CM, Chuang ST and Grunstein MM. 1992. Maturation of inositol 1,4,5-trisphosphate receptor binding in rabbit tracheal smooth muscle. *Am. J. Physiol.* 263: L501-L505.
15. Schramm CM and Grunstein MM. 1992. Assessment of signal transduction mechanisms regulating airway smooth muscle contractility. *Am. J. Physiol.* 262: L119-L139.
16. Bezprozvanny I, Watras J and Ehrlich BE. 1991. Bell-shaped calcium-response curves of Ins(1,4,5)P_3- and calcium-gated channels from endoplasmic reticulum of cerebellum. *Nature*. 351: 751-754.
17. Komalavilas P and Lincoln TM. 1996. Phosphorylation of the inositol 1,4,5-trisphosphate receptor. Cyclic GMP-dependent protein kinase mediates cAMP and cGMP dependent phosphorylation in the intact rat aorta. *J. Biol. Chem.* 271: 21933-21938.
18. Haug LS, Jensen V, Hvalby O, Walaas SI and Ostvold AC. 1999. Phosphorylation of the inositol 1,4,5-trisphosphate receptor by cyclic nucleotide-dependent kinases in vitro and in rat cerebellar slices in situ. *J. Biol. Chem.* 274: 7467-7473.
19. Prakash YS, Kannan MS, Walseth TF and Sieck GC. 1998. Role of cyclic ADP-ribose in the regulation of $[Ca^{2+}]i$ in porcine tracheal smooth muscle. *Am. J. Physiol.* 274: C1653-C1660.
20. Kuemmerle JF and Makhlouf GM. 1995. Agonist-stimulated cyclic ADP ribose.

Endogenous modulator of Ca^{2+}-induced Ca^{2+} release in intestinal longitudinal muscle. *J. Biol. Chem.* 270: 25488-25494.

21. Li PL, Tang WX, Valdivia HH, Zou AP and Campbell WB. 2001. cADP-ribose activates reconstituted ryanodine receptors from coronary arterial smooth muscle. *Am. J. Physiol. Heart Circ. Physiol.* 280: H208-H215.
22. Guse AH. 1999. Cyclic ADP-ribose: a novel Ca^{2+}-mobilising second messenger. *Cell. Signal.* 11: 309-316.
23. Guse AH. 2000. Cyclic ADP-ribose. *J. Mol. Med.* 78: 26-35.
24. Lee HC, Galione A, and Walseth TF. 1994. Cyclic ADP-ribose: metabolism and calcium mobilizing function. *Vit. Horm.* 48: 199-257.
25. Galione A. 1994. Cyclic ADP-ribose, the ADP-ribosyl cyclase pathway and calcium signalling. *Mol. Cell. Endocrinol.* 98: 125-131.
26. Rusinko N and Lee HC. 1989. Widespread occurrence in animal tissues of an enzyme catalyzing the conversion of NAD^{+} into a cyclic metabolite with intracellular Ca^{2+}-mobilizing activity. *J. Biol. Chem.* 264: 11725-11731.
27. Lee HC, Graeff RM and Walseth TF. 1997. ADP-ribosyl cyclase and CD38. Multi-functional enzymes in Ca^{+2} signaling. *Adv. Exp. Med. Biol.* 419: 411-419.
28. Ferrero E, Saccucci F and Malavasi F. 2000. The making of a leukocyte receptor: origin, genes and regulation of human CD38 and related molecules. *Chem. Immunol.* 75: 1-19.
29. Lee HC and Aarhus R. 1991. ADP-ribosyl cyclase: an enzyme that cyclizes NAD^{+} into a calcium-mobilizing metabolite. *Cell Regul.* 2: 203-209.
30. Glick DL, Hellmich MR, Beushausen S, Tempst P, Bayley H and Strumwasser F. 1991. Primary structure of a molluscan egg-specific NADase, a second-messenger enzyme. *Cell Regul.* 2: 211-218.
31. White TA, Johnson S, Walseth TF, Lee HC, Graeff RM, *et al.* 2000. Subcellular localization of cyclic ADP-ribosyl cyclase and cyclic ADP- ribose hydrolase activities in porcine airway smooth muscle. *Biochim. Biophys. Acta.* 1498: 64-71.
32. Chini EN, de Toledo FG, Thompson MA and Dousa TP. 1997. Effect of estrogen upon cyclic ADP ribose metabolism: beta-estradiol stimulates ADP ribosyl cyclase in rat uterus. *Proc. Natl. Acad. Sci. USA.* 94: 5872-5876.
33. Graeff RM, Walseth TF, Fryxell K, Branton WD and Lee HC. 1994. Enzymatic synthesis and characterizations of cyclic GDP-ribose. A procedure for distinguishing enzymes with ADP-ribosyl cyclase activity. *J. Biol. Chem.* 269: 30260-30267.
34. de Toledo FG, Cheng J and Dousa TP. 1997. Retinoic acid and triiodothyronine stimulate ADP-ribosyl cyclase activity in rat vascular smooth muscle cells. *Biochem. Biophys. Res. Commun.* 238: 847-850.
35. Graeff RM, Franco L, De Flora A and Lee HC. 1998. Cyclic GMP-dependent and -independent effects on the synthesis of the calcium messengers cyclic ADP-ribose and nicotinic acid adenine dinucleotide phosphate. *J. Biol. Chem.* 273: 118-125.
36. Yu JZ, Zhang DX, Zou AP, Campbell WB and Li PL. 2000. Nitric oxide inhibits Ca^{2+} mobilization through cADP-ribose signaling in coronary arterial smooth muscle cells. *Am. J. Physiol. - Heart Circ. Physiol.* 279: H873-H881.
37. Dogan S, White TA, Deshpande DA, Murtaugh MP, Walseth TF and Kannan MS. 2002. Estrogen increases CD38 gene expression and leads to differential regulation of adenosine diphosphate (ADP)-ribosyl cyclase and cyclic ADP-ribose hydrolase activities in rat myometrium. *Biol. Reprod.* 66: 596-602.
38. Howard M, Grimaldi JC, Bazan JF, Lund FE, Santos-Argumedo L, *et al.* 1993. Formation and hydrolysis of cyclic ADP-ribose catalyzed by lymphocyte antigen CD38. *Science* 262: 1056-9
39. Malavasi F, Funaro A, Roggero S, Horenstein A, Calosso L and Mehta K. 1994. Human CD38: a glycoprotein in search of a function. *Immunol. Today.* 15: 95-97.
40. De Flora A, Guida L, Franco L and Zocchi E. 1997. The CD38/cyclic ADP-ribose system: a topological paradox. *Int. J. Biochem. Cell Biol.* 29: 1149-1166.

41. Mehta K, Shahid U and Malavasi F. 1996. Human CD38, a cell-surface protein with multiple functions. *FASEB J.* 10: 1408-1417.
42. Li PL, Zou AP and Campbell WB. 1998. Regulation of KCa-channel activity by cyclic ADP-ribose and ADP-ribose in coronary arterial smooth muscle. *Am. J. Physiol.* 275: H1002-H1010.
43. Li N, Teggatz EG, Li PL, Allaire R and Zou AP. 2000. Formation and actions of cyclic ADP-ribose in renal microvessels. *Microvasc. Res.* 60: 149-159.
44. Wilson HL, Dipp M, Thomas JM, Lad C, Galione A and Evans AM. 2001. ADP-ribosyl cyclase and cyclic ADP-ribose hydrolase act as a redox sensor. a primary role for cyclic ADP-ribose in hypoxic pulmonary vasoconstriction. *J. Biol. Chem.* 276: 11180-11188.
45. Prakash YS, Kannan MS and Sieck GC. 1997. Nitric oxide inhibits ach-induced intracellular calcium oscillations in porcine tracheal smooth muscle. *Am. J. Physiol.* 272: L588-L596.
46. Clapper DL, Walseth TF, Dargie PJ and Lee HC. 1987. Pyridine nucleotide metabolites stimulate calcium release from sea urchin egg microsomes desensitized to inositol trisphosphate. *J. Biol. Chem.* 262: 9561-9568.
47. Galione A, Lee HC and Busa WB. 1991. Ca^{2+}-induced Ca^{2+} release in sea urchin egg homogenates: modulation by cyclic ADP-ribose. *Science* 253: 1143-1146.
48. Lee HC. 1996. Modulator and messenger functions of cyclic ADP-ribose in calcium signaling. *Recent Prog. Horm. Res.* 51: 355-388.
49. Kannan MS, Fenton AM, Prakash YS and Sieck GC. 1996. Cyclic ADP-ribose stimulates sarcoplasmic reticulum calcium release in porcine coronary artery smooth muscle. *Am. J. Physiol.* 270: H801-H806.
50. Murray RK, Fleischmann BK and Kotlikoff MI. 1993. Receptor-activated Ca influx in human airway smooth muscle: use of calcium imaging and perforated patch-clamp techniques. *Am. J. Physiol.* 264: C485-C490.
51. Iizuka K, Yoshii A, Dobashi K, Horie T, Mori M and Nakazawa T. 1998. $InsP_3$, but not novel Ca^{2+} releasers, contributes to agonist-initiated contraction in rabbit airway smooth muscle. *J. Physiol.* 511: 915-933.
52. Geiger J, Zou AP, Campbell WB and Li PL. 2000. Inhibition of cADP-ribose formation produces vasodilation in bovine coronary arteries. *Hypertension* 35: 397-402.
53. Franco L, Bruzzone S, Song P, Guida L, Zocchi E, *et al.* 2001. Extracellular cyclic ADP-ribose potentiates ACh-induced contraction in bovine tracheal smooth muscle. *Am. J. Physiol. -Lung Cell Mol. Physiol.* 280: L98-L106.
54. Dipp M and Evans AM. 2001. Cyclic ADP-ribose is the primary trigger for hypoxic pulmonary vasoconstriction in the rat lung in situ. *Circ. Res.* 89: 77-83.
55. Galione A, White A, Willmott N, Turner M, Potter BV and Watson SP. 1993. cGMP mobilizes intracellular Ca^{2+} in sea urchin eggs by stimulating cyclic ADP-ribose synthesis. *Nature* 365: 456-459.
56. Willmott N, Sethi JK, Walseth TF, Lee HC, White AM and Galione A. 1996. Nitric oxide-induced mobilization of intracellular calcium via the cyclic ADP-ribose signaling pathway. *J. Biol. Chem.* 271: 3699-3705.
57. Graeff, R. M., Franco, L., De Flora, A., and Lee, H. C. 1998. Cyclic GMP-dependent and -independent effects on the synthesis of the calcium messengers cyclic ADP-ribose and nicotinic acid adenine dinucleotide phosphate. *J. Biol. Chem.* 273: 118-125.
58. Takasawa S, Nata K, Yonekura H and Okamoto H. 1993. Cyclic ADP-ribose in insulin secretion from pancreatic beta cells. *Science* 259: 370-373.
59. Morita K, Kitayama S and Dohi T. 1997. Stimulation of cyclic ADP-ribose synthesis by acetylcholine and its role in catecholamine release in bovine adrenal chromaffin cells. *J. Biol. Chem.* 272: 21002-21009.
60. Higashida H, Egorova A, Higashida C, Zhong ZG, Yokoyama S, *et al.* 1999. Sympathetic potentiation of cyclic ADP-ribose formation in rat cardiac myocytes. *J. Biol. Chem.* 274: 33348-33354.

61. Higashida H, Zhang JS, Hashii M, Shintaku M, Higashida C and Takeda Y. 2000. Angiotensin II stimulates cyclic ADP-ribose formation in neonatal rat cardiac myocytes. *Biochem. J.* 352: 197-202.
62. Higashida H. 1997. ADP-ribosyl cyclase coupled with receptors via G proteins. *FEBS Lett.* 418: 355-356.
63. Higashida H, Yokoyama S, Hashii M, Taketo M, Higashida M, *et al.* 1997. Muscarinic receptor-mediated dual regulation of ADP-ribosyl cyclase in NG108-15 neuronal cell membranes. *J. Biol. Chem.* 272: 31272-31277.
64. Hotta T, Asai K, Fujita K, Kato T and Higashida H. 2000. Membrane-bound form of ADP-ribosyl cyclase in rat cortical astrocytes in culture. *J. Neurochem.* 74: 669-675.
65. Tohgo A, Takasawa S, Noguchi N, Koguma T, Nata K, *et al.* 1994. Essential cysteine residues for cyclic ADP-ribose synthesis and hydrolysis by CD38. *J. Biol. Chem.* 269: 28555-28557.
66. Han MK, Cho YS, Kim YS, Yim CY and Kim UH. 2000. Interaction of two classes of ADP-ribose transfer reactions in immune signaling. *J. Biol. Chem.* 275: 20799-20805.
67. Galione A, White A, Willmott N, Turner M, Potter BV and Watson SP. 1993. cGMP mobilizes intracellular Ca^{2+} in sea urchin eggs by stimulating cyclic ADP-ribose synthesis. *Nature* 365: 456-459.
68. Morita K, Kitayama S and Dohi T. 1997. Stimulation of cyclic ADP-ribose synthesis by acetylcholine and its role in catecholamine release in bovine adrenal chromaffin cells. *J. Biol. Chem.* 272: 21002-21009.
69. Willmott N, Sethi JK, Walseth TF, Lee HC, White AM and Galione A. 1996. Nitric oxide-induced mobilization of intracellular calcium via the cyclic ADP-ribose signaling pathway. *J. Biol. Chem.* 271: 3699-3705.
70. Wilson HL and Galione A. 1998. Differential regulation of nicotinic acid-adenine dinucleotide phosphate and cADP-ribose production by cAMP and cGMP. *Biochem. J.* 331: 837-843.
71. Kannan MS, Prakash YS, Johnson DE and Sieck GC. 1997. Nitric oxide inhibits calcium release from sarcoplasmic reticulum of porcine tracheal smooth muscle cells. *Am. J. Physiol.* 272: L1-L7.
72. Kannan MS and Johnson DE. 1995. Modulation of nitric oxide-dependent relaxation of pig tracheal smooth muscle by inhibitors of guanylyl cyclase and calcium activated potassium channels. *Life Sciences.* 56: 2229-38
73. Kannan MS and Johnson DE. 1992. Nitric oxide mediates the neural nonadrenergic, noncholinergic relaxation of pig tracheal smooth muscle. *Am. J. Physiol.* 262: L511-L514.
74. Bolotina VM, Najibi S, Palacino JJ, Pagano PJ, and Cohen RA. 1994. Nitric oxide directly activates calcium-dependent potassium channels in vascular smooth muscle. *Nature* 368: 850-853.
75. Bialecki RA and Stinson-Fisher C. 1995. KCa channel antagonists reduce NO donor-mediated relaxation of vascular and tracheal smooth muscle. *Am. J. Physiol.* 268: L152-L159.
76. Perkins WJ, Pabelick C, Warner DO and Jones KA. 1998. cGMP-independent mechanism of airway smooth muscle relaxation induced by S-nitrosoglutathione. *Am. J. Physiol.* 275: C468-C474.
77. Feletou M and Vanhoutte PM. 2000. Endothelium-dependent hyperpolarization of vascular smooth muscle cells. *Acta Pharmacol. Sin.* 21: 1-18.
78. Kuemmerle JF, Murthy KS and Makhlouf GM. 1998. Longitudinal smooth muscle of the mammalian intestine. A model for Ca^{2+} signaling by cADPR. *Cell Biochem. Biophys.* 28: 31-44.
79. Makhlouf GM and Murthy KS. 1997. Signal transduction in gastrointestinal smooth muscle. *Cell Signal.* 9: 269-276.
80. Amrani Y and Panettieri RA Jr. 1998. Cytokines induce airway smooth muscle cell hyperresponsiveness to contractile agonists. *Thorax.* 53: 713-716.

81. Howarth PH, Bradding P, Quint D, Redington AE and Holgate ST. 1994. Cytokines and airway inflammation. *Ann. N. Y. Acad. Sci.* 725: 69-82.
82. Bradding P, Roberts JA, Britten KM, Montefort S, Djukanovic R, *et al.* 1994. Interleukin-4, -5, and -6 and tumor necrosis factor-alpha in normal and asthmatic airways: evidence for the human mast cell as a source of these cytokines. *Am. J. Respir. Cell. Mol. Biol.* 10: 471-480.
83. Elias JA, Wu Y, Zheng T and Panettieri R. 1997. Cytokine- and virus-stimulated airway smooth muscle cells produce IL-11 and other IL-6-type cytokines. *Am. J. Physiol.* 273: L648-L655.
84. Kips JC, Tavernier J and Pauwels RA. 1992. Tumor necrosis factor causes bronchial hyperresponsiveness in rats. *Am. Rev. Respir. Dis.* 145: 332-336.
85. Renzetti LM, Paciorek PM, Tannu SA, Rinaldi NC, Tocker JE, *et al.* 1996. Pharmacological evidence for tumor necrosis factor as a mediator of allergic inflammation in the airways. *J. Pharmacol. Exp. Ther.* 278: 847-853.
86. Thomas PS, Yates DH and Barnes PJ. 1995. Tumor necrosis factor-alpha increases airway responsiveness and sputum neutrophilia in normal human subjects. *Am. J. Respir. Crit. Care Med.* 152: 76-80.
87. Amrani Y, Panettieri RA Jr. Frossard N and Bronner C. 1996. Activation of the TNF alpha-p55 receptor induces myocyte proliferation and modulates agonist-evoked calcium transients in cultured human tracheal smooth muscle cells. *Am. J. Respir. Cell. Mol. Biol.* 15: 55-63.
88. Amrani Y, Martinet N and Bronner C. 1995. Potentiation by tumour necrosis factor-alpha of calcium signals induced by bradykinin and carbachol in human tracheal smooth muscle cells. *Br. J. Pharmacol.* 114: 4-5.
89. Amrani Y and Bronner C. 1993. Tumor necrosis factor alpha potentiates the increase in cytosolic free calcium induced by bradykinin in guinea-pig tracheal smooth muscle cells. *C. R. Acad. Sci. III.* 316: 1489-1494.
90. Amrani Y, Krymskaya V, Maki C and Panettieri RA Jr. 1997. Mechanisms underlying TNF-alpha effects on agonist-mediated calcium homeostasis in human airway smooth muscle cells. *Am. J. Physiol.* 273: L1020-L1028.
91. Meszaros LG, Bak J and Chu A. 1993. Cyclic ADP-ribose as an endogenous regulator of the non-skeletal type ryanodine receptor Ca^{2+} channel. *Nature* 364: 76-79.
92. Perez CF, Marengo JJ, Bull R and Hidalgo C. 1998. Cyclic ADP-ribose activates caffeine-sensitive calcium channels from sea urchin egg microsomes. *Am. J. Physiol.* 274: C430-C439.
93. Lokuta AJ, Darszon A, Beltran C and Valdivia HH. 1998. Detection and functional characterization of ryanodine receptors from sea urchin eggs. *J. Physiol.* 510: 155-164.
94. Lee HC, Aarhus R and Graeff RM. 1995. Sensitization of calcium-induced calcium release by cyclic ADP-ribose and calmodulin. *J. Biol. Chem.* 270: 9060-9066.
95. Lee HC, Aarhus R, Graeff R, Gurnack ME and Walseth TF. 1994. Cyclic ADP ribose activation of the ryanodine receptor is mediated by calmodulin. *Nature* 370: 307-309.
96. Timerman AP, Onoue H, Xin HB, Barg S, Copello J, *et al.* 1996. Selective binding of FKBP12.6 by the cardiac ryanodine receptor. *J. Biol. Chem.* 271: 20385-20391.
97. Lam E, Martin MM, Timerman AP, Sabers C, Fleischer S, *et al.* 1995. A novel FK506 binding protein can mediate the immunosuppressive effects of FK506 and is associated with the cardiac ryanodine receptor. *J. Biol. Chem.* 270: 26511-26522.
98. Mayrleitner M, Timerman AP, Wiederrecht G and Fleischer S. 1994. The calcium releasechannel of sarcoplasmic reticulum is modulated by FK-506 binding protein: effect of FKBP-12 on single channel activity of the skeletal muscle ryanodine receptor. *Cell Calcium* 15: 99-108.
99. Timerman AP, Ogunbumni E, Freund E, Wiederrecht G, Marks AR and Fleischer S. 1993. The calcium release channel of sarcoplasmic reticulum is modulated by FK-506-binding protein. Dissociation and reconstitution of FKBP-12 to the calcium release

Author Index

Subject Index

MIX
Papier aus verantwortungsvollen Quellen
Paper from responsible sources
FSC® C105338

If you have any concerns about our products,
you can contact us on
ProductSafety@springernature.com

In case Publisher is established outside the EU,
the EU authorized representative is:
Springer Nature Customer Service Center GmbH
Europaplatz 3, 69115 Heidelberg, Germany

Printed by Libri Plureos GmbH
in Hamburg, Germany